Electrical Engineering Materials

Electrical Engineering Materials

Editor:

Dharani Sabba, Ph.D.

ArclerPress

www.arclerpress.com

Electrical Engineering Materials

Editor: Dharani Sabba, Ph.D.

708 3rd Avenue, 6th Floor
New York
NY 10017
United States of America
www.arclerpress.com

ISBN: 978-1-68094-392-4

Library of Congress Control Number: 2016951067

About the Editor

Dharani Sabba, Ph.D.

Dharani Sabba worked as a Research Fellow in Energy Research Institute of Nanyang Technological University (ERI@N), Singapore, on investigating novel semiconductor and photoactive materials for opto-electronic applications. She also obtained her Ph.D in Materials Science from Nanyang Technological University. Her research interests encompass development and synthesis of novel nanomaterials, investigation of their properties and their application in renewable energy sector.

Contents

List of Contributors

M. H. Ugur
Marmara University, Department of Chemistry, 34722 Istanbul, Turkey

R. D. Toker
Marmara University, Department of Chemistry, 34722 Istanbul, Turkey

A. Gungor
Marmara University, Department of Chemistry, 34722 Istanbul, Turkey

N. Kayaman-Apohan
Marmara University, Department of Chemistry, 34722 Istanbul, Turkey

T. Y. Kim
Institute of Materials Science, University of Connecticut, 97 North Eagleville Road, Storrs CT 06226, USA

W. J. Kim
Department of Materials Science, Korea University, 5-1 Anam-dong, Seongbuk-ku, Seoul 136-701, South Korea

T. H. Lee
Department of Materials Science, Korea University, 5-1 Anam-dong, Seongbuk-ku, Seoul 136-701, South Korea

J. E. Kim
Department of Materials Science, Korea University, 5-1 Anam-dong, Seongbuk-ku, Seoul 136-701, South Korea

K. S. Suh
Department of Materials Science, Korea University, 5-1 Anam-dong, Seongbuk-ku, Seoul 136-701, South Korea

B. W. Chieng
Chemistry Department, Faculty of Science, Universiti Putra Malaysia, 43400 UPM Serdang, Selangor, Malaysia

N. A. Ibrahim
Chemistry Department, Faculty of Science, Universiti Putra Malaysia, 43400 UPM Serdang, Selangor, Malaysia

W. M. Z. Wan Yunus
Chemistry Department, Faculty of Science, Universiti Putra Malaysia, 43400 UPM Serdang, Selangor, Malaysia

Edward Song
School of Electrical Engineering and Computer Science, Louisiana State University, Baton Rouge, LA 70803, USA

Jin-Woo Choi
School of Electrical Engineering and Computer Science, Louisiana State University, Baton Rouge, LA 70803, USA
Center for Advanced Microstructures and Devices, Louisiana State University, Baton Rouge, LA 70803, USA

Rajni Garg
Ian Wark Research Institute, University of South Australia, Mawson Lakes Campus, 5095 Adelaide, Australia

Naba K. Dutta
Ian Wark Research Institute, University of South Australia, Mawson Lakes Campus, 5095 Adelaide, Australia

Namita Roy Choudhury
Ian Wark Research Institute, University of South Australia, Mawson Lakes Campus, 5095 Adelaide, Australia

Fabiola Navarro-Pardo
División de Estudios de Posgrado e Investigación, Instituto Tecnológico de Querétaro, 76000 Santiago de Querétaro, QRO, Mexico
Centre Énergie Matériaux Télécommunications, Institut National de la Recherche Scientifique, Varennes, QC, Canada J3X 1S2

Ana L. Martinez-Hernandez
División de Estudios de Posgrado e Investigación, Instituto Tecnológico de Querétaro, 76000 Santiago de Querétaro, QRO, Mexico

Carlos Velasco-Santos
División de Estudios de Posgrado e Investigación, Instituto Tecnológico de Querétaro, 76000 Santiago de Querétaro, QRO, Mexico

Fang-Hsing Wang
Department of Electrical Engineering and Graduate Institute of Optoelectronic Engineering, National Chung Hsing University, Taichung 402, Taiwan

Kun-Neng Chen
Department of Electrical Engineering, Kun-Shan University, Tainan 710, Taiwan

Chao-Ming Hsu
Department of Mechanical Engineering, National Kaohsiung University of Applied Science, Kaohsiung 807, Taiwan

Min-Chu Liu
Department of Electrical Engineering and Graduate Institute of Optoelectronic Engineering, National Chung Hsing University, Taichung 402, Taiwan

Cheng-Fu Yang
Department of Chemical and Materials Engineering, National University of Kaohsiung, Kaohsiung 81141, Taiwan

Hongru Ma
School of Chemistry and Chemical Engineering, Shihezi University, Shihezi 832003, China

Jinfeng Zeng
School of Chemistry and Chemical Engineering, Shihezi University, Shihezi 832003, China

Steven Harrington
Tianjin International Center of Nanoparticle and Nanosystems, Tianjin University, Tianjin 300072, China

Lei Ma
Tianjin International Center of Nanoparticle and Nanosystems, Tianjin University, Tianjin 300072, China

Mingze Ma
School of Chemistry and Chemical Engineering, Shihezi University, Shihezi 832003, China

Xuhong Guo
School of Chemistry and Chemical Engineering, Shihezi University, Shihezi 832003, China
State Key Laboratory of Chemical Engineering, East China University of Science and Technology, Shanghai 200237, China
Key Laboratory for Green Processing of Chemical Engineering of Xinjiang Bingtuan, Shihezi University, Shihezi 832003, China
Engineering Research Center of Materials–Oriented Chemical Engineering of Xinjiang Bingtuan, Shihezi University, Shihezi 832003, China

Yanqing Ma
School of Chemistry and Chemical Engineering, Shihezi University, Shihezi 832003, China
Key Laboratory for Green Processing of Chemical Engineering of Xinjiang Bingtuan, Shihezi University, Shihezi 832003, China
Engineering Research Center of Materials–Oriented Chemical Engineering of Xinjiang Bingtuan, Shihezi University, Shihezi 832003, China

Man-Ning Lu
Department of Chemical Engineering, Tatung University, No. 40, Section 3, Chung Shan North Road, Taipei City 104, Taiwan
Department of Chemical Engineering, National Tsing-Hua University, Hsinchu 300, Taiwan

Chin-Yu Chang
Department of Chemical Engineering, Tatung University, No. 40, Section 3, Chung Shan North Road, Taipei City 104, Taiwan

Tzu-Chien Wei
Department of Chemical Engineering, National Tsing-Hua University, Hsinchu 300, Taiwan

Jeng-Yu Lin
Department of Chemical Engineering, Tatung University, No. 40, Section 3, Chung Shan North Road, Taipei City 104, Taiwan

Lu Yue
Jiangsu Collaborative Innovation Center for Ecological Building Materials and Environmental Protection Equipment, Yancheng Institute of Technology, Yancheng 224051, China
Key Laboratory for Advanced Technology in Environmental Protection of Jiangsu Province, Yancheng Institute of Technology, Yancheng 224051, China

Zhiqiang Zhang
Jiangsu Collaborative Innovation Center for Ecological Building Materials and Environmental Protection Equipment, Yancheng Institute of Technology, Yancheng 224051, China
Key Laboratory for Advanced Technology in Environmental Protection of Jiangsu Province, Yancheng Institute of Technology, Yancheng 224051, China

Yanyan Ma
Jiangsu Collaborative Innovation Center for Ecological Building Materials and Environmental Protection Equipment, Yancheng Institute of Technology, Yancheng 224051, China
Key Laboratory for Advanced Technology in Environmental Protection of Jiangsu Province, Yancheng Institute of Technology, Yancheng 224051, China

Wenhui Zhang
Jiangsu Collaborative Innovation Center for Ecological Building Materials and Environmental Protection Equipment, Yancheng Institute of Technology, Yancheng 224051, China
Key Laboratory for Advanced Technology in Environmental Protection of Jiangsu Province, Yancheng Institute of Technology, Yancheng 224051, China

Trong Tung Nguyen
College of Television Vietnam, Thuong Tin, Hanoi 158500, Vietnam
School of Engineering Physics, Hanoi University of Science and Technology, No.1 Daicoviet, Hanoi 112400, Vietnam

Ngoc Huyen Duong
School of Engineering Physics, Hanoi University of Science and Technology, No.1 Daicoviet, Hanoi 112400, Vietnam

Shanyue Zhao
School of Material and Science Engineering, Jiangsu University, Zhenjiang 212013, China

Yinqun Hua
School of Material and Science Engineering, Jiangsu University, Zhenjiang 212013, China
School of Mechanics Engineering, Jiangsu University, Zhenjiang 212013, China

Ruifang Chen
School of Mechanics Engineering, Jiangsu University, Zhenjiang 212013, China

Jian Zhang
School of Material and Science Engineering, Jiangsu University, Zhenjiang 212013, China

Ping Ji
School of Material and Science Engineering, Jiangsu University, Zhenjiang 212013, China

B. Conings
Instituut voor Materiaalonderzoek, Universiteit Hasselt, Wetenschapspark 1, 3590 Diepenbeek, Belgium

L. Baeten
Instituut voor Materiaalonderzoek, Universiteit Hasselt, Wetenschapspark 1, 3590 Diepenbeek, Belgium

T. Jacobs
Instituut voor Materiaalonderzoek, Universiteit Hasselt, Wetenschapspark 1, 3590 Diepenbeek, Belgium

R. Dera
Instituut voor Materiaalonderzoek, Universiteit Hasselt, Wetenschapspark 1, 3590 Diepenbeek, Belgium

J. D'Haen
Instituut voor Materiaalonderzoek, Universiteit Hasselt, Wetenschapspark 1, 3590 Diepenbeek, Belgium

J. Manca
Instituut voor Materiaalonderzoek, Universiteit Hasselt, Wetenschapspark 1, 3590 Diepenbeek, Belgium

H.-G. Boyen
Instituut voor Materiaalonderzoek, Universiteit Hasselt, Wetenschapspark 1, 3590 Diepenbeek, Belgium

Minghan Chen
School of Electronic Science and Technology, Faculty of Electronic Information and Electrical Engineering, Dalian University of Technology, Dalian 116024, China

Huichao Zhu
School of Electronic Science and Technology, Faculty of Electronic Information and Electrical Engineering, Dalian University of Technology, Dalian 116024, China

Xiaogan Li
School of Electronic Science and Technology, Faculty of Electronic Information and Electrical Engineering, Dalian University of Technology, Dalian 116024, China

Jun Yu
School of Electronic Science and Technology, Faculty of Electronic Information and Electrical Engineering, Dalian University of Technology, Dalian 116024, China

Haitao Cai
School of Electronic Science and Technology, Faculty of Electronic Information and Electrical Engineering, Dalian University of Technology, Dalian 116024, China

Xiaotong Quan
School of Electronic Science and Technology, Faculty of Electronic Information and Electrical Engineering, Dalian University of Technology, Dalian 116024, China

Kaili Wang
School of Electronic Science and Technology, Faculty of Electronic Information and Electrical Engineering, Dalian University of Technology, Dalian 116024, China

Jiaqi Zhang
School of Electronic Science and Technology, Faculty of Electronic Information and Electrical Engineering, Dalian University of Technology, Dalian 116024, China

Antonella Gentile
Department of Physics and Astronomy-University of Catania, via S. Sofia 64, 95123 Catania, Italy
MATIS IMM-CNR, via S. Sofia 64, 95123 Catania, Italy

Francesco Ruffino
Department of Physics and Astronomy-University of Catania, via S. Sofia 64, 95123 Catania, Italy
MATIS IMM-CNR, via S. Sofia 64, 95123 Catania, Italy

Maria Grazia Grimaldi
Department of Physics and Astronomy-University of Catania, via S. Sofia 64, 95123 Catania, Italy
MATIS IMM-CNR, via S. Sofia 64, 95123 Catania, Italy

Vu Binh Nam
Department of Mechanical Engineering, Gachon University, Seongnam 13120, Korea

Daeho Lee
Department of Mechanical Engineering, Gachon University, Seongnam 13120, Korea

Luiza A. Mercante
National Laboratory for Nanotechnology in Agribusiness (LNNA), Embrapa Instrumentation, 13560-970 São Carlos, SP, Brazil

Vanessa P. Scagion
National Laboratory for Nanotechnology in Agribusiness (LNNA), Embrapa Instrumentation, 13560-970 São Carlos, SP, Brazil
Center for Exact Sciences and Technology, Federal University of São Carlos (UFSCar), 13565-905 São Carlos, SP, Brazil

Adriana Pavinatto
National Laboratory for Nanotechnology in Agribusiness (LNNA), Embrapa Instrumentation, 13560-970 São Carlos, SP, Brazil

Rafaela C. Sanfelice
National Laboratory for Nanotechnology in Agribusiness (LNNA), Embrapa Instrumentation, 13560-970 São Carlos, SP, Brazil

Luiz H. C. Mattoso
National Laboratory for Nanotechnology in Agribusiness (LNNA), Embrapa Instrumentation, 13560-970 São Carlos, SP, Brazil

Daniel S. Correa
National Laboratory for Nanotechnology in Agribusiness (LNNA), Embrapa Instrumentation, 13560-970 São Carlos, SP, Brazil
Center for Exact Sciences and Technology, Federal University of São Carlos (UFSCar), 13565-905 São Carlos, SP, Brazil

Vishakha Kaushik
Nanobio Device Laboratory, School of Electrical and Electronic Engineering, Yonsei University, 50 Yonsei-ro, Seodaemun-Gu, Seoul 120-749, Korea

Jaehong Lee
Nanobio Device Laboratory, School of Electrical and Electronic Engineering, Yonsei University, 50 Yonsei-ro, Seodaemun-Gu, Seoul 120-749, Korea

Juree Hong, Seulah Lee
Nanobio Device Laboratory, School of Electrical and Electronic Engineering, Yonsei University, 50 Yonsei-ro, Seodaemun-Gu, Seoul 120-749, Korea

Sanggeun Lee
Nanobio Device Laboratory, School of Electrical and Electronic Engineering, Yonsei University, 50 Yonsei-ro, Seodaemun-Gu, Seoul 120-749, Korea

Jungmok Seo
Nanobio Device Laboratory, School of Electrical and Electronic Engineering, Yonsei University, 50 Yonsei-ro, Seodaemun-Gu, Seoul 120-749, Korea

Chandreswar Mahata
Nanobio Device Laboratory, School of Electrical and Electronic Engineering, Yonsei University, 50 Yonsei-ro, Seodaemun-Gu, Seoul 120-749, Korea

Taeyoon Lee
Nanobio Device Laboratory, School of Electrical and Electronic Engineering, Yonsei University, 50 Yonsei-ro, Seodaemun-Gu, Seoul 120-749, Korea

Preface

Research in electronic materials and its application in electrical engineering is rapidly progressing as it could pave way for advancements in electronics, optics, biotechnology, green energy generation as well as in medical diagnosis. In order to develop novel electronic materials and devices, it is pertinent to understand the underlying physics of these materials and devices at the micro and nano scale. This book is a compilation of research articles and reviews related to the application of various materials such as polymers, metal oxides, carbo nanotubes and graphene in the field of electrical engineering. It underlines the relationship between material composition, structure, properties and its application in real world.

Chapters 1-4 are compilations of research articles showcasing the electrical properties of polymer nanocomposites. The effect of composition of various kinds of polymer nanocomposites on electrical conduction and their application in chemical sensing is presented. While Chapters 5-8 focus on relatively "hot" engineering materials such as graphene and carbon nanotubes. In particular, Chapter 5 focuses on the engineering of graphene's work function which plays a pivotal role in electrical applications. Whereas Chapter 7 discusses the role of graphene as cathode material in one of the sub-categories of the third generation solar cells, commonly known as dye sensitized solar cells.

As materials at nanoscale exhibit unique properties, the effect of nanostructuring on the electrical properties of various metal oxides such as zinc oxide, titanium oxide and copper oxide is discussed in Chapters 9-14. While Chapters 15-17 focus on the application of metal nanoparticles and nanowires in transparent flexible conductors and electronic tongue. We come to conclusion of this book with the last Chapter 18 featuring the latest updates on the application of electrical materials for e-textiles: energy harvesting and storage.

This book will be a useful resource of materials to help students, engineers and researchers working in the field of electrical engineering and material science to extend their existing knowledge. The collection of articles and review articles is well versed and the reader can keep themselves abreast with the latest developments in the field of electrical engineering materials through this book.

Editor
Dharani Sabba, Ph.D.

CHAPTER 1

PREPARATION AND CHARACTERIZATION OF NOVEL THERMOSET POLYIMIDE AND POLYIMIDE-PEO DOPED WITH $LICF_3SO_3$

M. H. Ugur, R. D. Toker, A. Gungor and N. Kayaman-Apohan

Marmara University, Department of Chemistry, 34722 Istanbul, Turkey

ABSTRACT

This paper deals with the synthesis and characterization of a new type of anhydrous ionic conducting lithium doped membranes consist of polyimide (PI), poly (ethylene oxide) (PEO) and lithium trifluoromethanesulfonate ($LiCF_3SO_3$) for solid polymer electrolyte (SPE). For this purpose, different molar ratios of lithium salt (Li-salt) solution are added into poly (amic acid) (PAA) intermediate prepared from the reaction of 3,3!,4,4!-benzophenon tetracarboxylic dianhydride (BTDA) and 4,4!-oxydianiline (ODA). PEO is incorporated into PAA since it forms more stable complexes and possess high ionic conductivities. Then, Li-salt containing PAA solutions are imidized by thermal process. The effect of interaction between host polymer and Li-salt is characterized by FT-IR (Fourier Transform Infrared) spectroscopy and SEM (scanning electron micrsocopy). The conductivities of Li-salt and PEO containing PI composite membranes are in the range of 10^{-7}–10^{-5} S·cm^{-1}. The conductivity increases with incorporation of PEO. Thermogravimetric analysis results reveal that the PI/PEO/LiCF3SO3 composite polymer electrolyte membranes are thermally stable up to 500°C.

INTRODUCTION

Polyimides (PIs) belong to the important class of polymeric materials known as 'high performance' polymers due to their high thermo-oxidative stability [1], their exceptional mechanical properties, good film forming ability and superior chemical resistance [2–6]. Although, high performance polyimides are widely used in the microelectronics industry because the imide rings may provide permittivities equivalent to those of high dielectric constant to materials [7]. Work on polymer/salt mixtures has increased during recent years due to the growing interest in these systems as fast ion conducting systems. Several types of lithium ion conducting polymer electrolytes have been prepared as perfect homogenous mixtures of the components, namely, PEO as the polymer matrix which is capable of solving different alkali metal salts, and lithium trifluoromethanesulfonate ($LiCF_3SO_3$), lithium tetrafluoroborate ($LiBF_4$) and lithium perchlorate ($LiClO_4$) as a lithium salt [8]. However, these electrolytes show satisfactory ionic conductivity only at temperatures above 70°C. The conventional belief has been that the high degree of local order ('crystallinity') is what makes the ionic conductivity too low at ambient temperatures. Therefore, much attention has been devoted to the task of increasing the amorphous content of the PEO electrolyte; either by using large-anion lithium salts, by adding liquid plasticizers or ceramic fillers to the polymer

Citation: M. H. Ugur, R. D. Toker, A. Gungor, N. Kayaman-Apohan, Preparation and characterization of novel thermoset polyimide and polyimide-peo doped with $LiCF_3SO_3$, DOI 10.3144/expresspolymlett.2014.15.

[9–12]. When the electrolytic component is in the form of a dry polymer matrix, it consist of a high molecular weight homo or copolymer, which is cross-linkable or non-cross-linkable and includes a heteroatom in its repeating unit such as oxygen or nitrogen for example, in which an alkali metal salt is dissolved such as LiCF3SO3, LiClO4, LiTFSI etc. In the literature, ion-conducting polyimide membranes are selected as promising candidates for separating the electrodes because such a membrane has the mechanical stability to prevent shorts on rough handling, the chemical stability to prevent decomposition reactions that organic liquid electrolytes are prone to suffer at elevated temperatures, and the ability to resist penetration of lithium dendrites [13]. In the literature, there are only a few studies similar to the work that has been carried out by several research groups [14–22]. For instance, U.S. Patent 5888672 issued to Gustafson and Antonucci [14] disclosure a battery wherein each of the anode, cathode and electrolyte layer is based upon soluble, amorphous and thermoplastic polyimide. In another study, Johnson et al. [15] explored to comprise a solid, thermoset polyimide matrix doped with a lithium salt (LiTFSI). Wensley prepared several optically clear polyimide based electrolytes and determined the minimum required amount of lithium per mole of imide ring in soluble polyimide for rechargeable batteries [16]. The aim of the present work is to investigate the effects of Li-salt and PEO on the ionic conductivity and morphological behavior of PI based polymer electrolyte.

EXPERIMENTAL

Materials

3,3!,4,4!-Benzophenone tetracarboxylic dianhydride (BTDA, Aldrich, Steinheim-Germany), 4,4!- Oxydianiline (ODA, Aldrich, Steinheim-Germany), Poly(ethylene oxide) (average M_w = 100.000 g·mol^{-1}, powder, Aldrich, Steinheim-Germany), Lithium trifluoromethanesulfonate ($LiCF_3SO_3$, 99.995%-Aldrich, Steinheim-Germany), Dimethylacetamide (DMAc, Aldrich, Steinheim-Germany) were dried under vacuum before being used and stored in home-designed glove box (Ercom Kompresör, Kartal/"stanbul, Turkey) until use.

Preparation of lithium salt containing polyimide membrane

ODA and DMAc were placed into a 100 mL threeneck flask under nitrogen purge. The mixture was stirred until the solution was clear. BTDA in equal molar amount to that of ODA was added to the solution and stirred for 24 h at room temperature to give a viscous, transparent and yellow solution of PAA (Figure 1) which afforded approximately 20% solid by weight. PAA solutions containing different amounts of $LiCF_3SO_3$ salt were converted to polyimide by thermal imidization technique heating one hour each at 80–100–150–200–250°C ranges applied onto glass plates (F1–F4) (Figure 2). Polyimides having long PEO moieties in the main chain (F5 and F6) have been prepared by heating overnight at relatively low temperatures to prevent the degradation of polyether chain (#150°C).

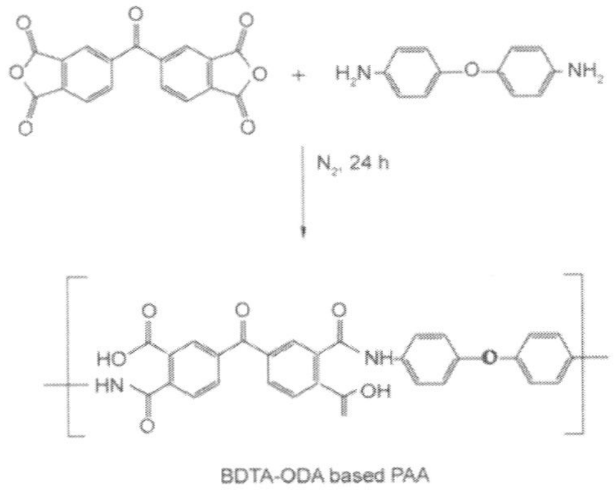

Figure 1: Shematic representation of synthesis BTDAODA based PAA.

Characterization

FT-IR spectrum was recorded on Perkin Elmer Spectrum 100 ATR-FTIR spectrophotometer (Perkin Elmer, Waltham, MA, USA). SEM imaging of the films were performed on Philips XL30 ESEM-FEG/ EDAX (Philips, Eindhoven, The Netherlands). The specimens were prepared for SEM by freeze fracturing in liquid nitrogen and applying a gold coating. The thermal stability of the samples was tested in the temperature range of 30–800°C by thermogravimetric analysis (TGA) using a PerkinElmer STA 6000 instrument (PerkinElmer, Waltham, MA, USA) under air atmosphere at a heating rate of 10°C/min. Differential scanning calorimetry (DSC) measurements were carried out with a Perkin Elmer Pyris Diamond (PerkinElmer, Shelton, CT, USA). The samples were analyzed under a nitrogen atmosphere in the temperature range –100 to 400ºC at a heating rate of 10ºC/min and cooling rate of 100ºC/min. Standard tensile stress-strain experiments were performed at room temperature on a Materials Testing Machine Z010/TN2S (Zwick GmbH&Co. KG, Ulm, Germany) using a crosshead speed of 5 mm·min^{-1}. The mechanical properties of the PI based polymer electrolytes were determined by standard tensile stress–strain tests in order to measure tensile strength and elongation at break. The specimen dimensions were 60.00 mm in length and 8.00 mm in width. Four parallel measurements were carried out for each sample.

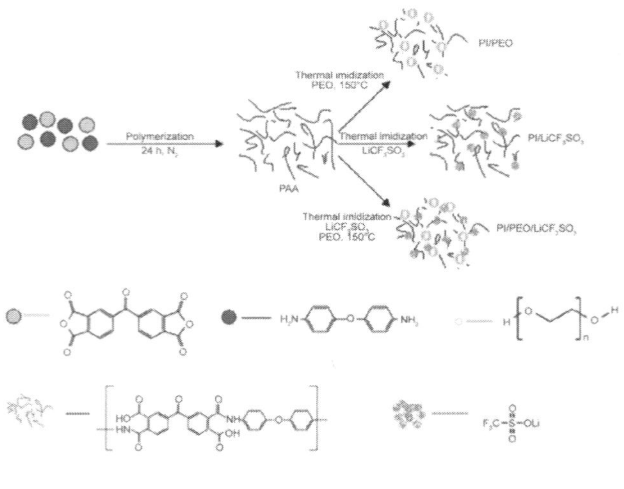

Figure 2: Shematic representation of preparation of Li-salt-doped polyimide membrane.

Ionic conductivity measurements were recorded using a Gamry Potentiostat/Galvanostat/ZRA (Gamry Series G 750, Warminster, PA, USA) with Gamry Framework software system EIS300. EIS spectra were analyzed using Echem Analyst 5.67 software. The polymer electrolyte (PE) was placed between two SS (type 304, 0.025 mm thick, Alfa Aesar, Karlsruhe, Germany) electrodes in a cell. The thickness of the membrane was measured before and after the EIS measurement, to assure a constant thickness throughout the experiment. Ionic conductivity was measured under an argon atmosphere using a potantiostat/galvanostat and a home-design glove box with conductivity cell. The frequency ranged from 40 Hz to 100 mHz at a perturbation voltage of 5 mV. The ionic conductivity values of the polymer electrolyte systems are calculated from the intercept of real part of the complex impedance plot, which is resistance of the film and known area using the Equation (1)

$$\sigma = \left(\frac{1}{R}\right)\cdot\left(\frac{L}{A}\right) \quad (1)$$

where σ is conductivity; L/A is geometrical factor, L is the thickness of the electrolyte film; R is resistance of the electrolyte film; and A is the area of the film.

Linear Sweep Voltammetry (LSV) measurement was recorded using a Gamry Potentiostat/Galvanostat/ZRA (Gamry Series G 750, Warminster, PA, USA) with Gamry Framework Software System PHE200 (Physical Electrochemistry Software, Warminster, PA, USA). LSV experiment was performed to investigate the electrochemical stability window of the polymer electrolyte (F5) employing SS as working electrode and lithium foil (Sigma-Aldrich, Steinheim-Germany) as reference and counter electrodes. Cell assembly was carried out in argon atmosphere inside the glove box. The scanning rate was 1 $mV\cdot s^{-1}$ and the potential ranged from open circuit potential to 5.0 V (vs. Li/Li^+).

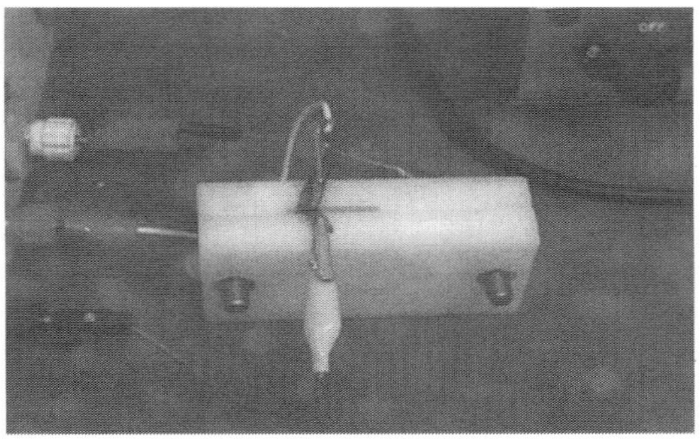

Figure 3: Picture of the conductivity cell.

All the electrochemical experiments were carried out in an Argon atmosphere in a home-designed glove box at room temperature (Figure 3).

RESULTS AND DISCUSSIONS

FT-IR

The structures of membranes were investigated by FTIR. To evaluate the molecular interactions between PI and PEO, the FTIR spectrum of pure PEO was also displayed in the insert of Figure 4. Figure 4 shows the FTIR spectra of (a) membrane F0, (b) membrane F1, (c) membrane F5, (d) membrane F6 and (e) pure PEO. In all these work, the complexation is confirmed by the appearance of new peaks, frequency shifts, changes in intensities or shape of existing peaks in the FTIR spectra. The chemical bonding between PEO chain, $LiCF_3SO_3$, PEO-$LiCF_3SO_3$ and PI structure was discussed below.

The FT-IR spectroscopy indicated that complexation has occurred between polymer matrix and salt. Figure 4 curve a is FT-IR spectra for polyimide film without the addition of a Li salt or any solvent. The characteristic absorption bands of the imide groups near 1776 (asym. imide carbonyl stretching), 1713 (sym. imide carbonyl stretching) and 1371 cm^{-1} (imide C–N stretching) were observed in the FTIR spectrum after thermal imidization of the poly (amic acid). Meanwhile, the characteristic absorption of the amide carbonyl at 1640 cm^{-1} did not appear in the spectrum, indicating that the imidization reaction is complete. Strong bands in the range of 1300 cm^{-1} (imide C–N stretching) and 744 cm^{-1} (imide C–H bending) are also observed in the spectrum. Other absorption bands at 1617 and 1492 cm^{-1} are aromatic C=C stretching and benzene ring vibrations, respectively. Moreover, the spectra of membrane F0 show asymmetric stretching ether (C–O–C) bands at 1228, 1162 and 1089 and aromatic C–H bending bands between 980 to 832 cm^{-1} [23].

On the other hand, Figure 4 curve b shows FT-IR spectrum of PI-$LiCF_3SO_3$ membrane. The resulting polyimide electrolyte shows strong peaks at about 1652 and 1590 cm^{-1}. Previously it was reported that, polyimide electrolyte shows a very strong doublet at about 1670 and 1640 cm^{-1}. These peaks are the evidence for the complex between lithium salt and the imide rings of the polyimide Slight shifts observed in these peaks could be attributed to the use of different salts and different polyimide materials. Also it can be seen that, the characteristic polyimide peaks at 1776 and 1713 cm^{-1} are shifted to 1770 and 1717 cm^{-1}, respectively. The other bands at 1256 and 1289 cm^{-1} related sym. CF_3 stretching and asymm. SO_3 stretching modes. Also the complexation of the $LiCF_3SO_3$ salt with PI was confirmed by the presence of new peak at 662 cm^{-1} corresponding to free SO_3^- ion. Furthermore it indicates an interaction/complexation between polyimide material and Li-salt. It is known that Li^+ is a very good complex forming species and therefore, it may possibly form some strong coordinate bond with the imide nitrogen atom or the imide carbonyl

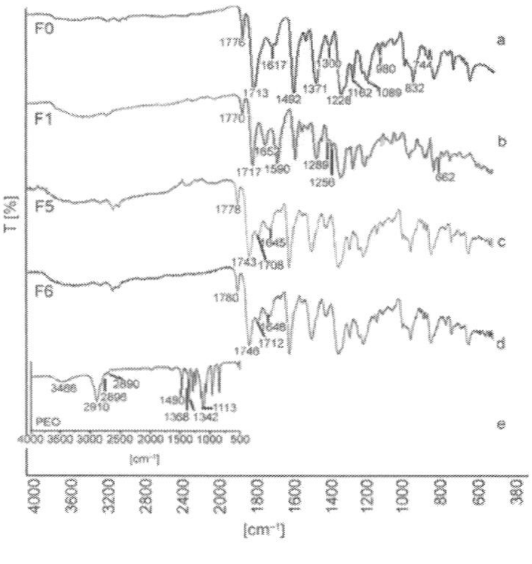

Figure 4: FT-IR spectra of a) PI and b) PI-$LiCF_3SO_3$ c) PIPEO and d) PI-PEO-$LiCF_3SO_3$ e) pure PEO.

As can be seen in Figure 4 curve e, in the pure PEO spectrum a large broad band appears centered at 3466 cm^{-1}. Pure PEO shows a large, broad band of CH_2 stretching at 2910 cm^{-1}. However, the band is split into two at 2918 and 2884 cm^{-1} corresponding to asymmetric and symmetric CH2 stretching, respectively 27. Other bands at 1490, 1368–1342 and 1113 cm^{-1} are CH_2 scissoring, CH_2 wagging, C–O–C stretching modes 28, 29.

The existence of chemical interchain between PEO with PI was proved by FTIR study. As it can be seen in Figure 4 curves c and d, the band at 1743 and 1746 cm^{-1} corresponds to the esther carbonyl groups. The presence of ester bonds resulted in a shift of the peak of the imide carbonyl groups at around 1708 (curve c) and 1712 cm^{-1} (curve d). In these spectra, the peaks at 1645 and 1648 cm^{-1} were attributed to the amide carbonyl groups due to partially imidization of membranes F5 and F6 which were imidized at low temperatures.

Morphology properties of polyimide membranes

The morphologies of the PI, PI-PEO and their composite polyelectrolyte membranes were studied by scanning electron microscopy (SEM) as shown in Figure 5. As seen in Figure 5a, the fractured surface of the PI membrane is dense and homogeneous. Since the thermal imidization was performed by stepwise heating up to 250°C, microporosity was not observed on the membrane. However, the surface morphology of the Li salt containing membrane is rough compared with neat PI film as seen in the SEM image (Figure 5b). In Figure 5c the random distribution of PEO structure in PI matrix can be seen. Since the imidization was performed at low temperature to protect the polyether chain from thermal degradation, course voids appeared in the membrane due to the low rate of solvent evaporation. The characteristic semi crystalline morphology of PEO chain was undetectable in the PI-PEO membrane. When only PEO or $LiCF_3SO_3$ or PEO/ $LiCF_3SO_3$ mixture was incorporated into PI matrix rough morphology was obtained (Figure 5b–5d).

a)

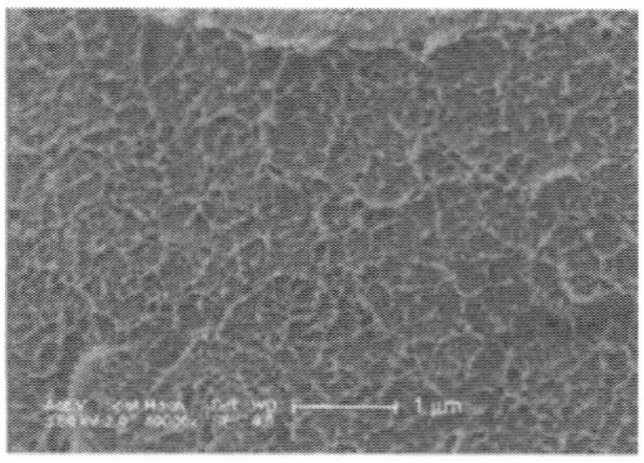

b)

c)

d)

Figure 5: SEM cross-section images of a) PI (F0), b) PI-$LiCF_3SO_3$ (F1), c) PI-PEO-$LiCF_3SO_3$ (F5), d) PI-PEO (F6) and e) pure PEO.

Thermal and mechanical properties of polyimide based polymer electrolytes

Thermo-oxidative stabilities of the polymers were determined by thermogravimetric analysis in air. Figure 6 shows TGA thermograms of polyimides. The dynamic TGA experiments were run from 30 to 800°C, at a heating rate of 10°C/min under air atmosphere. The first stage of decomposition started at about 250°C due to the degradation of organic group of $LiCF_3SO_3$ salt (–CF_3–SO_3 bridge) and PEO (C–O–C), which were the weakest linkages along the main chain. The second stage of decomposition started from 450°C corresponding to the degrading of the aromatic imido groups; this is typical for aromatic polyimides in general 30. In addition, although PEO decomposition temperature is much lower than polyimide, TGA curves of PEO containing polyimide membranes (F5–F6) have approximately same thermal properties. This result is explained by seconder interaction between molecules and also incorporating very low concentrations of PEO.

The TG and derivative thermogravimetric analysis (DTG) curves of PI based membranes under air indicate three stages of thermal decomposition. However, the resulting curve, shown in Figure 6 stable up PI based electrolytes approximately not less than 50% of their mass, which was illustrated in all of the PI membranes, had excellent stability. The glass transition temperature (T_g) of neat PI (F0) was found as 297.4 °C. When Li salt was added, the T_g of the composite film decreased to 271.6°C. Thus, it can be said that the Li salt particles decrease the chain interactions in the PI matrix by settling between polymer chains. On the other hand, in thermal imidization conditions (150°C for membranes F5 and F6) when PEO and the Li salt were incorporated into PI matrix, T_g was further reduced to 211.7°C. Partly remained solvent's effect is related to a weakening of the dipole-dipole interactions between the polymer chains. This result in a reduction of the T_g value. Therefore, this helps to soften the polymer backbone and increase its segmental motion. Relatively, this situation was attributed to the increased free volume at the molecular level due to presence of flexible PEO chains.

It is known that the mechanical properties of the SPEs are as important as the ionic conductivity at room temperature for their practical applications. Elongation at break and tensile strength of polymer electrolyte membranes are displayed in Table 2. For F1 to F4 membranes, decreasing Li content in the membrane decreases the elongation and tensile modules in the same manner. This behaviour was attributed decreasing the level of the hydrogen bond the carbonyl oxygen. When membranes F5 and F6 were compared, it is clearly observable that the addition of Li salt within the PEO containing PI membrane caused negligible changes. When membranes F2 and F5 were compared, it is clearly observable that the addition of PEO causes drastic reduction in the tensile strength and elongation at break. This could be attributed to the etheric structure of PEO. Finally, as can be seen in Table 2, raising the glass transition temperature of polymer electrolytes improves ionic conductivity but leads to a decrease in mechanical properties.

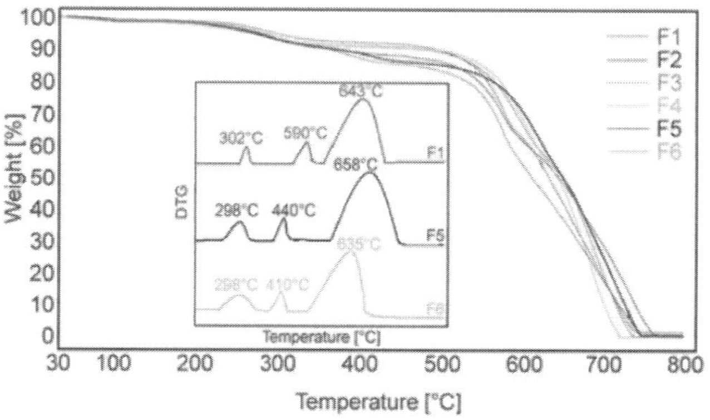

Figure 6: TGA thermograms of F1–F6 films recorded at a heating rate of 10°C/min for the polyimides containing various PEO, $LiCF_3SO_3$ and $PEOLiCF_3SO_3$ contents. Inset Figure: The DTG curve shows stepwise decomposition of F1, F5 and F6 membranes.

Ionic conductivity

The ionic conductivity of a polymer electrolyte depends on the concentration of ion carriers and on their mobility. The ionic conductivity in polymer electrolytes is assumed to occur by Lewis acid-base interactions between the cations and the polymer solvent. It is also generally accepted that the ionic conduction occurs mainly in the amorphous component of polymer electrolytes above their T_g with the chain segment mobility playing a critical role in the conductivity mechanism [31].

In the case of solvent free polymer electrolytes, the motions of the polymer host are responsible for the ionic mobility; ions move only if polymer segments undergo fairly large-amplitude motions 32–34. related to T_g. Polymer electrolytes show fast ionic conduction above their T_g where they are largely comprised of amorphous phases. Thus, a low T_g polymer like PEO (poly (ethylene oxide)); T_g, –50 to –57°C [35] has become an important polymer for solvent free electrolytes, and amorphization of this polymer is being researched 36, 37. as a way to increase its ionic conductivity.

In this study, the conductivity of the polymer electrolyte membranes was calculated from the measured resistance of the film for the known area and thickness of the polymer film. Table 1 and Figure 7 summarize the compositions and Nyquist diagrams of the thermoset PI based polymer electrolytes. Firstly, the ionic conductivity was studied as a function of Li salt concentration within PI matrix (F1– F4). As can be seen in Table 1, the Li^+ ion conductivity changed according to O./Li. ratio where O. is the number of mole of oxygen atoms in polyimide structure from membrane F1 to membrane F4. At lower O./Li. ratio, ion conductivity significantly decreased. This can be explained by the increased amount of Li^+ ion concentration in thermoset polyimide materials which was limited the mobility of polymer chains 24. In other words, further increase of Li ion concentration results in reduction of conductivity, explained by an increase in transient ionic crosslinks within polymer network, causing a decrease in chain segment mobility. It has also been suggested that at high salt content formation of ion aggregates will contribute to a decrease in conductivity 38. Hence, membrane F4 showed highest ionic conductivity of $8.7{\cdot}10^{-6}$ $S{\cdot}cm^{-1}$ compared the other membranes (F1, F2 and F3) at room temperature.

Table 1: The conditions, Li^+ ion conductivity values and O./Li. ratios of membranes

	F1	F2	F3	F4	F5	F6
Metod	**Thermal imidization**	**Thermal imidization**	**Thermal imidization**	**Thermal imidization**	**Thermal imidization**	**Thermal imidization**
Process temperature [°C]	**80–100–150–200–250**	**80–100–150–200–250**	**80–100–150–200–250**	**80–100–150–200–250**	**80–100–150**	**80–100–150**
Codes	$PI/LiCF_3SO_3$	$PI/LiCF_3SO_3$	$PI/LiCF_3SO_3$	$PI/LiCF_3SO_3$	$PI/PEO/LiCF_3SO_3$	PI/PEO
*[O]/[Li] ratio	1:1	2:1	4:1	8:1	1:1	–
PEO [mole]	–	–	–	–	10^{-6}	10^{-6}
Iomic conductivity [$S{\cdot}cm^{-1}$]	$9.2{\cdot}10^{-7}$	$1.08{\cdot}10^{-6}$	$4.44{\cdot}10^{-6}$	$8.7{\cdot}10^{-6}$	$2.04{\cdot}10^{-5}$	–

*[O] mole number of oxygen atoms is polyimide structure

Table 2: Thermal and mechanical properties of novel PI based thermoset polymer electrolytes

Sample code	Tensile strength [N/mm^2]	Elongation at break [%]	Max weight loss temperature [°C]	The glass transition temperature [°C]
F0	130.80	5.40	668	297.4
F1	64.53	1.07	643	271.6
F2	33.52	0.84	660	231.3
F3	17.80	0.60	659	220.8
F4	16.64	0.42	650	213.2
F5	16.80	0.41	658	211.7
F6	17.68	0.40	635	205.8

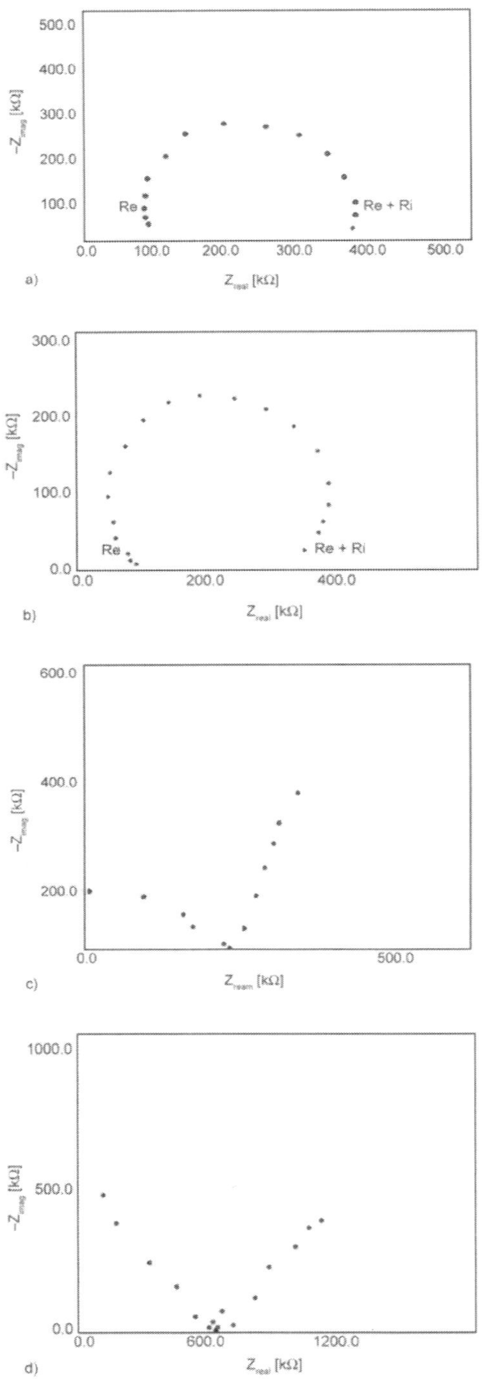

Figure 7: Impedance diagrams of ion-conductive polyimide membranes a) PI/PEO/$LiCF_3SO_3$ (F5) b) PI/$LiCF_3SO_3$ (F4) c) PI/$LiCF_3SO_3$ (F3) d) PI/$LiCF_3SO_3$ (F2).

On the other hand, the polymer–salt complexes are formed by complexes between salts of alkali metals and polymer containing solvating heteroatoms such as O, N, S, etc. The most common examples are complexation between PEO and alkali metal salts 39. The ether oxygens in PEO are hard Lewis bases, thus they have low polarizability and high electronegativity. They coordinate well to hard Lewis acids, which in general are small cations, e.g., Li^{+1}, Na^{+1} and Mg^{+2}. These cations then, in turn, form salts with low lattice energies together with large, e.g., PF_6^-, $CF_3SO_3^-$ and $(CF_3SO_3)_2N^-$. These larger anions can sometimes also have a plasticizing effect on the polymer 40. In addition, the increase in amorphous content is probably due to the large $CF_3SO_3^-$ anion that reduces the crystallinity of the complex. It can be suggested the increase in conductivity in our system is due to the amorphizing properties of the large $CF_3SO_3^-$ ion present in the system. The spectroscopically 'free' ions are believed to be responsible for ionic charge transport in polymer electrolyte. They can be temporarily complexed or attached to the polymer chain (Li^+ can attach to oxygen atoms on PEO) during segmental motion of the polymer and then hop to the next site. The fraction of 'free' ions will indicate the effectiveness of various electrolyte components in increasing charge concentration and subsequent ion conduction 41. Thus, the increasing ionic conductivity of PI-PEO-$LiCF_3SO_3$ film suggested that, even a small amount of PEO added to PI-$LiCF_3SO_3$ membrane, increases the free volume and decreases the T_g which leads to increase molecular mobility. Hence this will cause to increase conductivity. In the PI-PEO-$LiCF_3SO_3$ electrolyte system (F5) the composition with the highest conductivity is recorded which reaches $2.04 \cdot 10^{-5}$ $S \cdot cm^{-1}$ at room temperature.

Linear sweep voltammetry (LSV)

Electrochemical measurements showed that the cells prepared from using a stainless steel (SS) electrode, a lithium counter, and a lithium reference electrodes had a wide range electrochemical stability. Thus, the high stability of the PI based electrolyte membrane (F5) may be attributed to the absence of impurities, which is a welcome feature because it permits their use in high-voltage battery applications. Figure 8 shows the current-voltage response of the solid polymer membrane obtained in the potential range between open circuit and 5.0 V vs. Li at room temperature. As it can be seen, the plateau is very flat and straight; this very low residual current level prior to breakdown voltage, with no peaks in the 2.0–3.5 V voltage range. So, it confirms the high purity of the SPE membrane was obtained. In addition, the onset of the current during anodic scan, which is representative of the decomposition of the electrolyte, indicates an anodic break-down voltage of approximately 4.3 V vs. Li. So, membrane F5 showed an appreciable anodic breakdown voltage [42].

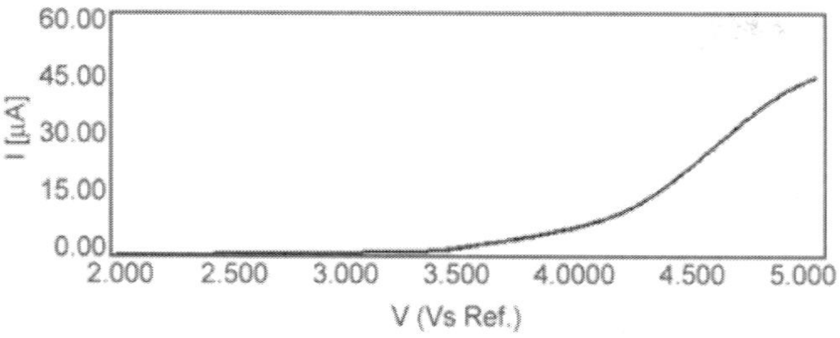

Figure 8: Electrochemical stability window at room temperature for PI based polymer electrolyte (F5), potential scan rate: 1 mV·s^{-1}.

CONCLUSIONS

Novel PI-based composite polymer electrolytes were prepared by the incorporation of $LiCF_3SO_3$ and PEO. Characterizations were carried out by FTIR, TGA, DSC, mechanical properties, SEM, impedance measurements and LSV. The thermal imidization technique was adopted to prepare polymer electrolyte membranes. BTDA and ODA based polyimide synthesized by step-growth polymerization method. In order to improve the conductivity behaviour of membranes, we have been focused on introducing the PEO and $LiCF_3SO_3$ into the polymer backbone. FT-IR measurements showed the existence of the complex formation between lithium salt and imide ring of the polyimide. F1 and F5 films showed an ionic conductivity of $9.2 \cdot 10^{-7}$ and $2.04 \cdot 10^{-5}$ S·cm^{-1} respectively at room temperature which is acceptable value for lithium rechargeable batteries. However, this conductivity is still low for practical use in lithium-polymer or lithium-metal batteries. The final solid polymer electrolyte is not soluble in the liquid electrolyte and also stable over a wide range of temperatures.

ACKNOWLEDGEMENTS

This work was supported by Marmara University, Commission of Scientific Research Project under grant Project FEN-E-090113-0006.

REFERENCES

1. Critchley J. P., Knight G. J., Wright W. W.: Heat-resistant polymers: Technologically useful materials. Plenum Press, New York (1983).
2. Mittal K. L.: Polyimides: Synthesis, characterization and applications. Plenum Press, New York (1984).
3. Bessonov M. T., Koton M. M., Kudryavtsev V. V., Laius L. A.: Polyimides: Thermally stable polymers. Consultants Bureau, New York (1987).
4. Wilson D., Stenzenberger H. D., Hergenrother P. M.: Polyimides. Blackie, Glasgow (1990).
5. Ghosh M. K., Mittal K. L.: Polyimides: Fundamentals and applications. Marcel Dekker, New York (1996).
6. Yamanaka K., Jikei M., Kakimoto M-A.: Preparation and properties of hyperbranched aromatic polyimides via polyamic acid methyl ester precursors. Macromolecules, 33, 6937–6944 (2000). DOI: 10.1021/ma000725n
7. Negi Y. S., Suzuki Y-I., Kawamura I., Hagiwara T., Takahashi Y., Iijima M., Kakimoto M-A., Imai Y.: Synthesis and properties of polyimides based on 2,2- bis
8. 4-(4-aminophenoxy)phenyl.propane and 2,2-bis
9. 4-(4-aminophenoxy)phenyl.hexafluoropropane. Journal of Polymer Science Part A: Polymer Chemistry, 30, 2281–2284 (1992). DOI: 10.1002/pola.1992.080301026
10. Vallee A., Brouilette D., Pugh J. K., Guindy W. W. J.: Polyimide-based lithium metal battery. U.S. Patent 7824808 B2, USA (2010).
11. Benrabah D., Sanchez J-Y., Armand M.: Synthesis and electrochemical characterization of a new family of lithium salts. Solid State Ionics, 60, 87–92 (1993). DOI: 10.1016/0167-2738(93)90279-C
12. Silva M. M., Barros S. C., Smith M. J., MacCallum J. R.: Study of novel lithium salt-based, plasticized polymer electrolytes. Journal of Power Sources, 111, 52–57 (2002). DOI: 10.1016/S0378-7753(02)00229-X
13. Croce F., Appetecchi G. B., Persi L., Scrosati B.: Nanocomposite polymer electrolytes for lithium batteries. Nature, 394, 456–458 (1998). DOI: 10.1038/28818
14. Reiche A., Weinkauf A., Sander B., Rittig F., Fleischer G.: Alternating copolymers for novel polymer electrolytes: The electrochemical properties. Electrochimica Acta, 45, 1327–1334 (2000). DOI: 10.1016/S0013-4686(99)00340-0
15. Meyer W. H.: Polymer electrolytes for lithium-ion batteries. Advanced Materials, 10, 439–448 (1998). DOI: 10.1002/(SICI)1521-4095(199804)10:63.0.CO;2-I
16. Gustafson S., Antonucci J. T.: Polyimid battery. U.S. Patent 5888672 A, USA (1999).
17. Johnson L. G., Allie L. A., Muller J. R.: Solid, lithiumsalt-doped, thermoset polyimide polymer electrolyte and electrochemical cell employing same. U.S. Patent 20130011745, USA (2013).
18. Wensley C. G., Vallée A., Brouillette D., Gustafson S.: Polyimide matrix electrolyte. U.S. Patent 7198870 B2, USA (2007).
19. Tigelaar D. M., Palker A. E., Meador M. A. B., Bennett W. R.: Synthesis and compatibility of ionic liquid containing rod-coil polyimide gel electrolytes with lithium metal electrodes. Journal of the Electrochemical Society, 155, 768–774 (2008). DOI: 10.1149/1.2967723
20. Pak Y. S., Xu G.: Ionic transport measurements of LiCF3SO3 doped polyimide-diaminobenzenesulfonic acid copolymer. Solid State Ionics, 67, 165–169 (1993). DOI: 10.1016/0167-2738(93)90322-T
21. Tian S. B., Pak Y. S., Xu G.: Polyimide-polysiloxanesegmented copolymers as high-temperature polymer electrolytes. Journal of Polymer Science Part B: Polymer Physics, 32, 2019–2023 (1994). DOI: 10.1002/polb.1994.090321208
22. Park J-H., Kim J-S., Shim E-G., Park K-W., Hong Y. T., Lee Y-S., Lee S-Y.: Polyimide gel polymer electrolyte-nanoencapsulated LiCoO2 cathode materials for high-voltage Li-ion batteries. Electrochemistry Communications, 12, 1099–1102 (2010). DOI: 10.1016/j.elecom.2010.05.038

23. Gustafson S., Wensley C. G.: Polyimide matrix electrolyte. U.S. Patent 20040229127, USA (2004).
24. Wensley C. G., Gustafson S., Nelson C. R., Singleton R. W., Vallee A., Brouillette D.: Polyimide matrix electrolyte and improved batteries therefrom. U.S. Patent 7129005, USA (2006).
25. Davenas J., Xu X. L., Boiteux G. Sage D.: Relation between structure and electronic properties of ion irradiated polymers. Nuclear Instruments and Methods in Physics Research Section B: Beam Interactions with Materials and Atoms, 39, 754–763 (1989). DOI: 10.1016/0168-583X(89)90891-4
26. Wensley C. G., Vallee A., Brouillette D., Gustafson S.: Polyimide matrix electrolyte. U.S. Patent 7198870 B2, USA (2007).
27. de Zea Bermudez V., Alcácer L., Acosta J. L., Morales E.: Synthesis and characterization of novel urethane cross-linked ormolytes for solid-state lithium batteries. Solid State Ionics, 116, 197–209 (1999). DOI: 10.1016/S0167-2738(98)00346-4
28. Cotton F. A., Wilkinson G.: Advanced inorganic chemistry. Wiley, New York (1988).
29. Ramesh S., Yuen T. F., Shen C. J.: Conductivity and FTIR studies on PEO–LiX
30. X: CF3SO3 –, SO4 2–. polymer electrolytes. Spectrochimica Acta Part A: Molecular and Biomolecular Spectroscopy, 69, 670–675 (2008). DOI: 10.1016/j.saa.2007.05.029
31. Rocco A. M., da Fonseca C. P., Pereira R. B.: A polymeric solid electrolyte based on a binary blend of poly (ethylene oxide), poly(methyl vinyl ether-maleic acid) and LiClO4. Polymer, 43, 3601–3609 (2002). DOI: 10.1016/S0032-3861(02)00173-8
32. Tang Z., Wang J., Chen Q., He W., Shen C., Mao X-X., Zhang J.: A novel PEO-based composite polymer electrolyte with absorptive glass mat for Li-ion batteries. Electrochimica Acta, 52, 6638–6643 (2007). DOI: 10.1016/j.electacta.2007.04.062
33. Dhara M. G., Banerjee S.: Fluorinated high-performance polymers: Poly(arylene ether)s and aromatic polyimides containing trifluoromethyl groups. Progress in Polymer Science, 35, 1022–1077 (2010). DOI: 10.1016/j.progpolymsci.2010.04.003
34. Berthier C., Gorecki W., Minier M., Armand M. B., Chabagno J. M., Rigaud P.: Microscopic investigation of ionic conductivity in alkali metal salts-poly(ethylene oxide) adducts. Solid State Ionics, 11, 91–95 (1983). DOI: 10.1016/0167-2738(83)90068-1
35. Wakihara M.: Recent developments in lithium ion batteries. Materials Science and Engineering R: Reports, 33, 109–134 (2001). DOI: 10.1016/S0927-796X(01)00030-4
36. Ratner M. A., Shriver D. F.: Ion transport in solventfree polymers. Chemical Reviews, 88, 109–124 (1988). DOI: 10.1021/cr00083a006
37. Papke B. L., Ratner M. A., Shriver D. F.: Conformation and ion-transport models for the structure and ionic conductivity in complexes of polyethers with alkali metal salts. Journal of the Electrochemical Society, 129, 1694–1701 (1982). DOI: 10.1149/1.2124252
38. Brandup J., Immergut E. H., Grulke E. A., Abe A., Bloch D. R.: Polymer handbook. Wiley, New York (1999).
39. Le Nest J. F., Callens S., Gandini A., Armand M.: A new polymer network for ionic conduction. Electro - chimica Acta, 37, 1585–1588 (1992). DOI: 10.1016/0013-4686(92)80116-4
40. Nishimoto A., Watanabe M., Ikeda Y., Kohjiya S.: High ionic conductivity of new polymer electrolytes based on high molecular weight polyether comb polymers. Electrochimica Acta, 43, 1177–1184 (1998). DOI: 10.1016/S0013-4686(97)10017-2
41. Angell C. A., Xu K., Zhang S-S., Videa M.: Variations on the salt-polymer electrolyte theme for flexible solid electrolytes. Solid State Ionics, 86, 17–28 (1996). DOI: 10.1016/0167-2738(96)00088-4
42. Chandrasekhar V.: Polymer solid electrolytes: Synthesis and structure. Advances in Polymer Science, 135, 139–205 (1998). DOI: 10.1007/3-540-69191-X_2
43. Armand M., Gorecki W., Andreani R.: Perfluorosulphonimide salts as solute for polymer electrolytes. in 'Second International Symposium on Polymer Electrolytes' (ed.: Scrosati B.) Elsevier, London, 91–97 (1990).
44. Tang C., Hackenberg K., Fu Q., Ajayan P. M., Ardebili H.: High ion conducting polymer nanocomposite electrolytes using hybrid nanofillers. Nano Letters, 12, 1152–1156 (2012). DOI: 10.1021/nl202692y
45. Appetecchi G. B., Croce F., Dautzenberg G., Mastragostino M., Ronci F., Scrosati B., Soavi F., Zanelli A., Alessandrini F., Prosini P. P.: Composite polymer electrolytes with improved lithium metal electrode interfacial properties I. Elechtrochemical Properties of Dry PEO-LiX. Journal of the Electrochemical Society, 145, 4126–4132 (1998). DOI: 10.1149/1.1838925.

CHAPTER 2

ELECTRICAL CONDUCTION OF POLYIMIDE FILMS PREPARED FROM POLYAMIC ACID (PAA) AND PRE-IMIDIZED POLYIMIDE (PI) SOLUTION

T. Y. Kim[1], W. J. Kim[2], T. H. Lee[2], J. E. Kim[2], K. S. Suh[2]

[1]Institute of Materials Science, University of Connecticut, 97 North Eagleville Road, Storrs CT 06226, USA

[2]Department of Materials Science, Korea University, 5-1 Anam-dong, Seongbuk-ku, Seoul 136-701, South Korea

ABSTRACT

Electrical conduction characteristics in two different polyimide films prepared by the imidization of polyamic acid (PAA) and pre-imidized polyimide (PI) solution were investigated. It is found that the current density of the polyimide film from PAA was higher than that of the polyimide film from PI at the same electric field, even though the conduction mechanism in both polyimide films follows the ionic hopping model. The hopping distance was calculated to be 2.8 nm for PAA type and 3.2 nm for PI type polyimide film. It is also found that the decay rate of the residual electrostatic charges on the polyimide films becomes faster in the PAA type than in the PI type polyimide film.

INTRODUCTION

Thin films of polymeric materials have been used widely in the field of microelectronics for insulation, isolation and passivation. Particularly, polyimides are growing important especially in various applications because of their excellent thermal and electrical properties [1–4]. One of the most important applications of polyimide films is used as alignment layers in liquid crystal display (LCD) device, which enable liquid crystal to orient along its designated direction [5, 6]. For this application, polyimide films can be simply prepared by coating either polyamic acid or pre-imidized polyimide solution onto glass substrates where thin film transistor (TFT) and color filter (CF) patterns are already prepared. With the thermal curing process at a high temperature thin films of polyimides are formed on the glass substrate and are subjected to the rubbing process for creating an alignment path [1–6].

However, since polyimides are one of the typical polymers having high electrical resistance, the electrostatic charges can be easily created on the surface by rubbing it with polyester cloth. Once electrostatic charges are formed on the polyimide layers, they last for some time during the manufacturing process, creating their own electric field between the two alignment layers. Then, liquid crystal director responds to the electric field from the electrostatic charges on the polyimide films, and this leads to unwanted light-transmission modulation of the liquid crystal that can be recognized as ghost images on the LCD panel [6, 7]. Even though this phenomenon is one of the major problems frequently occurring in the LCD manufacturing process, its mechanism has not been well understood. Since this phenomenon is believed to be related to the electrostatic charge-retention properties of polyimide films, the measurement of the electrical properties for polyimide films prepared from different precursors become necessary. Here, we present the comparison of the electrical

Citation: T. Y. Kim, W. J. Kim, T. H. Lee, J. E. Kim, K. S. Suh, Electrical conduction of polyimide films prepared from polyamic acid (PAA) and pre-imidized polyimide (PI) solution, DOI: 10.3144/expresspolymlett.2007.60.

properties in polyimide films which were separately prepared from polyamic acid and preimidized polyimide solution. The electrostatic charge susceptibilities of the two polyimide films are discussed in terms of their charge transport behavior.

EXPERIMENTAL DETAILS

The experiments reported here were performed on polyimide films which were separately prepared from two different types of precursors, namely polyamic acid (PAA) and pre-polymerized polyimide (PI) solution. The polyimide films were prepared on a glass substrate by a solvent casting method and pre-baked at 80°C on a heating plate, then subjected to a thermal curing in a vacuum oven at 180°C for 90 min. Then, the polyimide films were taken as free standing films and compression molded using a hot press at 200°C. The film thickness was controlled by the concentration of precursor solution and pressure to be 100 μm. Silver was vacuum evaporated through a mask to the surface of the polyimide films to form electrodes.

The apparatus used to measure the conduction currents was composed of a power supply (Keithley 247), electrometer (Keithley 617), and test cell. The test cell contained a measuring electrode, the counter electrode, and a guard ring. The contact area of measuring electrode was 6.202 cm^2. The temperature of 50–100°C was controlled in an air convection oven. All experiments were continued for a time period long enough to reach a steadystate current. For the evaluation of electrostatic charge susceptibility, two sets of normally dark mode LCD panels were fabricated using PAA type and PI type polyimide films as an alignment layer. Then, 1 kV of DC voltages were applied across the LCD panels and the brightness of light through LCD panel was measured by luminance calorimeter (BM-7, Topcon) on the voltage switching-off.

RESULTS AND DISCUSSION

Figure 1 shows a typical J–E plot of two different polyimide films prepared by the imidization of polyamic acid and pre-imidized polyimide solutions. A classical way to identify the conduction mechanism is to examine linearity in J–E curves. The plot consists of two regions, in the lower region of the applied electric field, the slope is around 1, and in the higher region, the slope is around 2. This means that both of polyimide films follow the ohmic conduction in the lower range of electric field, whereas non-ohmic conduction mechanisms are involved in higher range of electric field. There are three different types of electronic conduction mechanisms, namely, space charge limited conduction (SCLC), Shottky conduction, Poole-Frenkel conduction, and one of these mechanisms would be dominant in higher range of electric field [8, 9].

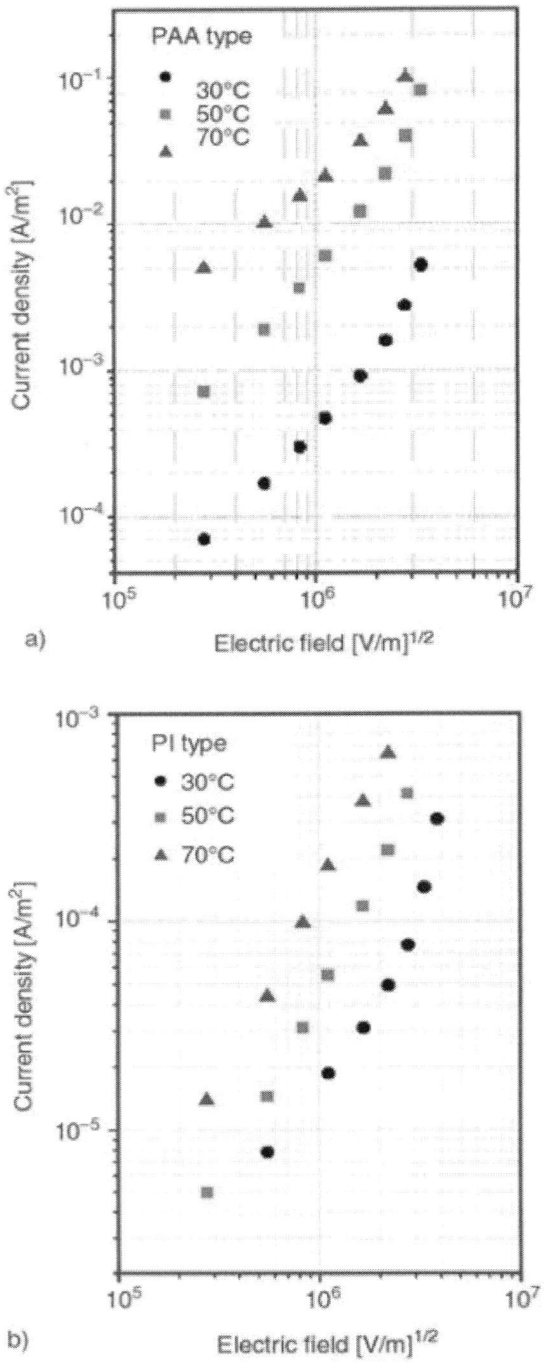

Figure 1: J–E curves of polyimide films: a) polyamic acid type and b) pre-imidized polyimide type.

When J is proportional to E^n , one can say that the sample obeys the SCLC mechanism. In this case, the slope of 2 represents the trap free SCLC, whereas the slope of 2–6 represents the trap modulated SCLC [8–12]. In Figure 1, the slope of 2 in higher field could be an indication that the samples follow the trap-free SCLC mechanism. However, for SCLC mechanism, the thickness dependence of the space charge limited current follows the relation of $J \propto d^{-n}$, where n is a parameter which depends on the trap distribution and is equal to or greater than 3 [8, 9].

Figure 2 shows the thickness dependence of the current density and the slopes of PAA- and PI-type polyimide samples are found to be 0.41 and –0.04, respectively. These values are much less than that required for space charge limited conduction so that SCLC conduction mechanism is ruled out. The conduction mechanisms of the polyimide films were further examined in terms of Schottky and Poole-Frenkel mechanisms which can be expressed by the Equation (1) [10–13]:

$$J = AT^2 \exp\left(\frac{\beta E^{\frac{1}{2}} - \phi}{kT}\right) \qquad (1)$$

where J, E, φ, k, and β are the current density, the electric field, the barrier height between the interfaces, the Boltzmann constant, and coefficient, respectively. The β coefficient is also given by the Equation (2):

$$\beta = \left(\frac{e^3}{a\pi\varepsilon\varepsilon_0}\right)^{\frac{1}{2}} \qquad (2)$$

where a = 1 for Poole-Frenkel effect and a = 4 for Schottky emission, ε is the dielectric constant and ε_0 is the permittivity of free space.

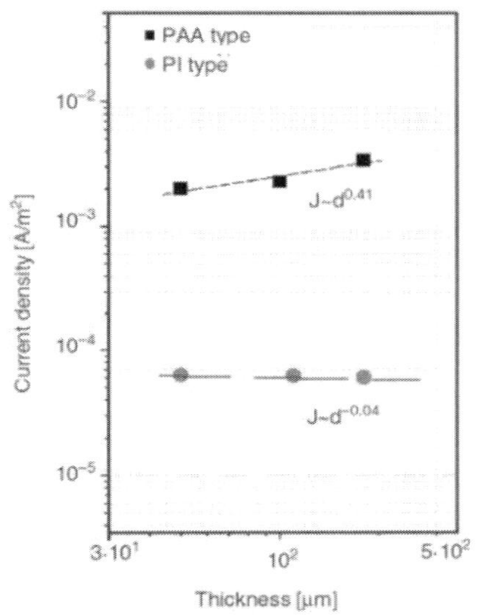

Figure 2: Thickness dependence of current density at 300 V in polyimide films.

According to Equation (1), a plot of logJ versus $E^{1/2}$ should yield a straight line if either of the above mentioned two mechanisms is dominant in polyimide films. The β coefficients are first calculated from the slope of logJ vs $E^{1/2}$ plots, and the dielectric constants are then estimated from these coefficients using Equation (2). If these dielectric constants agree with the values cited in the literature, it can be said that the samples follows the corresponding conduction mechanism.

The dielectric constants derived from the slope in Figure 3a and 3b range from 0.6 to 0.9 in the case of Schottky mechanism, whereas those from literatures range from 3.9 to 4.0 [14, 15]. A comparison of these values indicates that Schottky conduction mechanism is not the case for the polyimide films because dielectric constant of less than 1.0 is not conceivable. Since the dielectric constants by a Poole-Frenkel mechanism is twice of those by a Schottky mechanism, the Poole-Frenkel mechanism can be automatically excluded.

Measurement of complex impedance is one of the ways to determine whether the conduction mechanism is electronic or not. The measurement of complex impedance in the frequency range of 1 kHz– 5 MHz showed a semicircle for both types of polyimide samples, which implies that the dominant charge carriers are rather ions than electrons. When the ions existing in the polyimide samples are thermally activated and participate in the conduction process, the current density J is expressed in terms of ionic hopping conduction model as Equation (3) [16, 17]:

$$J = n\alpha\upsilon \exp\left(-\frac{U}{kT}\right)\exp\left(\frac{eE\alpha}{2kT}\right) \qquad (3)$$

where e, n, α, υ, U, k, and T are the charge of electron, carrier density, hopping distance, attempt to escape frequency, activation energy, Boltzmann constant, and absolute temperature.

Based on the ionic hopping model, Figure 4 compares the current density of two different types of polyimide films with respect to the applied electric field. It is clearly seen that the PAA type polyimide has the higher current density than PI type polyimide sample at the same electric field. In addition, since exp(–U/kT) term in Eqation (3) is independent of the electric field and can be treated as a constant with respect to electric field, the hopping distance α can be estimated from the slope of logJ vs E plot and the hopping distances for the PAA type and PI type polyimide were found to be 2.8 and 3.2 nm for 50°C, respectively. This is an indication that the ionic hopping motion is facilitated rather in PAA type than in PI type sample.

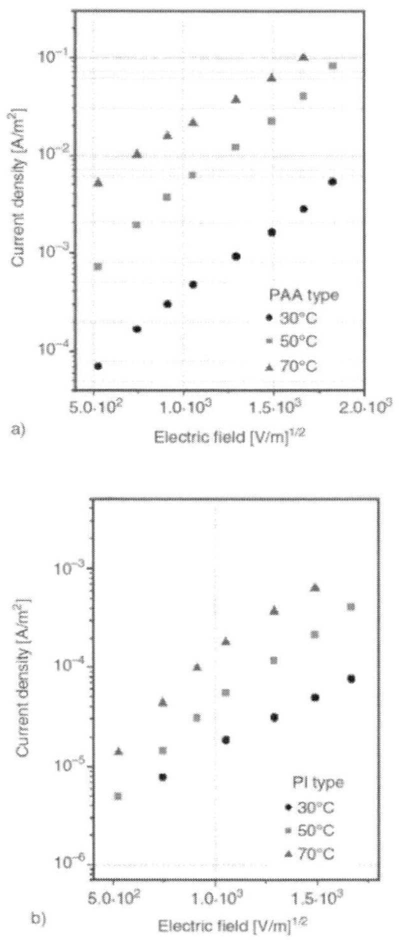

Figure 3: Plot of logJ vs $E^{1/2}$: a) polyamic acid type and b) pre-imidized polyimide type.

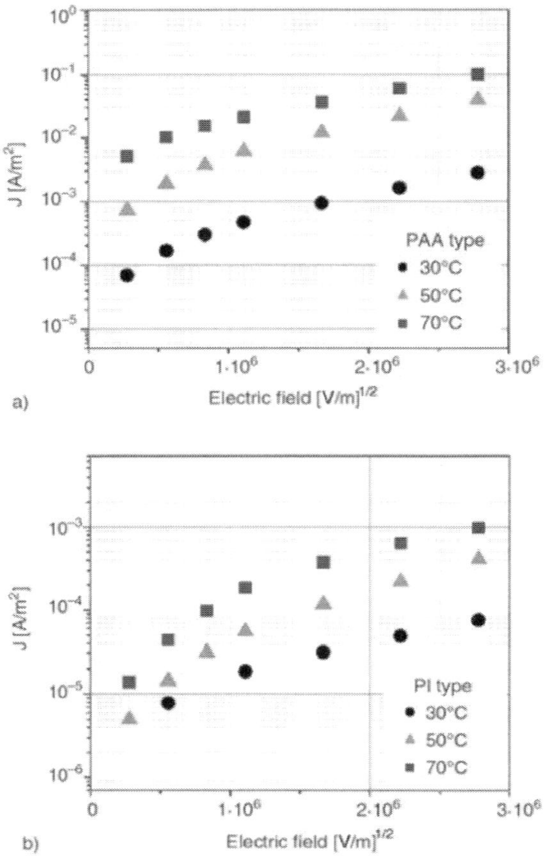

Figure 4: Plots of logJ vs E: a) polyamic acid type and b) pre-imidized polyimide type.

In order to verify that ionic hopping is more favorable in PAA type than in PI type sample, the activation energies of the two different samples were calculated from the slope of the ln[J/sinh(eEα/2kT)] vs 1/T plot in Figure 5. A linear relationship was observed in this figure, which further confirms that the conduction mechanism did not change over the corresponding temperature ranges. The activation energy were found to be 0.16 eV for PAA type polyimide and 0.19 eV for PI type polyimide and this corresponds with the results that the ionic conduction is much more favorable in PAA type polyimide.

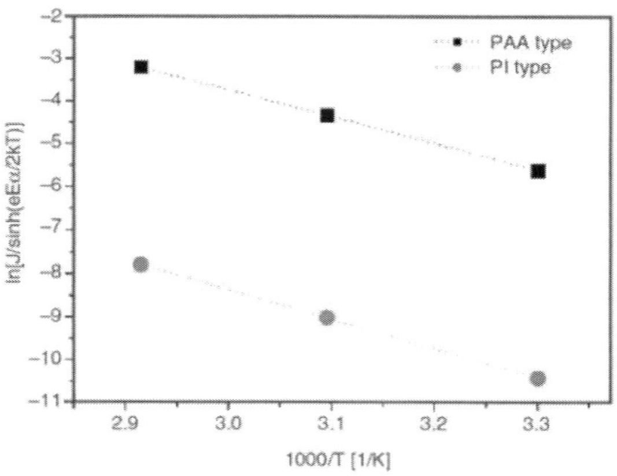

Figure 5: Temperature dependence of current densities in polyimide films.

Since ionic hopping conduction is known to originate from the ionic impurities such as polar and partially dissociable molecules, it is speculated that there are more impurities that contribute to the ionic conduction in PAA type sample [18, 19]. However, the sources of ionic conduction in PAA type polyimide are not identified at this moment.

In order to examine whether the ionic conductivity of polyimide films affects the electrostatic susceptibility of the LCD panel, two different sets of LCD panels were fabricated using PAA type and PI type polyimide films as an alignment layers and a DC voltage of 1 kV was applied across the LCD panels. When a voltage is applied across the normally dark LCD panels, the liquid crystal molecules respond instantaneously with an electric field, so that optical transmission increases. When the applied voltage is switched off, the optical transmission decreases as the liquid crystal molecules are restored to their original state. When charge accumulates on the alignment film, liquid crystal molecules are exposed to a residual voltage after the applied voltage is switched off. The slow decay of this residual voltage is expected to cause a slow change in the alignment of the liquid crystal molecules, and thus a slow change in the optical transmission.

Figure 6 compares the luminance of two LCD panels as a function of time. For the LCD panel with an alignment layer of PAA type polyimide film, the luminance decrease sharply on the removal of voltage application, whereas the slow change was observed in the case of the LCD panel with PI type polyimide film. This means that the residual charges remain for a longer period of time for PI type alignment film and these residual charges keep the liquid crystal light switch with a long relaxation time. In such case, the displayed image is retained after the applied voltage is switched off and recognized as a sticking image that is highly undesirable in the LCD manufacturing process. On the other hand, PAA type polyimide film shows higher ionic conductivity so that the residual charges can dissipate along or through the film or the charges can be neutralized by the mobile ions in PAA type polyimide film, enabling the liquid crystal molecule to be restored to its original state with a less probability of causing a sticking image.

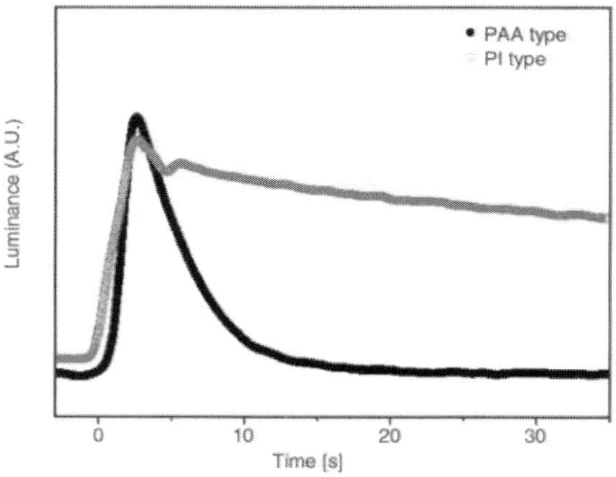

Figure 6: Comparison of the luminance with respect to time for LCD panels with PAA type and PI type polyimide films.

CONCLUSIONS

Polyimide films prepared from two different types of precursors are studied in terms of electrical conduction and electrostatic susceptibility. Although both polyimide films prepared from polyamic acid (PAA type) and pre-imidized polyimide (PI type) solution were governed by the thermally activated ionic hopping process, the polyimide film from PAA showed the current density higher than that from PI type solution at the same electric field. It is also found that the hopping distance and activation energy for PAA type sample was calculated to be 2.8 nm and 0.16 eV, while 3.2 nm and 0.19 eV for PI type sample.

Experiments for evaluating electrostatic susceptibility of the LCD panels incorporating two different polyimide alignment layers revealed that the residual charges on the PI type polyimide film remained over a time period of greater than 30 s, whereas the residual charges on the PAA type polyimide films decays rapidly. This implies that the charge accumulation is reduced for PAA type polyimide films so that image sticking problem can be minimized.

REFERENCES

1. Ghosh M. K.: Polyimides: Fundamentals and applications. Marcel Dekker, New York (1996).
2. Feger C., Khojasteh M. M., Htoo M.: Advances in polyimide science and technology. CRC Press, London (1993).
3. Mittal K. L.: Polyimides: Synthesis, characterization, and applications. Plenum Press, New York (1984).
4. Bessonov M. I., Zubkov V. A.: Polyamic acids and polyimides: Synthesis, transformations, and structure. CRC Press, London (1993).
5. Hasegawa M., Horie K.: Photophysics, photochemistry, and optical properties of polyimides. Progress in Polymer Science, 26, 259–335 (2001).
6. Hoogboom J., Rasing T., Rowan A. E., Nolte R. J. M.: LCD alignment layers. Controlling nematic domain properties. Journal of Materials Chemistry, 16, 1305–1314 (2006).
7. De Vleeschouwer H., Bougrioua F., Pauwels H.: Importance of ion transport in industrial LCD applications. Molecular Crystals and Liquid Crystals, 360, 29–39 (2001).
8. Lamb D. R.: Electrical conduction mechanisms in thin insulating films. Methuen, London (1967).
9. Blythe A. R.: Electrical properties of polymers. Cambridge University Press, New York (1979).
10. Nevin J. H., Summe G. L.: DC conduction mechanisms in thin polyimide films. Microelectronics and Reliability, 21, 699–705 (1981).
11. Suh K. S., Nam J. H., Lim K. J.: Electrical conduction in polyetherimide. Journal of Applied Physics, 80, 6333–6335 (1996).
12. Sessler G. M., Hahn B., Yoon D. Y.: Electrical conduction in polyimide films. Journal of Applied Physics, 60, 318–326 (2006).
13. Hill R. M.: Poole-Frenkel conduction in amorphous solids. Philosophical Magazine, 23, 59–86 (1971).
14. Simpson J. O., St Clair A. K.: Fundamental insight on developing low dielectric constant polyimides. Thin Solid Films, 308, 480–485 (1997).
15. Brandrup J., Immergut E. H.: Polymer handbook. Wiley, New York (1975).
16. Das-Gupta D.: Conduction mechanisms and high-field effects in syntheticinsulating polymers. Dielectrics and Electrical Insulation, IEEE Transactions on, 4, 149–156 (1997).
17. Mott N. F., Gurney R. W.: Electronic Processes in Ionic Crystals. Clarendon Press, Oxford 1940.
18. Bellucci F., Khamis I., Senturia, S. D., Latanision R. M.: Moisture effects on the electrical conductivity of kapton polyimide. Journal of the Electrochemical Society, 137, 1778–1784 (2006).
19. Lee Y. K., Murarka S. P.: Study on electrical characteristics of fluorinated polyimide film. Journal of Materials Science, 33, 4105–4109 (1998).

CHAPTER 3

EFFECT OF ORGANO-MODIFIED MONTMORILLONITE ON POLY (BUTYLENE SUCCINATE)/ POLY(BUTYLENE ADIPATE-CO-TEREPHTHALATE) NANOCOMPOSITES

B. W. Chieng, N. A. Ibrahim, W. M. Z. Wan Yunus

Chemistry Department, Faculty of Science, Universiti Putra Malaysia, 43400 UPM Serdang, Selangor, Malaysia

ABSTRACT

The composite material based on poly(butylene succinate) (PBS), poly(butylene adipate-co-terephthalate) (PBAT) and organo-modified montmorillonite (OMMT) were prepared by melt blending technique and characterized. Sodium montmorillonite (Na-MMT) was successfully modified by octadecylammonium (ODA) and dimethyldioctadecylammonium (DDOA) salts to become OMMT through cation exchange technique which is shown by the increase of basal spacing of clay by XRD. The addition of the OMMT to the PBS/PBAT blends produced nanocomposites which is proved by XRD and TEM. Tensile tests showed increase in tensile strength and modulus which is attributed to the existence of strong interactions between PBS/PBAT and clay, particularly with OMMT. Highest tensile strength of nanocomposite was observed at 1 wt% of OMMT incorporated. TGA study showed that the thermal stability of the blend increased after the addition of clays. SEM micrographs of the fracture surfaces show that the morphology of the blend becomes homogeneous and smoother with presence of OMMT.

INTRODUCTION

Nanocomposites based on biodegradable polyester and layered silicate have been extensively investigated because of their potential applications in the fields related to environmental protection. Biodegradable polymers are degradable polymers in which the degradation results from the action of naturally occurring microorganism such as bacteria, fungi and algae. Although biodegradable polymers have started making inroads into commercial applications, there is still lot of potential in the improvement of their properties for commercial large scale application.

Montmorillonite is one of the commercial layered silicates and has been widely used as reinforcing filler in the preparation of nanocomposites. However, the hydrophilic nature of montmorillonite limits its compatibility with organophilic polymers. Thus, chemical modification has been developed to make montmorillonite compatible with organophilic polymer. This can be readily achieved by modifying the montmorillonite with alkylammonium salts to form organo-montmorillonite (OMMT).

Poly(butylene succinate) (PBS) is a biodegradable aliphatic thermoplastic which synthesized by polycondensation of 1,4-butanediol with succinic acid. PBS possesses many desirable properties including biodegradability, melt processability, and thermal and chemical resistance. It can be processed in the field of textiles into melt-blown, multifilament, monofilament, flat and split yarn [1]. However, other properties such as gas

Citation: B. W. Chieng, N. A. Ibrahim, W. M. Z. Wan Yunus, Effect of organo-modified montmorillonite on poly(butylene succinate)/ poly(butylene adipate-co-terephthalate) nanocomposites, DOI: 10.3144/expresspolymlett.2010.51.

barrier properties, melt viscosity for further processing and so on, are frequently insufficient for various end-use applications [2]. Poly(butylene adipate-co-terephthalate) PBAT, trade name ECOFLEX® is an aliphatic/aromatic copolyester based on the monomers 1,4- butanediol, adipic acid and terephthalic acid. It is fully biodegradable polymer and is resistant to water. Typical applications are packaging films, agricultural films and compose bags.

Blending two or more polymers is an economic technique for producing new polymer materials at low cost and desirable properties combination for specific end use. They can be developed much more quickly than new polymers and require much less capital investment. However, thermodynamic immiscibility of polymers had limited the use of polymer blends in industry. These immiscible blends are characterized by a two phase morphology, narrow interphase, poor physical and chemical interactions across the phase boundaries, and poor mechanical properties. Many studies have focused on improving the compatibility between polymer components, either by adding third component, or by inducing a chemical reaction which leads to the modification of polymer interface [3]. To improve the properties of PBS, many researchers introduce clays to the PBS system [4–6]. Researchers also tried to improve the properties of PBS by blending with other polymers [7]. Jacob et al. [8] have studied the compatibility between PBS and PBAT. They revealed that this blend is immiscible but only one T_g was observed due to the close proximity of individual T_g's of both PBS and PBAT. Other reports have shown that OMMT can act as compatibilizer for immiscible polymer blends [9–15] which exhibit remarkable improvement in their properties compared to the pristine polymer or the conventional composite [16, 17]. The interlayer cations between the silicate layers such as Na^+ are exchanged with organic cations to increase the organonophilicity of the clay layers. This lowers the interfacial energy and improves wetting by the polymer matrix [16].

In this study, the OMMT was used as compatibilizer to improve the miscibility of the PBS/PBAT blends. The PBS/PBAT composites were prepared by the melt blending technique and the effect of OMMT on physical, mechanical and thermal properties was investigated. The compatibility between the PBS and PBAT was studied by examining the fracture surface of the tensile test specimens under SEM.

EXPERIMENTAL

Materials

Poly(butylene succinate) PBS, was supplied by Showa Highpolymer Co. Ltd. (Japan) under tradename 'BIONOLLE', grade #1020. The PBS pellets were dried in an oven at 60°C for 24 hours before processing. The poly(butylene adipate-co-terephthalate) PBAT, trade name ECOFLEX® F BX 7011 was supplied by BASF. PBAT was supplied in pellet form and used as received. Properties of BIONOLLE and ECOFLEX are summarized in Table 1. Na-MMT (Kunipia F) with cation exchange capacity of 119 meq/100 g was obtained from Kunimine Ind. Co. Japan and used as received. Octadecylamine and dimethyldioctadecylammonium bromide was purchased from Acros Organics and used as an organic modifier of montmorillonite.

Table 1: Properties of BIONOLLE and ECOFLEX

Properties	BIONOLLE	ECOFLEX
Melting point	116°C	110°C
Glass transition	–31°C	–29°C
Melt mass-flow rate (190°C/2.16 kg)	20–29 g/10 min	2.7–4.9 g/10 min
Molecular weight (M_w)	~40 000	40 000
Crystallinity	30–50%	~20%

Preparation of OMMT

The organo-modified montmorillonite (OMMT) was prepared according to the published method with slight modification [18]. The octadecylammonium-montmorillonite (ODA-MMT) was prepared by adding a dispersed 20.00 g of sodium montmorillonite in 800 ml of distilled water at 80°C into a solution of 13.476 g octadecylamine and 4.81 ml concentrated hydrochloric acid in 200 ml hot distilled water. The resultant suspension was vigorously stirred for 1 hour. The precipitate was repeatedly filtered and washed with hot distilled water until no chloride ion was detected with 0.1 N $AgNO_3$ solution. It was then dried at 60°C for 24 hours. The OMMT was ground with a mortar and sieved into particles of size less than 100 μm which will then used for the preparation of nanocomposites. Similar procedure was adopted to prepare dimethyldioctadecylammonium-montmorillonite (DDOA-MMT) with DDOA as intercalation agent.

Preparation of poly(butylene succinate) (PBS)/ poly(butylene adipate-co-terephthalate) (PBAT)/clay nanocomposites

Nanocomposites containing a thermoplastic blend and clay were produced by melt compounding. The blend composition was kept constant (PBS 70 wt% + PBAT 30 wt%), whereas the clay content was varied between 0 and 7 wt%. PBS, PBAT and clay were manually premixed in a container and fed into Brabender at 130°C and rotor speed of 50 rpm. The residence time was maintained at 10 minutes for all the preparation. The products were then compression molded into sheets of 1 mm thickness by an electrically heated hydraulic press with a force of 1500 kN at 125°C for 10 minutes. The sample sheets were then use for further characterization.

X-ray diffraction (XRD)

X-ray diffraction measurement was carried out by using a Shimadzu XRD 600 X-ray diffractometer with CuK_α radiation (λ = 1.542 Å) operated at 30 kV and 30 mA. Data were collected within the range of scattering angles (2θ) of 2 to 10° at the rate of 2°/min. The basal spacing of the clay was derived from the peak position (d_{001} reflection) in the XRD diffractogram according to the Bragg equation ($\lambda = 2d\sin\theta$).

Fourier Transform Infrared (FTIR)

Fourier transform infrared spectra were recorded using a Spectrum BX Perkin Elmer using KBR disk method in the range of frequency 280 to 4000 cm^{-1} at 25°C. FTIR spectrum was recorded using transmission method. For each spectrum, 128 consecutive scans with 4 cm^{-1} resolution were averaged.

Tensile properties

Tensile properties test were carried out by using Instron 4302 series IX. The samples were cut into the dumbbell shape follow the ASTM D638 (type V) standard. Average thickness and average width of the gauge section of each specimen were calculated using three measurements of the thickness and width respectively. A digital micrometer, Mitutoyo (Japan) with an accuracy of 0.001 mm was used in the measurements. Load of 1.0 kN was applied at constant crosshead speed of 10 mm/min at room temperature. Tensile strength and tensile moduli were evaluated from the stress-strain data. Seven specimens were tested and the average of the values was taken.

Thermogravimetric Analysis (TGA)

Thermogravimetric analysis was performed using a Perkin Elmer TGA7. The weight of the samples used was about 10.0 mg and were heated from 35 to 600°C at the heating rate of 10°C/min. The analysis was carried out in nitrogen atmosphere with nitrogen flow rate of 20 ml/min. The weight loss of samples were recorded and plotted as the function of temperature.

Scanning Electron Microscopy (SEM)

Scanning Electron Microscopy was done using JEOL scanning microscope JSM-6400, Japan operated at 20 kV. Fracture surfaces were obtained from plain strain fracture tensile tested specimens. The fracture surfaces were sputter coated with gold. A selective dissolution of polyester in dimethylformamide (DMF) was used to distinguish phases. The sample was held in DMF for 10 minutes, removed and dried. The surface of the blend was examined for changes in its phase morphology.

Transmission Electron Microscopy (TEM)

Transmission Electron Microscopy image were obtained using a Hitachi H-700 TEM operated at an accelerating voltage of 100 kV to observe the nanoscale structures of the composites. All samples were ultrathin-sectioned using a microtome equipped with a diamond knife. The samples were collected in a trough filled with water and placed on a 200 mesh copper grid.

RESULTS AND DISCUSSION

Characterization of OMMT

Fourier Transform Infrared (FTIR)

FTIR spectra of Na-MMT and OMMT (ODA-MMT and DDOA-MMT) are illustrated in Figure 1. For the unmodified montmorillonite, Na-MMT, the presence of broad band at 3400 cm^{-1} indicates the presence of the free water molecules vibration.

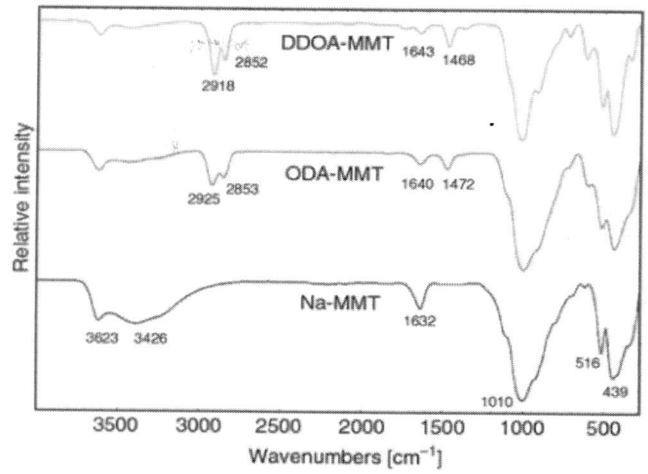

Figure 1: FTIR spectra of Na-MMT, ODA-MMT and DDOA-MMT.

This less intense band is ascribed to water hydrogen bonded to other water molecules within the interlayer of the montmorillonite. These molecules are involved with the structure of the hydration sphere of the Na^+ cation within the montmorillonite interlayer. It also shows that the hydrophilic nature of Na-MMT and water present in the Na-MMT. The infrared spectrum of Na-MMT exhibited four unique characteristic peaks which correspond to the O–H stretching (3623 cm^{-1}), Si–O stretching (1010 cm^{-1}), Al–O stretching (516 cm^{-1}) and Si–O bending (439 cm^{-1}). The OMMT presents three new peaks in the FTIR spectrum compared to NaMMT. The Bands around 2918 and 2852 cm^{-1} are attributed to the C–H asymmetric and symmetric stretching vibrations of surfactant, respectively.

The band around 1468 cm^{-1} is assigned to the CH_2 methylene (scissoring) vibration. The existence of these three new peaks in OMMT confirm the intercalation of alkylammonium ion in the interlayer spaces of montmorillonite [19]. This can further be confirmed with XRD observation.

On the other hand, the OMMT also become more organophilic as shown by the reduction in intensity of free O–H stretching peak. Furthermore, the H–O–H bending vibration of the water molecules is sensitive to the clay surface property transformation from hydrophilic to organophilic. The shifting of the H–O–H bending (1632 cm^{-1}) to a higher frequency (~1640 cm^{-1}), indicationg that H_2O is less hydrogen bonded and total sorbed water content decreases. Hydrophobic cations in organoclay will weaken the hydrogen bonding of the water. For the weaker hydrogen bond, lower energy is required for H–O–H bending vibration and thus a higher frequency of the H–O–H can be observed [20]. IR spectrum of OMMT keeps all the characteristic absorption bands of the host structure Na-MMT, it indicates that surfactants are intercalated into the silicate layer of Na-MMT. Therefore the clay was successfully modified to become OMMT.

X-ray diffraction analysis (XRD)

The changes of basal spacing of the resulted materials can reflect the intercalation of surfactant cation into montmorillonite interlayer spaces [21]. The natural montmorillonite, Na-MMT has 1.24 nm of basal spacing at $2\theta = 7.12°$ before it was modified with different surfactants. This basal spacing value is similar to the value obtained from the XRD measurement of previous work [22]. The values of 2θ and basal spacing of clays are summarized in Table 2. The XRD patterns for the clays are illustrated in Figure 2. After being modified by the surfactant ODA and DDOA, the basal spacing of the clay increase from 1.24 to 2.83 and 3.20 nm respectively. These indicated that the exchange cation is intercalated into the galleries of silicate layers after it is exchanged with the sodium ion. Thus montmorillonite is successfully modified by the surfactant and OMMT is formed.

Table 2: Basal spacing of Na-MMT, ODA-MMT and DDOA-MMT

Clay	Exchange cations	2θ [°]	Basal spacing [nm]
Na-MMT	Na^+	7.12	1.24
ODA-MMT	$CH_3(CH_2)_{17}NH_3^+$	3.12	2.83
DDOA-MMT	$[CH_3(CH_2)_{17}]_2N(CH_3)_2^+$	2.76	3.20

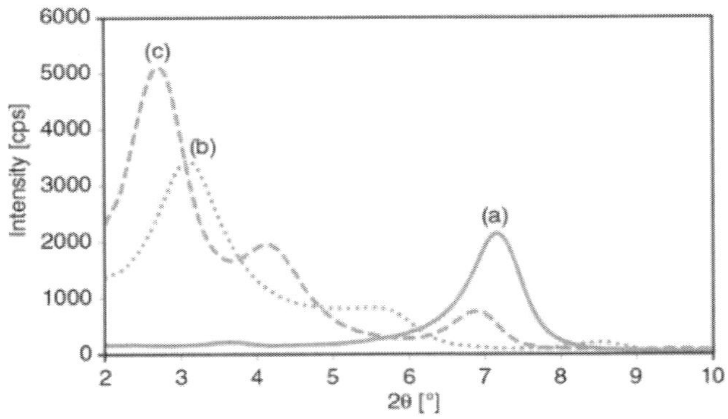

Figure 2: XRD patterns of (a) Na-MMT, (b) ODA-MMT and (c) DDOA-MMT.

There are three peaks observed for DDOA-MMT. The three spacings are attributed to the arrangement of the DDOA surfactant molecule in the clay layers and the presence of unmodified clay. The peak at 2θ = 6.88° (1.28 nm) is a similar space to the Na-MMT, so this spacing is attributed to unmodified Na-MMT. The peak at 2θ = 4.14° (2.13 nm) may be due to the molecular arrangement with the surfactant molecules laying flat to the surface, and peak at2θ = 2.76° (3.20 nm) is attributed to the surfactant molecules at 90° to the surface [23].

On the other hand, ODA-MMT shows only two peaks in the XRD pattern [24]. The peak at 2θ = 3.12° (2.83 nm) corresponds to the basal spacing of ODA-MMT. Another peak at 2θ = 5.75° (1.54 nm) attributed to the arrangement of surfactant molecules parallel to the surface of clay [23].

Characterization of nanocomposites

X-ray diffraction analysis

The nanostructure of layered silicate nanocomposites is established using XRD analysis at low angles (2θ < 10°). Figure 3 shows the XRD pattern of Na-MMT and PBS/PBAT/Na-MMT composites at different clay loadings. Due to the hydrophilic nature of Na-MMT, it is hardly intercalated into the polymer blend. The basal spacing of composites with the clay loading of 1 and 3 wt% could not be detected, which may be due to low clay content or absence of any ordered layer structure. This lack of intergallery clay diffraction is due to the exfoliation and random distribution of the clay platelets within the polymer blend [25]. When 5 and 7 wt% of Na-MMT is added to the polymer blend, the diffraction angle of Na-MMT is shifted from 7.12° (1.24 nm) to lower angle 6.48° (1.36 nm) and 6.58° (1.34 nm) respectively. The intensity of the characteristic peak of Na-MMT was reduced significantly, but almost retains its position, indicating the formation of a conventional or phase separated composites [26]. Such structure was expected since polymer matrices have no interaction with the highly hydrophilic Na-MMT surface.

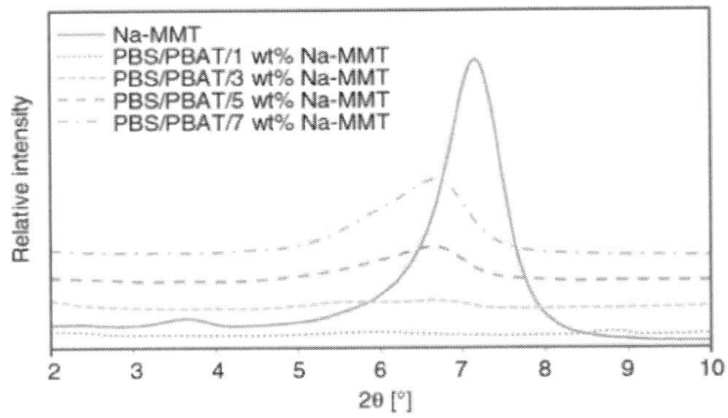

Figure 3: XRD patterns of Na-MMT and PBS/PBAT composites at different Na-MMT clay loadings.

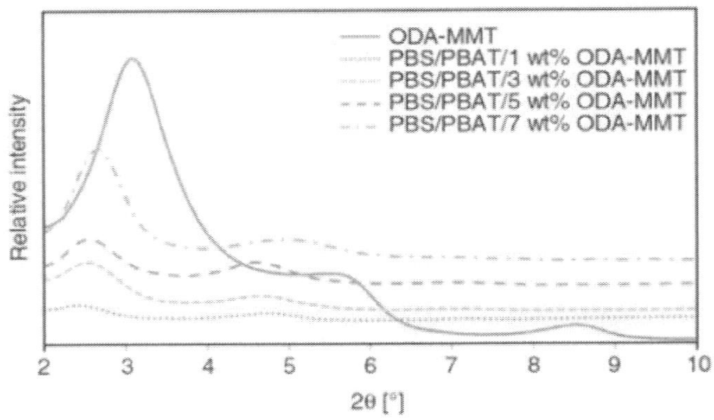

Figure 4: XRD patterns of ODA-MMT and PBS/PBAT composites at different ODA-MMT clay loadings.

The intercalation of the polymer chains would increase the interlayer spacing compared with that of clay, which would shift the peak towards a lower angle. Figure 4 shows the XRD pattern of ODAMMT and PBS/PBAT/ODA-MMT nanocomposites. ODA-MMT shows a characteristic peak at diffraction angle of 3.12° corresponding to basal spacing 2.83 nm. The increases in the basal spacing of ODA-MMT from 2.83 to 3.59, 3.45, 3.44 and 3.31 nm when 1, 3, 5 and 7 wt% of ODA-MMT respectively incorporated into PBS/PBAT, indicated that the polymer chain is successfully intercalated into the clay layers forming intercalated type nanocomposites.

Figure 5 shows the XRD pattern of DDOA-MMT and PBS/PBAT/DDOA-MMT composites. DDOAMMT shows a sharp peak at diffraction angle of 2.76° corresponding to 3.20 nm of basal spacing.

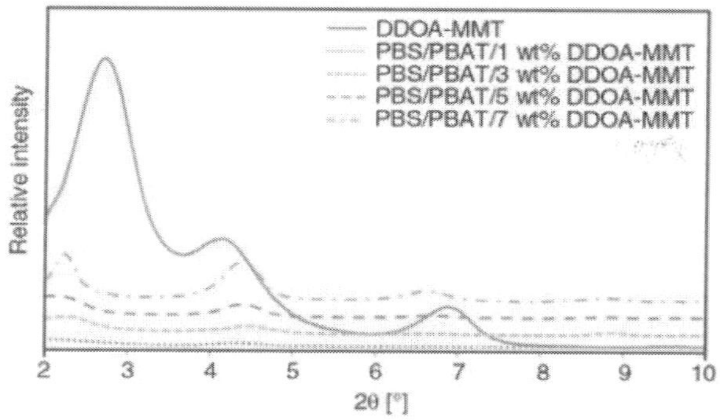

Figure 5: XRD patterns DDOA-MMT and PBS/PBAT composites at different DDOA-MMT clay loadings.

Table 3: Basal spacing of PBS/PBAT composites at various clay loadings

Sample	2θ [°]	Basal spacing [nm]
Na-MMT	7.12	1.24
PBS/PBAT/1 wt% Na-MMT	–	–
PBS/PBAT/3 wt% Na-MMT	–	–
PBS/PBAT/5 wt% Na-MMT	6.48	1.36
PBS/PBAT/7 wt% Na-MMT	6.58	1.34
ODA-MMT	3.12	2.83
PBS/PBAT/1 wt% ODA-MMT	2.46	3.59
PBS/PBAT/3 wt% ODA-MMT	2.56	3.45
PBS/PBAT/5 wt% ODA-MMT	2.57	3.44
PBS/PBAT/7 wt% ODA-MMT	2.67	3.31
DDOA-MMT	2.76	3.20
PBS/PBAT/1 wt% DDOA-MMT	–	–
PBS/PBAT/3 wt% DDOA-MMT	2.31	3.81
PBS/PBAT/5 wt% DDOA-MMT	2.28	3.88
PBS/PBAT/7 wt% DDOA-MMT	2.24	3.93

The increases in the basal spacing of DDOA-MMT from 3.20 to 3.81, 3.88, and 3.93 nm when 3, 5 and 7 wt% of DDOA-MMT respectively incorporated into PBS/PBAT, indicated that the polymer chain is successfully intercalated into the clay layers. The existence of sharp peaks shows that all the composites still retain an ordered structure. For the case of PBS/PBAT/1 wt%, the characteristic peak of the original clay disappeared, indicating that the clay was fully exfoliated in the blend matrix. Thus, it could be said that it forming the exfoliated type of nanocomposite and will be confirmed by TEM. Basal spacing of all nanocomposites was summarized in Table 3.

Transmission Electron Microscopy (TEM)

The dispersion of layered silicates in the composites were observed by TEM as shown in the Figure 6, which presents PBS/PBAT composites with 1 wt% of (a) Na-MMT and (b) DDOA-MMT. The dark lines are the cross section of intercalated silicate layers and the bright areas are the matrix.

In PBS/PBAT/Na-MMT, the clay persists as tactoids of agglomerates throughout the polymer matrix as shown in Figure 6a. This incomplete dispersal of the reinforcing phase inhibits ideal surface contact between the polymer and clay, creating large regions of clay tactoids in the composite. This observation is in agreement with the XRD result which shows little shifting in the diffraction peak of Na-MMT for PBS/PBAT/Na-MMT composites and formed as immiscible or phase separated microcomposites. PBS/PBAT/ODA-MMT (Figure 6b) shows that the layers of clay were disordered and the PBS/PBAT matrix was intercalated into the gallery of the clay. It is consistent with the XRD result shows that a large increase in basal spacing of the clay and formed intercalated type of nanocomposites.

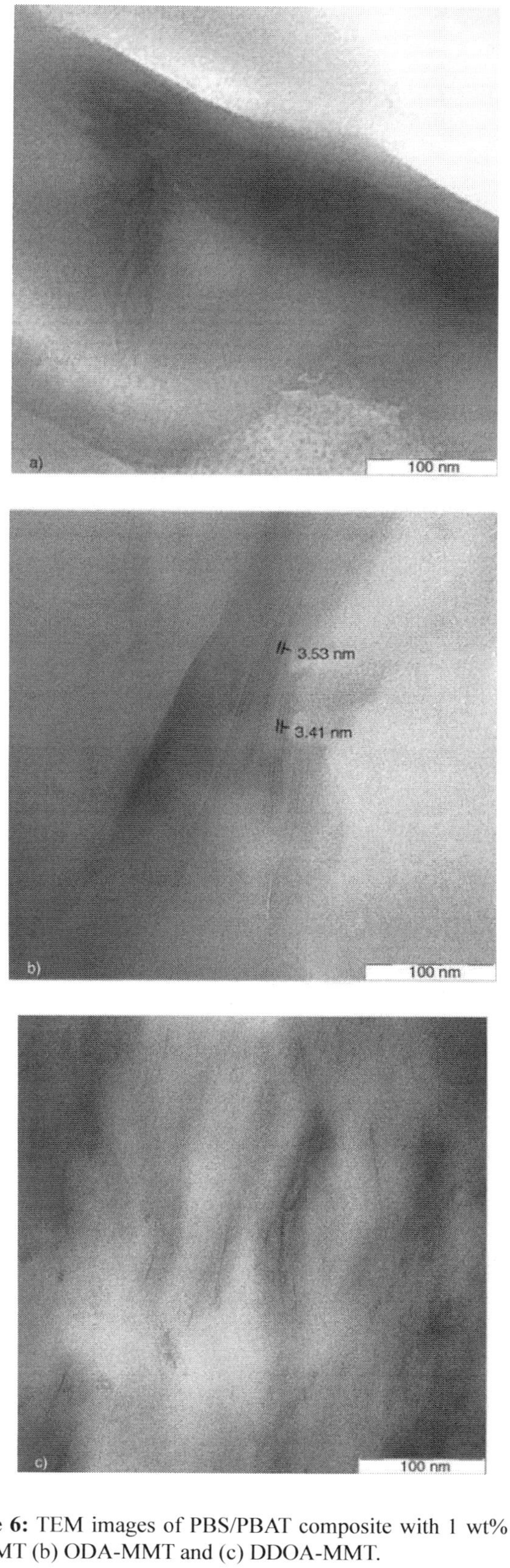

Figure 6: TEM images of PBS/PBAT composite with 1 wt% of (a) Na-MMT (b) ODA-MMT and (c) DDOA-MMT.

On the other hand, TEM image of PBS/PBAT/ DDOA-MMT shown in Figure 6c demonstrate single silicate layers homogenously dispersed in the polymer matrix, indicates the formation of a fully exfoliated nanocomposite. This finding shows that DDOA-MMT exhibits better dispersion of clay layers in the PBS/PBAT matrix, and is in accordance with the XRD observation shown in Figure 5, supports the formation of an exfoliated type nanocomposite. Because of that, better homogeneity and tensile properties were achieved for the PBS/PBAT/DDOA-MMT which will be proved by SEM and tensile tests.

Tensile roperties

The tensile properties of PBS/PBAT/Na-MMT and PBS/PBAT/OMMT composites at various contents of clays are shown in Figure 7 and Figure 8, respectively. The tensile strength of pristine PBS/PBAT blend is 21.20 MPa. Tensile strength of PBS/ PBAT/Na-MMT decreased as the clay content increases. This is due to the hydrophilic nature of Na-MMT which is incompatible with the hydrophobic polymer matrix. However, the reduction could also be attributed to poor dispersion and agglomeration of Na-MMT in the blends as shown in Figure 6a.

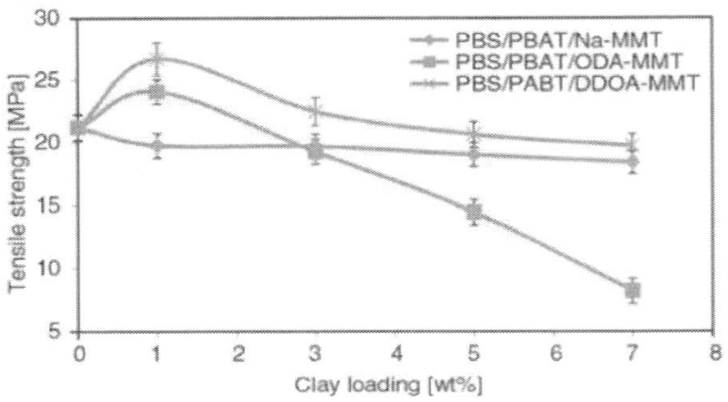

Figure 7: Tensile strength of PBS/PBAT composites.

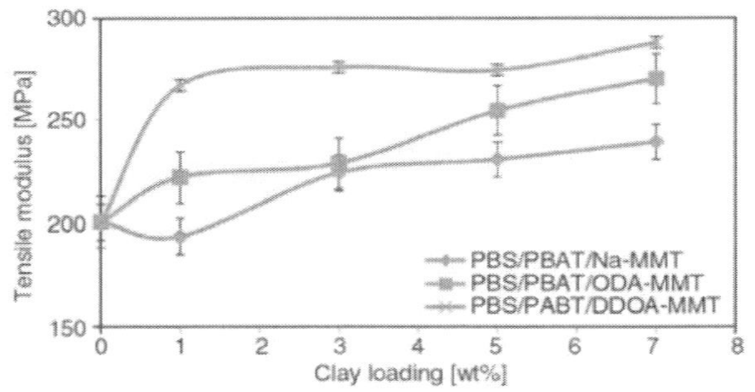

Figure 8: Tensile modulus of PBS/PBAT composites.

It is clearly seen that the highest tensile strength was achieved at 1 wt% of OMMT clay loading. This suggests that OMMT acts as reinforcing filler, which increases interaction at the phase boundaries upon the addition of OMMT. Strong interphase interaction between the matrix and the dispersed phase is believed reduce the stress concentration point when tensile load is applied. The OMMT is able to act as reinforcing filler due to its high aspect ratio and platelet structure. However, when the amount of clay is above 1 wt%, only a part of the clay locates in the interfacial area, and the excess is dispersed in the matrix affecting its homogeneity and consequently the tensile strength of the blends [27, 28]. Some of the clay is present in the form tactoids, remained partially intercalated and stacked, which weakened the reinforcing effect [29].

Thus, the reinforcing effect is reduced for nanocomposites with higher clay loading owing to poor dispersion of clay and hence unable to transfer stress efficiently. When comparing the OMMT as filler for PBS/PBAT blend, it is observed that at 1 wt% clay loading the reinforcing effect of DDOA-MMT is more noticeable.

Incorporation of clay into PBS/PBAT blend also increases the tensile modulus [30]. Clay has been found to be efficient in stiffening polymers [31]. The tensile modulus of PBS/PBAT/ODA-MMT increased up to 269.9 MPa, whereas the tensile modulus of PBS/PBAT/DDOA-MMT increased up to 287.7 MPa. The increase in tensile modulus is mainly due to the high aspect ratio and rigidity of clay layers. However, the tensile modulus increase in PBS/PBAT/OMMT is higher than PBS/PBAT/ Na-MMT at all clay loadings. This indicates that OMMT is more compatible with polymer blend compared to Na-MMT, and hence increase the stiffness of the nanocomposites. Once again the OMMT act as reinforcing filler better than the unmodified clay, Na-MMT. The enhancement of tensile modulus was reasonably ascribed to the constraint of the polymer chains by their interaction with the clay surfaces [32].

Thermogravimetric Analysis (TGA)

Generally, introduction of clay into polymeric matrices can improve their thermal stabilities since the clay can hinder the permeability of volatile degradation products out of the materials. The dispersed clay generates a barrier which delays the release of thermal degradation products in comparison the pristine polymer [22]. Figures 9 and 10 show the TG and DTG thermograms respectively of PBS/PBAT blend and PBS/PBAT composites. The weight loss of the blend and nanocomposites due to degradation is monitored as a function of temperature. The characteristic thermal parameters selected were onset temperature, which is the initial weight loss temperature, and maximum degradation temperature, which is the highest thermal degradation rate temperature. The results are summarized in the Table 4.

The PBS/PBAT blend shows the onset temperature of 215.3°C, which increased to 223.6, 237.5 and 234.0°C when Na-MMT, ODA-MMT and DDOA MMT clays respectively incorporated into the blend. The incorporation of clay into polymer matrix was found to enhance the thermal stability. The improved thermal stability attributed to an ablative reassembling of the silicate layers which may occur on the surface of the nanocomposites creating a physical protective barrier on the surface of the material which hinder the diffusion of volatiles and assist the formation of char after decomposition [33]. However, the OMMT gave higher thermal stability compared to the hydrophilic Na-MMT indicating better interaction between PBS/PBAT matrix and OMMT. The maximum degradation temperature (T_{max}) also increased by compounding the PBS/PBAT with OMMT.

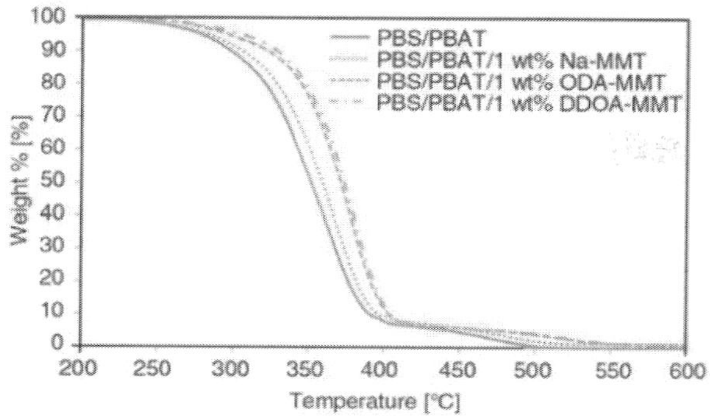

Figure 9: TG thermograms of PBS/PBAT and PBS/PBAT composites.

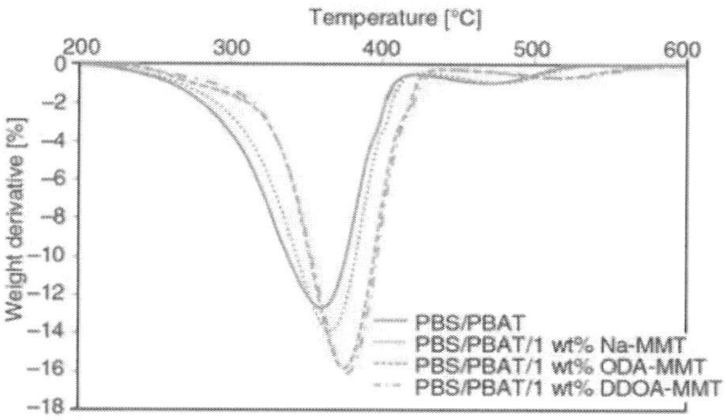

Figure 10: DTG thermograms of PBS/PBAT and PBS/PBAT composites.

Table 4: TG results of PBS/PBAT and PBS/PBAT composites

Sample	Onset temperature [°C]	Maximum degradation temperature T_{max} [°C]
PBS/PBAT	215.3	360.4
PBS/PBAT/Na-MMT	223.6	366.6
PBS/PBAT/ODA-MMT	237.5	375.4
PBS/PBAT/DDOA-MMT	234.0	377.0

Scanning Electron Microscopy (SEM)

The fracture surface of the nanocomposites was examined by scanning electron microscope to study the morphology of the surface. Figure 11 shows SEM micrographs of fractured surface of PBS/ PBAT blend and its nanocomposites at magnification of 1000×.

According to Jacob et al. [8], if the blend components have different melt viscosities, the morphology of the resulting blend will show the finely dispersed phase of the component with lower viscosity. Higher viscosity will lead to coarse dispersion of phases in a spherical domain. However in the case of blends containing higher concentrations of PBS, a reversal of the continuous phase was visible. Thus for PBS/ PBAT blend, the PBAT with higher viscosity formed the continuous phase as shown by red arrows in the Figure 11a. This can further confirmed by etching the PBS/PBAT blend with dimethylformamide (DMF). PBS is soluble in DMF, while PBAT is insoluble. The DMF was used to dissolve the soluble phase. Figure 12 represents the micrograph of the PBS/ PBAT blend after etching with DMF. The micrograph clearly showed that the PBS phase dissolved in the DMF and PBAT phase stayed intact during the etching process. However, PBS/ PBAT nanocomposites dissolved in DMF solvent during the same etching process. This means that the PBS/PBAT blend was a miscible blend after incorporation of OMMT.

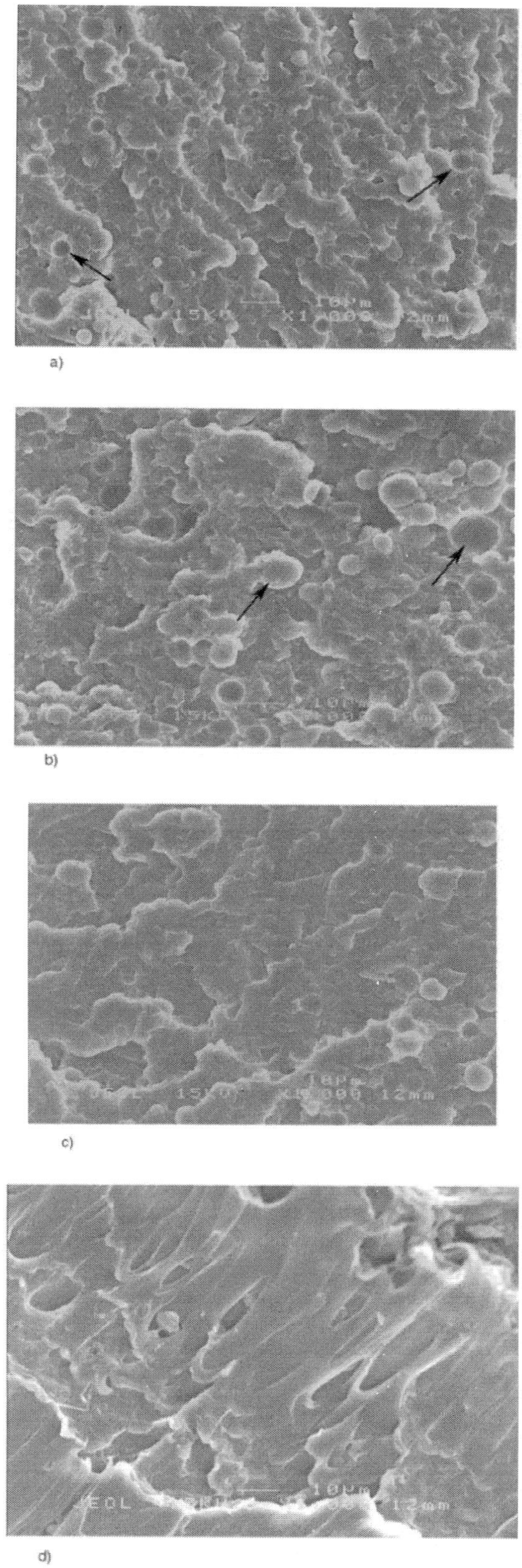

Figure 11: SEM images of (a) PBS/PBAT, and PBS/PBAT composite with 1 wt% of (b) Na-MMT, (c) ODA-MMT and (d) DDOA-MMT.

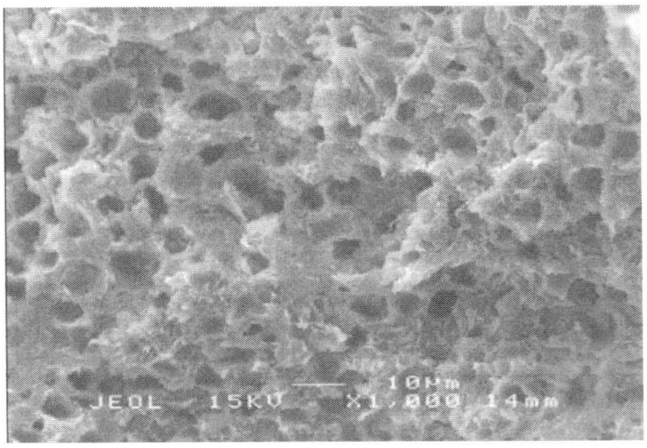

Figure 12: SEM micrograph of PBS/PBAT blend after etching with DMF.

This finding reveals that OMMT can improve the miscibility of the blend.

When OMMT was incorporated into PBS/PBAT, there is a reduction of the domain size of the dispersed phase (PBS) as shown by black arrows which is related to a reduction of the coalescence rate of the dispersed phase and a compatibilizing effect of the OMMT at the interface between the two phases as shown in Figure 11c and 11d. Conversely, there often appears to be an increase in the dispersed phase domain size when OMMT is located in the dispersed phase [9, 34]. So, the OMMT is located in the PBAT continuous phase in this PBS/PBAT blend system. According to Martins et al. [35], when the OMMT platelets are located in the continuous phase instead of the dispersed phase would lead to higher stiffness and this is in agreement with the tensile modulus results.

Some small void could be observed in PBS/PBAT/ Na-MMT (Figure 11b) compared to pristine PBS/ PBAT which indicates that the incompatibility of Na-MMT with the PBS/PBAT matrix. The presence of OMMT influences the morphological structural changes of the blend. As shown in Figure 11c and 11d, the PBS/PBAT phases have affinities to make nanolevel incorporation with OMMT. It indicates a good degree of mixing between the polymer phases with addition of OMMT, since this composite totally dissolved in DMF after addition of OMMT. As a result, the nanocomposite showed more homogeneous and single phase morphology compared to unmodified clay. In other word, the OMMT induces miscibility between PBS and PBAT.

CONCLUSIONS

OMMTs were successfully prepared from Na-MMT via cation exchange method by using octadecylammonium and dimethyldioctadecylammonium cation as surfactant. Intercalated and exfoliated types of PBS/PBAT/OMMT nanocomposites were successfully prepared by melt blending as proven by XRD and TEM. The tensile strength of PBS/PBAT/ OMMT nanocomposites was improved by reinforcing effects of OMMT and gives the maximum value at 1 wt% OMMT loading. Thermal stability was enhanced by incorporation of clay into PBS/ PBAT matrices, particularly OMMT. SEM micrographs showed that presence of OMMT in the PBS/PBAT blend improved the morphology of the compound. SEM micrographs also supported the finding of compatibility for the PBS/PBAT blend with OMMT through the homogeneous distribution which gives an improvement in mechanical properties.

REFERENCES

1. Fujimaki T.: Processability and properties of aliphatic polyesters, 'BIONOLLE', synthesized by polycondensation reaction. Polymer Degradation and Stability, 59, 209–214 (1998). DOI: 10.1016/S0141-3910(97)00220-6
2. Shih Y., Wang T., Jeng R., Wu J., Teng C.: Biodegradable nanocomposites based on poly(butylene succinate)/ organoclay. Journal of Polymers and the Environment, 15, 151–158 (2007). DOI: 10.1007/s10924-007-0055-6
3. Chen G-X., Kim H-S., Kim E-S., Yoon J-S.: Compatibilization-like effect of reactive organoclay on the poly(l-lactide)/poly(butylene succinate) blends. Polymer, 46, 11829–11836 (2005). DOI: 10.1016/j.polymer.2005.10.056
4. Okamoto K., Ray S. S., Okamoto M.: New poly(butylene succinate)/layered silicate nanocomposites. II. Effect of organically modified layered silicates on structure, properties, melt rheology, and biodegradability. Journal of Polymer Science Part B: Polymer Physics, 41, 3160–3172 (2003). DOI: 10.1002/polb.10708
5. Chen G-X., Kim E-S., Yoon J-S.: Poly(butylene succinate)/twice functionalized organoclay nanocomposites: Preparation, characterization, and properties. Journal of Applied Polymer Science, 98, 1727–1732 (2005). DOI: 10.1002/app.22264
6. Someya Y., Nakazato T., Teramoto N., Shibata M.: Thermal and mechanical properties of poly(butylene succinate) nanocomposites with various organo-modified montmorillonites. Journal of Applied Polymer Science, 91, 1463–1475 (2004). DOI: 10.1002/app.13366
7. Yu L., Dean K., Li L.: Polymer blends and composites from renewable resources. Progress in Polymer Science, 31, 576–602 (2006). DOI: 10.1016/j.progpolymsci.2006.03.002
8. Jacob J., Ramaswamy M., Mrinal B.: Evaluation of compatibility and properties of biodegradable polyester blends. Journal of Polymer Science Part A: Polymer Chemistry, 40, 2003–2014 (2002). DOI: 10.1002/pola.10297
9. Khatua B. B., Lee D. J., Kim H. Y., Kim J. K.: Effect of organoclay platelets on morphology of nylon-6 and poly(ethylene-ran-propylene) rubber blends. Macromolecules, 37, 2454–2459 (2004). DOI: 10.1021/ma0352072
10. Voulgaris D., Petridis D.: Emulsifying effect of dimethyldioctadecylammonium-hectorite in polystyrene/poly(ethyl methacrylate) blends. Polymer, 43, 2213–2218 (2002). DOI: 10.1016/S0032-3861(02)00039-3
11. Sinha Ray S., Pouliot S., Bousmina M., Utracki L. A.: Role of organically modified layered silicate as an active interfacial modifier in immiscible polystyrene/polypropylene blends. Polymer, 45, 8403–8413 (2004). DOI: 10.1016/j.polymer.2004.10.009

12. Si M., Araki T., Ade H., Kilcoyne A. L. D., Fisher R., Sokolov J. C., Rafailovich M. H.: Compatibilizing bulk polymer blends by using organoclays. Macromolecules, 39, 4793–4801 (2006). DOI: 10.1021/ma060125+

13. Hong J. S., Namkung H., Ahn K. H., Lee S. J., Kim C.: The role of organically modified layered silicate in the breakup and coalescence of droplets in PBT/PE blends. Polymer, 47, 3967–3975 (2006). DOI: 10.1016/j.polymer.2006.03.077

14. Su Q., Feng M., Zhang S., Jiang J., Yang M.: Melt blending of polypropylene-blend- polyamide 6-blendorganoclay systems. Polymer International, 56, 50–56 (2007). DOI: 10.1002/pi.2109

15. Gelfer M. Y., Song H. H., Liu L., Hsiao B. S., Chu B., Rafailovich M., Si M., Zaitsev V.: Effects of organoclays on morphology and thermal and rheological properties of polystyrene and poly(methyl methacrylate) blends. Journal of Polymer Science Part B: Polymer Physics, 41, 44–54 (2003). DOI: 10.1002/polb.10360

16. Sinha Ray S., Okamoto M.: Polymer/layered silicate nanocomposites: A review from preparation to processing. Progress in Polymer Science, 28, 1539–1641 (2003). DOI: 10.1016/j.progpolymsci.2003.08.002

17. Giannelis E. P.: Polymer layered silicate nanocomposites. Advanced Materials, 8, 29–35 (1996). DOI: 10.1002/adma.19960080104

18. Sharif J., Wan Yunus W. M. Z., Khairul Z. H. M. D., Mansor H. A.: Preparation and properties of radiation crosslinked natural rubber/clay nanocomposites. Polymer Testing, 24, 211–217 (2005). DOI: 10.1016/j.polymertesting.2004.08.008

19. Bala P., Samantaraya B. K., Srivastava S. K.: Synthesis and characterization of Na-montmorillonite-alkylammonium intercalation compounds. Materials Research Bulletin, 35, 1717–1724 (2000). DOI: 10.1016/S0025-5408(00)00368-8

20. Xue W., He H., Zhu J., Yuan P.: FTIR investigation of CTAB-Al-montmorillonite complexes. Spectrochimica Acta Part A: Molecular and Biomolecular Spectroscopy, 67, 1030–1036 (2007). DOI: 10.1016/j.saa.2006.09.024

21. Xi Y., Frost R. L., He H.: Modification of the surfaces of Wyoming montmorillonite by the cationic surfactants alkyl trimethyl, dialkyl dimethyl, and trialkyl methyl ammonium bromides. Journal of Colloid and Interface Science, 305, 150–158 (2007). DOI: 10.1016/j.jcis.2006.09.033

22. Agag T., Koga T., Takeichi T.: Studies on thermal and mechanical properties of polyimide-clay nanocomposites. Polymer, 42, 3399–3408 (2001). DOI: 10.1016/S0032-3861(00)00824-7

23. Zhou Q., Frost R. L., He H., Xi Y., Zbik M.: TEM, XRD, and thermal stability of adsorbed paranitrophenol on DDOAB organoclay. Journal of Colloid and Interface Science, 311, 24–37 (2007). DOI: 10.1016/j.jcis.2007.02.039

24. Someya Y., Sugahara Y., Shibata M.: Nanocomposites based on poly(butylene adipate-co-terephthalate) and montmorillonite. Journal of Applied Polymer Science, 95, 386–392 (2005). DOI: 10.1002/app.21333

25. Wu Z., Zhou C., Qi R., Zhang H.: Synthesis and characterization of nylon 1012/clay nanocomposite. Journal of Applied Polymer Science, 83, 2403–2410 (2002). DOI: 10.1002/app.10198

26. Paul D. R., Robeson L. M.: Polymer nanotechnology: Nanocomposites. Polymer, 49, 3187–3204 (2008). DOI: 10.1016/j.polymer.2008.04.017

27. Sathe S. N., Devi S., Srinivasa Rao G. S., Rao K. V.: Relationship between morphology and mechanical properties of binary and compatibilized ternary blends of polypropylene and nylon 6. Journal of Applied Polymer Science, 61, 97–107 (1996). DOI: 10.1002/(SICI)1097-4628(19960705)61:13.0.CO;2-X

28. Sathe S. N., Srinivasa Rao G. S., Rao K. V., Devi S.: The effect of composition on morphological, thermal, and mechanical properties of polypropylene/nylon-6/polypropylene-g-butyl acrylate blends. Polymer Engineering and Science, 36, 2443–2450 (1996). DOI: 10.1002/pen.10642

29. Ferrigno T. H.: Principles of filler selection and use. in 'Handbooks of fillers for plastic' (eds.: H. S. Kartz, J. V. Milewski) Van Nostrand Reinhold, New York, 6–81 (1987).

30. Reichert P., Nitz H., Klinke S., Brandsch R., Thomann R., Mühaupt R.: Poly(propylene)/organoclay nanocomposite formation: Influence of compatibilizer functionality and organoclay modification. Macromolecular Materials and Engineering, 275, 8–17 (2000). DOI: 10.1002/(SICI)1439-2054(20000201)275:13.0.CO;2-6

31. Wang J., Pyrz R.: Prediction of the overall moduli of layered silicate-reinforced nanocomposites- Part I: Basic theory and formulas. Composites Science and Technology, 64, 925–934 (2004). DOI: 10.1016/S0266-3538(03)00024-1

32. Shelley J. S., Mather P. T., DeVries K. L.: Reinforcement and environmental degradation of nylon-6/clay nanocomposites. Polymer, 42, 5849–5858 (2001). DOI: 10.1016/S0032-3861(00)00900-9

33. Gilman J. W., Jackson C. L., Morgan A. B., Harris R., Manias E., Giannelis E. P., Wuthenow M., Hilton D., Phillips S. H.: Flammability properties of polymerlayered-silicate nanocomposites. Polypropylene and polystyrene nanocomposites. Chemistry of Materials, 12, 1866–1873 (2000). DOI: 10.1021/cm0001760

34. Kontopoulou M., Liu Y., Austin J. R., Parent J. S.: The dynamics of montmorillonite clay dispersion and morphology development in immiscible ethylenepropylene rubber/polypropylene blends. Polymer, 48, 4520–4528 (2007). DOI: 10.1016/j.polymer.2007.05.068

35. Martins C. G., Larocca N. M., Paul D. R., Pessan L. A.: Nanocomposites formed from polypropylene/ EVA blends. Polymer, 50, 1743–1754 (2009). DOI: 10.1016/j.polymer.2009.01.059

CHAPTER 4

CONDUCTING POLYANILINE NANOWIRE AND ITS APPLICATIONS IN CHEMIRESISTIVE SENSING

Edward Song[1] and Jin-Woo Choi[1,2]

[1]School of Electrical Engineering and Computer Science, Louisiana State University, Baton Rouge, LA 70803, USA

[2]Center for Advanced Microstructures and Devices, Louisiana State University, Baton Rouge, LA 70803, USA

ABSTRACT

One dimensional polyaniline nanowire is an electrically conducting polymer that can be used as an active layer for sensors whose conductivity change can be used to detect chemical or biological species. In this review, the basic properties of polyaniline nanowires including chemical structures, redox chemistry, and method of synthesis are discussed. A comprehensive literature survey on chemiresistive/conductometric sensors based on polyaniline nanowires is presented and recent developments in polyaniline nanowire-based sensors are summarized. Finally, the current limitations and the future prospect of polyaniline nanowires are discussed.

INTRODUCTION

The accurate detection and quantification of chemical and biological species are of great importance in many areas including health care, environmental monitoring, and recently, defense against biological warfare. The key features of an effective sensor are that, in addition to having high sensitivity and selectivity, the detection method must be simple, rapid, and the sensor must be multiplexed for parallel detection of multiple target analytes [1]. However, existing technologies have not been able to demonstrate such sensing capability.

Nanomaterial-based sensors seem to be promising in this regard and have received great attention in recent years due to many unique properties that nanoscale materials offer. The main advantages of nanoscale materials include high surface-to-volume ratio leading to fast reaction speed and sensitivity, ease of miniaturization, minimum power consumption, and low cost as a result of the small volume of required reagents.

Chemiresistive or conductometric sensors have attracted a great deal of attention because of their simple structure and their ability to be miniaturized to give high density and high throughput sensor arrays. Chemiresistive or conductometric sensors measure the change in the conductance or resistance of the sensing material caused by its interaction with the target analytes. Chemiresistive sensors have the benefit of being simple in configuration as well as in signal measurement.

One dimensional nanomaterial-based chemiresistors can achieve high sensitivity and fast response while preserving simplicity and high density. In particular, nanotubes, nanowires, or nanorods are ideal for this configuration [2]. First of all, the small cross-sectional area of the nanowires maximizes the current response along the axial direction of the wires creating a large conductance change. Secondly, the large surface area of the nanowires improves the sensitivity of the nanowire-based sensor by increasing the chance of target analytes reacting

Citation: Edward Song and Jin-Woo Choi, Conducting Polyaniline Nanowire and Its Applications in Chemiresistive Sensing, doi: 10.3390/nano3030498.

with the surface of the nanowire. Thirdly, the direct conversion of the chemical change into an electrical signal greatly simplifies the device configuration. Finally, the nanoscale of the sensing material enables the development of high density, high throughput, and individually addressable sensor arrays for simultaneous multi-analyte detection. Reported results show that the large surface area of 1-dimensional nanowire network-based sensors improved the sensitivity compared to the conventional 2-dimensional films [3,4]. Recently, label-free detection of analytes using chemiresistive nanowire-based sensors has gained much interest due to its high sensitivity and real-time monitoring with a fast transduction mechanism [1,5,6]. The use of nanowires has been applied in biosensors as well [7].

Because of these advantages that the 1-dimensional nanowires provide in sensing capabilities, many types of nanowires have been studied as a sensing material such as carbon nanotubes [8,9,10], silicon nanowires [1,11,12], platinum and gold nanowires, and metal oxide nanowires [13]. However, fabrication of these nanowires is expensive and often requires harsh conditions [14]. Moreover, the inability to grow these nanowires in a site-specific location has limited their application.

Conducting polymer nanowires (CPNWs) have recently emerged as an attractive alternative to metal and semiconducting nanowires for their large conductivity change, flexibility, and ease of synthesis [15,16]. Furthermore, the CPNWs can be synthesized site-specifically at the desired location [16,17,18,19,20].

The discovery of the first organic polymer that possesses metallic conductivity, polyacetylene, in 1977 [21] has opened up possibilities for the development of a new class of materials known as intrinsically conducting polymers, or simply conducting polymers. In 2000, the Nobel Prize for Chemistry was awarded to the three scientists who discovered polyacetylene (Alan G. MacDiarmid, Hideki Shirakawa, and Alan J. Heeger) "for the discovery and development of conducting polymers". Since then, a surge of research effort has been directed towards the development, synthesis, and characterization of new types of conducting polymer materials. As a result, many different kinds of conducting polymers have been discovered and synthesized. Some of the most well-known examples are polyaniline, polypyrrole, polythiophene, poly(3,4-ethylenedioxythiophene) (PEDOT), to name a few. In particular, polyaniline is one such organic polymer which has been known for over 150 years but was revisited in the 1980s and discovered that it possesses an electron conducting nature. Because of its tendency to be synthesized in a 1-dimensional nanofiber-like morphology, combining its environmental stability [22,23] and reversible redox chemistry, it has emerged as an excellent candidate to be used as a material for various nanoscale applications, particularly for chemical and biological sensors [24,25,26].

POLYANILINE NANOWIRES: AN OVERVIEW

Polyaniline is one of the oldest known conducting polymers [27] and has been extensively reviewed [28,29,30,31,32]. It has been the most studied conducting polymer closely followed by polypyrrole [33]. First discovered in the 19th century, polyaniline was originally known as "aniline black" [29,30]. It was later found to be electrically conductive in nature and many researchers began to closely examine the properties of this material. Polyaniline was initially grown as thin films but later it was discovered that, under certain conditions, it can be grown in the form of an interwoven nanowire network. It has been reported that polyaniline has an intrinsic nature to grow 1-dimensionally [34] which is not the case for other types of conducting polymers such as polypyrrole or polythiophene. Its unique ability to easily form 1-dimensional nanostructures, including wires, rods, tubes, and ribbons presents many advantages in nanoscale devices [4]. Polyaniline has been studied for use in a wide range of applications [35] such as chemical sensors, battery electrodes [36], supercapacitors [37], fuel cells [38], display devices [39], separation membranes [40] and anticorrosion coatings [41].

Protonation of the polymer, also known as doping, is the mechanism by which polyaniline becomes electrically conductive. However its conductivity is influenced by various factors including its electrochemical redox state, pH, humidity, temperature, and the type of anions in the solution [42]. The conductivity of polyaniline, in its pressed-pellet form, is typically in the range 2–10 S/cm [43].

To use polyaniline nanowires as chemiresistive sensors in the solution phase, a thorough knowledge of the polymer, especially its chemical backbone structures, conductivity and its dependence on the redox state, pH, and synthesis method among other points is needed. Therefore, in this section, a review covering the above mentioned topics is presented.

Chemical Structure and Electrochemical Properties

The two most important factors determining the chemical structure of polyaniline are the redox state and the doping level. Polyaniline mainly has three distinguishable oxidation states, namely, the fully reduced (leucoemeraldine), the half oxidized (emeraldine), and the fully oxidized (pernigraniline) state, with virtually an infinite number of possible oxidation states existing in between. Therefore, in principle, polyaniline can exist in a continuum of oxidation states ranging from a completely reduced to a completely oxidized form. A general chemical structure of polyaniline is shown in Figure 1 where the polymer chain consists of two types of repeating units, namely, the reduced and the oxidized unit. The degree of oxidation is described by the variable x whose value is between 0 and 1, and it represents the fraction of the two repeating units.

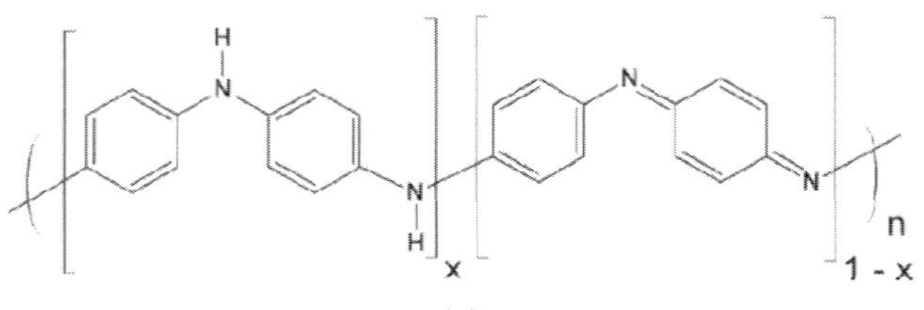

(a)

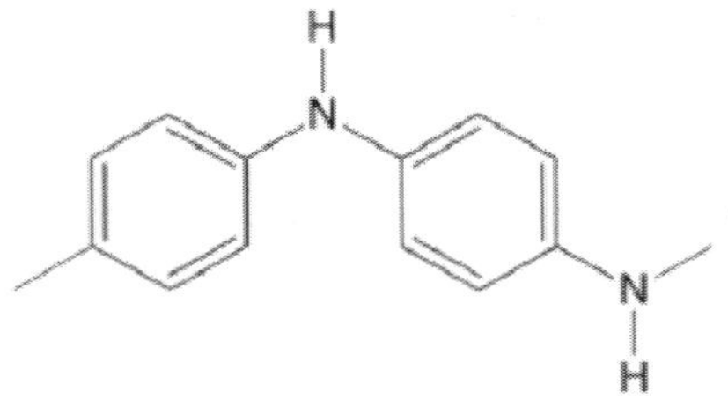

(b)

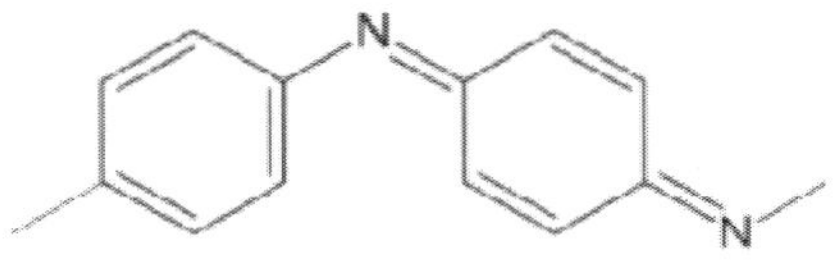

(c)

Figure 1: A general chemical structure of polyaniline and its repeating units. (**a**) A general chemical structure of polyaniline, (**b**) Reduced repeating unit, and (**c**) oxidized repeating unit.

It follows therefore that leucoemeraldine, emeraldine, and pernigraniline refer to the chemical formula where $x = 1$, 0.5, and 0, respectively. The reduced repeating unit contains only the amine nitrogen atoms whereas the oxidized repeating unit is made up of imine nitrogen atoms only. The imine nitrogen atoms, which do not have hydrogen bonds, can be protonated in an acidic environment where protons can be attached to these nitrogen atoms to generate radical cations [44,45]. The degree of protonation strongly depends on the oxidation state of polyaniline and the pH of the aqueous solution in which the polymer is immersed. It was originally thought that protonation occurred exclusively on the imine nitrogen atoms, however, experimental results show that some amine nitrogen atoms can also be protonated to give NH_2^+ groups even if all the imines are not protonated [28,46,47,48]. An unprotonated polyaniline is known as a base whereas the protonated polymer is called a salt. The three different redox states of polyaniline in their base forms and the corresponding salts are illustrated in Figure 2.

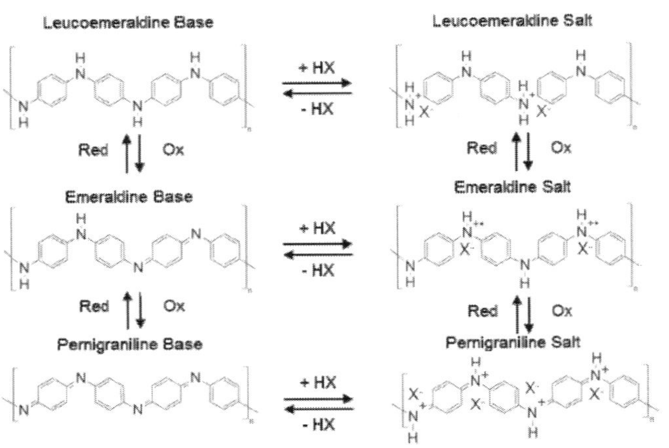

Figure 2: Electrochemical redox states and the corresponding doped form of polyaniline.

The electrochemical behavior and the redox states of polyaniline are typically studied using cyclic voltammetry (CV). The CV response of polyaniline has been well documented [31,49,50] and some of the key results are described here. The electrochemical behavior of polyaniline is dependent on many parameters including the applied potential, the choice of material and the surface area of the electrodes, composition of the electrolyte, and temperature, among others. A typical CV curve of a synthesized polyaniline is shown in Figure 3.

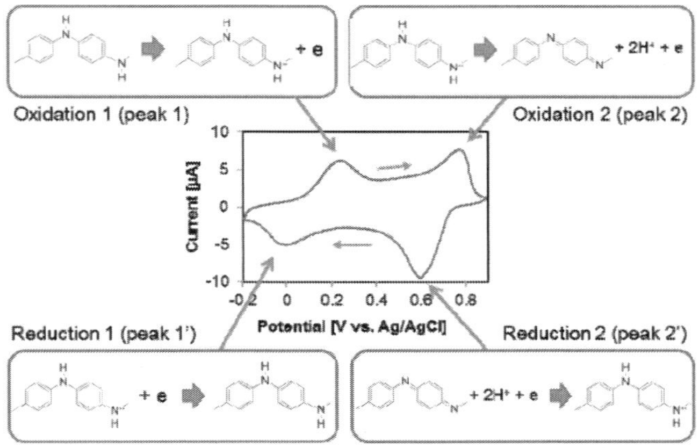

Figure 3: A typical cyclic voltammetry (CV) curve of polyaniline in HCl (pH 1) showing two sets of redox couples. The direction of potential scan is shown with the arrows.

The CV curve shows two sets of distinct redox activity as indicated by the two pairs of anodic and cathodic current peaks. The first set of a redox couple which appears between 0 and 0.25 V *vs.* silver/silver chloride reference electrode (Ag/AgCl) is associated with the conversion of the fully reduced leucoemeraldine base to the partially oxidized emeraldine, and the second set of redox current peaks occurring between 0.6 and 0.8 V *vs.* Ag/AgCl pertains to the conversion of emeraldine to the fully oxidized pernigraniline form. The potential of the first redox couple (peaks 1 and 1') is largely independent of the pH whereas the potential for the second redox couple (peaks 2 and 2') is strongly dependent on the pH value. This indicates that protons are involved in the redox reaction associated with the second peaks while the redox reaction related to the first anodic peak does not require protons as part of the reaction. Another point to note is that polyaniline is more easily oxidized in less acidic solutions, and this can be experimentally verified where the peaks 2 and 2' shift to the left as the pH increases. Irreversible degradation also occurs at the second oxidation peak especially in strong acids. A linear relationship has been observed between peak height of the redox current and scan rate in a solution containing aniline and sulfuric acid solution which is indicative of an electron transfer limited process [51].

Using the CV data and the electrochemistry, the structural formula for polyaniline as it goes through the redox process can be interpreted. The following reactions have been proposed by Focke *et al.* [42] and a more detailed account can be found in [31]. Since the peak position of the first redox process in the CV plot is independent of pH, no proton is involved in the reaction. Hence, the reduction reaction can be described as the following:

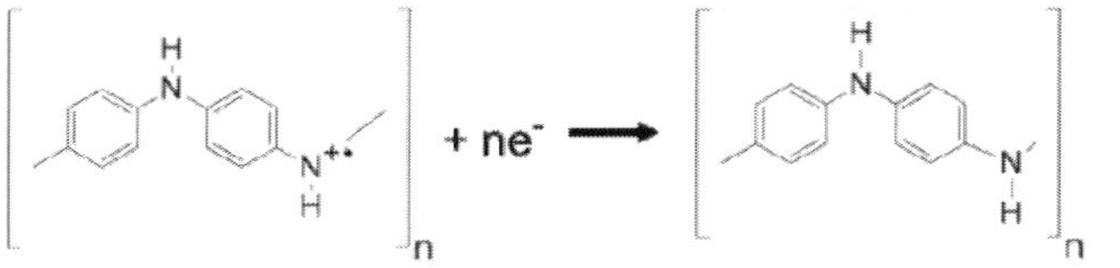

with $E_0 = 0.1$ V *vs.* Ag/AgCl. However, the peak position of the second redox process is highly dependent upon the pH of the solution. The second oxidation peak and its corresponding reduction peak tend to move to more negative values at a rate of approximately 120 mV per pH unit as the pH is increased [31]. Since the peak position moves as a function of the proton concentration in the solution, it can be expected that protons are involved in the reaction as shown below:

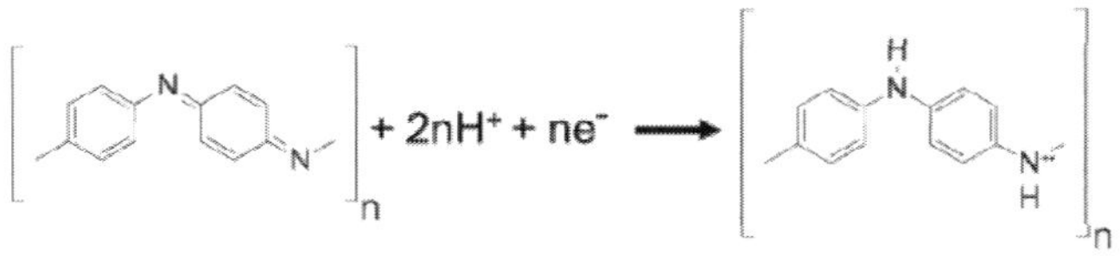

The doping mechanism of polyaniline is also unique among other conducting polymers. Most conducting polymers undergo redox doping processes during which the number of electrons associated with the polymer backbone changes. However, polyaniline can be doped through a non-redox process where the number of electrons in the polymer backbone structure remains unchanged [52], making the doping process simpler.

There are two ways that polyaniline can be doped, one is through acid treatment which is a non-redox doping process, and the other is by oxidation of leucoemeraldine to emeraldine which is a redox process. Upon proton doping, radical cations are formed at the imine nitrogen atoms and these charge carriers are believed to be responsible for the electronic conduction in polyaniline [53]. Hence, the majority charge carriers in polyaniline are holes.

Electronic Conduction

Since the delocalized positively charged free radicals are the main source of conduction in polyaniline, it can be expected that maximum conductivity is exhibited when the number of radical cations in the polymer chains is maximized. This is in agreement with the fact that both leucoemeraldine and pernigraniline, neither of which have free radicals in their backbone structure, are completely insulating. Hence, the conduction current *vs.* the electrochemical potential relationship shows a bell shaped curve as illustrated in Figure 4. The graph also confirms that the most conductive form of polyaniline is the fully protonated, half-oxidized emeraldine salt form, and the conductivity decreases as the polymer is deprotonated or the oxidation state changes toward either a fully oxidized or a fully reduced state. Over-oxidation of polyaniline by applying potentials beyond +0.6 V *vs.* Ag/AgCl should be avoided because it causes the irreversible formation of quinonediimine structures which are electrochemically inactive [42,54,55].

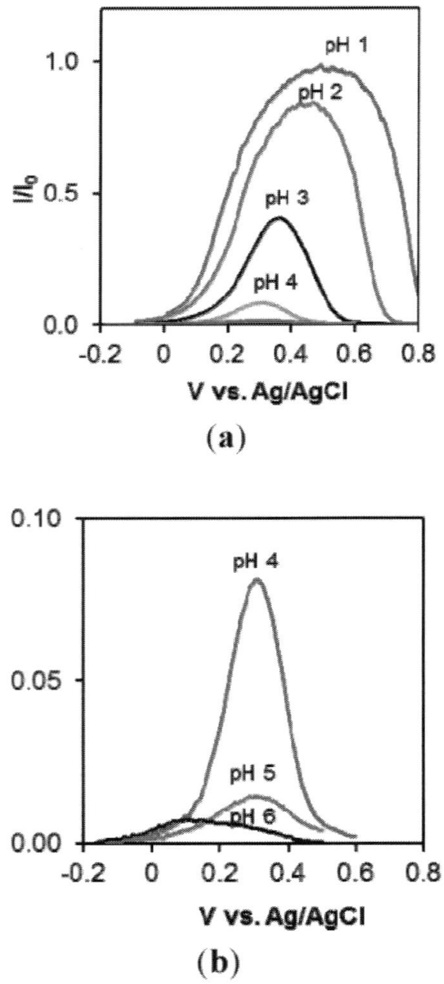

Figure 4: Conductance current *vs.* potential of polyaniline in various pH solutions: pH range 1–4 (**a**) and pH range 4–6 (**b**). I_0 indicates maximum current observed.

Although a true mechanism of electron transport is still under debate, many theories have been proposed to explain the electronic conduction of polyaniline. It is generally accepted that polyaniline nanowires consist of pockets of conductive grains embedded in an insulating region as illustrated in Figure 5a. Since the conductive grains are separated by an insulating medium, electrons must gain sufficient energy to be able to overcome the insulating barrier and "hop" into the nearest neighboring conductive grain. Some models also take into consideration electronic conduction through internanotubular contact as illustrated in Figure 5b.

To further elucidate this phenomenon, many different models to describe the electron transport in polyaniline nanowires have been proposed, including granular-rod model [56], 3-dimensional variable range hopping (3D VRH) [57,58], 1-dimensional variable range hopping (1D VRH) with interchain coupling [59], Efros-Schklovskii (E-S) hopping conduction [60], and charging energy limited tunneling (CELT) [61]. A recent survey indicates that the true nature of the conduction mechanism must be examined on a mesoscopic scale [62].

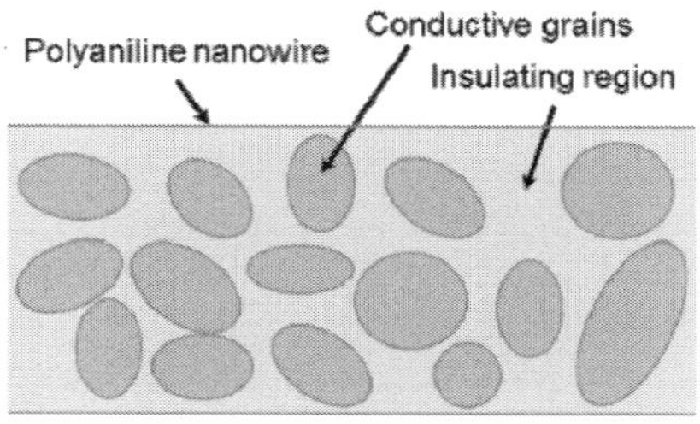

(a)

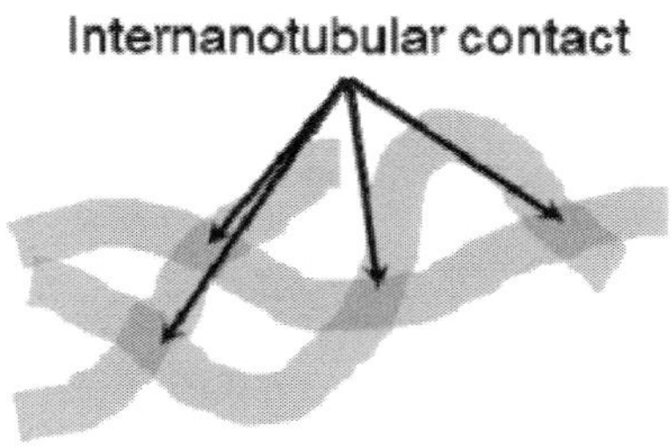

(b)

Figure 5: The electronic conduction path of the polyaniline nanowires: (**a**) internanotubular contacts between polyaniline nanowires, and (**b**) conductive granular region encapsulated in the insulating region of the nanowire. The drawings were adopted from [60].

SYNTHESIS OF POLYANILINE NANOWIRES

There are many techniques available for synthesizing 1-dimensional polyaniline nanowires. The two most common techniques are chemical synthesis where polyaniline nanowires are formed by chemical oxidation of aniline monomers, and electrochemical polymerization where aniline monomers are electrochemically oxidized and polymerized on the surface of the anode of the electrochemical cell. Although other methods such as electrospinning [63], enzyme assisted growth [64], and DNA template method [65] exist, in this review, emphasis will be placed on the chemical and the electrochemical polymerization methods.

Various shapes and forms of polyaniline nanowires have been obtained, however the mechanism of the formation of such nanowires has not yet been fully elucidated [66,67,68]. Although the exact growth mechanism is uncertain, there have been many reports suggesting the probable growth mechanisms based on experimental results [29,49,69,70].

Chemical Synthesis

The classical chemical synthesis involves the direct oxidation of aniline monomers by chemical oxidants [29]. Polyaniline nanofibers can be obtained by mixing aniline monomers with a strong oxidizing agent in an acidic environment [71,72]. The most commonly used oxidant in chemical synthesis is ammonium peroxydisulfate (APS) [73]. However, using APS creates a harsh synthetic condition due to its strong oxidizing nature. Therefore, more environmentally benign synthesis methods that use a relatively mild oxidant, such as hydrogen peroxide with catalysts, instead of APS have been reported [64,73]. Depending on the growth condition, chemical synthesis can yield various shapes and forms of polyaniline nanostructures such as irregularly shaped agglomerates, granular particles, and elongated nanofibers. It has been experimentally observed that, during the early stage of polymerization, only 1-dimensional nanofibers were observed due to the homogeneous nucleation of polyaniline molecules. However, as further polymerization proceeds, preferential growth on previously formed nanowires by heterogeneous nucleation occurs, resulting in irregularly shaped particles. Therefore, suppression of this 'secondary growth' that takes place during the later stage of polymerization is likely to be the key to growing directional nanowires.

Two novel synthesis methods that yield highly directional polyaniline nanowires with controllable diameters have been suggested: the first method is interfacial polymerization [74,75], and the second method is rapid mixing technique [72,76]. In the interfacial synthesis method, the polyaniline polymerization only occurs at the boundary of two immiscible organic/aqueous solutions. The synthesized polyaniline nanowires at the interface of the two solutions migrate into the bulk of the aqueous phase, thereby avoiding further polymerization. Figure 6 illustrates the early, the intermediate, and the final stages of the interfacial polymerization process. The downside of this technique is that the yield of nanofibers produced is generally too low to be mass produced on a large scale. On the basis of the experimental results that a fast reaction rate promotes homogeneous nucleation, and that conducting the reaction at a raised temperature (for example, 60 °C) produces less granular particles, a rapid mixing method was suggested. By quickly mixing aniline monomers with an appropriate portion of the oxidant at room temperature or higher, one can obtain large quantities of highly directional polyaniline nanofibers. The diameter of the nanowires is dependent on the type of acid used in the polymerization process. The average diameter of nanowires synthesized with HCl is approximately 30 nm while those obtained with camphorsulfonic acid (CSA) are roughly 50 nm, and $HClO_4$ yielded an average diameter of 120 nm [74].

Since the nanofiber morphology of polyaniline can be intrinsically grown, templates are not necessary for 1-D polyaniline synthesis although there have been reports that use porous templates for the synthesis of nanowires [77,78]. There are other techniques available that yield the growth of polyaniline nanowires without the aid of templates including oligomer-assisted polymerization, seeding polymerization, falling pH polymerization, and dilute polymerization [79].

(a)

(b)

(c)

Figure 6: Interfacial polymerization process with 0.32 M of aniline in chloroform (bottom layer) interfacing 0.08 M of ammonium peroxydisulfate (top layer): (**a**) 1 min, (**b**) 5 min, and (**c**) 10 min of the reaction time after the reaction started.

Electrochemical Synthesis

Polyaniline can be synthesized by anodic oxidation of aniline monomers through an inert electrode [29]. One of the key advantages of the electrochemical method is that this technique allows direct deposition of polyaniline nanostructures onto metal electrodes in a simple and cost effective way which ensures good ohmic contact between the electrode and the polymer. The nanowires are also addressable since the location of polymer growth is designated by the shape and pattern of the working electrode. Electropolymerization of polyaniline can be categorized into three types: potentiostatic, galvanostatic, and potentiodynamic growth.

Potentiostatic Growth

In the potentiostatic method, polyaniline nanowires can be grown by applying a constant oxidative potential to the anode of the electrochemical cell, which causes polymerization of aniline on the surface of the anode. Polyaniline can be grown on a variety of metallic surfaces including platinum, gold, stainless steel, iron, copper, zinc, ITO, graphite, and glassy carbon among many others [30,35]. The electrolyte solution is generally a mixture of a strong acid (such as 1 M H_2SO_4, HCl, or $HClO_4$) and aniline monomer with a concentration in the area of 0.05 M. The potential under which polyaniline nanowires can be grown is between 0.7 and 0.8 V *vs.* Ag/AgCl. The oxidation of the aniline monomers occurs at around 0.7 V. These oxidized species are thought to be in radical forms that quickly become *p*-aminodiphenylamine dimers. These formed dimers are much more easily oxidized than aniline monomers (around 0.2 V), and can be further oxidized to form longer chains of polyaniline. Therefore, once these dimers are formed, polymerization of aniline can proceed even at potentials lower than 0.7 *vs.* Ag/AgCl as long as the electrode surface is preliminarily covered with small amounts of polyaniline [80]. Typical polyaniline nanowires grown under the potentiostatic method are shown in Figure 7.

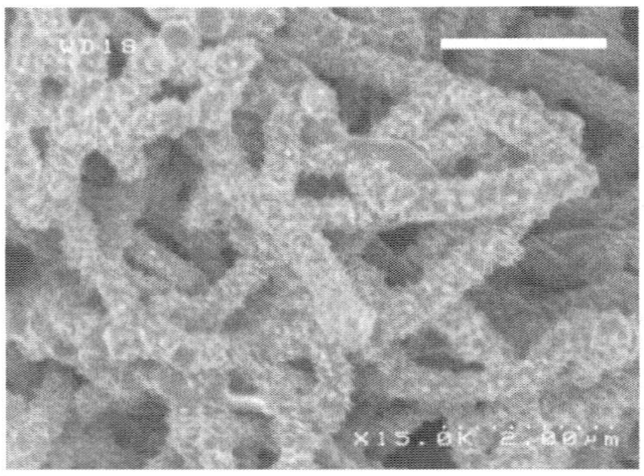

(**a**)

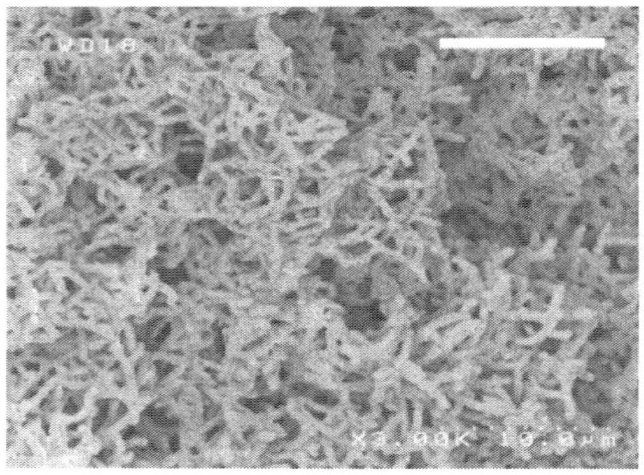

(**b**)

Figure 7: SEM images of the electrochemically synthesized polyaniline nanowires. Scale bars are 2 μm (**a**) and 10 μm (**b**).

Galvanostatic Growth

Galvanostatic growth is a method of electrochemical polymerization where the current flow in the working electrode is maintained at a constant value. A well-established low current polymerization technique involves a three step galvanostatic growth method [81]. In the first step, the current density is fixed to 0.08 mA/cm^2 for 0.5 h which generates particles to act as nucleation sites for growing extended polymer nanostructures. This process is followed by the second step where the current density is reduced to 0.04 mA/cm^2 for

3 h. During this stage, continued polymerization occurs at a slower growth rate. Finally in the third step, the current density is further reduced to 0.02 mA/cm^2 for 3 h, which continues to elongate the nanostructures into wires. Typical diameters of these nanowires are 50 to 70 nm with approximately 0.8 μm in length. For producing longer nanowires, higher current density which promotes faster growth rate is required. Further, galvanostatically grown polyaniline is less stable due to the formation of quinoid structures which contribute to the final formation of the degradation product, *p*-benzoquinone, which prevents further growth of the polymer [82,83].

Potentiodynamic Growth

Potentiodynamic growth involves scanning the potential of the working electrode linearly from initial to final value in a forward and reverse direction repeatedly until the desired amount of polymer has been deposited [84,85]. The sweeping voltage technique for the production of polyaniline is useful for elucidating basic aspects of the polymer growth and the redox mechanism [51]. It has been reported that continuously cycling the potential of the working electrode produces a more uniformly deposited polyaniline film compared to that formed at constant potential and promotes better adhesion to the electrode surface [86].

Other Growth Methods

Other methods of polyaniline synthesis include the electrospinning technique [63,87,88] and the template-guided method. The hard template approach was proposed by Martin [77,78,89,90,91] and has been reviewed in detail [92]. DNAs have also been used as a template for the synthesis of polyaniline [65,93,94]. Horseradish peroxidases have been used as catalysts for the oxidation of aniline monomers in the presence of hydrogen peroxide to synthesize polyaniline [64,95]. Wang *et al.* have used an *in situ* electrochemical growth method to produce polyaniline and polypyrrole nanowires within a microfluidic device [16].

POLYANILINE NANOWIRE-BASED CHEMIRESISTIVE SENSORS

The applications for polyaniline nanowires are being actively explored by many researchers due to a number of useful properties such as direct and easy deposition on the electrode, control of thickness, high surface area, long term environmental stability, and tunable electrical conductivity [25]. Although our focus is on the development of a chemiresistive sensor that measures the change in conductivity of polyaniline caused by a reaction of interest, other detection mechanisms are also briefly discussed followed by a general survey of existing polyaniline-based chemiresistive sensors.

Potentiometric, Amperometric, and Conductometric Sensors

A potentiometric sensor measures the change in potential caused by a chemical reaction that separates electric charge. In general, the potential difference is in a logarithmic relationship with the activity of the target analyte as given by the Nernst equation. Examples of potentiometric sensors include ion-selective field effect transistors (ISFETs) [96,97,98], ion-selective electrodes (ISEs), and conventional pH meters. If there is a large difference between the activity and the concentration of the species of interest, then the potentiometric sensor may give inaccurate results. Amperometric sensors measure the current generated at the surface of the electrode as a result of electron transfer from the oxidation or reduction reaction of the electroactive species, typically at a constant applied potential. If the applied potential is not a constant but contains a change in potential, such as a step change, a pulsed voltage, or a linear sweep, then these types of sensing are referred to as voltammetric sensors. The current must vary linearly with respect to the concentration of the analyte in order to have accurate readings. Moreover, if two or more species in the sample solution have similar redox potentials then the generated electrochemical current may contain a substantial amount of interference which may compromise the selectivity of the sensor. Conductometric or resistive sensors measure the change in resistivity of the transducing material due to chemical reactions. Regardless of the type and the detection mechanism of the sensors, the important considerations inherent to all devices include, but are not limited to, sensitivity (limit of detection), selectivity (minimum false negative/positive signals), response time (fast reading), stability (short term and long term), and hysteresis (same measurement regardless of the past history of the sensor). Although polyaniline has been applied in all three types of sensors, to be able to truly make use of the tunable conductivity nature of polyaniline as a transducing mechanism, polyaniline must be used as a conductometric sensor. Chemiresistors and chemically sensitive field effect transistors (chemFETs) are the two main examples of conductometric sensors. In the following sections, polyaniline nanowire based sensors are reviewed. Table 1 summarizes the chemical [3,15,17,99,100,101,102,103,104,105,106,107,108,109,110] and biological [94,111,112] sensors that utilize polyaniline nanowires as sensing materials.

Gas Sensors

The importance of a reliable and accurate gas monitoring system is well understood not only from the safety but also from the environmental standpoint. Most commercially available gas sensors are based on metal oxide semiconducting materials (such as tin oxide) operating at high temperature to increase sensitivity. The use of conducting polymers as an alternative to inorganic semiconducting materials for the gas sensitive layer offers many advantages such as, low cost, ease of synthesis, tunable conductivity, fast response due to porosity of the material [113], and high sensitivity at room temperature. Polypyrrole was one of the first conducting polymers to be applied for gas sensing materials [114]. However, its low sensitivity, slow response, lack of adsorption-desorption reversibility has limited its use. Hence, the search for a more reliable and better performing material was extended to polyaniline. Huang *et al.* developed a chemically synthesized polyaniline nanofiber-based gas sensor and studied its response to 100 ppm of HCl and NH_3 vapor [3,75,115]. Upon exposure to HCl vapor the resistance of the polyaniline nanofibers reduced through the doping process, and exposure to NH_3 had a dedoping effect which increased the resistance of the polymer [99]. They also demonstrated that the nanofiber based sensor responded much faster than the conventional film to both doping and dedoping due to the highly porous morphology of the nanofiber film with the small diameter of the fibers resulting in faster diffusion of the gas molecules. Similar work was done in [17] where a polyaniline nanowire framework was formed to bridge the electrode junctions which acted as a resistive sensor, and its responses to HCl, NH_3, and ethanol vapors were demonstrated. A polyaniline chemiresistors have also been developed for various gas sensors [15,100,101,116] as well as humidity sensors [102,103]. Combining polyaniline nanowires with other nanomaterials such as carbon nanotubes [104,117,118], graphenes [105], and gold nanoparticles [106] to form a composite layer has been shown to improve the sensitivity as well as the carrier mobility of the gas sensing material.

Separation of a gas species of interest from a mixture of gases is essential in improving the performance of a gas sensor in terms of sensitivity, selectivity, and protection of the sensing material from unwanted molecules. One of the most promising techniques to achieve the separation of gas is a membrane-based gas separation method. Separation of gas generally requires a membrane with pore sizes of the order of 3 nm or less [119]. Polyaniline-based membranes have been reported to be the most effective since their pore size can be controlled via the doping process either before or after the synthesis of the membrane. Polyaniline is an ideal choice for this application due to its simple doping/dedoping chemistry and its stability in air. The earliest work on polyaniline-based membranes for gas separation was reported in [40] where the porosity of the membrane was tailored by controlling the level of doping as well as the size of the dopant counterions. In a more recent work, composite membranes such as polyaniline/carbon nanotube [120], polyaniline-polysulfone [121], and polyaniline/polypropylene [122] have also been reported to have improved permeability and selectivity.

Table 1: Summary of chemical and biological sensors based on polyaniline nanowires and their derivatives

Sensor Type	Sensing Material	Diameter (nm)	Analyte	Detection Limit (LOD)	Response Time	References
Surface acoustic wave	Polyaniline/In_2O_3	90	H_2, NO_2, CO	~ 2 ppm	~ 30 s	[107]
Surface acoustic wave	Polyaniline	30–50	H_2	0.06%	~ 100 s	[108,109]
Amperometric	Polyaniline	60–80	nitrite	5×10^{-8} M	~ 5 s	[110]
Amperometric	Polyaniline/Au nanoparticles	30–50	glucose	5×10^{-7} M	~ 5 s	[111]
Chemiresistive	Polyaniline/CSA	~100	H_2	<1%	-	[15]
Chemiresistive	Polyaniline	40–80	NH_3, HCl, EtOH	~0.5 ppm	~100 s	[17]
Chemiresistive	Polyaniline/Au nanoparticles	80	H_2S, CH_3SH	~1 ppm	~20 s	[101]
Chemiresistive	Polyaniline	30–40	CO	~1 ppm	~100 s	[100]
Chemiresistive	Polyaniline	335	NH_3	0.5 ppm	~75 s	[99]
Chemiresistive	Polyaniline	30–120	HCl, NH_3, N_2H_4, $CHCl_3$, CH_3OH	100 ppm	2~200 s	[3]
Chemiresistive	SWCNT/Polyaniline	15	NO_2, H_2S	500 ppb	~10 min	[104]
Chemiresistive	Graphene/Polyaniline	25–50	H_2	-	~1 min	[105]
Chemiresistive	Polyaniline/Au nanoparticles	250–320	H_2S	0.1 ppb	<2 min	[106]
Chemiresistive	polyaniline	100	humidity	-	~1 min	[102]
Chemiresistive	Polyaniline/PVB/PEO	100	humidity	-	~50 s	[103]
Target-guided formation method	Polyaniline	-	microRNA	5 fM	30–60 min	[94]
Labeled direct charge transfer	polyaniline	~200	*Bacillus cereus*	~10 CFU/mL	-	[112]

Biosensors

The development of biosensors has emerged as a topic of great importance due to their applications in clinical diagnostics, environmental monitoring, food safety, and defense against biological warfare. An affinity based molecule recognition method which uses the specific interactions of biomolecules such as antibodies to antigen binding, DNA hybridization, enzyme catalysis, is the most common method to ensure high specificity and selectivity of a biosensor. One of the earliest biosensors was developed by Clark and Lyons [123] who used glucose oxidase immobilized on a semi-permeable membrane, and glucose was monitored by measuring the oxygen consumed by the enzyme catalytic reaction. The glucose sensor is one of the most commercially successful electrochemical biosensors existing today. Most glucose sensors employ an amperometric measuring technique [111,124], however, this technology has major limitations if the sensor is to be made smaller or if lower concentrations of analyte are to be measured. One of the strategies to overcome this limitation was the use of microelectrochemical transistors [125]. These devices make use of conductive polymers whose conductivity changes by several orders of magnitude upon oxidation or reduction [126,127]. This property of conducting polymers can be used to amplify signals transduced by electrochemical reactions. Therefore, polyaniline is an excellent candidate for biosensor applications in this regard [5,24,25]. This idea has been realized with a glucose sensor by attaching or embedding enzymes such as glucose oxidase to the polyaniline film [128,129,130].

A nanoscale glucose sensor was implemented by bridging a 20–60 nm electrode gap with polyaniline [131]. Pal *et al.* used polyaniline nanowires as direct charge transfer (DCT) electrical transducers for the detection of foodborne pathogen, *Bacillus cereus* [112]. In this method, polyaniline nanowires are attached to each of the antibodies and when these modified antibodies bind to antigen to form a sandwich complex via lateral flow immunoassay method, the nanowires form a bridge between two open electrodes to give resistive measurements [112,132,133,134]. Forzani *et al.* developed a hybrid amperometric and conductometric chemFET-based sensor that can detect neurotransmitter dopamine even in the presence of ascorbic acid, an interference, whose concentration can be much higher than the analyte itself [135]. Polyaniline chemiresistors have also been applied as biosensors for the detection of microRNAs [94,136,137], urea [138,139], *E. coli* [140], and H_2O_2 [141].

CURRENT LIMITATIONS OF POLYANILINE-BASED SENSORS

Although polyaniline nanowires have made great progress as chemiresistive sensors, the current limitations and challenges facing polyaniline nanowires as chemical sensors must be addressed in order for the technology to make further advancement.

One of the main limitations of polyaniline is the loss of conductivity in neutral and high pH environment. Since polyaniline requires a large amount of protons attached to the polymer to be electrically conducting, it is a very poor conductor when the pH is greater than 5, which significantly limits its application. This is especially critical for biosensors where most enzymatic and cellular activities are pH sensitive, and function properly only in neutral pH environment, typically between pH 6 and 8. Therefore, preventing the loss of protons bound to the polyaniline structure in neutral pH solutions is the key to maintaining conductivity in such an environment. There have been attempts to achieve this by attaching negatively charged anions to the polymer which will attract positively charged protons. MacDiarmid and Epstein developed a “self-doped” polyaniline which contains negatively charged sufonate groups covalently bound to the aromatic rings of polyaniline [142,143,144]. Such self-doped polyaniline can also be synthesized by electro-copolymerization of aniline with its ring-substituted derivatives such as aminobenzenesulfonate [145,146]. Another method to achieve increased conductivity at neutral pH is the polymerization of aniline with large molecular weight organic acids. In this case, large molecular sized anions are trapped within the polyaniline matrix, thereby maintaining the polymer’s electronegativity in order to attract protons. Camphorsulfonic acid [147,148,149], dodecylbenzenesulfonic acid (DBSA) [150,151], and poly(2-acrylamido-2-methyl-1-propanesulphonic acid) (PAMPSA) [152,153,154] were reported to be the most effective organic acids used for this purpose. Furthermore, the use of such organic acids enhances solubility of the polymer and improves solution processability. Long chain polymeric anions such as polyacrylic acid (PAA) [155,156] and polystyrene sulfonate (PSS) [157,158] have also been incorporated during the polymerization. These aforementioned techniques have extended the polymer’s electroactivity at the neutral pH environment, however its conductivity has only slightly improved. Further improvements of its conductivity in the physiological pH range are required in order to implement reliable biosensors.

Exposing polyaniline to elevated pH solutions such as pH 5 or higher causes an irreversible conductivity degradation of the polymer. Some of the possible causes for this conductivity degradation are structural damage due to mechanical stretching or twisting of the polymer chain caused by the electrostatic charge of the dopant, loss of anions to counter balance the positive charge gained by proton adsorption [159], and the production of quinone-hydroquinone couples [53,81]. Therefore, minimizing or compensating for the conductivity degradation is crucial in developing a repeatable polyaniline-based sensor and this still remains to be solved.

The conductivity of polyaniline is known to possess hysteresis, which is illustrated in the current versus potential sweep characteristics plot [127,131,160,161] where the current response to the potential sweep in the positive direction is different from that to the reverse sweep of the potential. This existence of hysteresis is more closely related to the level of doping than the electrochemical potential or the pH of the polymer [47] and this apparent hysteresis or “memory effect” has been attributed to structural relaxations [162]. Hence, a sensor calibration step is required to eliminate hysteresis in measurements. A simple way to calibrate is to deprotonate the polymer in strong base solutions. However, this method is inconvenient and can also accelerate the conductivity degradation as previously discussed. Therefore, a convenient and reliable self-calibration method for the development of a hysteresis-free polyaniline-based sensor is required. The authors have recently reported a potential self-calibration mechanism of a polyaniline-based chemical sensor by way of electrochemical resetting of the polymer [163]. The doping level of the polyaniline nanowires can be initialized by changing the redox state of the polymer to the fully oxidized form, and by returning the electrochemical potential to the initial redox state, the hysteresis can be minimized for subsequent pH measurement. Further characterization and improvement of the concept of electrochemical self-calibration mechanisms for reproducible and low hysteresis pH sensing is currently being investigated.

One of the main drawbacks of the polyaniline-based chemiresistive sensor, especially in gas sensing, is its relatively slow response time. As indicated in Table 1, the response times of polyaniline-based sensors are of the order of tens of seconds. For practical use, the response time must be reduced preferably to less than several seconds. Polyaniline composited with other nanomaterials such as carbon nanotubes, graphenes, or gold nanoparticles seems promising in this regard and has been shown to improve the response time as discussed in the previous section. Another issue with a polyaniline-based chemical sensor is its longevity. Due to natural degradation and loss of conductivity over time, the practical application of such devices is constrained by limited shelf life. These concerns must be addressed in the near future before any realistic commercial development can be made.

CONCLUSIONS

In summary, we have reviewed the basic principles and properties of polyaniline nanowires and their use in various sensor applications. The conductivity of polyaniline can be changed by a few orders of magnitude by varying its electrochemical potential and the solution pH. This unique property can be utilized for various chemiresistive or conductometric sensors. Moreover, the large surface area of the polyaniline nanowires makes this an ideal material for applications in high sensitivity gas sensors and biosensors. Conducting polyaniline nanowires have advantages over other metal or semiconducting nanowires for their low cost, ease of synthesis, and for the ability to locally or site-specifically fabricate the nanowires. However, some disadvantages may include fragility and structural degradation under harsh environments. The polyaniline nanowire-based sensors have been reported to have improved sensitivity and response time due to their nanoscale morphology. Some of the immediate challenges regarding polyaniline nanowires include improving their conductivity in the physiological pH range, preventing or minimizing the conductivity degradation, and minimizing the hysteresis effect. Much work is currently underway to address these issues with some areas already showing signs of success, and the number of applications for polyaniline nanowires is expected to increase in the future.

REFERENCES

1. Patolsky, F.; Zheng, G.; Lieber, C.M. Nanowire-based biosensors. *Anal. Chem.* 2006, *78*, 4260–4269.
2. Bangar, M.A.; Shirale, D.J.; Purohit, H.J.; Chen, W.; Myung, N.V.; Mulchandani, A. Single conducting polymer nanowire chemiresistive label-free immunosensor for cancer biomarker. *Anal. Chem.* 2009, *81*, 2168–2175.
3. Virji, S.; Huang, J.; Kaner, R.B.; Weiller, B.H. Polyaniline nanofiber gas sensors: Examination of response mechanisms.*Nano Lett.* 2004, *4*, 491–496.
4. Zhang, D.; Wang, Y. Synthesis and applications of one-dimensional nano-structured polyaniline: An overview. *Mater. Sci. Eng. B* 2006, *134*, 9–19.
5. Mulchandani, A.; Myung, N.V. Conducting polymer nanowires-based label-free biosensors. *Curr. Opin. Biotechnol.*2011, *22*, 502–508.
6. Ramgir, N.S.; Yang, Y.; Zacharias, M. Nanowire-based sensors. *Small* 2010, *6*, 1705–1722.
7. Matlock-Colangelo, L.; Baeumner, A.J. Recent progress in the design of nanofiber-based biosensing devices. *Lab. Chip*2012, *12*, 2612–2620.
8. Iijima, S. Helical microtubules of graphitic carbon. *Nature* 1991, *354*, 56–58.
9. Wang, J. Carbon-nanotube based electrochemical biosensors: A review. *Electroanalysis* 2005, *17*, 7–14.
10. Jacobs, C.B.; Peairs, M.J.; Venton, B.J. Review: Carbon nanotube based electrochemical sensors for biomolecules. *Anal. Chim. Acta* 2010, *662*, 105–127.
11. Cui, Y.; Wei, Q.; Park, H.; Lieber, C.M. Nanowire nanosensors for highly sensitive and selective detection of biological and chemical species. *Science* 2001, *293*, 1289–1292.
12. Patolsky, F.; Lieber, C.M. Nanowire nanosensors. *Mater. Today* 2005, *8*, 20–28.
13. Yogeswaran, U.; Chen, S.-M. A review on the electrochemical sensors and biosensors composed of nanowires as sensing material. *Sensors* 2008, *8*, 290–313.
14. Sunkara, M.K.; Sharma, S.; Miranda, R.; Lian, G.; Dickey, E.C. Bulk synthesis of silicon nanowires using a low-temperature vapor–liquid–solid method. *Appl. Phys. Lett.* 2001, *79*, 1546–1548.
15. Virji, S.; Kaner, R.B.; Weiller, B.H. Hydrogen sensors based on conductivity changes in polyaniline nanofibers. *J. Phys. Chem. B* 2006, *110*, 22266–22270.
16. Wang, J.; Bunimovich, Y.L.; Sui, G.; Savvas, S.; Wang, J.; Guo, Y.; Heath, J.R.; Tseng, H.-R. Electrochemical fabrication of conducting polymer nanowires in an integrated microfluidic system. *Chem. Commun.* 2006, *29*, 3075–3077.
17. Wang, J.; Chan, S.; Carlson, R.R.; Luo, Y.; Ge, G.; Ries, R.S.; Heath, J.R.; Tseng, H.-R. Electrochemically fabricated polyaniline nanoframework electrode junctions that function as resistive sensors. *Nano Lett.* 2004, *4*, 1693–1697.
18. Kuhn, P.; Puigmartí-Luis, J.; Imaz, I.; Maspoch, D.; Dittrich, P.S. Controlling the length and location of in situ formed nanowires by means of microfluidic tools. *Lab Chip* 2011, *11*, 753–757.
19. Hou, S.; Wang, S.; Yu, Z.T.F.; Zhu, N.Q.M.; Liu, K.; Sun, J.; Lin, W.-Y.; Shen, C.K.-F.; Fang, X.; Tseng, H.-R. A hydrodynamically focused stream as a dynamic template for site-specific electrochemical micropatterning of conducting polymers. *Angew. Chem.* 2008, *120*, 1088–1091.
20. Puigmartí-Luis, J.; Schaffhauser, D.; Burg, B.R.; Dittrich, P.S. A microfluidic approach for the formation of conductive nanowires and hollow hybrid structures. *Adv. Mater.* 2010, *22*, 2255–2259.
21. Shirakawa, H.; Louis, E.J.; MacDiarmid, A.G.; Chiang, C.K.; Heeger, A.J. Synthesis of electrically conducting organic polymers: Halogen derivatives of polyacetylene, $(CH)_x$. *J. Chem. Soc. Chem. Commun.* 1977, *1977*, 578–580.
22. Macdiarmid, A.G.; Chiang, J.C.; Richter, A.F.; Epstein, A.J. Polyaniline: A new concept in conducting polymers. *Synth. Met.* 1987, *18*, 285–290.
23. Sergeyeva, T.A.; Lavrik, N.V.; Piletsky, S.A.; Rachkov, A.E.; El'skaya, A.V. Polyaniline label-based conductometric sensor for IgG detection. *Sens. Actuators B* 1996, *34*, 283–288.
24. Gerard, M.; Chaubey, A.; Malhotra, B.D. Application of conducting polymers to biosensors. *Biosens. Bioelectron.* 2002,*17*, 345–359.
25. Dhand, C.; Das, M.; Datta, M.; Malhotra, B.D. Recent advances in polyaniline based biosensors. *Biosens. Bioelectron.*2011, *26*, 2811–2821.
26. Lange, U.; Roznyatovskaya, N.V.; Mirsky, V.M. Conducting polymers in chemical sensors and arrays. *Anal. Chim. Acta*2008, *614*, 1–26.
27. Reddinger, J.; Reynolds, J. Molecular Engineering of π-Conjugated Polymers. In *Advances in Polymer Science*; Springer: Berlin, Heidelberg, Germany, 1999; Volume 145, pp. 57–122.
28. MacDiarmid, A.G.; Epstein, A.J. Polyanilines: A novel class of conducting polymers. *Faraday Discuss. Chem. Soc.* 1989,*88*, 317–332.
29. Syed, A.A.; Dinesan, M.K. Review: Polyaniline—A novel polymeric material. *Talanta* 1991, *38*, 815–837.
30. Geniès, E.M.; Boyle, A.; Lapkowski, M.; Tsintavis, C. Polyaniline: A historical survey. *Synth. Met.* 1990, *36*, 139–182.
31. Huang, W.-S.; Humphrey, B.D.; MacDiarmid, A.G. Polyaniline, a novel conducting polymer. Morphology and chemistry of its oxidation and reduction in aqueous electrolytes. *J. Chem. Soc. Faraday Trans. 1* 1986, *82*, 2385–2400.

32. Bhadra, S.; Khastgir, D.; Singha, N.K.; Lee, J.H. Progress in preparation, processing and applications of polyaniline.*Prog. Polym. Sci.* 2009, *34*, 783–810.

33. Stejskal, J.; Sapurina, I.; Trchová, M. Polyaniline nanostructures and the role of aniline oligomers in their formation.*Prog. Polym. Sci.* 2010, *35*, 1420–1481.

34. Tran, H.D.; Wang, Y.; D'Arcy, J.M.; Kaner, R.B. Toward an understanding of the formation of conducting polymer nanofibers. *ACS Nano* 2008, *2*, 1841–1848.

35. Gupta, V.; Miura, N. Large-area network of polyaniline nanowires prepared by potentiostatic deposition process. *Electrochem. Commun.* 2005, *7*, 995–999.

36. Macdiarmid, A.G.; Mu, S.-L.; Somasiri, N.L.D.; Wu, W. Electrochemical characteristics of "polyaniline" cathodes and anodes in aqueous electrolytes. *Mol. Cryst. Liq. Cryst.* 1985, *121*, 187–190.

37. Gupta, V.; Miura, N. High performance electrochemical supercapacitor from electrochemically synthesized nanostructured polyaniline. *Mater. Lett.* 2006, *60*, 1466–1469.

38. Zhiani, M.; Gharibi, H.; Kakaei, K. Performing of novel nanostructure MEA based on polyaniline modified anode in direct methanol fuel cell. *J. Power Sources* 2012, *210*, 42–46.

39. Kelly, F.M.; Meunier, L.; Cochrane, C.; Koncar, V. Polyaniline: Application as solid state electrochromic in a flexible textile display. *Displays* 2013, *34*, 1–7.

40. Anderson, M.R.; Mattes, B.R.; Reiss, H.; Kaner, R.B. Conjugated polymer films for gas separations. *Science* 1991, *252*, 1412–1415.

41. Chang, C.-H.; Huang, T.-C.; Peng, C.-W.; Yeh, T.-C.; Lu, H.-I.; Hung, W.-I.; Weng, C.-J.; Yang, T.-I.; Yeh, J.-M. Novel anticorrosion coatings prepared from polyaniline/graphene composites. *Carbon* 2012, *50*, 5044–5051.

42. Focke, W.W.; Wnek, G.E.; Wei, Y. Influence of oxidation state, pH, and counterion on the conductivity of polyaniline.*J. Phys. Chem.* 1987, *91*, 5813–5818.

43. Zhang, X.; Goux, W.J.; Manohar, S.K. Synthesis of polyaniline nanofibers by "nanofiber seeding". *J. Am. Chem. Soc.*2004, *126*, 4502–4503.

44. Stafström, S.; Brédas, J.L.; Epstein, A.J.; Woo, H.S.; Tanner, D.B.; Huang, W.S.; MacDiarmid, A.G. Polaron lattice in highly conducting polyaniline: Theoretical and optical studies. *Phys. Rev. Lett.* 1987, *59*, 1464–1467.

45. Heeger, A.J. Semiconducting and metallic polymers: The fourth generation of polymeric materials. *J. Phys. Chem. B*2001, *105*, 8475–8491.

46. Ray, A.; Richter, A.F.; MacDiarmid, A.G.; Epstein, A.J. Polyaniline: Protonation/deprotonation of amine and imine sites. *Synth. Met.* 1989, *29*, 151–156.

47. Nechtschein, M.; Genoud, F.; Menardo, C.; Mizoguchi, K.; Travers, J.P.; Villeret, B. On the nature of the conducting state of polyaniline. *Synth. Met.* 1989, *29*, 211–218.

48. McManus, P.M.; Cushman, R.J.; Yang, S.C. Influence of oxidation and protonation on the electrical conductivity of polyaniline. *J. Phys. Chem.* 1987, *91*, 744–747.

49. Genies, E.M.; Tsintavis, C. Redox mechanism and electrochemical behaviour or polyaniline deposits. *J. Electroanal. Chem. Interfacial Electrochem.* 1985, *195*, 109–128.

50. Geniès, E.M.; Lapkowski, M.; Penneau, J.F. Cyclic voltammetry of polyaniline: Interpretation of the middle peak. *J. Electroanal. Chem. Interfacial Electrochem.* 1988, *249*, 97–107.

51. Nunziante, P.; Pistoia, G. Factors affecting the growth of thick polyaniline films by the cyclic voltammetry technique.*Electrochim. Acta* 1989, *34*, 223–228.

52. MacDiarmid, A.G. "Synthetic metals": A novel role for organic polymers (Nobel lecture). *Angew. Chem. Int. Ed.* 2001,*40*, 2581–2590.

53. Focke, W.W.; Wnek, G.E. Conduction mechanisms in polyaniline (emeraldine salt). *J. Electroanal. Chem. Interfacial Electrochem.* 1988, *256*, 343–352.

54. Saheb, A.H.; Seo, S.S. UV-vis and Raman spectral analysis of polyaniline/gold thin films as a function of applied potential. *Anal. Lett.* 2011, *44*, 1206–1216.

55. Kobayashi, T.; Yoneyama, H.; Tamura, H. Electrochemical reactions concerned with electrochromism of polyaniline film-coated electrodes. *J. Electroanal. Chem. Interfacial Electrochem.* 1984, *177*, 281–291.

56. Li, Q.; Cruz, L.; Phillips, P. Granular-rod model for electronic conduction in polyaniline. *Phys. Rev. B* 1993, *47*, 1840–1845.

57. Li, W.; Wan, M. Porous polyaniline films with high conductivity. *Synth. Met.* 1998, *92*, 121–126.

58. Mott, N.F.; Davis, E.A. *Electronic Processes in Non-Crystalline Materials*; Oxford University Press: Oxford, UK, 2012.

59. Joo, J.; Long, S.M.; Pouget, J.P.; Oh, E.J.; MacDiarmid, A.G.; Epstein, A.J. Charge transport of the mesoscopic metallic state in partially crystalline polyanilines. *Phys. Rev. B* 1998, *57*, 9567–9580.

60. Ghosh, M.; Barman, A.; De, S.K.; Chatterjee, S. Crossover from Mott to Efros-Shklovskii variable-range-hopping conductivity in conducting polyaniline. *Synth. Met.* 1998, *97*, 23–29.

61. Sheng, P.; Abeles, B.; Arie, Y. Hopping conductivity in granular metals. *Phys. Rev. Lett.* 1973, *31*, 44–47.

62. Lin, Y.-F.; Chen, C.-H.; Xie, W.-J.; Yang, S.-H.; Hsu, C.-S.; Lin, M.-T.; Jian, W.-B. Nano approach investigation of the conduction mechanism in polyaniline nanofibers. *ACS Nano* 2011, *5*, 1541–1548.

63. Zhou, Y.; Freitag, M.; Hone, J.; Staii, C.; Johnson, A.T.; Pinto, N.J.; MacDiarmid, A.G. Fabrication and electrical characterization of polyaniline-based nanofibers with diameter below 30 nm. *Appl. Phys. Lett.* 2003, *83*, 3800–3802.

64. Liu, W.; Kumar, J.; Tripathy, S.; Senecal, K.J.; Samuelson, L. Enzymatically synthesized conducting polyaniline. *J. Am. Chem. Soc.* 1999, *121*, 71–78.

65. Ma, Y.; Zhang, J.; Zhang, G.; He, H. Polyaniline nanowires on Si surfaces fabricated with DNA templates. *J. Am. Chem. Soc.* 2004, *126*, 7097–7101.

66. Trchová, M.; Šeděnková, I.; Konyushenko, E.N.; Stejskal, J.; Holler, P.; Ćirić-Marjanović, G. Evolution of polyaniline nanotubes: The oxidation of aniline in water. *J. Phys. Chem. B* 2006, *110*, 9461–9468.

67. Zhang, L.; Zujovic, Z.D.; Peng, H.; Bowmaker, G.A.; Kilmartin, P.A.; Travas-Sejdic, J. Structural characteristics of polyaniline nanotubes synthesized from different buffer solutions. *Macromolecules* 2008, *41*, 8877–8884.

68. Chiou, N.-R.; Epstein, A.J. Polyaniline nanofibers prepared by dilute polymerization. *Adv. Mater.* 2005, *17*, 1679–1683.

69. Wei, Y.; Tang, X.; Sun, Y.; Focke, W.W. A study of the mechanism of aniline polymerization. *J. Polym. Sci. Part Polym. Chem.* 1989, *27*, 2385–2396.

70. Yang, H.; Bard, A.J. The application of fast scan cyclic voltammetry. Mechanistic study of the initial stage of electropolymerization of aniline in aqueous solutions. *J. Electroanal. Chem.* 1992, *339*, 423–449.

71. Li, D.; Huang, J.; Kaner, R.B. Polyaniline nanofibers: A unique polymer nanostructure for versatile applications. *Acc. Chem. Res.* 2009, *42*, 135–145.

72. Huang, J.; Kaner, R.B. Nanofiber formation in the chemical polymerization of aniline: A mechanistic study. *Angew. Chem.* 2004, *116*, 5941–5945.

73. Dias, H.V.R.; Wang, X.; Rajapakse, R.M.G.; Elsenbaumer, R.L. A mild, copper catalyzed route to conducting polyaniline. *Chem. Commun.* 2006, 976–978.

74. Huang, J.; Kaner, R.B. A general chemical route to polyaniline nanofibers. *J. Am. Chem. Soc.* 2004, *126*, 851–855.

75. Huang, J.; Virji, S.; Weiller, B.H.; Kaner, R.B. Polyaniline nanofibers: Facile synthesis and chemical sensors. *J. Am. Chem. Soc.* 2003, *125*, 314–315.

76. Qiang, J.; Yu, Z.; Wu, H.; Yun, D. Polyaniline nanofibers synthesized by rapid mixing polymerization. *Synth. Met.*2008, *158*, 544–547.

77. Martin, C.R. Template synthesis of electronically conductive polymer nanostructures. *Acc. Chem. Res.* 1995, *28*, 61–68.

78. Martin, C.R. Nanomaterials: A membrane-based synthetic approach. *Science* 1994, *266*, 1961–1966.

79. Li, G.; Zhang, C.; Li, Y.; Peng, H.; Chen, K. Rapid polymerization initiated by redox initiator for the synthesis of polyaniline nanofibers. *Polymer* 2010, *51*, 1934–1939.

80. Kitani, A.; Kaya, M.; Yano, J.; Yoshikawa, K.; Sasaki, K. "Polyaniline": Formation reaction and structure. *Synth. Met.*1987, *18*, 341–346.

81. Liang, L.; Liu, J.; Windisch, C.F., Jr.; Exarhos, G.J.; Lin, Y. Direct assembly of large arrays of oriented conducting polymer nanowires. *Angew. Chem. Int. Ed.* 2002, *41*, 3665–3668.

82. Choi, S.-J.; Park, S.-M. Electrochemical growth of nanosized conducting polymer wires on gold using molecular templates. *Adv. Mater.* 2000, *12*, 1547–1549.

83. Stilwell, D.E.; Park, S.-M. Electrochemistry of conductive polymers IV electrochemical studies on polyaniline degradation—Product identification and coulometric studies. *J. Electrochem. Soc.* 1988, *135*, 2497–2502.

84. Huang, L.; Wang, Z.; Wang, H.; Cheng, X.; Mitra, A.; Yan, Y. Polyaniline nanowires by electropolymerization from liquid crystalline phases. *J. Mater. Chem.* 2002, *12*, 388–391.

85. Mu, S.; Yang, Y. Spectral characteristics of polyaniline nanostructures synthesized by using cyclic voltammetry at different scan rates. *J. Phys. Chem. B* 2008, *112*, 11558–11563.

86. Diaz, A.F.; Logan, J.A. Electroactive polyaniline films. *J. Electroanal. Chem. Interfacial Electrochem.* 1980, *111*, 111–114.

87. Huang, Z.-M.; Zhang, Y.-Z.; Kotaki, M.; Ramakrishna, S. A review on polymer nanofibers by electrospinning and their applications in nanocomposites. *Compos. Sci. Technol.* 2003, *63*, 2223–2253.

88. Reneker, D.H.; Chun, I. Nanometre diameter fibres of polymer, produced by electrospinning. *Nanotechnology* 1996, *7*, 216–223.

89. Van Dyke, L.S.; Martin, C.R. Electrochemical investigations of electronically conductive polymers. 4. Controlling the supermolecular structure allows charge transport rates to be enhanced. *Langmuir* 1990, *6*, 1118–1123.

90. Martin, C.R. Membrane-based synthesis of nanomaterials. *Chem. Mater.* 1996, *8*, 1739–1746.

91. Parthasarathy, R.V.; Martin, C.R. Template-synthesized polyaniline microtubules. *Chem. Mater.* 1994, *6*, 1627–1632.

92. Jackowska, K.; Biegański, A.; Tagowska, M. Hard template synthesis of conducting polymers: A route to achieve nanostructures. *J. Solid State Electrochem.* 2008, *12*, 437–443.

93. Xiao, Y.; Kharitonov, A.B.; Patolsky, F.; Weizmann, Y.; Willner, I. Electrocatalytic intercalator-induced winding of double-stranded DNA with polyaniline. *Chem. Commun.* 2003, *2003*, 1540–1541.

94. Fan, Y.; Chen, X.; Trigg, A.D.; Tung, C.; Kong, J.; Gao, Z. Detection of microRNAs using target-guided formation of conducting polymer nanowires in nanogaps. *J. Am. Chem. Soc.* 2007, *129*, 5437–5443.

95. Liu, W.; Anagnostopoulos, A.; Bruno, F.F.; Senecal, K.; Kumar, J.; Tripathy, S.; Samuelson, L. Biologically derived water soluble conducting polyaniline. *Synth. Met.* 1999, *101*, 738–741.

96. Bergveld, P. Development of an ion-sensitive solid-state device for neurophysiological measurements. *IEEE Trans. Biomed. Eng.* 1970, *BME-17*, 70–71.

97. Bergveld, P. Thirty years of ISFETOLOGY: What happened in the past 30 years and what may happen in the next 30 years. *Sens. Actuators B* 2003, *88*, 1–20.

98. Bergveld, P. Development, operation, and application of the ion-sensitive field-effect transistor as a

tool for electrophysiology. *IEEE Trans. Biomed. Eng.* 1972, *BME-19*, 342–351.

99. Liu, H.; Kameoka, J.; Czaplewski, D.A.; Craighead, H.G. Polymeric nanowire chemical sensor. *Nano Lett.* 2004, *4*, 671–675.

100. Liu, C.; Noda, Z.; Sasaki, K.; Hayashi, K. Development of a polyaniline nanofiber-based carbon monoxide sensor for hydrogen fuel cell application. *Int. J. Hydrog. Energy* 2012, *37*, 13529–13535.

101. Liu, C.; Hayashi, K.; Toko, K. Au nanoparticles decorated polyaniline nanofiber sensor for detecting volatile sulfur compounds in expired breath. *Sens. Actuators B* 2012, *161*, 504–509.

102. Zeng, F.-W.; Liu, X.-X.; Diamond, D.; Lau, K.T. Humidity sensors based on polyaniline nanofibres. *Sens. Actuators B*2010, *143*, 530–534.

103. Lin, Q.; Li, Y.; Yang, M. Polyaniline nanofiber humidity sensor prepared by electrospinning. *Sens. Actuators B* 2012,*161*, 967–972.

104. Lim, J.-H.; Phiboolsirichit, N.; Mubeen, S.; Deshusses, M.A.; Mulchandani, A.; Myung, N.V. Electrical and gas sensing properties of polyaniline functionalized single-walled carbon nanotubes. *Nanotechnology* 2010, *21*, 075502:1–075502:7.

105. Al-Mashat, L.; Shin, K.; Kalantar-zadeh, K.; Plessis, J.D.; Han, S.H.; Kojima, R.W.; Kaner, R.B.; Li, D.; Gou, X.; Ippolito, S.J.; *et al.* Graphene/polyaniline nanocomposite for hydrogen sensing. *J. Phys. Chem. C* 2010, *114*, 16168–16173.

106. Shirsat, M.D.; Bangar, M.A.; Deshusses, M.A.; Myung, N.V.; Mulchandani, A. Polyaniline nanowires-gold nanoparticles hybrid network based chemiresistive hydrogen sulfide sensor. *Appl. Phys. Lett.* 2009, *94*, 083502:1–083502:3.

107. Sadek, A.Z.; Wlodarski, W.; Shin, K.; Kaner, R.B.; Kalantar-zadeh, K. A layered surface acoustic wave gas sensor based on a polyaniline/In_2O_3 nanofibre composite. *Nanotechnology* 2006, *17*, 4488–4492.

108. Sadek, A.Z.; Baker, C.O.; Powell, D.A.; Wlodarski, W.; Kaner, R.B.; Kalantar-zadeh, K. Polyaniline nanofiber based surface acoustic wave gas sensors-effect of nanofiber diameter on H_2 response. *IEEE Sens. J.* 2007, *7*, 213–218.

109. Arsat, R.; Yu, X.F.; Li, Y.X.; Wlodarski, W.; Kalantar-zadeh, K. Hydrogen gas sensor based on highly ordered polyaniline nanofibers. *Sens. Actuators B* 2009, *137*, 529–532.

110. Wang, H.; Yang, P.-H.; Cai, H.-H.; Cai, J. Constructions of polyaniline nanofiber-based electrochemical sensor for specific detection of nitrite and sensitive monitoring of ascorbic acid scavenging nitrite. *Synth. Met.* 2012, *162*, 326–331.

111. Xian, Y.; Hu, Y.; Liu, F.; Xian, Y.; Wang, H.; Jin, L. Glucose biosensor based on Au nanoparticles–conductive polyaniline nanocomposite. *Biosens. Bioelectron.* 2006, *21*, 1996–2000.

112. Pal, S.; Alocilja, E.C.; Downes, F.P. Nanowire labeled direct-charge transfer biosensor for detecting Bacillus species.*Biosens. Bioelectron.* 2007, *22*, 2329–2336.

113. Li, G.; Martinez, C.; Janata, J.; Smith, J.A.; Josowicz, M.; Semancik, S. Effect of morphology on the response of polyaniline-based conductometric gas sensors: Nanofibers *vs.* thin films. *Electrochem. Solid-State Lett.* 2004, *7*, H44–H47.

114. Nicolas-Debarnot, D.; Poncin-Epaillard, F. Polyaniline as a new sensitive layer for gas sensors. *Anal. Chim. Acta* 2003,*475*, 1–15.

115. Huang, J.; Virji, S.; Weiller, B.H.; Kaner, R.B. Nanostructured polyaniline sensors. *Chem. Eur. J.* 2004, *10*, 1314–1319.

116. Sadek, A.Z.; Wlodarski, W.; Kalantar-Zadeh, K.; Baker, C.; Kaner, R.B. Doped and dedoped polyaniline nanofiber based conductometric hydrogen gas sensors. *Sens. Actuators A* 2007, *139*, 53–57.

117. Liao, Y.; Zhang, C.; Zhang, Y.; Strong, V.; Tang, J.; Li, X.-G.; Kalantar-zadeh, K.; Hoek, E.M.V.; Wang, K.L.; Kaner, R.B. Carbon nanotube/polyaniline composite nanofibers: Facile synthesis and chemosensors. *Nano Lett.* 2011, *11*, 954–959.

118. Liao, Y.; Zhang, C.; Wang, X.; Li, X.-G.; Ippolito, S.J.; Kalantar-zadeh, K.; Kaner, R.B. Carrier mobility of single-walled carbon nanotube-reinforced polyaniline nanofibers. *J. Phys. Chem. C* 2011, *115*, 16187–16192.

119. Kaner, R.B. Gas, liquid and enantiomeric separations using polyaniline. *Synth. Met.* 2001, *125*, 65–71.

120. Liao, Y.; Yu, D.-G.; Wang, X.; Chain, W.; Li, X.-G.; Hoek, E.M.V.; Kaner, R.B. Carbon nanotube-templated polyaniline nanofibers: Synthesis, flash welding and ultrafiltration membranes. *Nanoscale* 2013, *5*, 3856–3862.

121. Guillen, G.R.; Farrell, T.P.; Kaner, R.B.; Hoek, E.M.V. Pore-structure, hydrophilicity, and particle filtration characteristics of polyaniline–polysulfone ultrafiltration membranes. *J. Mater. Chem.* 2010, *20*, 4621–4628.

122. Blinova, N.V.; Svec, F. Functionalized polyaniline-based composite membranes with vastly improved performance for separation of carbon dioxide from methane. *J. Membr. Sci.* 2012, *423–424*, 514–521.

123. Clark, L.C.; Lyons, C. Electrode systems for continuous monitoring in cardiovascular surgery. *Ann. N. Y. Acad. Sci.*1962, *102*, 29–45.

124. Wang, J. Electrochemical glucose biosensors. *Chem. Rev.* 2008, *108*, 814–825.

125. Bartlett, P.N.; Astier, Y. Microelectrochemical enzyme transistors. *Chem. Commun.* 2000, *2000*, 105–112.

126. White, H.S.; Kittlesen, G.P.; Wrighton, M.S. Chemical derivatization of an array of three gold microelectrodes with polypyrrole: Fabrication of a molecule-based transistor. *J. Am. Chem. Soc.* 1984, *106*, 5375–5377.

127. Ofer, D.; Crooks, R.M.; Wrighton, M.S. Potential dependence of the conductivity of highly oxidized polythiophenes, polypyrroles, and polyaniline: Finite windows of high conductivity. *J. Am. Chem. Soc.* 1990, *112*, 7869–7879.

128. Bartlett, P.N.; Birkin, P.R. A microelectrochemical enzyme transistor responsive to glucose. *Anal. Chem.* 1994, *66*, 1552–1559.

129. Bartlett, P.N.; Birkin, P.R.; Wang, J.H.; Palmisano, F.; De Benedetto, G. An enzyme switch employing direct electrochemical communication between horseradish peroxidase and a poly(aniline) film. *Anal. Chem.* 1998, *70*, 3685–3694.

130. Battaglini, F.; Bartlett, P.N.; Wang, J.H. Covalent attachment of osmium complexes to glucose oxidase and the application of the resulting modified enzyme in an enzyme switch responsive to glucose. *Anal. Chem.* 2000, *72*, 502–509.

131. Forzani, E.S.; Zhang, H.; Nagahara, L.A.; Amlani, I.; Tsui, R.; Tao, N. A conducting polymer nanojunction sensor for glucose detection. *Nano Lett.* 2004, *4*, 1785–1788.

132. Yuk, J.S.; Jin, J.-H.; Alocilja, E.C.; Rose, J.B. Performance enhancement of polyaniline-based polymeric wire biosensor.*Biosens. Bioelectron.* 2009, *24*, 1348–1352.

133. Muhammad-Tahir, Z.; Alocilja, E.C. A conductometric biosensor for biosecurity. *Biosens. Bioelectron.* 2003, *18*, 813–819.

134. Tahir, Z.M.; Alocilja, E.C.; Grooms, D.L. Polyaniline synthesis and its biosensor application. *Biosens. Bioelectron.* 2005,*20*, 1690–1695.

135. Forzani, E.S.; Li, X.; Tao, N. Hybrid amperometric and conductometric chemical sensor based on conducting polymer nanojunctions. *Anal. Chem.* 2007, *79*, 5217–5224.

136. Gao, Z.Q.; Rafea, S.; Lim, L.H. Detection of nucleic acids using enzyme-catalyzed template-guided deposition of polyaniline. *Adv. Mater.* 2007, *19*, 602–606.

137. Chang, H.; Yuan, Y.; Shi, N.; Guan, Y. Electrochemical DNA biosensor based on conducting polyaniline nanotube array. *Anal. Chem.* 2007, *79*, 5111–5115.

138. Castillo-Ortega, M.M.; Rodriguez, D.E.; Encinas, J.C.; Plascencia, M.; Méndez-Velarde, F.A.; Olayo, R. Conductometric uric acid and urea biosensor prepared from electroconductive polyaniline–poly(n-butyl methacrylate) composites.*Sens. Actuators B* 2002, *85*, 19–25.

139. Sangodkar, H.; Sukeerthi, S.; Srinivasa, R.S.; Lal, R.; Contractor, A.Q. A biosensor array based on polyaniline. *Anal. Chem.* 1996, *68*, 779–783.

140. Muhammad-Tahir, Z.; Alocilja, E.C. Fabrication of a disposable biosensor for *Escherichia coli* O157:H7 detection. *IEEE Sens. J.* 2003, *3*, 345–351.

141. Raffa, D.; Leung, K.T.; Battaglini, F. A microelectrochemical enzyme transistor based on an n-alkylated poly(aniline) and its application to determine hydrogen peroxide at neutral pH. *Anal. Chem.* 2003, *75*, 4983–4987.

142. Yue, J.; Epstein, A.J. Synthesis of self-doped conducting polyaniline. *J. Am. Chem. Soc.* 1990, *112*, 2800–2801.

143. Yue, J.; Epstein, A.J.; Macdiarmid, A.G. Sulfonic acid ring-substituted polyaniline, a self-doped conducting polymer.*Mol. Cryst. Liq. Cryst. Inc. Nonlinear Opt.* 1990, *189*, 255–261.

144. Ghenaatian, H.R.; Mousavi, M.F.; Kazemi, S.H.; Shamsipur, M. Electrochemical investigations of self-doped polyaniline nanofibers as a new electroactive material for high performance redox supercapacitor. *Synth. Met.* 2009,*159*, 1717–1722.

145. Thiemann, C.; Brett, C.M. Electropolymerisation and properties of conducting polymers derived from aminobenzenesulphonic acids and from mixtures with aniline. *Synth. Met.* 2001, *125*, 445–451.

146. Kamaraj, K.; Karpakam, V.; Sathiyanarayanan, S.; Venkatachari, G. Electrosysnthesis of poly(aniline-co-m-amino benzoic acid) for corrosion protection of steel. *Mater. Chem. Phys.* 2010, *122*, 123–128.

147. Lukachova, L.V.; Shkerin, E.A.; Puganova, E.A.; Karyakina, E.E.; Kiseleva, S.G.; Orlov, A.V.; Karpacheva, G.P.; Karyakin, A.A. Electroactivity of chemically synthesized polyaniline in neutral and alkaline aqueous solutions: Role of self-doping and external doping. *J. Electroanal. Chem.* 2003, *544*, 59–63.

148. Zhang, L.; Dong, S. The electrocatalytic oxidation of ascorbic acid on polyaniline film synthesized in the presence of camphorsulfonic acid. *J. Electroanal. Chem.* 2004, *568*, 189–194.

149. Lee, K.-H.; Park, B.J.; Song, D.H.; Chin, I.-J.; Choi, H.J. The role of acidic m-cresol in polyaniline doped by camphorsulfonic acid. *Polymer* 2009, *50*, 4372–4377.

150. Del Castillo-Castro, T.; Castillo-Ortega, M.M.; Villarreal, I.; Brown, F.; Grijalva, H.; Pérez-Tello, M.; Nuño-Donlucas, S.M.; Puig, J.E. Synthesis and characterization of composites of DBSA-doped polyaniline and polystyrene-based ionomers. *Compos. Part Appl. Sci. Manuf.* 2007, *38*, 639–645.

151. Haberko, J.; Bernasik, A.; Łużny, W.; Hasik, M.; Raczkowska, J.; Rysz, J.; Budkowski, A. Humidity and wetting effects in spin-cast blends of insulating polymers and conducting polyaniline doped with DBSA. *J. Appl. Polym. Sci.* 2012,*127*, 2354–2361.

152. Lyutov, V.V.; Ivanov, S.D.; Mirsky, V.M.; Tsakova, V.T. Polyaniline doped with poly(acrylamidomethyl propanesulphonic acid): Electrochemical behaviour and conductive properties in neutral solutions. *Chem. Pap.* 2013, *67*, 1002–1011.

153. Tarver, J.; Yoo, J.E.; Dennes, T.J.; Schwartz, J.; Loo, Y.-L. Polymer acid doped polyaniline is electrochemically stable beyond pH 9. *Chem. Mater.* 2009, *21*, 280–286.

154. Bayer, C.L.; Konuk, A.A.; Peppas, N.A. Development of a protein sensing device utilizing interactions between polyaniline and a polymer acid dopant. *Biomed. Microdevices* 2010, *12*, 435–442.

155. Raitman, O.A.; Katz, E.; Bückmann, A.F.; Willner, I. Integration of polyaniline/poly(acrylic acid) films and redox enzymes on electrode supports: An *in situ* electrochemical/surface plasmon resonance study of the bioelectrocatalyzed oxidation of glucose or lactate in the integrated bioelectrocatalytic systems. *J. Am. Chem. Soc.* 2002,*124*, 6487–6496.

156. Tang, Y.; Pan, K.; Wang, X.; Liu, C.; Luo, S. Electrochemical synthesis of polyaniline in surface-attached poly(acrylic acid) network, and its application to the electrocatalytic oxidation of ascorbic acid. *Microchim. Acta* 2010, *168*, 231–237.

157. Bonastre, A.M.; Sosna, M.; Bartlett, P.N. An analysis of the kinetics of oxidation of ascorbate at poly(aniline)-poly(styrene sulfonate) modified microelectrodes. *Phys. Chem. Chem. Phys.* 2011, *13*, 5365–5372.

158. Bartlett, P.N.; Wang, J.H. Electroactivity, stability and application in an enzyme switch at pH 7 of poly(aniline)-poly(styrenesulfonate) composite films. *J. Chem. Soc. Faraday Trans.* 1996, *92*, 4137–4143.

159. Łapkowski, M.; Vieil, E. Control of polyaniline electroactivity by ion size exclusion. *Synth. Met.* 2000, *109*, 199–201.

160. Lee, S.-Y.; Choi, G.-R.; Lim, H.; Lee, K.-M.; Lee, S.-K. Electronic transport characteristics of electrolyte-gated conducting polyaniline nanowire field-effect transistors. *Appl. Phys. Lett.* 2009, *95*, 013113:1–013113:3.

161. Lee, S.-Y.; Lee, S.-K.; Lim, H.; Choi, G.-R. Irreversible degradation behaviors of an electrolyte-gated polyaniline (PANI) nanowire field-effect transistor. *J. Korean Phys. Soc.* 2010, *57*, 1416–1420.

162. He, H.; Zhu, J.; Tao, N.J.; Nagahara, L.A.; Amlani, I.; Tsui, R. A conducting polymer nanojunction switch. *J. Am. Chem. Soc.* 2001, *123*, 7730–7731.

163. Song, E.; Choi, J.-W. An On-chip Chemiresistive Polyaniline Nanowire-based pH Sensor with Self-calibration Capability. In Proceedings of the 2012 Annual International Conference of the IEEE Engineering in Medicine and Biology Society (EMBC), San Diego, CA, USA, 28 August–1 September 2012; pp. 4018–4021.

CHAPTER 5

WORK FUNCTION ENGINEERING OF GRAPHENE

Rajni Garg, Naba K. Dutta and Namita Roy Choudhury

Ian Wark Research Institute, University of South Australia, Mawson Lakes Campus, 5095 Adelaide, Australia

ABSTRACT

Graphene is a two dimensional one atom thick allotrope of carbon that displays unusual crystal structure, electronic characteristics, charge transport behavior, optical clarity, physical & mechanical properties, thermal conductivity and much more that is yet to be discovered. Consequently, it has generated unprecedented excitement in the scientific community; and is of great interest to wide ranging industries including semiconductor, optoelectronics and printed electronics. Graphene is considered to be a next-generation conducting material with a remarkable band-gap structure, and has the potential to replace traditional electrode materials in optoelectronic devices. It has also been identified as one of the most promising materials for post-silicon electronics. For many such applications, modulation of the electrical and optical properties, together with tuning the band gap and the resulting work function of zero band gap graphene are critical in achieving the desired properties and outcome. In understanding the importance, a number of strategies including various functionalization, doping and hybridization have recently been identified and explored to successfully alter the work function of graphene. In this review we primarily highlight the different ways of surface modification, which have been used to specifically modify the band gap of graphene and its work function. This article focuses on the most recent perspectives, current trends and gives some indication of future challenges and possibilities.

SCOPE OF THE REVIEW

Graphite and its intercalation compounds have been studied for over 150 years; however, serious scientific investigation on graphene is of rather recent origin, from when Novoselov *et al.* [1] 2004 reported the facile synthesis of single layer/few layer graphene from graphite using mechanical exfoliation from graphite. Since this report on the unusual electronic properties of single layers of the graphene lattice, research interest on graphene increased exponentially over the past decade (Figure 1).

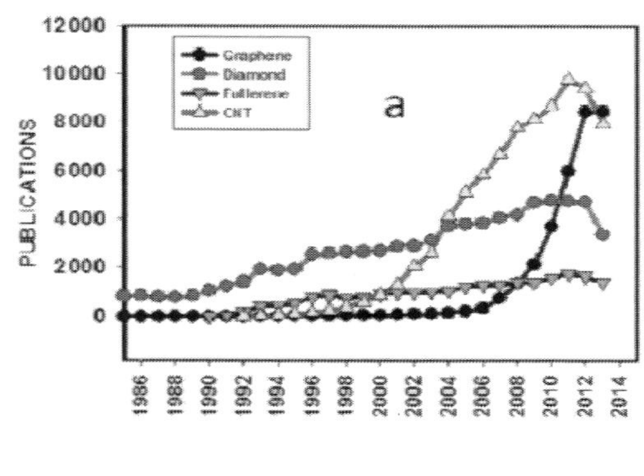

(a)

Citation: Rajni Garg, Naba K. Dutta and Namita Roy Choudhury, Work Function Engineering of Graphene, doi: 10.3390/nano4020267.

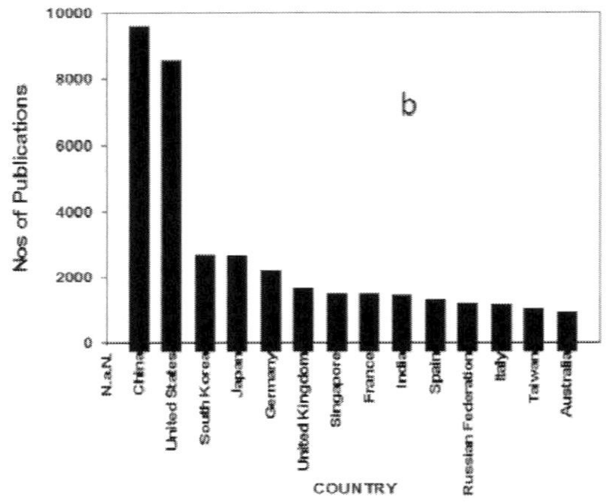

(b)

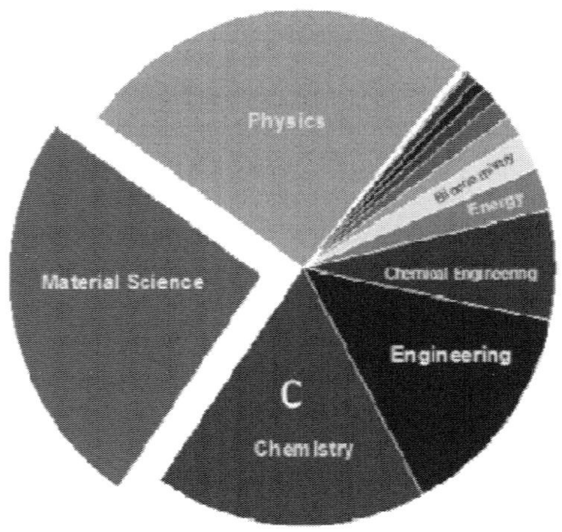

(c)

Figure 1: History of scientific publications on carbon-based materials: (**a**) Total annual number of publications for each carbon allotrope based on Scopus; (**b**) History of the number of publications in graphene from different countries; (**c**) History of the number of publications on grapheme based on different subject areas.

Andre Geim and Konstantin Novoselov share the 2010 Nobel Prize in Physics for their ground-breaking work on graphene. Figure 1a illustrates the annual number of articles published in refereed journals containing different carbon allotropes as key words. The recent dramatic growth of publications in graphene is remarkable with more than 8000 papers published in 2012 (~22 papers/day). Figure 1b shows that China and USA are the two countries that are pursuing research on graphene most vigorously (~25% of total publication each). Moreover, the aspects of graphene from physics and material science can be identified as the two major focused areas of research (Figure 1c). Although, a large number of review and prospective articles on different aspects of graphene are plentiful, there are however a limited number of critical reviews on graphene interfacial engineering and work function (WF) tuning. Most of the organic electronics, optoelectronics and printed electronics technologies require electrodes and conductors with an appropriately tuned work function to facilitate efficient charge transport. The band alignment of two different materials at the heterojunction is also governed by their respective WF. Therefore, in graphene-based electronic devices the WF of graphene under a given metal electrode is critical information for the realization of high-performance graphene-based interconnects. In this review, we focus our attention on the recent advances on different methods for graphene synthesis and modification and its effect on tuning of WF. Initially we discuss briefly the unique band structure and a short history of the synthesis and characteristics of graphene, as well as the exciting recent progresses.

GRAPHENE: A UNIQUE CARBON ALLOTROPE

The element carbon occurs in several allotropes, including crystalline, three-dimensional diamond, graphite and lonsdaleite; two dimensional graphene, one-dimensional nanotubes, zero dimensional fullerene (Figure 2) and several other non-crystalline forms [2,3]. These wide-ranging carbon allotropes exhibit extremes in the physical, chemical and morphological behaviour; which display the diversity of carbon's atomic structures, crystal chemistry and bonding differences. For example, in diamond each C-atom is sp^3 hybridized, the C–C–C bond angle is 109.5°, the C–C bond length is 1.54 Å; and forms basic tetrahedral units and a cubic unit cell. On the one hand diamond is considered as the hardest known substance. On the other hand, graphite is known to be one of the softest materials; and acts as an efficient solid lubricant. In graphite, graphene, fullerenes, carbon nanotubes, and several types of other amorphous and glassy carbons; the carbon atoms are in a planar three-coordination state that results from the sp^2 hybridization. In this trigonal coordination state typical C–C distance is ~1.42 Å and C–C–C angles are ~120°. Graphene (GR) is a flat single layer, with sp^2 hybridized carbon atoms tightly packed into a two dimensional honeycomb lattice structure; and it is one of the most exciting two dimensional materials being investigated today [4]. Graphene may also be considered as the basic building block for graphite materials of all other dimensions. It can be stacked into 3D graphite, rolled into 1D carbon nanotubes and wrapped into 1D fullerenes. The sp^2 hybridisation forms strong directed bonds and determines a honeycomb lattice structure for graphene; and the $p^z(\pi)$ orbitals form a delocalized π-system which regulates the conduction properties/charge mobility of graphite.

Graphene is the first example of a close-packed two dimensional (2D) crystalline material isolated in nature; and it is currently receiving unusual growth in research attention. The hexagonal arrangement of the carbon atoms in graphene can be reduced into two interpenetrating sub-lattices of carbon atoms with inversion symmetry between them (Figure 3) [5].

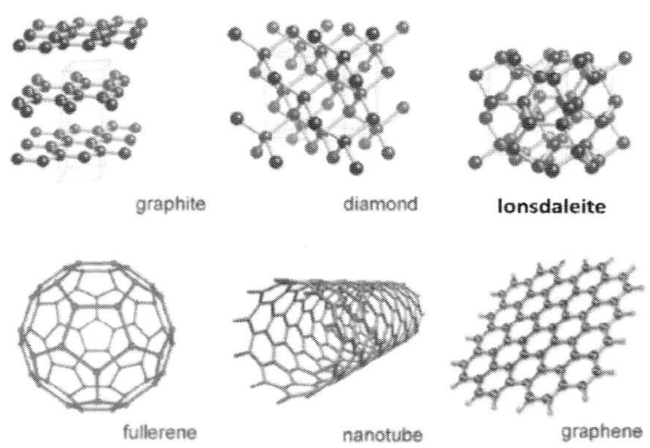

Figure 2: The common naturally-occurring sp^2 and sp^3 allotropes of carbon occur in different crystallographic forms. **Graphite**: Hex-

agonal; stacked flat layers of 3-coordinated sp^2 C. **Diamond**: Cubic; framework of 4-coordinated sp^3 C. **Lonsdaleite**: Hexagonal; framework of 4-coordinated sp^3 C. **Fullerenes**: Closed cage molecules sp^{2}C: C60, C70, C76, *etc.* Nanotubes cylindrical fibers of sp^2 C, single tubes or nested. **Graphene**: one-atom-thick graphitic layers with sp^2 bonding. Reprinted with permission from [3]. Copyright 2013 Mineralogical Society of America.

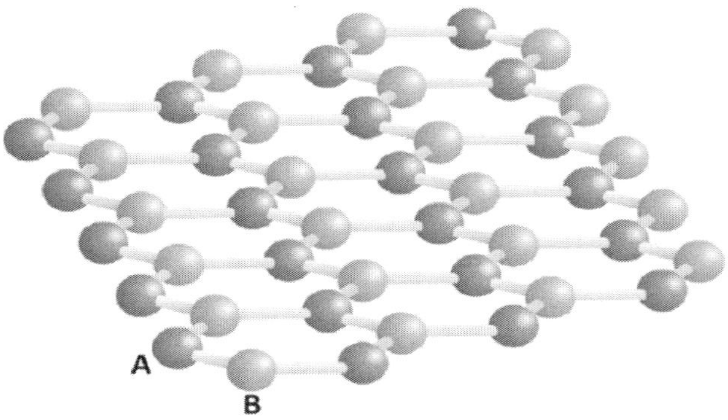

Figure 3: Schematic representation of the hexagonal arrangement of the carbon atoms in graphene. It can be reduced into two interpenetrating sub-lattices of carbon atoms with inversion symmetry between them. Atoms from different sub-lattices (**A** and **B**) are marked by different colors. Reprinted with permission from [6]. Copyright 2007 Elsevier.

GRAPHENE: A TWO DIMENSIONAL FORM OF CARBON WITH UNUSUAL BAND STRUCTURE AND CHARACTERISTICS

Graphene is a two-dimensional zero-band gap semiconductor with a unique energy diagram shown in Figure 4 [7]. This unique topology provides an unusual energy dispersion relation in graphene [8]. Graphene's unique electronic structure is characterized by the pointed valence and conduction bands that meet at a single point in momentum space (the Dirac crossing energy). Quantum mechanical hopping between the sub-lattices in graphene leads to the formation of two energy bands, and their intersection near the edge of the Brillion zone yields the conical energy spectrum near the Dirac points K and K› as represented in Figure 4 [7]. It has now been confirmed that the charge carriers in graphene can be described by the Dirac-like equation, rather than the usual Schrödinger equation [6]. Novoselov *et al.* [1] demonstrated the strong ambipolar electric field effect of GR with electrons and holes in concentrations up to 10^{13}/cm^2, and with room temperature mobility of ~10,000 cm^2/V·s.

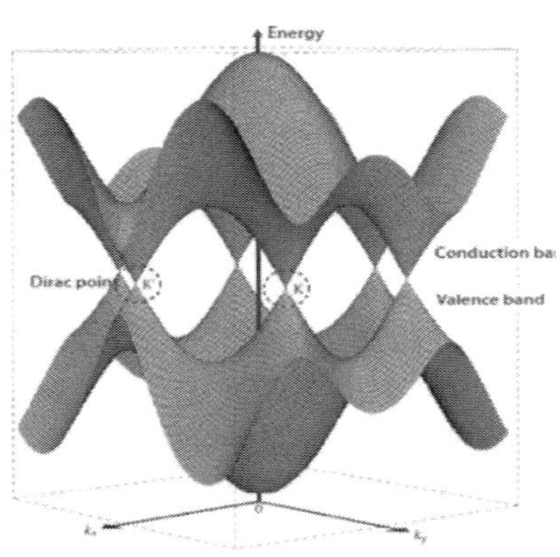

(a)

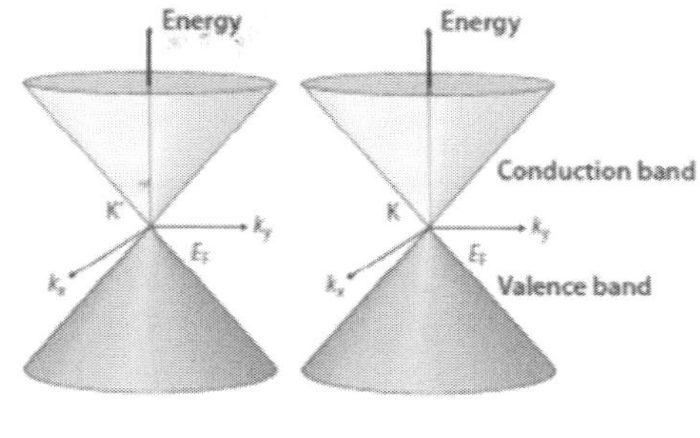

(b)

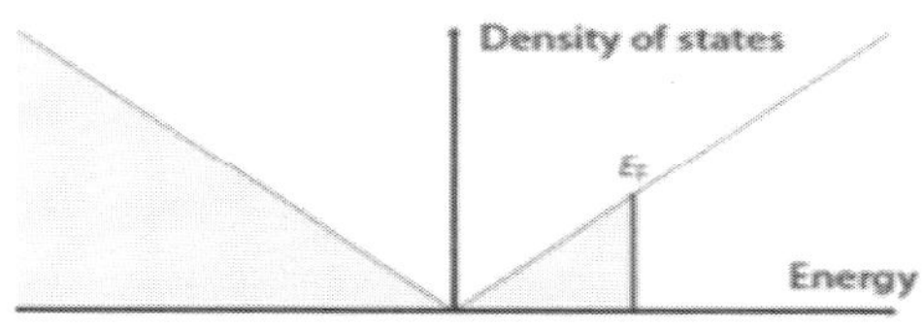

(c)

Figure 4: (**a**) Energy bands near the Fermi level in graphene. The first Brillouin zone of graphene is illustrated in the horizontal plane and labelled with some points of interest. The conduction and valence bands cross at points K and K'—the two non-equivalent corners of the zone, also known as the Dirac points; (**b**) Conic energy bands in the vicinity of the K and K' points; (**c**) Density of states near the Fermi level with Fermi energy E_F. Reprinted with permission from [7]. Copyright 2009 Nature publishing group.

It has also been demonstrated that through selective control of the carrier concentration in the graphene layers, the band structure can be easily tuned near the Dirac crossing [9]. The electronic structure, unique morphological and electronic characteristics of graphene; and its potential applications have been the subject of extensive theoretical and experimental investigations [10,11,12,13]. Summarily, graphene has many special attributes including the followings unique characteristics:

1. Graphene is a zero-band gap perfect 2D semi-conductor with a tiny overlap between valence and conduction bands [6];
2. By strictly confining electrons to two dimensions graphene displays an unusual fractional quantum Hall effect [14,15,16,17];
3. Graphene shows a strong ambipolar electric field effect with charge carrier concentrations up to 10^{13} cm^2 and room temperature mobility of ~10,000 cm^2·s^{-1} [6];
4. Graphene demonstrates transport via relativistic Dirac Fermions [18,19];
5. Graphene displays high thermal conductivity with a value of ~5000 Wmk^{-1} for a single layer sheet at room temperature [19];
6. Graphene exhibits high optical transparency with absorption of ~2.3% towards visible light [20];
7. Graphene is incredibly strong, mechanically (tensile strength of ~130 GPa, for a defect free single layer, Young's modulus of 1 Tpa, third order elastic stiffness ≈2 Tpa); while remaining highly flexible and very light (0.77 mg m^{-2}) and possesses a very high specific surface area (~2630 m^2·g^{-1}) [21].

GRAPHENE: POTENTIAL APPLICATIONS AND IMPORTANCE OF WORK FUNCTION

Despite its short history, graphene, graphene-related nanomaterials and their processing including their synthesis, direct liquid-phase exfoliation of graphite to produce single/few layered pristine graphene, doped graphene, graphene oxide, fluorographene, graphene, graphdiyne, and porous graphene have all been the subject of many recent reviews [22,23,24,25,26]. Allen *et al.* [27] presented a comprehensive review on graphene from the materials chemistry stand point. In-depth discussion focused on the physics aspect of graphene has been the subject of many recent perspectives and reviews [5,13,28,29]. The challenges and prospects of graphene based new energy materials specifically for supercapacitors, lithium ion batteries, water splitting, electrocatalysts for fuel cells, and solar cells have recently been reviewed by Sun *et al.* [30]. Recent advances in the field of graphene from the standpoint of electrochemistry have been presented by Chen *et al.* [31]. Some of the recent progresses in graphene synthesis, fundamental concepts and prominent applications in electronics and photonics have been summarized by Brownson *et al.* [32]. Compton and Nguyen [33] have highlighted the wide ranging functionalized materials that can be prepared using a bottom-up approach from bulk dispersions of graphene oxide, HRG, and graphene in various media. The opportunities and challenges of graphene-based hybrid materials for energy and sensing applications have been highlighted [34,35,36]. Graphene has also been promoted as a reinforcing filler in polymer composites and nano-composites, and is expected to be a less expensive substitute than carbon nanotubes [37]. In a recent review, Kuila *et al.* [38] discussed the different methods used for chemical functionalization of graphene and its importance in processing, properties and final applications. Yang *et al.* [39] reviewed the available different protocols for functionalization of graphene with special reference to its potential biomedical applications including drug delivery and multimodal imaging to preparation of bio-sensing devices. Craciun *et al.* [40] discussed the possibilities to engineer the electrical and optical properties in graphene through chemical functionalization of graphene. Such modifications have the potential to widen the applications of graphene devices on an industrial scale.

Due to its unusual electronic characteristics and distinct capabilities graphene has been used successfully to fabricate a number of simple electronic devices [10,41]. One of the major forces for the use of graphene in nanotechnology is to develop integrated circuits on a much smaller scale that is possible with current silicon-based complementary metal-oxide-semiconductor (CMOS) technology, and it has been identified as the most promising material for post-silicon electronics [42,43]. It has been predicted [44] that evolutionary miniaturization using silicon based technology will reach the fundamental limits of extreme physical size in the near future. Since graphene can exhibit room-temperature ballistic transport over mean free paths of up to 300 nm [6,45] its use has been suggested either as a channel material for the next generation of field-effect transistors (FET); or as a conductive sheet upon which nanometer scale devices may be patterned to create single electron or few electron transistors. Graphene is also an excellent candidate for ultra-high frequency transistors [26]. It has been identified as a novel electrode material with excellent stability, high transparency, flexibility and outstanding charge transfer mobility, which is a critical requirement for many optoelectronic devices such as: organic solar cells (OSCs), organic light-emitting diodes (OLEDs), and organic thin-film transistors (TFTs) [46,47,48,49,50]. The unique properties of graphene have also potential for use in gas sensors [51], supercapacitors [52] and printed electronics [35]. Recently, a CMOS compatible graphene photodetector covering all optical communication bands was also reported [53]. In all these practical device designs the WF of graphene is one of the most important considerations; moreover, WF is also dependent on the environment, such as adjacent substrates [54]. Another important consideration for both fundamental property measurements and practical device design is the contact resistance, and the choice of contact material. In graphene electronics, the contact resistance (Rc) between graphene and other metal limits the performance of the device such as the carrier mobility and ON-state current of graphene field effect transistors (FETs) [55]. The contact resistance suppresses the on-current, which is detrimental to high-frequency transistor performance; however, in optoelectronics this effect enhances photocurrent efficiency [56]. A lower WF can dramatically enhance the emitting current [57]. The WF of graphene electrodes is also critical to maximize energy conversion efficiency in OPVs. Besides good conductivity and transparency of the electrode, the performance and current density for semiconducting electronic devices such as light-emitting diodes and field-effect transistors is strongly dependent on the carrier injection efficiency through the contact between electrodes and semiconducting material layers. Graphene being a prospective conductive material with the desirable properties including engineered WF for efficient carrier injection, makes it highly attractive. Due to the critical importance, different approaches have been investigated to modulate the WF of graphene including employing an external electric field [58], reaction with organic and inorganic molecules [59], chemical modification of the surface [60], metal doping [61,62,63], substrate orientation [64], and a self-assembled monolayer formation [65]. Intercalation of different species such as hydrogen [66], fluorine [67], lithium [68], gold [69] and iron (III) chloride [70] has also been identified as a potential method to modulate the work function of few layer graphene [71,72]. However, controlling the WF of graphene, precisely, on demand has yet to be demonstrated.

GRAPHENE SYNTHESIS: RELATIONSHIP TO APPLICATIONS

Graphene was first successfully synthesized using mechanical exfoliation of graphite in 2004 by Novoselov *et al.* [1]. Such single/few layer graphene was used to elucidate the unique characteristics of graphene, which was the catalyst for the explosive growth in research on graphene. However, this method is unreliable, of low yield and basic; and it has very limited relevance to commercial high-end electronic applications. As the research interest and promise of large scale application of graphene has grown extensively, various practical methods have been attempted to synthesize high quality pristine graphene in large scale at low cost. The fundamental properties of graphene have been well investigated using high-quality graphene produced by "top-down" physical exfoliation [1,72] and solvation-assisted exfoliation of graphite [73]. Among the many approaches for single or few layer graphene synthesis, the most significant ones are the exfoliation of graphite [74,75,76,77], arc discharge of graphite in the presence of helium and hydrogen gases [78], laser based green synthesis, [79], chemical vapour deposition [35,80], low-cost liquid-phase exfoliation of graphite and graphene oxide reduction [81]. In Table 1, we summarize the synthesis methods employed; and comment on their maturity, advantages and disadvantages, and targeted use. Before the wider application of graphene, the most important challenge is to develop a facile and efficient method for controlled production of processable, large graphene sheets of the desirable properties for specific applications. An important challenge in the large-scale production of pristine graphene is to select/design the proper exfoliating reagents including the right solvents. The continuous and scalable large area synthesis of graphene by chemical vapour deposition (CVD) has reinforced its study in a broad range of research areas [82,83]. Recently, Colson *et al.* [84] demonstrated covalent organic framework thin films on single-layer graphene. The electronic characteristics of a planar covalent organic framework (COF) on graphene are investigated by means of dispersion-corrected density functional theory [85].

Ideally, the exfoliation protocol should involve the use of low-cost natural graphite powder as starting material, with commonly available solvents to assist that can be removed easily to produce pristine graphene. Various suitable chemical methods have been developed and identified for mass production of graphene and functional graphene using natural graphite as the starting material [86]. Among the different chemical methods, synthesis of solution processable graphite oxide as precursor of graphene-based materials is one of the most widely used protocols. In general, such methods involve chemical oxidation of natural graphite using various solution-based routes to synthesize hydrophilic graphite oxide followed by exfoliation (e.g., using ultrasonication) to form a single layer of few layer graphene oxide [87,88,89]. Thermal/chemical/electrochemical reduction of GO is normally used to produce reduced graphene rGO [86,90,91,92,93,94,95] (Figure 5).

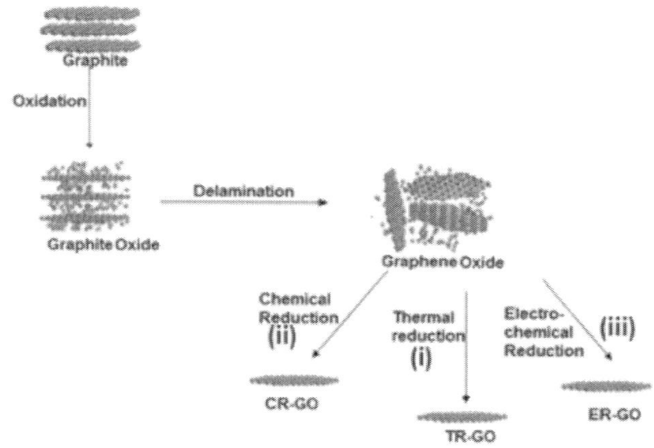

Figure 5: Schematic representation of the synthesis of single/few layer graphene from graphite: An oxidative treatment is initially performed to generate graphite oxide; which is followed by exfoliation to produce graphene oxide. Finally, **(i)** Thermal reduction; **(ii)** Chemical reduction or **(iii)** Electrochemical reduction of graphene oxide produces reduced graphene (r-GO).

Table 1: Different methods for synthesis of single or multilayer grapheme

Method	Precursor	Electronic quality	Advantage	Disadvantage	Commercialization	References
Mechanical Exfoliation	Graphite	High	Inexpensive and time saving method	Flakes randomly distributed, poor yield	Not scalable for commercialization	[76]
Arc discharge method	Graphite	High	Applicable to obtain Boron or nitrogen doped graphene	Cannot obtain pure graphene	Not scalable	[78]
Wet chemical synthesis such as Hummer, Brodie	Graphite	High	Transparent conductive film, useful to synthesis graphene based composites	Presence of oxygen impurities are not suitable for most of the electronic applications	Can be obtained in lab but not good enough for commercialization	[92]
Chemical vapour deposition	Hydrocarbons	High	Promising method that has all the above mentioned advantages	Transfer of graphene films deteriorates graphene quality and causes wrinkle formation	Possible	[80,82]
Solvothermal synthesis	Ethanol	Not available	Cheap and easily available precursor	Popcorn effect arises due to nucleation of sheets	Scalable	[96]
Epitaxial growth on metals	Ultrathin graphitic film	High	Single to multi layer graphene sheets can be obtained	Requires high temperature, expensive and difficult transfer process	Not feasible	[97]

During reduction GO, which is a non-conductive and hydrophilic layer of graphene with oxygen atoms attached to it, gets reduced to graphene. Currently, one of the most popular methods for synthesizing graphite oxide is Hummer's or a modified Hummer method [86,94]. These methods for graphene preparation allowed facile large scale synthesis, and observation of its unique properties including: flexibility, transparency, chemical robustness and roll to roll processability. This protocol has the potential not only to significantly reduce the cost of mass-produced graphene, but also allow simple, non-chemical, thermal or electrochemical conversion of GO powder to graphene film or powder respectively. It has also widely opened the opportunity for a variety of functionalization and hybridization of graphene/few layer graphene and to tune the final characteristics. It is a potential method for synthesis of large-scale graphene to be used in various industrial applications including organic solar cells, capacitors, sensors, and transparent electrodes. The thermal reduction of GO to produce graphene; however, is a complex phenomenon and involves multistep removal processes of the intercalated H_2O molecules, oxide groups of –COOH (carboxyl group), hydroxyl group and epoxy group. Therefore, the thermal reduction of GO and resultant GP needs to be investigated in great detail. Electrochemical reduction of GO has been successfully employed for graphene based electrochemical sensors and biosensors. Electrochemically reduced GO to rGO is the preferred method for reduction when rGO is to be used in different electrodes with different energy, electrochemical and electrochemical immunosensing platforms. [95,96,98,99,100] However, a much detailed understanding of the mechanisms of the exfoliation and reduction of graphite is essential to the design and discovery of efficient exfoliation systems with precise properties for specific applications.

WORK FUNCTION AND TUNING OF THE WORK FUNCTION OF GRAPHENE

The work function, Φ of any material can be defined as the energy required to remove an electron from the highest filled level in the Fermi distribution of a solid to vacuum (*i.e.*, stationary at a point in a field-free zone just outside the solid) at absolute zero [97,101]:

$$\Phi = V_{\text{vacuum}} - E_{\text{Fermi}} \quad (1)$$

Work function (WF) is a fundamental electronic property of any material, and provides understanding of the relative position of the Fermi surface levels. WF, determines the band alignment in the contact at the heterojunction to facilitate selective electron and hole transport [101]. WF of graphene and the ability to engineer it, is a very important factor in governing the application of graphene as an electrode material [102]. Graphene is attractive as an important electrode material for transparent conducting electrodes in many modern optoelectronic devices including liquid crystal displays (LCDs), organic light emitting diodes (OLEDs), touch screens, and solar cells. Graphene is a zero band gap material, and its conductivity cannot be turned off electronically as in conventional semiconductor materials. It is a critical issue in graphene-based nanoelectronic and optoelectronic devices to tune the WF of graphene layers, while preserving its unique band structure [103]. For example, in organic thin-film devices such as light-emitting diodes (LEDs); the selection of an electron injecting contact with a WF that matches the energy level of the lowest unoccupied molecular orbital (LUMO) of the organic layer is critical. Such a choice prevents the formation of an electron injection barrier. Facilitating the charge injection improves the device performance, and tuning the WF of electrode to match the HOMO (or valence band) and/or LUMO (conducting band) of the active layers is essential [104,105]. Several physical and chemical methods including depositing the dopant atoms [106], absorption of gas molecules [107] or use of aromatic compounds [108,109,110,111] have recently been proposed for tuning the WR of graphene. Yu *et al.* [112] proposed tuning of graphene WR by electric field effect. Gui [113] proposed and modelled the band structure engineering of GR by application of strain. The chemical doping and deposition of different noble metal nanoparticles on the graphene surface is one of the most effective methods to tailor the WF of graphene [114,115]. Various chemical functionalization strategies also introduce band gap opening in graphene which leads to a change in work function [116]. The band gap opening leads to a shift in the Fermi level, however, the extent to which the band gap of graphene can be tuned has not been investigated in detail; and little is known about the precise role of interface and surface adsorbents [117]. In the following sections we will review in some detail different surface modifications techniques that have been used to modify graphene/GO/r-GO in an attempt to engineer the WF of graphene.

Effect of Oxygen Functionalities on the Work Function of Graphene

The wet chemical method based on oxidation of natural graphite to graphite oxide (GO) followed by exfoliation to graphene oxide, and finally thermal/chemical/electrochemical reduction of GO to reduced graphene oxide (r-GO) is one of the most widely used protocols for large scale synthesis of graphene/r-GO. This method, however, introduces different oxygen containing functional groups to graphene. The type and concentration of such functionalities have a decisive influence on the WF of graphene. For example, incorporation of electron withdrawing groups such as –OH to the surface of r-GO increases the WF; whereas, the addition of electron donating group (e.g., $-CH_3$) to the surface decreases the WF [101]. Kumar *et al.* [118] recently thoroughly examined the impact of oxygen-containing chemical functionality on the WF of rGO using classical molecular dynamics simulation and density functional theory on a set of realistic rGO structures. From such an exercise, it was predicted that the presence of carbonyl groups on the rGO surface results in the largest impact on WF (6.7 eV) relative to all other groups. It has been demonstrated that the presence of such a group can induce a WF value of ~60% higher than that of pristine graphene (4.2 eV). On that basis they projected a significant tunability in the WF of rGO (up to ~2.5 eV) by altering the composition of the oxygen-containing functional groups (Figure 6). Mishra *et al.* [119] experimentally evaluated the WF of GO as a function of oxygen content. They employed contact surface potential difference (CPD) estimation as an indicator of WF using a scanning Kelvin probe method (SKPM). SKPM, is a scanning probe method, where, a vibrating capacitor is used to investigate the WF of metal and semiconductor surfaces at nano/micro level; and provides information about the electronic state of the local structures on the surface of a solid. In principle it is based on the detection and dynamic compensation of the electrostatic forces that arise between the micro/nano tip and the surface when they are electrically connected [119]. CPD measurement estimated WF indirectly in terms of relative surface contact and may be converted to an absolute value by the appropriate calibration. The CPD measurement between two different surfaces follows the relation:

$$V_{cPD} = \frac{\Phi_{HP} - \Phi_{Sample}}{e} \quad (2)$$

where Φ is the WF, and *e* is the elementary charge of electron [119].

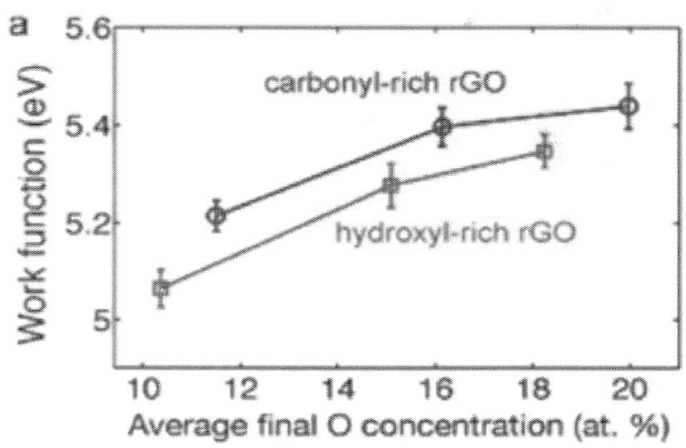

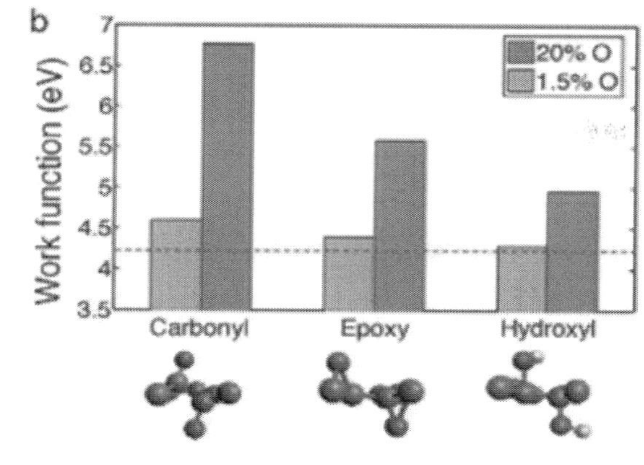

Figure 6: Work function tunability in rGO structures. (**a**) Calculated work function of carbonyl-rich and hydroxyl-rich rGO structures with different oxygen content; (**b**) The effect of individual functional groups on the work function of rGO, for two different total oxygen concentrations of 1.5% (for validation purpose) and of 20%. Change in Work function in rGO structures with respect to its functional groups. Reprinted with permission from [118]. Copyright 2013 American Chemical Society.

Figure 7 illustrates the variation of experimental CPD as a function of oxygen content for Go to r-GO. It is clear fromFigure 7b that CPD and oxygen content at different stages from GO follow a linear trend. These recent predictive methods and experimental study demonstrated the possibility of precise control of WF of rGO by tuning the oxygen-containing functionalities on the surface. However, further in-depth experimental investigation of the effect of different type and concentration of oxygen containing functionality is critical to fine tune the WF of graphene.

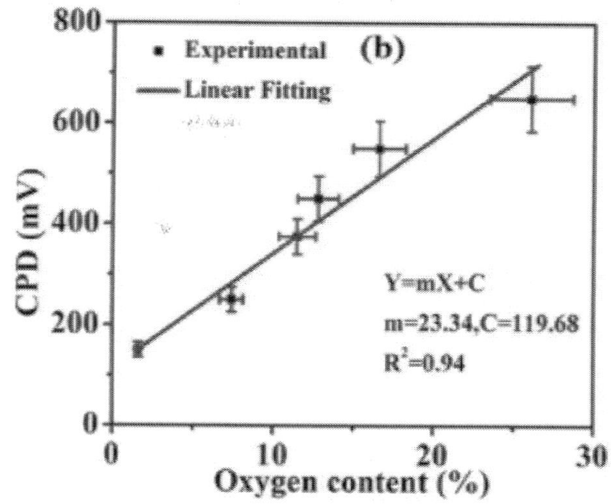

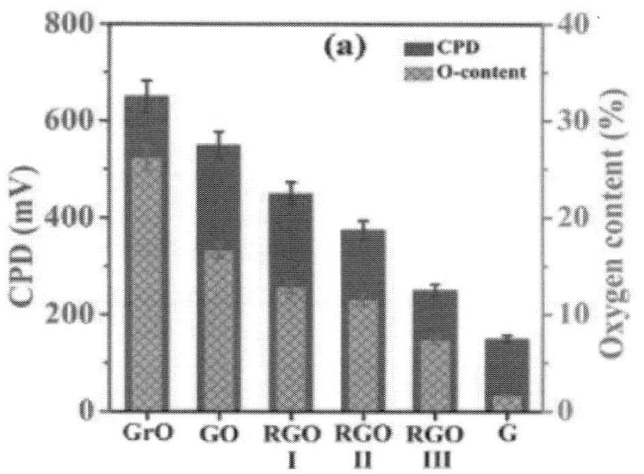

Figure 7: (**a**) Bar graph of contact surface potential difference (CPD) as a function of oxygen content at different stages in graphene synthesis; and (**b**) Linear fitting of CPD variation during decreasing oxygen content at different stages from graphite oxide to graphene. Reprinted with permission from [119]. Copyright 2013 American Chemical Society.

Work Function Engineering by Reduction of Graphene Oxide

An extensive amount of research effort has been focused on thermal [120,121], chemical [14,122,123], and electrochemical [124,125] reduction of solution processed GO to grapheme [126,127]. Reduction of graphene is nothing but removal of oxygen functionalities present on the GO surface, which will ultimately alter the WF of the rGO. However, actual chemical transformation during the reduction process is complicated and significantly dependent on the method used for reduction and the environmental conditions imposed during the reaction. Different reduction methods have different effects on the graphene surface, and consequently result in different WFs. As the presence of surface functionality is the most important factor to control WF of graphene; understanding the reduction mechanism for removal of different functional groups at different steps is very important. Gao *et al.* [128] reported the reduction of GO through simultaneous chemical and thermal reduction methods. They elucidated the mechanism of chemical reduction via hydrazine as decarbonylation, and thermal reduction at 900 °C as dehydroxylation of GO (Figure 8). Higher C to O ratio was observed as GO was progressively annealed from 500 to 900 °C.

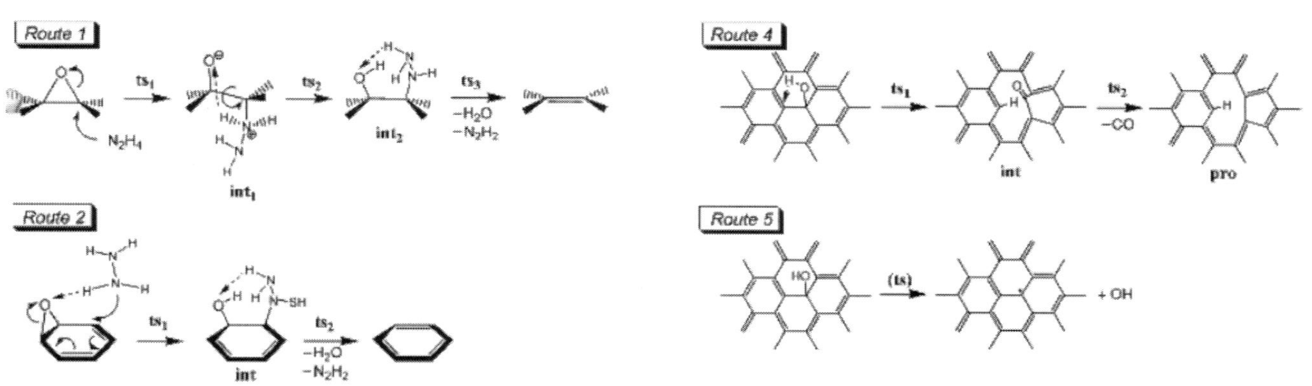

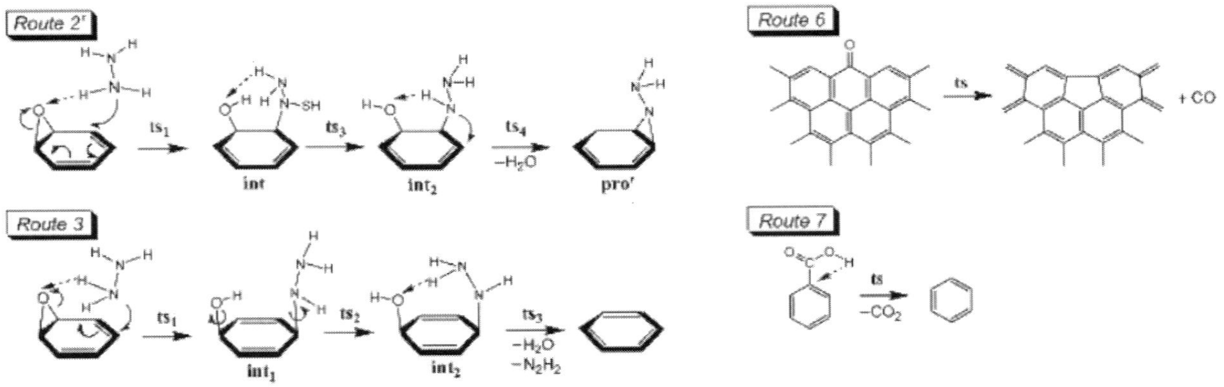

Figure 8: GO reduction mechanism. Routes1–3 and 2'represents the mechanism for hydrazine de-epoxidation of GO. Routes 4–5 represent the mechanism for thermal dehydroxylation for GO. Routes 6 and 7 represent the mechanism for thermal decarbonylation and thermal decarboxylation of GO. Reprinted with permission from [128]. Copyright 2009 American Chemical Society.

Hwang *et al.* [129] attempted to tune the WF of N-doped rGO film. N-doped reduced graphene was obtained by sequential chemical treatment of GO with hydrazine followed by thermal treatment. N-doping was performed to increase the electron density and thus further reduce WF. They observed that the hydrazine treatment lowered the oxygen containing groups; whereas, the thermal treatment removed the less stable N and O containing groups present at dangling bonds. The WF for N-doped graphene, reduced using hydrazine based pre-treatment, showed almost a constant value of WF (4.25 eV). However, N-doped films without pre-treatment exhibited variation in WF value within the range of 4.2–4.4 eV; and such rGOs exhibit higher conductivity, superior transmittance (~80%) and lower sheet resistance (Figure 9). Hydrothermal reduction techniques have been employed by Chieh *et al.* [130], to tune the WF of GO over a wide range from 5.72 to 4.43 eV. They examined the treated GO using X-ray absorption near edge structure (XANES), and confirmed progressive increase in the sp^2 to sp^3 ratio with respect to increase in the hydrothermal temperature. This observation is consistent with the increase in conductivity with increased amount of sp^2 species with a progressively higher treatment temperature of GO. The use of a combination of treatments provides opportunity to fine tune the WF for graphene for many applications including cathode material, to obtain highly efficient polymer light emitting diodes (PLEDs).

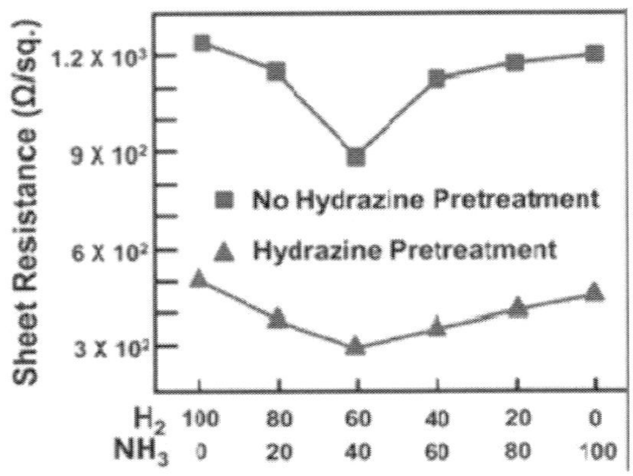

(b)

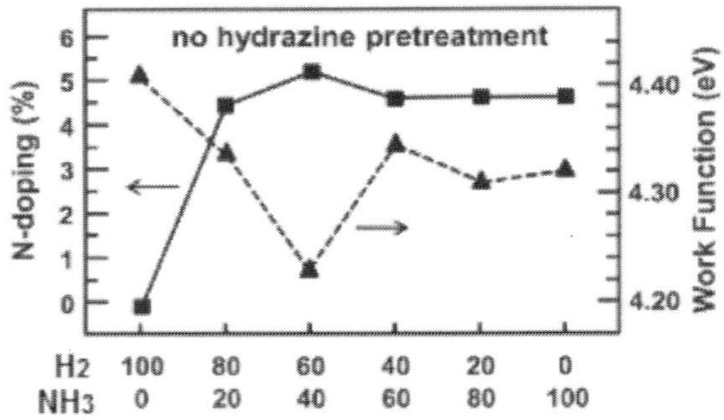

(c)

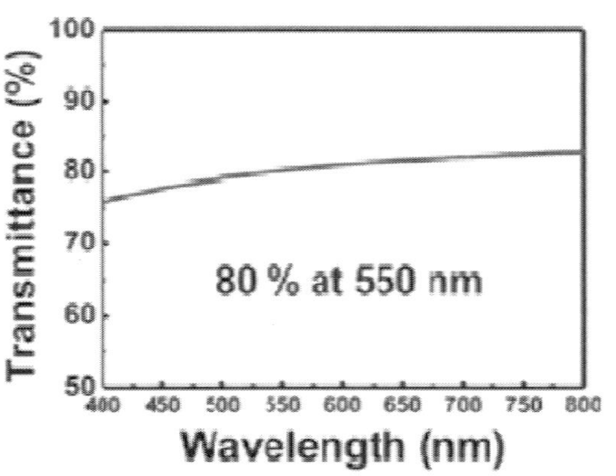

(a)

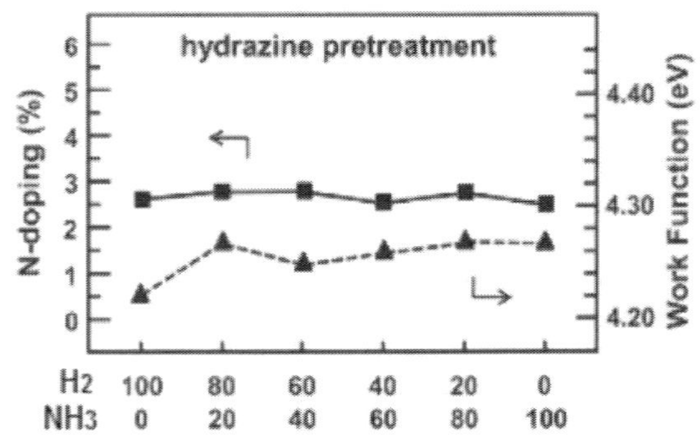

(d)

Figure 9: Effect of GO treatment on the properties of GO. (**a**) Shows increase in transmittance with respect to wavelength and (**b**) Shows decrease in sheet resistance with respect to hydrazine treatment, whereas (**c**) Shows percentage of N-doping and work function modulation with no hydrazine pre-treatment, and (**d**) With hydrazine pre-treatment plotted again H_2/NH_3 ratio. Reprinted with permission from [129]. Copyright 2011 American Chemical Society.

Liu *et al.* [131] fabricated a polymer solar cell using graphene as the hole transport layer (HTL) and P3HT(donor)/PCBM (acceptor) as active layer, and the reported V_{OC} was 0.6 V at 130 °C. However, they observed a decrease in open circuit voltage (V_{OC}) when GO was annealed to 230 °C. The V_{OC} value of a device is governed by the difference in WF between HOMO of the donor (P3HT (5.0 eV)) and LUMO of the acceptor (PCBM (3.7 eV)) material. The authors ascribed this change in V_{OC} with thermal treatment and the WF change in GO due to the removal of functional groups at higher temperature.

Work Function Engineering of GO Using Functionalization and Self-Assembled Monolayer

Functionalization of GO or graphene using various chemical approaches has been attempted to modulate WF [109]. Liu *et al.* [132] reported tuning of the WF of GO to match the WF of higher occupied molecular orbital (HOMO) of P3HT (5.0 eV) through synthesis of sulfated graphene oxide (GO–OSO_3H), and achieved a WF of 4.8 eV. To achieve such a goal, GO was functionalized (Figure 10) where the –OSO_3H group was attached to the –COOH groups present at the edge of GO. Such WF matching has the potential to enhance the conductivity of the charge transfer layer of organic solar cells. Yang *et al.* [133] demonstrated O_2 plasma treatment of GO to achieve a work function of 5.2 eV. Such chemically modified GO could be used as a hole transport layer in organic solar cells.

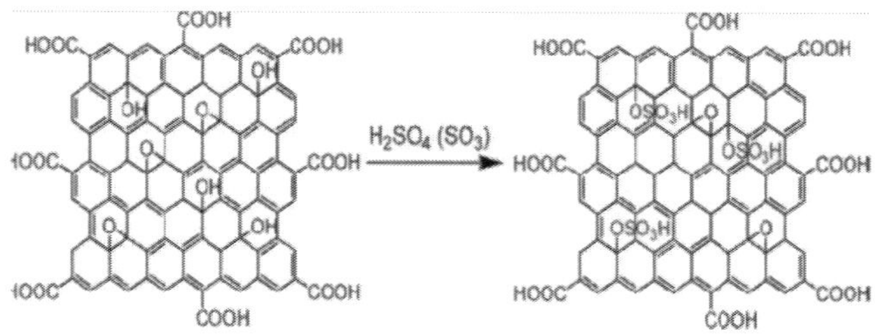

Figure 10: Formation of GO–OSO_3H. Reprinted with permission from [132]. Copyright 2012 American Chemical Society.

By considering the influence of functional groups and strong pi-pi conjugation (that leads to changes in Fermi level) between graphene and P3HT on WF, Wang *et al.* [134] attempted to tune the WR of graphene with a view to use solution-processable functionalized graphene as the acceptor material in OPV to replace PCBM (e.g., Phenyl-C61-butyric acid methyl ester). They used phenyl isocyanate-treated GO to obtain SPF-graphene to replace acceptor component. SPF-graphene can be easily dissolved in organic solvents such as 1,2-dichlorobenzene (DCB), which is also the solvent for P3HT. Such photovoltaic devices can be fabricated using common fabrication processes, and the active layer can be prepared by spin coating a P_3HT/SPF graphene solution in DCB. The WF difference between graphene and P_3HT is generally ~0.7 eV, which was further increased to 1.08 eV for SPF-graphene. The investigators confirmed the influence of WF on V_{OC} by subjecting SPF-graphene/P_3HT active layer to a range of annealing temperatures. They demonstrated that the V_{OC} value of the system was between 0.7 and 1.1 eV on annealing the system at 160 °C, whereas the value fell to 0.5 eV when the annealing temperature was increased to 210 °C. This change in WF was attributed to the elimination of functional groups at higher temperature, which led to a change in WF and thus V_{OC} [122]. Similar observation was reported by Wang *et al.* [135] when SPF-graphene was used as acceptor along with PCBM and P3OT as donor.

Liu *et al.* [136] fabricated an OPV device with P3HT/SPF-graphene as active layer; and they reported a short circuit current density J_{sc} of 4.0 mA·cm^{-2}, open circuit voltage V_{OC} of 0.72 V, and a solar power conversion efficiency of 1.1% under simulated AM1.5G, 100 mW illumination in air for such a device. This observation indicates that soluble graphene has the potential to be a promising acceptor material for OPVs; however, further research is critical to optimise the donor/acceptor interface engineering and work function tuning. Recently, Lin *et al.* [137] further reported findings, both theoretical and experimental, on the graphene/semiconductor heterojunction solar cells. They predicted that for such a Schottky junction solar cell a maximum theoretical conversion efficiency of ~9.2% could be achieved by controlling the graphene layer number, tuning the graphene WF and adding an antireflection film. They also experimentally demonstrated that solar cells based on modified graphene and Si pillar arrays can deliver enhanced cell performance with efficiencies of up to 7.7%.

Work Function Engineering Using Self-Assembled Monolayer (SAM) and Layer by Layer Technique

Kang *et al.* [138] investigated the influence of the tuning of the WF of rGO by self-assembled monolayer (SAM) formation on p-channel organic field-effect transistors (FETs). They showed that the use of (tridecafluoro-1,1,2,2-tetrahydrooctyl) trichlorosilane SAM modified GO (FTS-GO) display strong p-doped behavior relative to r-GO. FTS-rGO exhibited WF of 5.51 eV, a considerable increase when compared to the WF for r-GO of 4.9 eV. However, aminopropyl triethoxysilane (APTS) SAM modified rGO (APTS-rGO), exhibited WF of 4.31 eV, which is 0.63 eV lower than that of p-RGO (Figure 11). This tunability of WF for SAM functionalized rGO was utilized in developing source/drain electrodes in bottom-contact FETs. It is suggested that the APTS functionalization helps in neutralization of unintentionally doped p-rGO [138]. Layer-by-layer technique is a facile means of fabricating multilayer hybrid thin films. Kong *et al.* [139] employed LBL assembly to make a hybrid thin film of graphene and gold nanoparticles. They demonstrated that for such films electrons in the negatively charged (rGO) substrate participated in the reduction of metal (Au) cations. Such electrochemical transformation occurs because the reduction potential of rGO is much less than that of Au (+1.002 V). The significant variation in reduction potential allows reduction of Au cations by donating electrons from rGO to Au^{3+}. This material was found to be highly applicable to sensors and other electronic devices.

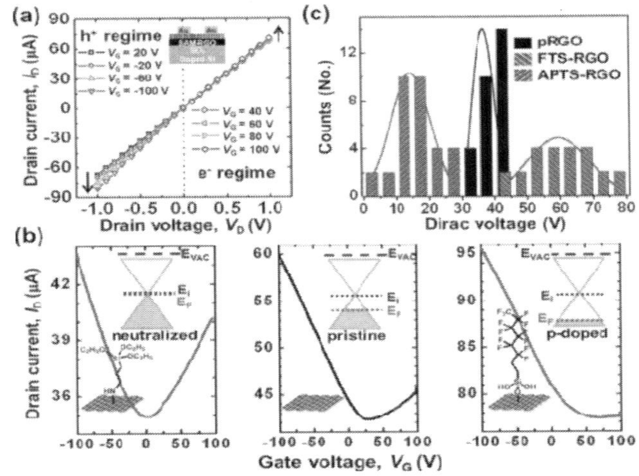

Figure 11: Work functions of graphene can be widely tuned using direct surface functionalization, which is demonstrated by self-assembled monolayers anchored onto the surfaces of the r-GO. Charge-transport characteristics of r-GO field-effect transistors (FETs) functionalized with the various self-assembled monolayers (SAMs). The inset of (**a**) shows the device configuration, where r-GO was used as an active layer. (**a**) Output characteristics of pristine r-GO FETs; (**b**) Transfer characteristics ($VD = -1$ V) of various r-GO FETs: APTS-rGO (**left**), pr-GO (**middle**), and FTS-r-GO (**right**). The insets show schematic band diagrams of SAM-functionalized r-GOs; (**c**) Comparative plots of Dirac voltages of the r-GO FETs. Reprinted with permission from [138]. Copyright 2013 Wiley-VCH.

Work Function Engineering of Graphene Using Noble Metal Doping

The band gap structure, and the type and the concentration of carriers in graphene—both electrons and holes—can be controlled by the introduction of metals or molecules on the graphene surface. This is easily possible because, zero-gap 2D semiconductor graphene has extreme sensitivity to molecular adsorbents. The relative position of the HOMO and LUMO of the adsorbents with respect to the Dirac point in pristine graphene determines the direction of charge transfer. In general GO and r-GOs synthesized by oxidation and reducing exfoliated graphite possess many defects including vacancies, oxide functionalities and substitutional nitrogen formed by harsh chemical treatments. The presence of these defects and the functionalities perturbs the π–π network of graphene, resulting in a loss of electrical conductivity of the GO/r-GO structure. However, it has been identified that GO/r-GO has the potential to recover conductivity through molecular doping by defect filling. Moreover, the presence of various oxygen containing functional groups on GO and rGO make them suitable platforms for the nucleation and growth of noble metal nanostructures, and synthesis of noble metal nanoparticle/graphene composites [140,141].

The controlled synthesis of hetero-structures based on Ag, Au, Pt with graphene has attracted significant attention in recent years due to the unique size and shape dependent properties including high catalytic activity, resulting in potential applications in chemical sensors, energy storage devices, catalysis, hydrogen storage, fuel cells, solar cells, electrochemical sensors and surface enhanced Raman scattering [140,141,142,143,144,145,146,147,148,149,150]. Huang *et al.* [127] recently presented a review on the current development of graphene-based composites including metal-graphene composites. Subrahmanyam *et al.* [151] examined the specific interaction of nanoparticles of metals such as Ag, Au, Pt and Pd with graphene using Raman spectroscopy and first-principles calculations. A significant charge-transfer interaction between graphene and the metal nanoparticles was demonstrated through both experimental and theoretical calculations. Lightcap and Kamat [152] discussed the potential of the graphene/semiconductor and graphene/metal-nanoparticle composites to function as efficient, multifunctional materials for energy conversion and storage. They emphasized that such advanced composites have the promise to integrate conversion and storage of solar energy, the detection and selective destruction of trace environmental contaminants, and to achieve single-substrate, multistep heterogeneous catalysis. In the next section, we attempt to discuss in some details various approaches used for tuning the WF of graphene-based noble metal nanostructures that can be useful as both electroactive and photoactive components in many devices.

Work Function Engineering on Graphene Based Gold (Au) Composite

Graphene is a promising nanoscale building block of new nanocomposites and can act as a support material for the dispersion of metal nanoparticles. Kim *et al.* [153] reported $AuCl_3$-doped graphene transparent conductive electrodes that were integrated in GaN-based ultraviolet (UV) light-emitting diodes (LEDs). They reported that the transmittance and the sheet resistance of the graphene electrode layer decreased with increase in the concentration of the $AuCl_3$ in solution. It was also demonstrated that *p*-type doping of the graphene by $AuCl_3$ dramatically improved the I-V characteristics and EL intensities. However, there is an optimal in the level of $AuCl_3$ due to the trade-off between the decrease in both transmittance and the sheet resistance of the graphene electrode with increase in $AuCl_3$. Choe *et al.* [154] investigated the WF of $AuCl_3$-doped graphene layers, and reported that the WF of p-doped graphene layers exposed to 20 mM $AuCl_3$ solutions increased from 4.42 to 5.12 eV. The p-doping occurs due to the electron transfer from the graphene layer to the Au nanoparticles. Shi *et al.* [60] also examined surface modification of graphene using Au dopant (e.g., $AuCl_3$) for specific range and demonstrated the upshift in WF of graphene film to ~0.5 eV (Figure 12). Cho *et al.* [155] demonstrated the efficiency of gold (Au) doped multi-layer graphene (MLG)/AuNP composite as a transparent conducting layer in near-ultraviolet light-emitting diodes (NUV-LEDs) (Figure 13). They reported that the use of thermally annealed Au-doped MLG enhanced the optical output power of NUV-LEDs by 34% relative to that of NUV-LEDs with a bare MLG. Au-doped MLG exhibited low sheet resistance and high current injection in the NUV-LED. A shift in WR of Au doped GR from 4.5 eV (undoped) to 4.9 eV (doped) was demonstrated, and a decrease in sheet resistance (*Rs*) from 500 to 90 Ω^{-2} was observed. The improved I-V characteristics of the NUV-LEDs with an Au-doped MLG layer can be attributed to the reduced sheet resistance of the Au-doped MLG films, and the decreased contact resistance between Au-doped MLG and the p-GaN layer.

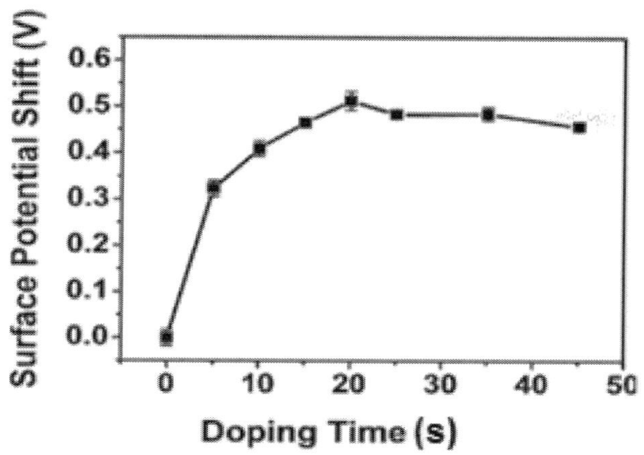

Figure 12: Change in WF with respect to doping time. Reprinted with permission from [60]. Copyright 2010 American Chemical Society.

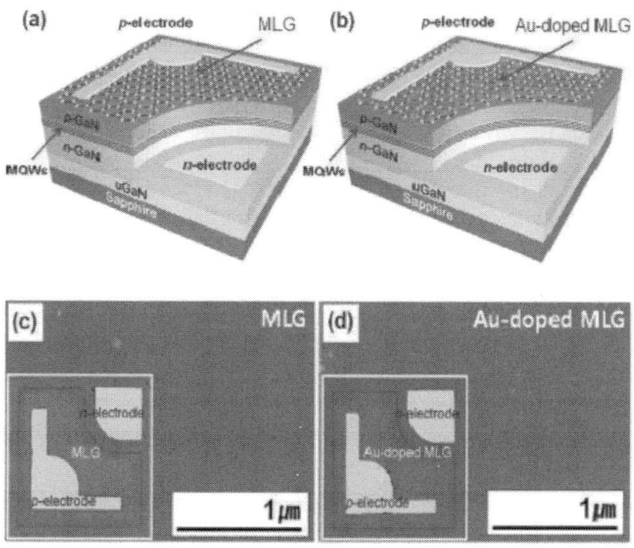

Figure 13: Schematic diagrams of near-ultraviolet light-emitting diodes (NUV-LEDs) with (**a**) multi-layer graphene (MLG) and (**b**) Au-doped MLG TCLs; (**c**) SEM images of MLG and (**d**) Au-doped MLG. Reprinted with permission from [155]. Copyright 2013 AIP Publishing.

Kwon *et al.* [156] tuned the WF of Au/graphene composite using different Au precursors containing different anions. They obtained WF of 4.3–4.6; 4.8; 5.0; 4.9 eV when Au $(OH)_3$; Au_2S; $AuBr_3$; $AuCl_3$ respectively were used as the precursors (Figure 14). They also demonstrated that thermal annealing of the doped graphene surface decreases the WF (Figure 14b). The decrease in WF value was attributed to the breakage of ionic bonding between Au cation and anion in each dopant doped cationic graphene. This observation indicates the sensitivity of the WF on surface engineering. Goncalves *et al.* [140] examined the role of oxygen moieties at GO and rGO surfaces on the gold nucleation and growth. They reported that the nucleation and growth mechanism depends strongly on the degree of oxygen functionalization on the graphene surface. No AuNP was reported to be observed at totally reduced graphene surfaces.

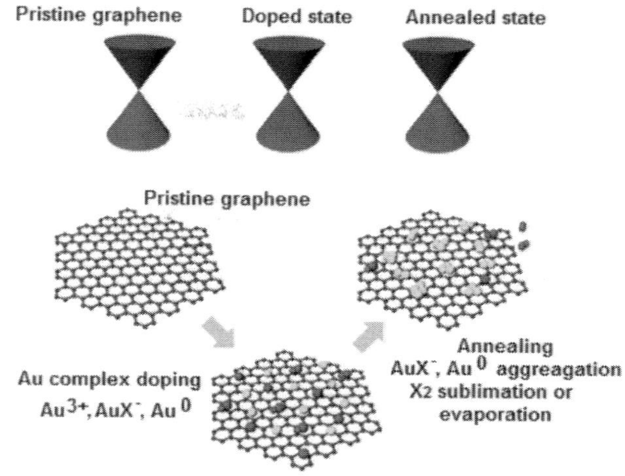

Figure 14: Schematic representation of the Dirac point state and morphological state of graphene according to sample treatment condition. Schematic representation of morphological state of graphene, before treatment, after treatment and after annealing are shown (left to right). Reprinted with permission from [156]. Copyright 2013 Royal Society of Chemistry.

A facile fabrication of macroporous gold films using graphene sheets as template was reported by Sun and Wu [157]. AuNP decorated graphene sheets were prepared using a one-pot simultaneous reduction of GO and gold precursor $HAuCl_4$by sodium citrate. Roy *et al.* [146] recently reported a highly efficient photocatalytic graphene-ZnO-Au nanocomposite using a simple hydrothermal method. This type of complex nano-composite has great potential in removing organic pollutants such as nitrobenzene. A rapid conversion of nitrobenzene to aniline within 60 s was demonstrated, and it was interpreted on the basis of WF tuning and electron-hole combination. It was implied that WF decreased via rapid electron transfer from ZnO to graphene (WFs of conduction and valence bands of ZnO are −4.05 eV and −7.25 eV respectively, and WF for graphene and Au are −4.42 eV and −4.70 eV) (Figure 15). It was hypothesised that the presence of solvent such as methanol (hydroxyl group) contributes to trapping holes, enables photo-generated electrons to catalyse reduction of nitrobenzene (NB) to aniline with a yield of 97.8%. Koo *et al.* [158] observed the growth of sub-nano sized Au clusters on r-GOs, which could reinforce the conductivity of the resulting r-GOs by defect filling. The resulting Au/r-GOs were reported to exhibit an improvement of bulk electrical conductivities and a reduced ratio of the intensity of the D band to that of the G band (ID/IG), relative to the r-GOs without Au nanoclusters. The decrease of the ID/IG was explained to be related to the filling of sub-nano sized Au clusters on the r-GOs owing to the enhancement of the flat geometry of the graphene nanosheets.

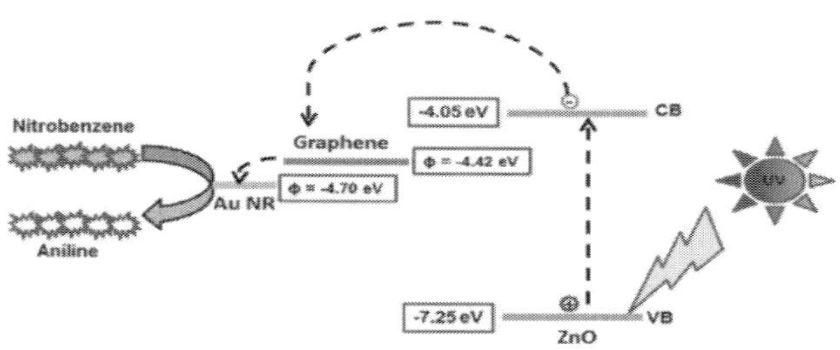

Figure 15: Schematic diagram showing the electron transfer mechanism from different energy levels of the Graphene-ZnO-Au

heterostructure for the photo-reduction of nitrobenzene. Reprinted with permission from [146]. Copyright 2013 American Chemical Society.

Work Function Engineering on Graphene Based Silver (Ag) Composite

For the synthesis of Ag-graphene nanostructure, $AgNO_3$ is the commonly used precursor, which can be easily reduced in the presence of different reducing agents including amines, $NaBH_4$ and ascorbic acid. [159,160]. Dutta *et al.* [161] reported synthesis of Ag-nanoparticle conjugated rGO nanosheets using dimethylformamide (DMF) as an efficient reducing agent that reduces both silver nitrate ($AgNO_3$) and graphene oxide (GO) in the reaction mixture. Shim *et al.* [162] demonstrated that in the case of GaN-based LEDs the WF of graphene could be tuned from 4.3 to 5.0 eV by introducing Ag, and such a device could produce uniform and stable light emission: 1.7 times higher than graphene only LEDs. Recently, the importance of WF was revealed during electron transfer from a Rhodamine B (RhB) dye to Ag/Graphene nanocomposites by Martínez-Orozco [163]. RhB (5.67 eV), after exciting electron, was converted to RhB* (3.08 eV) that acted as sensitizer in visible light. It was highlighted that RhB* injected electrons to the electron acceptor, *i.e.*, graphene (4.26 eV) to become a cationic radical RhB^+; which underwent self-degradation by trapping excited electrons of adsorbed oxygen. During this process, recombination of the injected electron and surface adsorbed RhB^+ might be possible, however, it was hypothesised it could be stopped by transferring injected electrons from graphene to Ag-nanoparticles leading to a better separation of electron (e^-) and hole (h^+). This electron transfer process that occurs on the basis of WF difference proves the importance of WF study in energy conversion devices.

Work Function Engineering for Graphene Based Platinum (Pt) Composite

Pt nanoparticles supported on a highly dispersed support have attracted much attention, and this is one of the most widely used catalyst materials for many chemical reactions and energy conversion devices including hydrogen oxidation and oxygen reduction reactions in fuel cell. Thus, significant efforts have been focused to combine graphene as a conductive support for Pt nanostructures. The surface area of graphene is exceptionally high and has also the potential to reduce carbon corrosion resulting in enhanced electrocatalytic performance and long term stability. Most of the techniques that were used to prepare graphene-Au or Ag composites and discussed above could be equally applied in general to fabricate graphene supported Pt nanomaterials. K_2PtCl_4 has been commonly used as a precursor that can be reduced on the surface of graphene to form Pt NPs [143,149,164,165,166]. Li *et al.* [165] synthesized Pt-nanoflowers loaded on rGO using a green approach for methanol electro-oxidation. Graphene supported Pt nanostructures have been synthesized that can be used as electrodes in fuel cells or other energy related devices [149,165,166,167]. In rGO-NP heterostructures, the electronic structure of GO/rGO is preserved by weak adsorption on these metals. However, even when the bonding is weak, the metal substrates caused the Fermi level of graphene to move away from the conical points, resulting in doping with either electrons or holes. The amount of doping in any specific case is dependent on the difference in WF of metal and graphene. Therefore, tuning of WF can be performed by doping graphene using different metal substrates such as Au, Ag, Pt, *etc.*, and a shift of WF has been observed from 5.54 to 4.74 eV, 4.92–4.24 eV, and 6.13–4.8 eV, respectively [61]. It is realized that graphene-supported noble metal nanostructures are promising electrodes and photoactive material for energy conversion and sensing applications [168]. However, significant further fundamental investigation on the effect of quality of interaction between the support and the NP on the WF is essential to make graphene a multipurpose material for energy conversion and sensing devices. Table 2 illustrates WF data for graphene based noble metal composites using a variety of precursors and their respective advantages and applications.

Table 2: Work function data for graphene based noble metal composites

S.No.	Modified Graphene	Method	Precursor	WF (eV)	Improved property	Application	Ref.
1	Graphene-ZnO-Au	Hydrothermal	Zn Acetate , $HAuCl_4$	Value not given	Efficiency = 3.5–4.5 fold more than ZnO	Removing pollutant (nitrobenzene)	[157]
2	Various Au dopants on graphene	Chemical doping	$AuBr_3$, Au_2S, $Au(OH)_3$, $AuCl_3$	5, 4.8, 4.6, 4.9 as doped and 4.5, 4.4, 4.55, 4.3 eV as annealed (w.r.t. the precursor)	Multiuse of graphene due to tunable WF property	Energy conversion devices and sensors	[159]
3	Au/graphene	Chemical doping	$AuCl_3$	Increase by 0.5 eV with increase in doping time	Tunable WF property	Optoelectronic devices	[60]
4	Ag/graphene	Photochemical silver functionalization	$AgNO_3$	Value not given	Suppressed e-h recombination	Efficient removal of hazardous materials	[161]
5	Au/Ag/ Pt-graphene	Graphene adsorption metal substrate	Au, Ag, Pt substrates were used	5.54–4.74, 4.92–4.24, 6.13–4.8 eV	Multipurpose modified graphene	Energy conversion devices	[168]

CONCLUSIONS

High-quality graphene has only been available for less than a decade; however, it has generated unprecedented excitement in the scientific community. The electronic structure, unique morphological and electronic characteristics of graphene and its potential applications have been the subject of extensive theoretical and experimental investigations. This advanced material has not only afforded the Nobel Prize to Andrei Geim and Konstantin Novoselov in 2010 "*for ground-breaking experiments regarding the two-dimensional material graphene*" but has also generated great interest to wide ranging industries including semiconductor, optoelectronics and printed electronics. Graphene has indeed been identified as a next-generation material with potential to replace traditional electrode materials in many electronic and optoelectronic devices. One of the most important themes of current research on graphene is to understand the interaction between graphene and its surrounding environment including the presence of optically and electrochemically active nanoparticles, and the consequences on the band gap structure and Fermi level. Such investigations are of significant fundamental interest from the quantum physics point of view, as well as being relevant for the process of sample production and fabrication of graphene-based electronic and optoelectronic devices. The influence of the presence of ad atoms, different functionalities and modifications on the WF of graphene, GO and r-GO has been studied to understand the different functionalization strategies to tune the WF of graphene. This effect of WF modulation has proved to be very useful in employing graphene as a multipurpose material for applications related to energy conversion devices such as an electrode, hole transporting layer (HTL), photoactive material. The surface functionalization of solution processable graphene has attracted significant attention as a potential method for the synthesis of graphene to be used in various industrial applications including photovoltaic cells, capacitors, sensors, and transparent electrodes. Recent innovations in the continuous and scalable large area synthesis of graphene using chemical vapour deposition (CVD) have reinforced its study in a broad range of research areas. The most critical technological challenge to be faced before a wider application of graphene is to develop a facile and efficient method for controlled production of processable single/few layer large graphene sheets of the desired properties, and tuning of the electronic properties including work function.

ACKNOWLEDGMENTS

This research was financially supported by the Australian Research Council (ARC) through Discovery Projects: DP120103537 and DP1092678.

REFERENCES

1. Novoselov, K.S.; Geim, A.K.; Morozov, S.V. Electric field effect in atomically thin carbon films. *Science* 2004, *306*, 666–669.
2. Hazen, R.M.; Downs, R.T.; Jones, A. Carbon mineralogy and crystal chemistry. *Rev. Mineral. Geochem.* 2013, *75*, 7–46.
3. Oganov, A.R.; Hemley, R.J.; Hazen, R.M. Structure, bonding and mineralogy of carbon at extreme conditions. *Rev. Mineral. Geochem.* 2013, *75*, 47–77.
4. Rao, C.N.R.; Biswas, K.; Subrahmanyam, K.S.; Govindaraj, A. Graphene, the new nanocarbon. *J. Mater. Chem.* 2009, *19*, 2457–2469.
5. Katsnelson, M.I. Graphene: Carbon in two dimensions. *Mater. Today* 2007, *10*, 20–27.
6. Katsnelson, M.I.; Novoselov, K.S. Graphene: New bridge between condensed matter physics and quantum electrodynamics. *Solid State Commun.* 2007, *143*, 3–13.
7. Ando, T. The electronic properties of graphene and carbon nanotubes. *NPG Asia Mater.* 2009, *1*, 17–21.
8. Wallace, P.R. The band theory of graphite. *Phys. Rev.* 1947, *71*, 622–634.
9. Ohta, T.; Bostwick, A.; Seyller, T. Controlling the electronic structure of bilayer graphene. *Science* 2006, *313*, 951–954.
10. Bonaccorso, F.; Sun, Z.; Hasan, T. Graphene photonics and optoelectronics. *Nat. Photonics* 2010, *4*, 611–622.
11. Nardecchia, S.; Monte, F.; Ferrer, M. Three dimensional macroporous architectures and aerogels built of carbon nanotubes and/or graphene: Synthesis and applications. *Chem. Soc. Rev.* 2013, *42*, 794–830.
12. Castro Neto, A.H.; Guinea, F.; Peres, N.M.R.; Novoselov, K.S.; Geim, A.K. The electronic properties of graphene. *Rev. Mod. Phys.* 2009, *81*, 109–162.
13. Avouris, P.; Dimitrakopoulos, C. Graphene: Synthesis and applications. *Mater. Today* 2012, *15*, 86–97.
14. Zhang, Y.; Tan, Y.W.; Stormer, H.L.; Kim, P. Experimental observation of the quantum hall effect and Berry's phase in graphene. *Nature* 2005, *438*, 201–204.
15. Novoselov, K.S.; McCann, E.; Morozov, S.V.; Falko, V.I.; Katsnelson, M.I.; Zeitler, U.; Jiang, D.; Schedin, F.; Geim, A.K. Unconventional quantum hall effect and Berry's phase of 2pi in bilayer graphene. *Nat. Phys.* 2006, *2*, 177–180.
16. Novoselov, K.S.; Jiang, Z.; Zhang, Y.; Morozov, S.V.; Stormer, H.L.; Zeitler, U.; Maan, J.C.; Boebinger, G.S.; Kim, P.; Geim, A.K. Room-temperature quantum hall effect in graphene. *Science* 2007, *315*, 1379.
17. Novoselov, K.S.; Geim, A.K.; Morozov, S.V.; Jiang, D.; Katsnelson, M.I.; Grigorieva, I.V.; Dubonos, S.V.; Firsov, A.A. Two dimensional gas of massless Dirac fermion in graphene. *Nature* 2005, *438*, 197–200.
18. Zhou, S.Y.; Gweon, G.H.; Graf, J.; Fedorov, A.V. First direct observation of Dirac fermions in graphite. *Nat. Phys.* 2006,*2*, 595–599.
19. Chen, S.; Wu, Q.; Mishra, C. New graphene-related materials on the horizon. *Nat. Mater.* 2012, *11*, 203–207.
20. Nair, R.R.; Blake, P.; Grigorenko, A.N. Fine structure constant defines visual transparency of graphene. *Science* 2008,*320*, 1308.
21. Lee, C.; Wei, X.; Kysar, J.W.; Hone, J. Measurement of the elastic properties and intrinsic strength of mono-layer graphene. *Science* 2008, *321*, 385–388.
22. Du, W.; Jiang, X.; Zhub, L. From graphite to graphene: Direct liquid phase exfoliation of graphite to produce single and few layered pristine graphene. *J. Mater. Chem. A* 2013, *1*, 10592–10606.
23. Krishnan, D.; Kim, F.; Luo, J. Energetic graphene oxide: Challenges and opportunities. *Nano Today* 2012, *7*, 137–152.
24. Geim, A.K. Graphene: Status and prospects. *Science* 2009, *324*, 1530–1534.
25. Tang, Q.; Zhou, Z.; Chen, Z. Graphene-related nanomaterials: Tuning properties by functionalisation. *Nanoscale* 2013,*5*, 4541–4583.
26. Zhu, Y.; Murali, S.; Cai, W. Grpahene and graphene oxide: Synthesis, properties and applications. *Adv. Mater.* 2010,*22*, 3906–3924.

27. Allen, J.M.; Tung, C.V.; Kaner, R. Honeycomb carbon: A review on graphene. *Chem. Rev.* 2010, *110*, 132–145.
28. Geim, A.K.; Novoselov, K.S. The rise of graphene. *Nat. Mater.* 2007, *6*, 183–199.
29. Das Sarma, S.; Geim, A.; Kim, P.; MacDonald, A. Exploring graphene—Recent research advances—Foreword. *Solid State Commun.* 2007, *143*, 1–2.
30. Sun, Y.; Wu, Q.; Shi, G. Graphene based new energy materials. *Energy Environ. Sci.* 2011, *4*, 1113–1132.
31. Chen, D.; Tanga, L.; Li, J. Graphene based material in electrochemistry. *Chem. Soc. Rev.* 2010, *39*, 3157–3180.
32. Brownson, D.; Kampouris, K.D. Graphene electrochemistry: Fundamental concepts through to prominent applications. *Chem. Soc. Rev.* 2012, *41*, 6944–6976.
33. Compton, C.O.; Nguyen, S. Graphene oxide, highly reduced graphene oxide, and graphene: Versatile building blocks for carbon-based materials. *Small* 2010, *6*, 711–723.
34. Cui, S.; Mao, S.; Lu, G.; Chen, J. Graphene coupled with nanocrystals: Opportunities and challenges for energy and sensing application. *J. Phys. Chem. Lett.* 2013, *4*, 2441–2454.
35. Weiss, N.; Zhou, H.; Liao, L.; Liu, Y.; Jiang, S. An emerging electronic material. *Adv. Mater.* 2012, *24*, 5782–5825.
36. Wasse, J.K.; Kaner, R.B. Exploring the synthesis and application of graphene. *Mater. Today* 2010, *13*, 52–59.
37. Stankovich, S.; Dikin, D.A.; Dommett, G.H.B.; Kohlhaas, K.M.; Zimney, E.J.; Stach, E.A.; Piner, R.D.; Nguyen, S.T.; Ruoff, R.S. Graphene based composite materials. *Nature* 2006, *442*, 282–286.
38. Kuila, T.; Bose, S.; Mishra, A.K.; Khanra, P.; Hoon, K.N.H.; Lee, J. Chemical functionalization of graphene and its applications. *Prog. Mater. Sci.* 2012, *57*, 1061–1105.
39. Yang, K.; Feng, L.; Hong, H.; Cai, W.; Liu, Z. Preparation and functionalization of graphene nanocomposites for biomedical applications. *Nat. Protoc.* 2013, *8*, 2392–2403.
40. Craciun, M.F.; Khrapach, I.; Barnes, M.D.; Russo, S. Properties and applications of chemically functionalized graphene. *J. Phys. Condens. Matter* 2013, *25*, 423201:1–423201:42.
41. Bunch, J.S.; van der Zande, A.M.; Verbridge, S.S.; Frank, I.W.; Tanenbaum, D.M.; Parpia, J.M.; Craighead, H.G. Electromechanical resonators from graphene sheets. *Science* 2007, *315*, 490–493.
42. Staley, N.; Wang, H.; Puls, C.; Forster, J.; Jackson, T.N.; McCarthy, K.; Clouser, B.; Liu, Y. Lithography free fabrication of graphene devices. *Appl. Phys. Lett.* 2007, *90*, 143518:1–143518:3.
43. Gilje, S.; Han, S.; Wang, K.L.; Kaner, R.B. Graphene: Calling all chemists. *Nano Lett.* 2007, *7*, 3394–3398.
44. Moore, G.E. Cramming more components onto integrated circuits. *Electronics* 1965, *38*, 114–117.
45. Son, Y.W.; Cohen, M.L.; Louie, S.G. Energy gap in graphene nanoribbons. *Phys. Rev. Lett.* 2006, *97*, 216803:1–216803:4.
46. Wang, X.; Zhi, L.; Mullen, K. Transparent, conductive graphene electrodes for dye-sensitized solar cells. *Nano Lett.*2008, *8*, 323–327.
47. Zhu, Y.; Cai, W.; Piner, R.D.; Velamakanni, A.; Ruoffa, R.S. Transparent self-assembled films of reduced graphene oxide platelets. *Appl. Phys. Lett.* 2009, *95*, 103104:1–103104:3.
48. Becerril, H.A.; Mao, J.; Liu, Z.; Stoltenberg, R.M.; Bao, Z.; Chen, Y. Evaluation of solution-processed reduced graphene oxide films as transparent conductors. *ACS Nano* 2008, *2*, 463–470.
49. Shin, H.J.; Kim, K.K.; Benayad, A.; Yoon, S.; Park, H. Efficient reduction of graphite oxide by sodium borohydride and its effect on electrical conductance. *Adv. Funct. Mater.* 2009, *19*, 1987–1992.
50. Watcharotone, S.; Dikin, D.; Stankovich, S. Graphene-silica composite thin films as transparent conductors. *Nano Lett.*2007, *7*, 1888–1892.
51. Robinson, J.T.; Perkins, F.K.; Snow, E.S.; Wei, Z.; Sheehan, P.E. Reduced graphene oxide molecular sensors. *Nano Lett.*2008, *8*, 3137–3140.
52. Wang, D.W.; Li, F.; Zhao, J.; Ren, W.; Chen, Z.G.; Tan, J. Fabrication of graphene/polyaniline composite paper via anodic electropolymerization for high-performance flexible electrode. *ACS Nano* 2009, *3*, 1745–1752.
53. Andreas, P.; Humer, M.; Furchi, M.M.; Bachmann, D.; Guider, R.; Fromherz, T.; Mueller, T. CMOS-compatible graphene photodetector covering all optical communication bands. *Nat. Photonics* 2013, *7*, 892–896.
54. Zhou1, S.Y.; Gweon1, G.H.; Fedorov, A.V. Substrate induced bandgap opening. *Nat. Mater.* 2007, *6*, 770–775.
55. Song, S.M.; Park, J.; Sul, O.J. Determination of Work function of graphene under a metal electrode and its role in contact resistance. *Nano Lett.* 2012, *12*, 3887–3892.
56. Xia, F.; Mueller, T.; Mojarad, R.; Avouris, P. Photocurrent Imaging and efficient photon detection in a graphene transistor. *Nano Lett.* 2009, *9*, 1039–1044.
57. Eda, G.; Unalan, H.E.; Rupesinghe, N.; Chhowalla, M. Graphene based composite thin films. *Appl. Phys. Lett.* 2008, *93*, 233502:1–233502:3.
58. Yu, Y.-J.; Zhao, Y.; Ryu, S.; Brus, L.E.; Kim, K.S.; Kim, P. Tuning the graphene work function by electric field effect.*Nano Lett.* 2009, *9*, 3430–3434.
59. Loh, K.P.; Bao, Q.; Ang, P.K.; Yang, J. The chemistry of graphene. *J. Mater. Chem.* 2010, *20*, 2277–2289.
60. Shi, Y.; Kim, K.K.; Reina, A.; Hofmann, M.; Li, L.-J.; Kong, J. Work function engineering of graphene electrode via chemical doping. *ACS Nano* 2010, *4*, 2689–2694.
61. Giovannetti, G.; Khomyakov, P.A.; Brocks, G.; Karpan, V.M.; van den Brink, J.; Kelly, P.J. Doping graphene with metal contacts. *Phys. Rev. Lett.* 2008, *101*, 026803:1–026803:4.

62. Benayad, A.; Shin, H.J.; Park, H.K.; Yoon, S.M. Controlling work function of reduced graphite oxide with Au-ion concentration. *Chem. Phys. Lett.* 2009, *475*, 91–95.

63. Wang, B.; Gunther, S.; Wintterlin, J. Peridoicity, work function and reactivity of graphene on Ru(001) from first principles. *New J. Phys.* 2010, *12*, 043041:1–043041:15.

64. Murata, Y.; Starodub, E.; Kappes, B.B.; Ciobanu, C.V.; Bartelt, N.C.; McCarty, K.F.; Kodambaka, S. Orientation-dependent work function of graphene on Pd(111). *Appl. Phys. Lett.* 2010, *97*, 143114:1–143114:3.

65. Park, J.; Lee, W.H.; Huh, S.; Sim, S.H.; Kim, S.B.; Cho, K.; Hong, B.H.; Kim, K.S. Work-function engineering of graphene electrodes by self-assembled monolayers for high-performance organic field-effect transistors. *J. Phys. Chem. Lett.* 2011, *2*, 841–845.

66. Elias, D.C.; Nair, R.R.; Mohiuddin, T.M.G.; Morozov, S.V.; Blake, P.; Halsall, M.P.; Ferrari, A.C.; Boukhvalov, D.W.; Katsnelson, M.I.; Geim, A.K.; *et al.* Control of graphene's properties by reversible hydrogenation: Evidence for graphane. *Science* 2009, *323*, 610–613.

67. Cheng, Y.C.; Kaloni, T.P.; Huang, G.S.; Schwingenschlögl, U. Origin of the high p-doping in F intercalated graphene on SiC. *Appl. Phys. Lett.* 2011, *99*, 053117:1–053117:3.

68. Virojanadara, C.; Watcharinyanon, S.; Zakharov, A.A.; Johansson, L.I. Epitaxial graphene on SiC and Li intercalation.*Phys. Rev. B* 2010, *82*, 205402:1–205402:6.

69. Premlal, B.; Cranney, M.; Vonau, F.; Aubel, D.; Casterman, D.; De Souza, M.M.; Simon, L. Surface intercalation of gold underneath a graphene monolayer on SiC(0001) studied by scanning tunneling microscopy and spectroscopy. *Appl. Phys. Lett.* 2009, *94*, 263115:1–263115:3.

70. Khrapach, I.; Withers, F.; Bointon, T.H.; Polyushkin, D.K.; Barnes, W.L.; Russo, S.; Craciun, M.F. Novel highly conductive and transparent graphene-based conductors. *Adv. Mater.* 2012, *24*, 2844–2849.

71. Zhao, W.; Tan, P.H.; Liu, J.; Ferrari, A.C. Intercalation of few-layer graphite flakes with $FeCl_3$: Raman determination of fermi level, layer by layer decoupling, and stability. *J. Am. Chem. Soc.* 2011, *133*, 5941–5946.

72. McAllister, M.J.; Li, J.; Adamson, D.H.; Schniepp, H.C.; Aksay, I.A. Single sheet functionalized graphene by oxidation and thermal expansion of graphite. *Chem. Mater.* 2007, *19*, 4396–4406.

73. Lotya, M.; Hernandez, Y.; King, P.J. Solvation assisted exfoliation of graphite. *Chem. Soc.* 2009, *131*, 3611–3620.

74. Kim, K.S.; Zhao, Y.; Jang, H.; Lee, S.Y. Pattern growth of graphene films for stretchable transparent electrodes. *Nature*2009, *457*, 706–710.

75. Obraztsov, A.N. Chemical vapour deposition: Making graphene on a large scale. *Nat. Nano* 2009, *4*, 212–213.

76. Cai, M.; Thorpe, D.; Adamson, D.H.; Schniepp, H.C. Methods of graphite exfoliation. *J. Mater. Chem.* 2012, *22*, 24992–25002.

77. Chen, L.; Hernandez, Y.; Feng, X.; Müllen, K. From nanographene and graphene nanoribbons to graphene sheets: Chemical synthesis. *Angew. Chem. Int. Ed.* 2012, *51*, 7640–7654.

78. Wu, C.; Dong, G.; Guan, L. Production of graphene sheets by a simple helium arc-discharge. *Phys. E* 2010, *42*, 1267–1271.

79. Kumar, P. Laser flash synthesis of graphene and its inorganic analogues: An innovative breakthrough with immense promise. *RSC Adv.* 2013, *3*, 11987–12002.

80. Liao, C.D.; Lu, Y.Y.; Tamalampudi, S.R.; Cheng, H.C.; Chen, Y.T. Chemical vapor deposition synthesis and raman spectroscopic characterization of large-area graphene sheets. *J. Phys. Chem. A* 2013, *117*, 9454–9461.

81. Kuila, T.; Mishra, A.K.; Khanra, P.; Kim, N.H.; Lee, J.H. Recent advances in the efficient reduction of graphene oxide and its application as energy storage electrode materials. *Nanoscale* 2013, *5*, 52–71.

82. Li, X.S.; Cai, W.; An, J.; Kim, S.; Nah, J.; Yang, D.; Piner, R.; Velamakanni, A.; Jung, I.; Tutuc, E.; *et al.* Large-area synthesis of high-quality and uniform graphene films on copper foils. *Science* 2009, *324*, 1312–1314.

83. Lin, Y.M.; Dimitrakopoulos, C.; Jenkins, K.A.; Farmer, D.B.; Chiu, H.-Y.; Grill, A.; Avouris, P. 100-GHz transistors from wafer-scale epitaxial graphene. *Science* 2010, *327*, 662.

84. Colson, J.W.; Woll, A.R.; Mukherjee, A.; Spitler, E.L.; Shields, V.B.; Spencer, M.G.; Park, J.; Dichtel, W.R. Oriented 2D.*Science* 2011, *332*, 228–231.

85. Gunasinghe, R.N.; Reuven, D.G.; Suggs, K.; Wang, X.-Q. Filled and empty orbital interactions in a planar covalent organic framework on graphene. *J. Phys. Chem. Lett.* 2012, *3*, 3048–3052.

86. Park, S.; Ruoff, S.R. Chemical methods for the production of graphene. *Nanotechnology* 2009, *4*, 217–224.

87. Brodie, C. On the atomic weight of graphite. *Phil. Trans. R. Soc. A* 1859, *149*, 249–259.

88. Hummers, W.S.; Offerman, R.E. Preparation of graphite oxide. *J. Am. Chem. Soc.* 1958, *80*, 1339.

89. Hirata, M.; Gotou, T.; Horiuchi, S.; Fujiwara, M.; Ohba, M. Progress towards monodisperse single-walled carbon nanotubes. *Carbon* 2004, *42*, 2929–2937.

90. Wei, G.; Lawrence, A.; Li, B. New insights into the structure and reduction of graphite oxide. *Nat. Chem.* 2009, *1*, 403–408.

91. Jeong, H.K.; Lee, Y.P.; Jin, M.H.; Kim, E.S.; Bae, J.J.; Lee, Y.H. Thermal stability of graphite oxide. *Chem. Phys. Lett.*2009, *470*, 255–258.

92. Dreyer, D.R.; Park, S.; Bielawski, C.W.; Ruoff, R.S. The chemistry of graphene oxide. *Chem. Soc. Rev.* 2010, *39*, 228–240.

93. Ambrosi, A.; Bonanni, A.; Sofer, Z.K.; Cross, J.S.; Pumera, M. Electrochemistry at chemically modified graphenes.*Chem. Eur. J.* 2011, *17*, 10763–10770.

94. Chua, C.K.; Pumera, M. Chemical reduction of graphene oxide: A synthetic chemistry viewpoint. *Chem. Soc. Rev.* 2014,*43*, 291–312.

95. Mao, S.; Pu, H.; Chen, J. Graphene oxide and its reduction: Modeling and experimental progress. *RSC Adv.* 2012, *2*, 2643–2662.

96. Choucair, M.; Thordarson, P.; Stride, J.A. Gram-scale production of graphene based on solvothermal synthesis and sonication. *Nat. Nano* 2009, *4*, 30–33.

97. Zangwill, A.; Vvedensky, D.D. Novel growth mechanism of epitaxial graphene on metals. *Nano Lett.* 2011, *11*, 2092–2095.

98. Shao, Y.; Wang, J.; Wu, H. Graphene based electrochemical sensors and biosensors. *Electroanalysis* 2010, *22*, 1027–1036.

99. Monsur, H.; Hyejin, P.; Sung, D. An electrochemically reduced graphene oxide-based electrochemical immunosensing platform for ultrasensitive antigen detection. *Anal. Chem.* 2012, *84*, 1871–1878.

100. Shao, Y.; Wang, J.; Engelhard, M. Facile and controllable electrochemical reduction of graphene oxide and its applications. *J. Mater. Chem.* 2010, *20*, 743–748.

101. Peng, X.; Tang, F.; Copple, A. Engineering the work function of armchair graphene nanoribbons using strain and functional species: A first principles study. *J. Phys. Condens. Matter* 2012, *24*, 075501:1–075501:10.

102. Jiang, J.; Krauss, T.D.; Brus, L. Electrostatic force microscopy characterization of trioctylphosphine oxide self-assembled monolayers on graphite. *J. Phys. Chem. B* 2000, *104*, 11936–11941.

103. Shin, H.J.; Choi, W.M.; Choi, D.; Han, G.H. Control of electronic structure of graphene by various dopants and their effects on a nanogenerator. *J. Am. Chem. Soc.* 2010, *132*, 15603–15609.

104. Nduwimana, A.; Wang, X.-Q. Energy gaps in supramolecular functionalized graphene nanoribbons. *ACS Nano* 2012,*6*, 1011–1017.

105. Williams, M.D.; Samarakoon, D.K.; Hessc, D.W.; Wang, X.-Q. Tunable bands in biased multilayer epitaxial grapheme.*Nanoscale* 2012, *4*, 2962–2967.

106. Singh, A.K.; Ahmad, M.; Singh, V. Tailoring the electrical properties of graphene layers by molecular doping. *ACS Appl. Mater. Interfaces* 2013, *5*, 5276–5281.

107. Lee, J.; Novoselov, K.S.; Shin, H.S. Interaction between metal and graphene: Dependence on the layer number of graphene. *ACS Nano* 2011, *5*, 608–612.

108. Wang, X.R.; Li, X.L.; Zhang, L. N-doping of graphene through electrothermal reactions with ammonia. *J. Sci.* 2009,*324*, 768–771.

109. Georgakilas, V.; Otyepka, M.; Bourlinos, A.B. Functionalization of graphene: Covalent and non-covalent approaches.*Chem. Rev.* 2012, *112*, 6156–6214.

110. Peimyoo, N.; Yu, T.; Shang, J.; Cong, C.; Yang, H. Thickness-dependent azobenzene doping in mono- and few-layer graphene. *Carbon* 2012, *50*, 201–208.

111. Dong, X.; Long, Q.; Wei, A.; Huang, W. The electrical properties of graphene modified by bromophenyl groups derived from a diazonium compound. *Carbon* 2012, *50*, 1517–1522.

112. Yu, Y.J.; Zhao, Y.; Ryu, S.; Brus, L.E.; Kim, K.S. Tuning the graphene work function by electric field effect. *Nano Lett.*2009, *10*, 3430–3434.

113. Gui, G.; Li, J.; Zhong, J. Band structure engineering of graphene by strain: First-principles calculations. *Phys. Rev. B*2008, *78*, 075435:1–075435:3.

114. Zhang, Z.; Huang, H.; Yang, X.; Zang, L. Tailoring electronic properties of graphene by π–π stacking with aromatic molecules. *J. Phys. Chem. Lett.* 2011, *2*, 2897–2905.

115. Singh, A.K.; Iqbal, M.W.; Singh, V.K.; Iqbal, M.Z. Molecular n-doping of chemical vapour deposition. *Mater. Chem.*2012, *22*, 15168–15174.

116. Eda, G.; Lin, Y.-Y.; Mattevi, C. Blue photoluminescence from chemically derived graphene oxide. *Adv. Mater.* 2010, *22*, 505–509.

117. Matis, B.R.; Burgess, J.S.; Bulat, F.A.; Friedman, A.L.; Houston, B.H.; Baldwin, J.W. Surface doping and band gap tunability in hydrogenated graphene. *ACS Nano* 2012, *6*, 17–22.

118. Kumar, P.V.; Bernardi, M.; Grossman, J.C. The impact of functionalization on the stability, work function, and photoluminescence of reduced graphene oxide. *ACS Nano* 2013, *7*, 1638–1645.

119. Mishra, M.; Joshi, R.K.; Ojha, S. Role of oxygen in the work function modification at various stages of chemically synthesized graphene. *J. Phys. Chem. C* 2013, *117*, 19746–19750.

120. Wong, C.H.; Ambrosi, A.; Pumera, M. Thermally reduced graphenes exhibiting a close relationship to amorphous carbon. *Nanoscale* 2012, *4*, 4972–4977.

121. Chen, W.; Yan, L. Preparation of graphene by a low-temperature thermal reduction at atmosphere pressure. *Nanoscale*2010, *2*, 559–563.

122. Stankovich, S.; Dikin, D.A.; Piner, R.D.; Kohlhaas, K.A.; Kleinhammes, A.; Jia, Y.; Wu, Y.; Nguyen, S.T.; Ruoff, R.S. Synthesis of graphene-based nanosheets via chemical reduction of exfoliated graphite oxide. *Carbon* 2007, *45*, 1558–1565.

123. Navarro, C.; Weitz, R.T.; Bittner, A.M.; Scolari, M.; Mews, A.; Burghard, M.; Kern, K. Electronic transport properties of individual chemically reduced graphene oxide sheets. *Nano Lett.* 2007, *7*, 3499–3503.

124. Xiong, Z.D.; Chen, Y.; Zhong, S.X.; Wei, Y. Electrochemical reduction of graphene oxide films: Preparation, characterization and their electrochemical properties. *Chin. Sci. Bull.* 2012, *57*, 3045–3050.

125. Wang, Z.; Zhou, X.; Zhang, J.; Boey, F.; Zhang, H. Direct electrochemical reduction of single-layer graphene oxide and subsequent functionalization with glucose oxidase. *J. Phys. Chem. C* 2009, *113*, 14071–14075.

126. Cooper, D.R.; D'Anjou, B.; Ghattamaneni, N. Experimental review of graphene. *ISRN Condens. Matter Phys.* 2012, *2012*, 501686:1–501686:56.

127. Huang, X.; Qi, X.; Boey, F.; Zhang, H. Graphene-based composites. *Chem. Soc. Rev.* 2012, *41*, 666–686.

128. Gao, X.; Jang, J.; Nagase, S. Hydrazine and thermal reduction of graphene oxide: Reaction mechanisms,

product structures, and reaction design. *J. Phys. Chem. C* 2009, *114*, 832–842.

129. Hwang, J.O.; Park, J.S.; Choi, D.S. Workfunction-tunable, n-doped reduced graphene transparent electrodes for high-performance polymer light-emitting diodes. *ACS Nano* 2011, *6*, 159–167.

130. Chieh, Y.; Yan, W.; Chen, C.W. Work Function Evolution of Graphene Oxide by Utilizing Hydrothermal Treatment. In Proceedings of the 2010 8th International Vacuum Electron Sources Conference and Nanocarbon (IVESC), Nanjing, China, 14–16 October 2010; p. 552.

131. Liu, X.; Kim, H.; Guo, L.J. Optimization of thermally reduced graphene oxide for an efficient hole transport layer in polymer solar cells. *Org. Electron.* 2013, *14*, 591–598.

132. Liu, J.; Xue, Y.; Dai, L. Sulfated graphene oxide as a hole-extraction layer in high-performance polymer solar cells. *J. Phys. Chem. Lett.* 2012, *3*, 1928–1933.

133. Yang, D.; Zhou, L.; Chen, L.; Zhao, B.; Zhang, J.; Li, C. Chemically modified graphene oxides as a hole transport layer in organic solar cells. *Chem. Commun.* 2012, *48*, 8078–8080.

134. Wang, J.; Wang, Y.; He, D.; Wu, H.; Wang, H.; Zhou, P.; Fu, M.; Jiang, K.; Chen, W. Organic photovoltaic devices based on an acceptor of solution-processable functionalized graphene. *J. Nanosci. Nanotechnol.* 2011, *11*, 9432–9438.

135. Wang, H.; He, D.; Wang, Y.; Liu, Z.; Wu, H.; Wang, J. Organic photovoltaic devices based on graphene as an electron-acceptor material and P3OT as a donor material. *Phys. Status Solidi* 2011, *208*, 2339–2343.

136. Liu, Q.; Liu, Z.; Zhang, X.; Zhang, N.; Yang, L.; Yin, S.; Chen, Y. Organic photovoltaic cells based on an acceptor of soluble graphene. *Appl. Phys. Lett.* 2008, *92*, 223303:1–223303:3.

137. Lin, Y.; Li, X.; Xie, D.; Feng, T.; Chen, Y.; Song, R.; Tian, H.; Ren, T.; Zhong, M.; Wang, K. Graphene/semiconductor heterojunction solar cells with modulated antireflection and graphene work function. *Energy Environ. Sci.* 2013, *6*, 108–115.

138. Kang, B.; Lim, S.; Lee, W.H.; Jo, S.B.; Cho, K. Work-function-tuned reduced graphene oxide via direct surface functionalization as source/drain electrodes in bottom-contact organic transistors. *Adv. Mater.* 2013, *25*, 5856–5862.

139. Kong, B.S.; Geng, J.; Jung, H.T. Layer-by-layer assembly of graphene and gold nanoparticles by vacuum filtration and spontaneous reduction of gold ions. *Chem. Commun.* 2009, *16*, 2174–2176.

140. Goncalves, G.; Marques, P.A.A.P.; Granadeiro, C.M.; Nogueira, H.I.S.; Singh, M.K.; Grácio, J. Surface modification of graphene nanosheets with gold nanoparticles: The role of oxygen moieties at graphene surface on gold nucleation and growth. *Chem. Mater.* 2009, *21*, 4796–4802.

141. Li, J.; Yang, J.; Yang, Z.; Li, Y.; Yu, S.; Xu, Q.; Hu, X. Graphene-Au nanoparticles nanocomposite film for selective electrochemical determination of dopamine. *Anal. Methods* 2012, *4*, 1725–1728.

142. Zhang, Y.; Liu, S.; Wang, L.; Qin, X.; Tian, J.; Lu, W.; Chang, G.; Sun, X. One-pot green synthesis of Ag nanoparticles-graphene nanocomposites and their applications in SERS, H_2O_2, and glucose sensing. *RSC Adv.* 2012, *2*, 538–545.

143. He, H.K.; Gao, C. Graphene nanosheets decorated with Pd, Pt, Au, and Ag nanoparticles: Synthesis, characterization, and catalysis applications. *Sci. Chin. Chem.* 2011, *54*, 397–404.

144. Wang, P.; Liu, Z.G.; Chen, X.; Meng, F.L.; Liu, J.H.; Huang, X.J. UV irradiation synthesis of an Au-graphene nanocomposite with enhanced electrochemical sensing properties. *J. Mater. Chem. A* 2013, *1*, 9189–9195.

145. Liu, N.; Chen, X.; Ma, Z. Ionic liquid functionalized graphene/Au nanocomposites and its application for electrochemical immunosensor. *Biosens. Bioelectron.* 2013, *48*, 33–38.

146. Roy,P.;Periasamy,A.P.;Liang,C.T.;Chang,H.T.Synthesis of graphene-ZnO-Au nanocomposites for efficient photocatalytic reduction of nitrobenzene. *Environ. Sci. Technol.* 2013, *47*, 6688–6695.

147. Lu, G.; Li, H.; Liusman, C.; Yin, Z.; Wu, S.; Zhang, H. Surface enhanced Raman scattering of Ag or Au nanoparticle-decorated reduced graphene oxide for detection of aromatic molecules. *Chem. Sci.* 2011, *2*, 1817–1821.

148. Huang, X.; Li, S.; Wu, S.; Huang, Y.; Boey, F.; Gan, C.L.; Zhang, H. Graphene oxide-templated synthesis of ultrathin or tadpole-shaped Au nanowires with alternating hcp and fcc domains. *Adv. Mater.* 2012, *24*, 979–983.

149. Gu, H.; Yang, Y.; Tian, J.; Shi, G. Photochemical synthesis of noble metal (Ag, Pd, Au, Pt) on graphene/ZnO multihybrid nanoarchitectures as electrocatalysis for H_2O_2 reduction. *ACS Appl. Mater. Interfaces* 2013, *5*, 6762–6768.

150. Hou, J.; Shao, Y.; Ellis, M.A.; Moore, R.; Yi, B. Graphene-based electrochemical energy conversion and storage: Fuel cells, supercapacitors and lithium ion batteries. *Phys. Chem. Chem. Phys.* 2011, *13*, 15384–15402.

151. Subrahmanyam, K.S.; Manna, A.; Pati, S.K.; Rao, C.N.R. Graphene. *Chem. Phys. Lett.* 2010, *497*, 70–75.

152. Lightcap, I.V.; Kamat, P.V. Graphitic design: Prospects of graphene-based nanocomposites for solar energy conversion, storage, and sensing. *Acc. Chem. Res.* 2012, *46*, 2235–2243.

153. Kim, B.J.; Yang, G.; Kim, H.Y.; Hyeon, K.B.; Mastro, M.A.; Hite, J.K.; Eddy, C.R., Jr.; Ren, F.; Pearton, S.J.; Kim, J. GaN-based ultraviolet light-emitting diodes with $AuCl_3$-doped graphene electrodes. *Opt. Express* 2013, *21*, 29026–29030.

154. Choe, M.; Cho, C.Y.; Shim, J.P.; Park, W.; Lim, S.K.; Hong, W.K.; Lee, B.H.; Lee, D.S.; Park, S.J.; Lee, T. Au nanoparticle-decorated graphene electrodes for GaN-based optoelectronic devices. *Appl. Phys. Lett.* 2012, *101*, 031115:1–031115:5.

155. Cho, C.-Y.; Choe, M.; Lee, S.J.; Hong, S.H.; Lee, T.; Lim, W.; Kim, S.T.; Park, S.J. Near-ultraviolet light-emitting diodes with transparent conducting layer of gold-doped multi-layer graphene. *J. Appl. Phys.* 2013, *113*, 113102:1–113102:5.

156. Kwon, K.C.; Kim, B.J.; Lee, J.-L.; Kim, S.Y. Effect of anions in Au complexes on doping and degradation of graphene. *J. Mater. Chem. C* 2013, *1*, 2463–2469.

157. Sun, S.; Wu, P. Easy fabrication of macroporous gold films using graphene sheets as a template. *ACS Appl. Mater. Interfaces* 2013, *5*, 3481–3486.

158. Koo, H.Y.; Lee, H.-J.; Noh, Y.Y.; Lee, E.S.; Kim, Y.H.; Choi, W.S. Gold nanoparticle-doped graphene nanosheets: Sub-nanosized gold clusters nucleate and grow at the nitrogen-induced defects on graphene surfaces. *J. Mater. Chem.* 2012,*22*, 7130–7135.

159. Jeon, E.K.; Seo, E.; Lee, E.; Lee, W.; Um, M.K.; Kim, B.S. Mussel-inspired green synthesis of silver nanoparticles on graphene oxide nanosheets for enhanced catalytic applications. *Chem. Commun.* 2013, *49*, 3392–3394.

160. Ran, X.; Sun, H.; Pu, F.; Ren, J.; Qu, X. Ag Nanoparticle-decorated graphene quantum dots for label-free, rapid and sensitive detection of Ag^+ and biothiols. *Chem. Commun.* 2013, *49*, 1079–1081.

161. Dutta, S.; Ray, C.; Sarkar, S. Silver nanoparticles decorated reduced graphene oxide (rGO) nanosheets: A platform for SERS-based low level detection of uranyl ion. *ACS Appl. Mater. Interfaces* 2013, *5*, 8724–8732.

162. Shim, J.P.; Kim, D.; Choe, M.; Lee, T.; Park, S.J.; Lee, D.S. A self-assembled Ag nanoparticle agglomeration process on graphene for enhanced light output in GaN-based LEDs. *Nanotechnology* 2012, *23*, 255201:1–255201:6.

163. Martínez-Orozco, R.D.; Rosu, H.C.; Lee, S.-W.; González, V. Understanding the adsorptive and photoactivity properties of Ag-graphene oxide nanocomposites. *J. Hazard Mater.* 2013, *263*, 52–60.

164. Yang, G.; Li, Y.; Rana, R.K.; Zhu, J.-J. Pt-Au/nitrogen-doped graphene nanocomposites for enhanced electrochemical activities. *J. Mater. Chem. A* 2013, *1*, 1754–1762.

165. Li, Y.; Gao, W.; Ci, L.; Wang, C.; Ajayan, P.M. Catalytic performance of Pt nanoparticles on reduced graphene oxide for methanol electro-oxidation. *Carbon* 2010, *48*, 1124–1130.

166. Chen, X.; Su, B.; Wu, G.; Yang, C.J.; Zhuang, Z.; Wang, X.; Chen, X. Platinum nanoflowers supported on graphene oxide nanosheets: Their green synthesis, growth mechanism, and advanced electrocatalytic properties for methanol oxidation. *J. Mater. Chem.* 2012, *22*, 11284–11289.

167. Chen, C.; Long, M.; Wu, H.; Cai, W. One-step synthesis of Pt nanoparticles/reduced graphene oxide composite with enhanced electrochemical catalytic activity. *Sci. Chin. Chem.* 2013, *56*, 354–361.

168. Tan, C.; Huang, X.; Zhang, H. Synthesis and applications of graphene-based noble metal nanostructures. *Mater. Today*2013, *16*, 29–36.

CHAPTER 6

CARBON NANOTUBE AND GRAPHENE BASED POLYAMIDE ELECTROSPUN NANOCOMPOSITES: A REVIEW

Fabiola Navarro-Pardo[1,2] Ana L. Martinez-Hernandez[1] and Carlos Velasco-Santos[1]

[1]División de Estudios de Posgrado e Investigación, Instituto Tecnológico de Querétaro, 76000 Santiago de Querétaro, QRO, Mexico

[2]Centre Énergie Matériaux Télécommunications, Institut National de la Recherche Scientifique, Varennes, QC, Canada J3X 1S2

ABSTRACT

Electrospinning is a unique and versatile technique to produce nanofibres; the facility to incorporate fillers has expanded its range of applications. This review gives a brief description of the process and the different polymers employed for obtaining nanofibres. Owing to the ability of fibrillation of polyamides, these polymers have resulted in a wide variety of interesting results obtained when using this technique; therefore these features are summarised. Additionally, because of the feasibility of incorporating carbon nanotubes and graphene in these nanofibres and the growing interest on these nanomaterials, this review focuses in the most common methods employed for their incorporation in electrospun polyamides. Several equipment setups used for the electrospinning of the nanofibres are explained. The outstanding electrical, optical, crystallinity, and mechanical properties obtained by a number of research groups are discussed. The potential applications of the resulting nanocomposites have also been explored.

INTRODUCTION

Electrospinning is a unique technique that can effectively produce fibres with diameters ranging from micrometers to several nanometres using a wide variety of materials [1–4]. This approach has been known since its patent was issued in 1924 by Formhals; however it was not until the last couple of decades that there was a revival on the interest of applying such technique in order to produce nanofibres. In this regard, Reneker and Chun promoted the interest of electrospinning and they have also provided a detailed review about the mechanism of electrospinning process including a diversity of polymers that were electrospun in their laboratory [1]. Teo and Ramakrishna have also offered a review on this process about different materials that have been electrospun, such as polymers, composites, ceramics, and metals [3]. Electrospinning of polymers has been very attractive because of the interesting characteristics when the diameters of the fibres are in the nanoscale [1–7].

Luo et al. reviewed the different insights found between the academia and the industry about the nanofibre research, providing also a comprehensive comparison of the conventional techniques for obtaining fibres [7]. Nanofibres are exceptional when compared to any other fibres due to the very large surface area to volume ratio (which can be as large as 10^3 times compared to a microfibre) which leads superior mechanical performance (e.g., stiffness and tensile strength), high porosity, high gas permeability, and small interfibrous

Citation: Fabiola Navarro-Pardo, Ana L. Martinez-Hernandez, and Carlos Velasco-Santos, "Carbon Nanotube and Graphene Based Polyamide Electrospun Nanocomposites: A Review," Journal of Nanomaterials, vol. 2016, Article ID 3182761, 16 pages, 2016. doi:10.1155/2016/3182761.

pore size [5–16]. Some authors have reviewed a variety of electrospun nanofibres and highlighted their potential for their use in filtration membranes, [5, 6], solar cells [6, 8], electronic devices [8], and biomedical applications [9, 17,18].

A number of polymers such as polyacrylonitrile [10], polyvinyl acetate [11], polyvinylidene fluoride/polypyrrole [12], poly-ε-caprolactone [13], polystyrene [14], polyethylene [15], and natural-synthetic hybrids like chitosan/starch/polyethylene terephthalate [16] have been used to obtain electrospun fibres. Huang and coworkers have made an extensive review of the processing parameters, features, modelling, and simulations related to a variety of polymers that have been electrospun into nanofibres [2]. Polyamides are suitable for electrospinning processing due to their polyelectrolytic behaviour in acid solution [19]. They are among the most used polymers for the successful electrospinning due to their mechanical properties, such as high tensile and impact strength, durability, and chemical and abrasion resistance [5, 19]. The outstanding properties of the nanofibres have attracted a lot of attention from many research groups due to the high production rate and the facility to incorporate reinforcing materials expanding the range of applications that can be obtained through this process.

Nanomaterials can impart remarkable structural and physical properties to the matrix as reinforcement agents [20–22]. Among the fillers used to reinforce nanofibres are metal nanoparticles [23–25], nanoclays [26–28], and carbon nanomaterials [29–38]. CNTs have been studied in the last two decades and since the raising of graphene a lot of research has been focused on these two carbon allotropes in polymer nanocomposites [21, 29,30]. CNTs are known to possess high aspect ratio, elastic modulus values of ~1 TPa, and tensile strength of 60–150 GPa [21, 29]. They have been observed to have thermal conductivity as high as 6000 W/cmK and electrical conductivity of 5000 S/cm [29]. The tensile strength of graphene is similar or slightly higher than CNTs (130 GPa); although CNTs show comparable mechanical properties to graphene, this 2D carbon filler has superior properties than those of the 1D carbon filler in certain aspects, such as thermal and electrical conductivity [30]. Because of their high strength and high aspect ratio carbon based nanocomposite fibres with extraordinary superior mechanical properties have been obtained [20, 31, 32]. Electrical [33, 34] and physical [35, 36] properties have also been enhanced by the incorporation of these nanomaterials. Yeo and Friend have reviewed several aspects of CNT based electrospun nanofibres [20]. However, there are no reviews on graphene based nanocomposites obtained from electrospinning.

This review focuses on CNT and graphene based polyamide nanofibres because of the above mentioned characteristics that carbon nanofillers can provide and the wide use of this polymer in electrospinning. In addition, it is important to highlight the different properties that can be obtained in the nanofibres taking into account different aspects for the processing of these carbon based nanocomposites. In this regard, we have summarised the different methods and setups used for the nanocomposite preparation and the effects provided in the morphological, optical, electrical, crystallisation, and mechanical properties of nanocomposite electrospun fibres. The prospective applications of these nanofibres are also presented.

NANOCOMPOSITE PREPARATION

Electrospun fibres are obtained from a polymer solution; when nanofillers are incorporated, they have to be dispersed in the liquid medium by vigorous stirring and/or sonication followed by the mixing of the dispersion with a polymer solution [29]. Pristine nanometric carbon is extremely difficult to disperse and align in a polymer matrix because it usually forms strong bundles due to van der Waals forces between adjacent carbon nanomaterials [37, 38]. Furthermore, CNTs and graphene have an atomically smooth nonreactive surface; the lack of interfacial bonding inhibits load transfer from the matrix to nanomaterial across the nanomaterial/polymer interface [64]. Therefore, the main challenges for obtaining carbon based nanocomposites with remarkable properties are the improvement of the dispersion, alignment, and interfacial adhesion of nanometric carbon within the matrix. Many works have been directed to overcome these shortcomings in carbon based nanocomposites [25, 37, 38, 64–69].

Electrospinning is an ideal route for aligning carbon nanomaterials [2, 65]; the different approaches and results obtained by several groups will be discussed in the next section. Functionalisation provides efficient stress transfer from the polymer matrix to the nanometric carbon by preventing aggregation of them, providing a better dispersion of the nanomaterials in the polymer matrix [37, 66, 67]. It also increases the polymer-nanomaterial physical contact and can form chemical bonds between the nanomaterial and the polymer matrix [66]. The functional groups at the surface of nanometric carbon make the strongest type of interfacial bonding with the polar polymer matrices [64]. Selected functional groups can broaden the properties of the carbon nanomaterials through the formation of donor-acceptor complexes with the graphitic structure, affording the tunability of electrical conductivity and optical properties [68]. Moreover, the additional organic moieties on the surface of carbon nanomaterial can improve the solubility of them to suit different solvents [68, 69].

Polyamides consist of methylene segments $(CH_2)_n$ separated by amide units (NH-CO). The presence of the polar groups makes functionalisation of carbon nanofillers a relevant approach for the preparation of electrospun fibres. Table 1 shows the different carbon nanomaterials used in polyamide electrospun fibres. According to the works summarised, the most commonly employed carbon nanofillers are those characterised by having oxygenated groups in their structure. Pristine carbon nanotubes typically contain carbonaceous impurities and metal catalyst particles [21, 29]. Purification of CNTs is typically achieved by oxidation using strong acids [21, 29, 66]. This process originates functional groups such as carbonyl, hydroxyl, and carboxyl on the graphitic surface [21, 69]. An extensively used strategy for obtaining graphene is the modification of graphite through oxidative routes for its subsequent exfoliation to produce GO [38]. The carboxylic acid groups at the sheet edges and epoxy and hydroxyl groups on the basal plane of GO sheets allow them to disperse in polar solvents [64]. Amino functionalised CNTs and graphene have also been incorporated into electrospun nanocomposites [31, 32, 44, 45, 70]. Jeong et al. showed that amino functionalised CNTs have better stability during the same period of time when compared to acid treated nanotubes; see Figure 1 [34, 45]. Our research group studied the influence of the dimensionality of both 1D and 2D carbon fillers on the dispersion of PA66; the capability of the CNTs to bend and the higher amount of NH_2 on their surface when compared to graphene sheets allowed the 1D nanotubes to interact with each other resulting in aggregation of them when incorporated into the polymer matrix [31]. Avila-Vega et al. incorporated nitroxide groups to GO using a new method for functionalising and exfoliating GO sheets in one step [41]. Table 1 also shows that other authors have employed techniques taking advantage of Van der Waals, π-π, CH-π, and other interactions; the adsorbed surfactants, biomolecules, or polymers provide repulsive and attractive forces creating stable dispersions [47–49, 63].

Table 1: Polyamide electrospun nanofibres reinforced with 1D and 2D carbon

Nanofiller	Functionalising agents	Polymer matrix	Solvent	Reference
P-MWCNTs	—	PA66	FA	[32]
O-MWCNTs	H_2SO_4/HNO_3	PA66	FA	[32]
A-MWCNTs	EDAC/DMA	PA66	FA	[31]
O-MWCNTs	Not specified	PA66	FA/DCM	[35, 39]
O-SWCNTs	H_2SO_4/HNO_3	PA6	FA	[40]
P-MWCNTs	—	PA6	FA	[41]
P-CNTs	—	PA6	HFIP	[42]
MWCNTs-OH	$KMnO_4$ in the presence of a phase transfer catalyst/dispersed in DMF	PA66	FA	[43]
O-MWCNTs	HNO_3/HCl dispersed in DMF	PA11	FA/DCM	[44]
A-MWCNTs	$SOCl_2$/EDA	PA66	FA	[34, 45]
O-MWCNTs	H_2SO_4/HNO_3	PA6	HFIP	[46]
O-MWCNTs	HNO_3/HCl dispersed in Triton X-100	PA610	HFIP	[47]
O-MWCNTs	HNO_3/HCl dispersed in DMF or Triton X-100 and sodium dodecyl sulfate	PA66	FA	[48]
P-MWCNTs	Dispersed in Triton X-100	PA6	HFIP	[49]
A-MWCNTs	$SOCl_2$/EDA	PA6	FA	[50]
P-MWCNTs	—	PA6	FA	[50]
O-MWCNTs	H_2SO_4/HNO_3	PA6	FA	[51]
Ac-MWCNTs	Friedel-Crafts acylation	PA6	FA	[52]
P-MWCNTs	—	PA6	FA	[52]
O-MWCNTs	Commercially obtained	PA6	HFIP	[53]
O-MWCNTs	Commercially obtained	PA6	Cresol/FA	[54]
RGO	Hydrazine	PA66	FA	[55]
NGO	Oxoammonium salts	PA6	HFIP	[41, 56]
RGO	Hydrazine	PA66	FA/DCM	[33, 57]
RGO	Hydrazine	PA66	FA	[57]
GnPs	Commercially obtained	PA6	HFIP	[58]
GO	Hummers method	PA66	FA	[32]
AGe	EDAC/DMA	PA66	FA	[31]
RGO	Hexamethylentetramine	PA66	FA	[32]
TiO_2 -RGO	Hydrothermal method	PA6	FA/AA	[59]
RGO	Hydrothermal method	PA6	FA/AA	[60]
GO	Hummers and Offeman's method	PA6	FA	[61]
RGO	Hydroiodic acid	PA6	FA	[62]
BSA-GO	Electrostatic self-assembly	PA6	FA	[63]

(a)

(b)

Figure 1: Images of the MWCNTs dispersed in a formic acid solution (sonication ~30 min). (a) Initial status and (b) after 44 h. Reprinted from [45], Copyright 2006, with permission from Elsevier.

A large number of investigations have employed a nanofiller dispersion followed by mixing with the polyamide solution [31–36, 39, 41, 42, 45, 50, 51, 55, 56, 58–60, 71–73]. Table 1 also shows that PA6 and PA66 are the most studied polymers and the solvents used for the electrospinning of the nanofibres are also included. Kim et al. reported that the ultrasonication of O-MWCNTs in DMF provides a dispersion of individual nanotubes which is stable and can be stored at room temperature for several months without precipitation [48]. According to Lala and coworkers, HFIP is the best solvent for dispersing P-MWCNTs effectively when compared to different solvents [49]. FA has also been used for dispersing CNTs and graphene in polyamides [31, 32, 55]. This organic solvent offers fine dispersion stability of the carbon nanomaterials and good solubility of polyamides [34, 45]. A combination of this solvent with others such as DCM and AA has also been used [33, 35, 39].

Another approach employed for obtaining these nanocomposites is by dip-coating the nanofibres previously obtained by electrospinning in a carbon nanomaterial dispersion [33, 40, 43, 48, 49, 57, 74, 75]. In this method, the nanometric carbon is adsorbed onto the surface of nanofibre mats as shown in Figure 2. Kim et al. studied the effect of two surfactants on the adsorption of O-MWCNTs; the nonionic surfactant resulted in a higher degree of adsorption of the nanotubes on the PA66 nanofibres [48]. This method is useful for obtaining transparent films [40]. Havel et al. adapted a similar method to deposit O-MWCNTs onto nanofibre mats allowing the formation of a thin conductive 2D network [44].

(c)

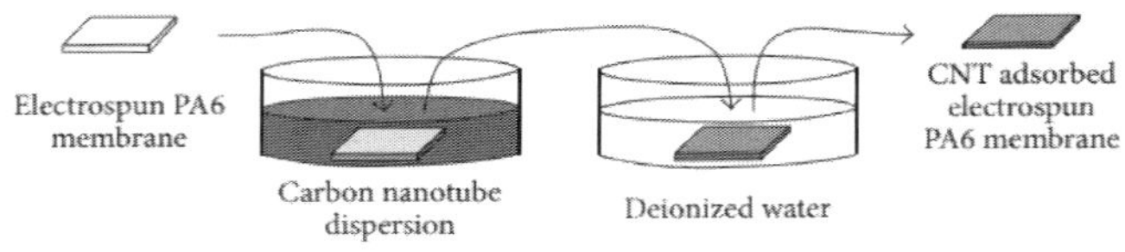

(d)

Figure 2: High-resolution SEM images of the nonwoven fibrous PA6 membranes (a) before and (b) after dip-coating in a dispersion of O-MWCNTs in water (0.05 wt.%) containing Triton X-100 surfactant (0.3 wt.%). (c) Images of nonwoven fibrous PA6 and the O-MWCNT-adsorbed nonwoven fibrous PA6 and (d) scheme of the simple processing technique used for producing the O-MWCNT-adsorbed nonwoven fibrous PA6 membranes. Reprinted from [48], Copyright 2005, with permission from John Wiley and Sons.

PVP was used to act as a modifier for improving the adsorption of GO sheets onto the nanofibres via strong π-πinteractions; the additional sites on the PVP-GO surface provided hydrogen bonding with neighbouring PA66 nanofibres; this route was also used for the reduction of GO by a combination of hydrazine vapour treatment and thermal annealing (350°C) [33]. Pant et al. employed hydrothermal treatment for the reduction of GO in PA6 nanofibres [60]. In addition, another work of these authors showed that these RGO/P6 fibres facilitated the deposition of TiO_2 nanoparticles either on the nanonets or on the graphene sheets present on the main nanofibres [59]. Other authors have adsorbed GO onto the nanofibres followed by the reduction of the graphene sheets [33, 55, 57, 61]. Figure 3 shows a method for fabricating yarns from PA6 electrospun fibres functionalised with BSA molecules (yellow dots, Figure 3(a)). The GO was wrapped via electrostatic self-assembly with BSA (Figure 3(b)), which served as an adhesive for improving the adsorption of GO sheets onto the textile, followed by a low-temperature chemical reduction of GO (Figure 3(c)) [62, 63]. Cruz-Silva et al. developed a new technique for obtaining electrospun PA fibres on top of freestanding GO film [76]. Wang et al. penetrated RGO nanosheets of different sizes onto PA66 nanofibres for constructing smooth conductive paths within polymer nanofibre fabrics [77].

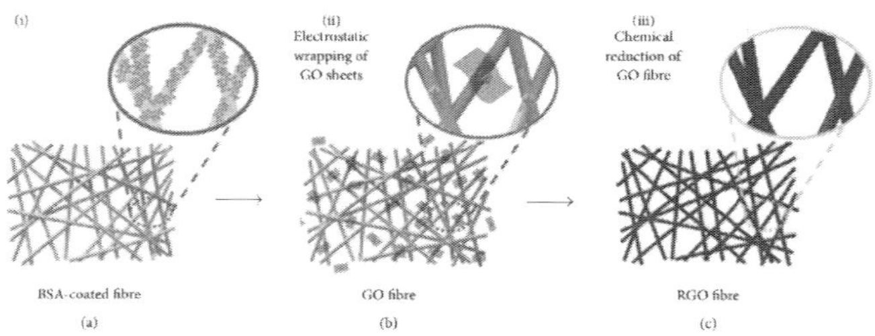

Figure 3: Schematic illustration of the three steps used to prepare RGO nanoyarns. Reprinted from [63], Copyright 2005, with permission from John Wiley and Sons.

An in situ polymerisation technique has also been used to prepare MWCNT/PA6 nanocomposites with subsequent electrospinning [52], showing better dispersion of the nanotubes when functionalised with amino groups. Kang and Jin also used a similar approach in order to incorporate O-MWCNTs into PA610 [47].

The type of method used for the incorporation of the carbon nanofiller in a matrix will have a repercussion in the application to which they will be destined. A good dispersion of the carbon filler inside the polymer is required for those applications where strength is needed [4, 17, 18, 78–82]. Thin, transparent, conducting films are critical for their use as electrodes in modern electronic devices [8, 22, 40, 57, 77]. For chemical sensors the interactions of the analyte with the carbon nanomaterials have suggested a better performance when they are surface adsorbed using a surfactant [49, 74, 75].

MORPHOLOGY OF THE ELECTROSPUN FIBRES

Electrospinning provides electrostatic stretching forces (whipping elongation) for overcoming any entanglement of nanomaterials [70]. A uniform distribution of carbon nanomaterials is critical for preventing nanofiller protrusion, notches, and beads across the fibre body [56, 70]. Polymer solution properties, such as molecular weight, solution viscosity, surface tension, solution conductivity, and dielectric constant, are critical factors that affect the electrospun fibre morphology [68, 83, 84]. Process parameters, such as applied voltage, polymer flow rate, and capillary-collector distance, also have diverse effects on the morphological appearance and average fibre diameter [14, 17, 35, 71, 84].

Different electrospinning setups have been made to control the alignment of the polyamide electrospun fibres [2, 17, 71]; among the different approaches developed are rotating drums [14, 34, 45], parallel plates [39], rotating discs [70], electrically rotating viscoelastic jets [10], grounded collector plates [31, 32, 55], and self-blending coelectrospinning [58, 85]. Applying other stretching forces can also enhance the alignment of the nanofillers along the fibre axis [70] and this is also a promising way to improve the molecular orientation degree in the fibre [70]. Parallel plates enabled the collection of aligned PA66 fibre arrays [39]. Aligned fibre bundles can also be obtained by a phase-inversion method. In this method the fibres are collected on a water bath; once they are floating on water they are transferred to a rotating drum and aligned in bundles of ~20 μm in diameter [6, 51]. Liu et al. obtained individual fibres of PA6/MWCNTs with diameters ranging from 200 to 300 nm using this technique [51]. This method has been used to fabricate microfiltration and ultrafiltration membranes [59, 70]. Moreover the manipulation of geometrical stretching of the electrospun nanofibres can favour the alignment of nanomaterials in the polymer matrix [70]. High-speed rotating collectors have also been used to collect and to align CNTs in the polymer matrix. A study showed that the mechanical properties are affected by the alignment of PA6 fibres and the enhanced orientation of MWCNTs by the take-up speed [46]. The alignment of fibres is important for engineering the nanostructure and it also provides the ability to twist the aligned fibre bundles for fabricating nanoyarns [63, 71]. Yarns of twisted electrospun MWCNT/PA6 composite nanofibres, ranging from 5 to 10 μm in diameter, have been produced by a two-disk mechanism where a first disk with controlled rotation imparts twist to a fibre bundle and then this is continuously wound on by a second disk with a constant linear speed [70].

The electrospinning setups have an influence on the morphology of the fibres; however, the content of the nanomaterials also affects the diameter of the nanofibres. These nanofillers produce changes in the physical properties of the solutions such as viscosity and electrical conductivity. Increasing the content of the carbon nanofiller provides a higher viscosity which in turn will produce thicker fibres. On the other hand, the electrical conductivity will also be higher and will favour the stretching of thinner fibres [32]. Due to these opposite behaviours some studies have shown variable fibre diameters as the loading of the nanofiller is increased [32,46, 50, 51]. Li et al. obtained PA6 nanofibres with decreasing diameter as the loading of GnPs was increased. Figure 4 shows that GnPs gave rise to the electrical conductivity and produced finer nanofibres [58]. There are other works that have shown the reduced diameter of the nanofibres as the nanofiller loading increased [42,55]. Addition of LA to PA6 for obtaining electrospun fibres produced nanofibres with flat and ribbon-shaped morphologies with wrinkled surfaces when compared to the cylindrical pure PA6 nanofibres. Furthermore, incorporation of CNTs decreased the fibre diameter, attributed to the increased electrical conductivity of the electrospun solutions [42]. Avila-Vega et al. found that the stability of the solutions also lead to effects on the homogeneity of the diameters of NGO/PA6 nanofibres [41]. The diameter of the nanofibres has been found to be related to the mechanical and thermal properties [35, 39].

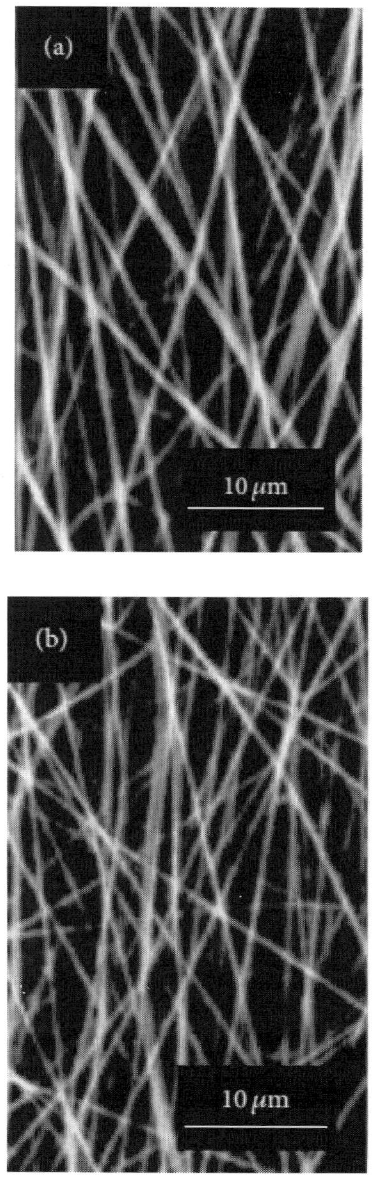

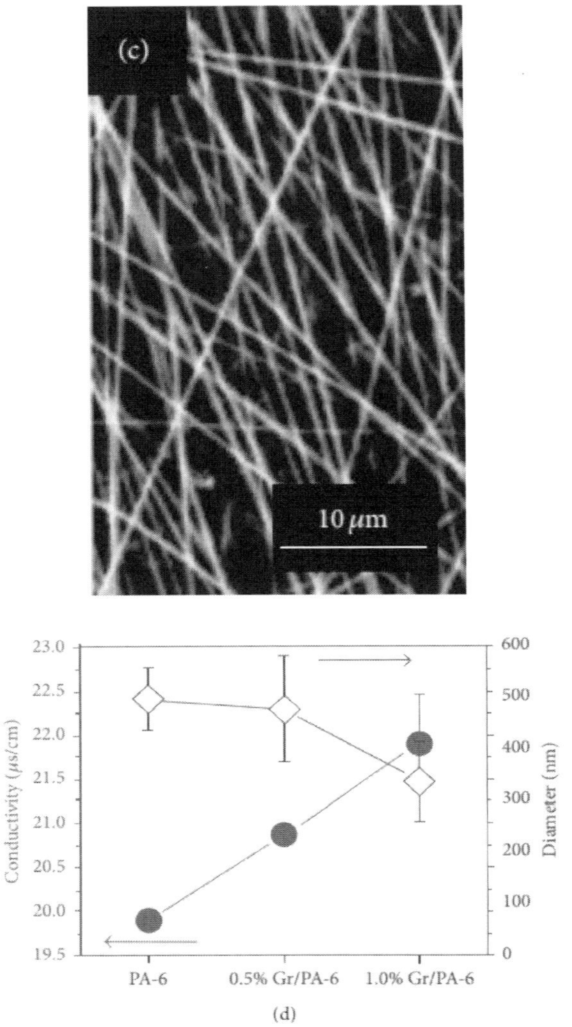

Figure 4: SEM images of as-spun (a) PA6 nanofibres, (b) 0.5% GnP/PA6 nanofibres, and (c) 1.0% GnP/PA-6 nanofibres. (d) Plot on the correlations of fibre diameters (black hollow diamond) and electrical conductivity (blue solid circle) versus PA6 solutions containing varied amount of GnP nanosheets. Reprinted from [58], Copyright 2013, with permission from Elsevier.

Additionally, in the last decade a variant of the commonly known electrospun nanofibres has been developed; this approach known as electrospinning/netting refers to the formation of nanofibres with diameters below 50 nm which connect with each other forming a spider-web-like morphology among the main nanofibres of larger diameter. Wang et al. have made a review on the recent advances when achieving this type of morphology, summarising several properties found in these type of bimodal diameter nanofibres [84]. Interestingly, PA6 was the first polymer to show this type of morphology; therefore there is valuable information about the parameters to obtain this interconnecting nets in electrospun polyamides; however, research regarding this type of morphology in nanofibres reinforced with carbon nanomaterials is limited. Pant et al. have reported that the incorporation of GO into PA6 allowed the formation of a large/scale uniform bimodal fibres with distributed thick (~200 nm) and thin (~13 nm) fibre diameter in spider-wave-like nanonets [36, 60]. The formation of spider-wave-like nanonets was attributed to the fast solvent degradation of PA6 caused by well dispersed GO sheets through the electrospinning solution [36]. Figure 5 displays TEM images obtained in our research group; the spider-net like structures were also found in PA66 electrospun fibres reinforced with oxidised MWCNTs. Furthermore, Figure 6 shows this type of morphology in PA66 nanofibres containing diverse 1D and 2D carbon nanomaterials also synthesised in our group, indicating that addition of carbon nanofillers has a favouring effect for the formation of netting among the main PA66 nanofibres. This can be related to the increased conductivity achieved in the solution by the incorporation of nanofillers. Electrospinning conditions can be found in [31, 32].

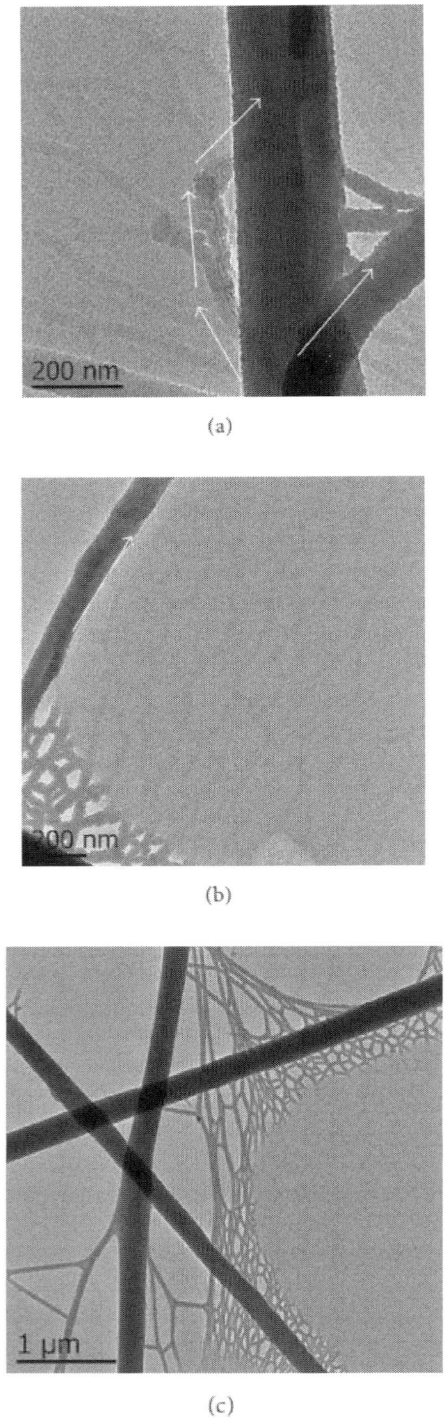

Figure 5: TEM images of 1 wt.% O-MWCNT/PA66 nanofibres showing (a) O-MWCNT oriented along the small diameter nanofibre, (b) O-MWCNT oriented in the large diameter nanofibre, and (c) bimodal diameter nanofibres.

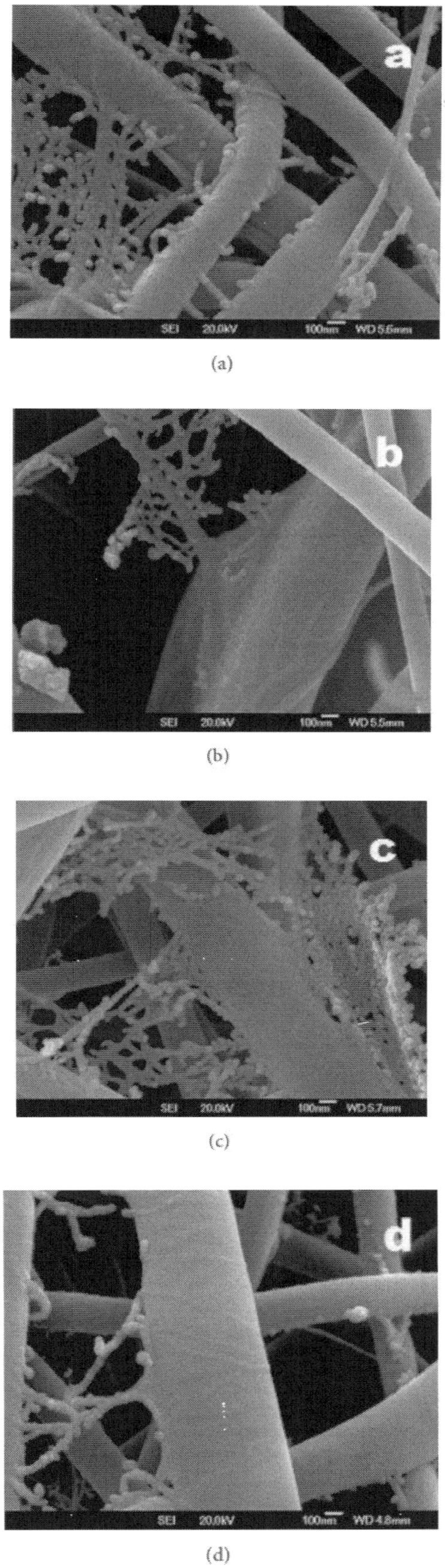

Figure 6: SEM images of (a) P-CNT/PA66 nanofibres, (b) A-CNT/PA66, (c) RGO/PA66, and (d) AG/PA66.

Pant and coworkers have explained that the increased conductivity of different pure PA6 solutions increased the ionization of polymer and therefore the fibres were characterised by these subnanofibres [86]. Furthermore, several authors have found that the electrospinning process usually favours the alignment of CNTs in the nanofibres [31, 32, 53, 70]. This behaviour is due to the converging nanoscale jet reducing the number of available orientations in the flow field [53]. NGO was aligned within the PA6 nanofibres meanwhile agglomerates of the nanofiller could hardly be aligned; therefore the NGO sheets protruded from the fibres; Figure 7 shows the NGO/PA6 nanofibres [56].

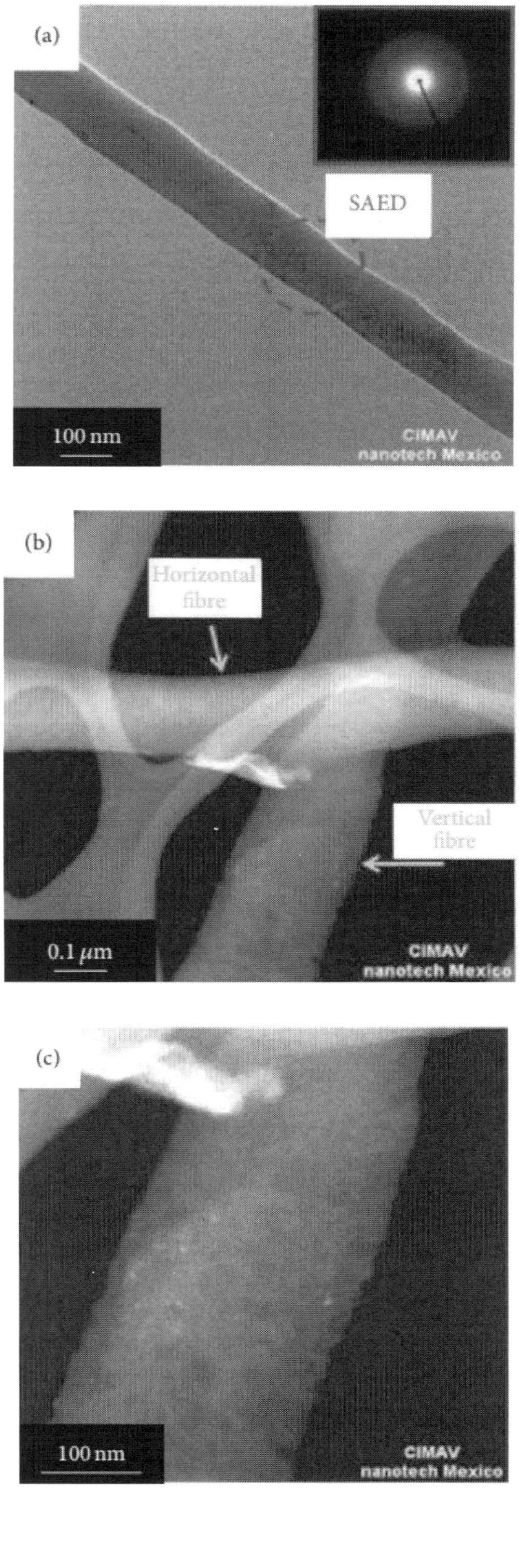

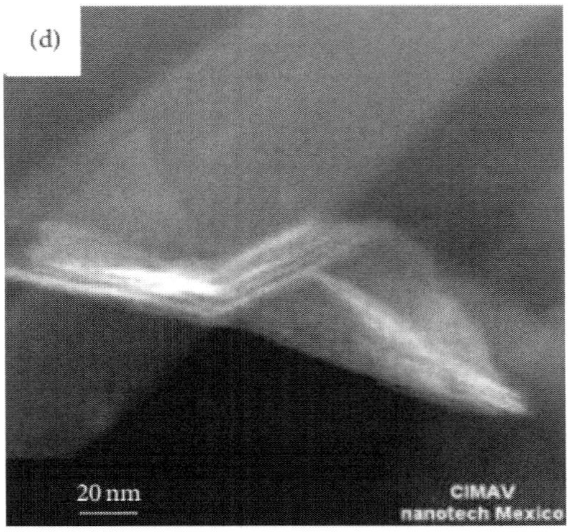

Figure 7: (a) TEM bright field image. The inset shows the SAED acquired in the circumscribed region by the red dotted circle. (b) STEM Z-contrast image showing two fibres containing NGO layers. Arrows indicate the horizontal and vertical fibres. Higher magnification Z-contrast images from (c) vertical fibre and (d) horizontal fibre. Reprinted from [56], Copyright 2010, with permission from Elsevier.

Zomer Volpato et al. found that the surface of PA6 electrospun fibres was rough with defects attributed to the fibre stretching and the presence of O-MWCNTs in the nanocomposite fibres resulted in greater surface roughness compared to the control nanofibres [53]. Yun et al. fabricated a RGO/PA6 fabric from randomly oriented nanocomposite fibres with diameters of 150–200 nm; numerous wrinkles were observed by high-resolution SEM, indicating the uniform wrapping of the RGO nanosheets onto PA nanofibres [63]. GnPs/PA6 nanofibres have also been used as reinforcement fillers in simultaneous electrospinning of PMMA [58]. The PMMA fibres were melted in situ to become the matrix, whereas the GnPs/PA6 nanofibres with higher melting point maintained their original morphology within the PMMA matrix; this method was used for achieving a good dispersion by using the graphene based nanofibres as dispersing carriers in the polymer matrix.

ELECTRICAL AND OPTICAL PROPERTIES

Nanometric carbon based films have attracted a lot of interest because of the outstanding electrical and mechanical properties of CNTs and graphene [40, 57, 61]. Using minimal amounts of CNTs or graphene in a polymer makes the resulting nanocomposites electrically conductive. Conductivity of carbon based nanocomposites increases markedly once the nanofiller content becomes high enough to percolate, that is, to form a connected network that acts as a conductive pathway through the nonconductive polymer matrix [33,34, 43, 49, 69]. In addition, percolation threshold increases significantly when processing methods cause nanofillers to align [69] and also polyamides can offer high transparency depending upon electrospinning parameters [33]. Consequently, electrospinning of polyamides reinforced with carbon nanomaterials offers the possibility to guide the nanofillers adsorbed on the nanofibres in order to obtain thin transparent conducting films [33, 40].

Bak et al. incorporated a transparent epoxy resin into a membrane made from O-SWCNT adsorbed on polyamide nanofibres [40]. Figure 8 shows that a content of 0.02 wt.% was sufficient for the formation of a percolated network of O-SWCNTs with electrical pathways, resulting in sheet resistance of 950 kΩ/sq for the nanofibres and the same behaviour was found in the epoxy resin-impregnated films which achieved higher sheet resistance than the membrane. This film transmitted 77.4% of the light at 550 nm. Another approach for enhancing the electrical conductivity of MWCNT/PA fibres was developed by Blasdel et al., where PPy was used for interconnecting MWCNT in the fibres; the resulting material acted as a reliable resistance temperature detector in the range of 25°C to 45°C [75].

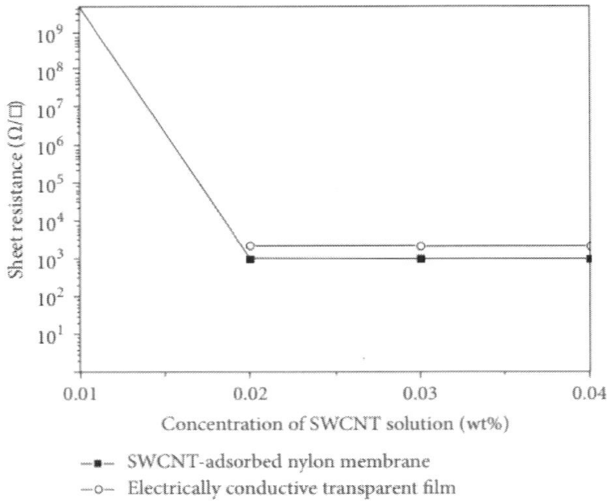

Figure 8: Sheet resistances of the O-SWCNT-absorbed PA6 membranes and the electrically conductive transparent films with respect to the O-SWCNT dispersion concentration. Reprinted from [40], Copyright 2013, with permission from Elsevier.

Adsorption of O-MWCNTs on PA11 fibres showed the highest coating of CNTs and thinnest electrospun films resulted in the best combination of electrical and optical properties; the results achieved were a sheet resistance of 440 kΩ/sq and 95% transmittance was achieved after melting the polymer nanofibres [44]. These properties were also studied in RGO/PA66 films as a function of electrospinning time. The surface electrical resistivity gradually decreased with increasing density of PA66 nanofibres and the transmittance decreased as the electrospinning time increased [33]. After annealing a film obtained from 120 s of electrospinning time and immersed in 0.05 wt.% PVP-GO solution exhibited a surface resistance of 8.6 kΩ/sq and 88% light transmittance [33]. Higher loadings of nanofillers have also been used [45, 51]. Liu et al. found that the conductivity of PA6 improved significantly as the loading of MWCNTs was increased up to a 3 wt.% [51]. Jeong et al. conducted I–V measurements of PA66 nanofibres containing different amounts of MWCNTs; they found that the current increased from 0.59 mA to 1.77 mA when the nanofiller loading was increased from 10 wt.% to 20 wt.% [45]. Nirmala et al. found further increases when adding Ag nanoparticles to MWCNTs/PA6 fibres [87].

Addition of GO or RGO into PA6 nanofibres boosted the electron conduction path. Furthermore the spider-wave structure which was a feature of these nanocomposites allowed the bridging between the PA6 nanofibres and the graphene sheets increasing the amounts of charge carriers and reducing the hopping distance of conduction electrons which favoured the percolation phenomenon [60]. The GO/PA6 nanofibres favoured the electrical conductivity resulting in a three times larger increment on this property than that of the pure PA6 nanofibres; additionally, GO reduction treatment produced an excellent enhancement in this property achieving values 300 times higher than the pure polymer mats [60]. In another work, chemical reduction of GO resulted in an increase of six orders of magnitude in the electrical conductivity when compared to the GO/PA6 nanofibres; this was attributed to the restoration of the conjugated network on the RGO sheets [63]. Wang et al. effectively penetrated RGO into PA66 fibres facilitating the construction of electron pathways within the fabric and promoting electrical conductivity [77].

CRYSTALLINITY PROPERTIES

Semicrystalline polymers are sensitive to shear and deformation producing changes in the polymorphism and crystallite orientation [88]. Polyamides are semicrystalline polymers; their molecular structure, morphology, and crystallinity are determined by hydrogen bonded sheets formed between near-neighbour amide groups; and these sheets are stacked together by Van der Waals forces between methylene chains [5, 89]. Stephens et al. found that electrospinning process modifies the chain conformation of polyamide backbone due to the high stress induced on the jet as the fibres are being formed [90]. Electrospun fibres consist of densely packed aligned lamellae and fibrillar structures. Surrounding the crystalline regions the amorphous phase consists of extended tie molecules [91]. The addition of a second phase is expected to favour crystallisation, reduce crystal size, and therefore impart better mechanical properties to the fibres [26, 31, 32, 35]. However, there is disagreement among diverse studies on carbon nanocomposites as to whether their presence alters the mechanism of crystal growth or the overall degree of crystallinity [69].

Several studies of electrospun CNT/polyamide nanocomposites showed the influence that nanotubes have on crystallisation of the polymer [31, 32, 35, 39, 46, 52, 92]. Jose et al. found that the structure of PA6 transforms from the single γ phase for pure polymer to a mixture of γ and α phases as the nanotube loading increased. This study also showed that the structure of the nanocomposite and the take-up speed of the obtained fibres were independent when comparing similar loadings of O-MWCNTs [46]. Saeed et al. found that electrospinning of P-MWCNTs and Ac-MWCNTs did not affect the crystal structure of PA6 upon functionalisation of nanotubes. On the other hand, this study showed that the shear force during electrospinning might favour the γ phase [52]. This phase is often associated with the formation of extended chain crystals and is typically obtained from a process involving elongational flow [88]. The addition of a small amount of O-MWCNTs induced crystallisation in PA66 fibres; however no significant difference in enthalpy values was seen when the loading of O-MWCNTs was higher than 1 wt.% [35, 39]. The presence of nanotubes in the fibres leads to smaller but larger number of crystals, explaining the increase of crystallinity in the reinforced fibres [35]. The crystallisation of PA66 electrospun fibres using A-MWCNTs and AGe as filler has also been investigated in our group. CNTs provided decrease in the crystallinity and higher crystal size when compared to graphene based nanofibres [32]. Lee et al. found that P-MWCNTs influenced the crystallinity degree of PA6 nanofibres [50]. In addition, the two crystalline peaks of PA6 downshifted from the theoretical value; this behaviour was attributed to the close-packing during the electrospinning. Cai and coworkers showed that the addition of LA produced a detriment on the crystallisation of PA6 nanofibres. However, the incorporation of P-CNTs could effectively promote the heterogeneous nucleation in the polymer [42]. Jeon et al. also showed the nucleating behaviour of MWCNTs on electrospun fibres, where pristine MWCNTs started nucleation at higher crystallisation temperatures when compared to O-MWCNTs [92].

The study on the crystallinity properties of graphene based nanofibres obtained from polyamides is limited. Albañil-Sanchez et al. found that increasing the concentration of RGO intensified and narrowed the crystalline reflections of PA66, suggesting the nucleating behaviour of these nanomaterial [55]. Similar results were found in our research group, by adding GO and AGe [31, 32]. Our results showed that an increase in the nanofiller content produced smaller crystal sizes in the nanocomposite fibres. In another experiment, addition of GnPs into PA6 modified the crystalline phase from α to γ, when compared to the pure polymer; this was attributed to high-speed electrospinning process along with the nucleating behaviour of the nanofiller [58]. Figure 9 shows that the crystal structure of PA transforms from γ (Figure 9(a)) to α upon hydrothermal treatment of GO (Figure 9(b)). The diffractogram also shows the presence of the different peaks of the anatase (A) and rutile (R) forms of TiO_2 and also both crystalline phases of PA6 [59].

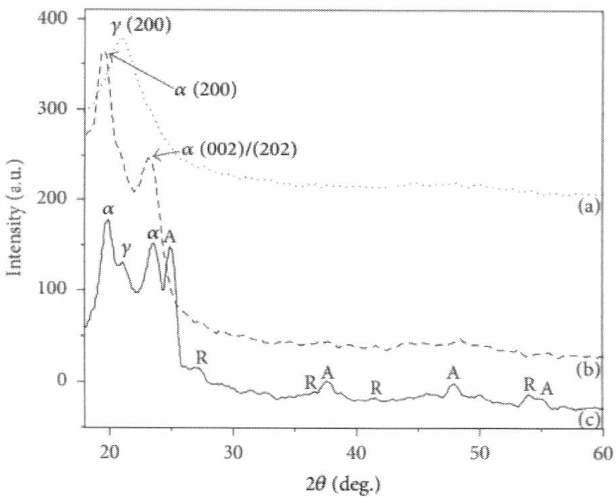

Figure 9: XRD patterns of (a) GO/PA6, (b) RGO/PA6, and (c) TiO_2-RGO/PA6 mats. Reprinted from [59], Copyright 2007, with permission from Elsevier.

MECHANICAL PROPERTIES

The alignment of carbon nanofillers by electrospinning enhances the axial mechanical and physical properties of the fibres [65]. Several researchers have reported systematic investigations of the effects of electrospinning parameters on fibre diameter, morphology, and the effect on the mechanical properties [33, 36, 39, 46]. A study using different collector speeds showed the influence on mechanical properties at different loadings of O-MWCNTs in PA6 fibres; the high drawing speed provided a fibre rearrangement at testing direction and therefore better load transfer. Combining high take-up speeds and 1 wt.% of O-MWCNTs increased the modulus 797% when compared to pure PA6 fibres, as seen in Figure 10 [46]. Baji et al. studied the influence of diameter on the mechanical properties. The tensile modulus and strength versus O-MWCNT content for each fibre diameter are presented in Figures 11(a) and 11(b), respectively. They concluded that the reinforcement effect of O-MWCNTs is controlled by the fibre diameter. In the fibres with smaller diameter, the size of the surface regions is comparable to the overall fibre diameter, while in the fibres with larger diameter, the surface regions are much smaller compared to the overall fibre diameter. Therefore, fibres of smaller diameter display improved strength and stiffness values [35]. O-MWCNT/PA6 nanofibres showed decreased stiffness when compared to the pure polymer nanofibres; this was attributed to the different architecture of the tested mats [53]. Bazbouz and Stylious found that the tensile test for nonwoven nanofibre mats was not suitable for reflecting the exact mechanical properties of nanofibres. They provided a more accurate characterisation by conducting the tensile tests on single nanofibres or even aligned nanofibre bundles [72].

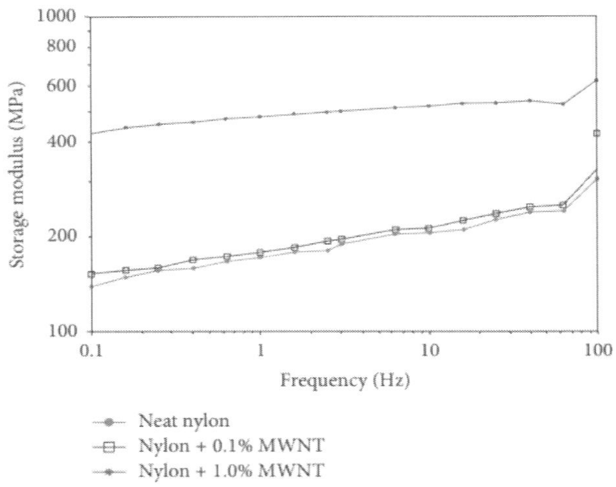

Figure 10: Effect of O-MWNT on storage modulus of the nonwoven aligned fibre mat at 6000 rpm. Reprinted from [46], Copyright 2011, with permission from Elsevier.

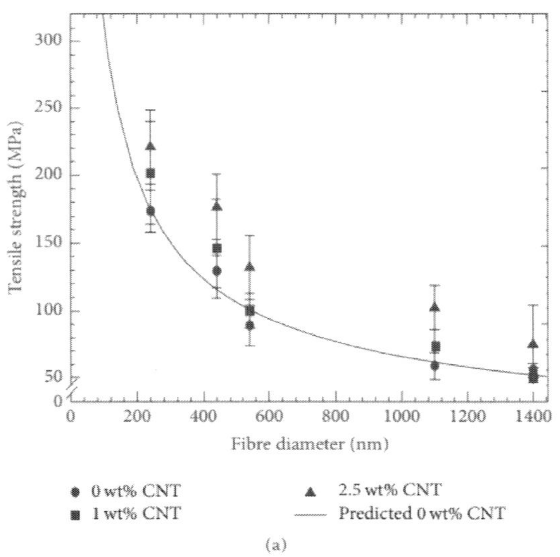

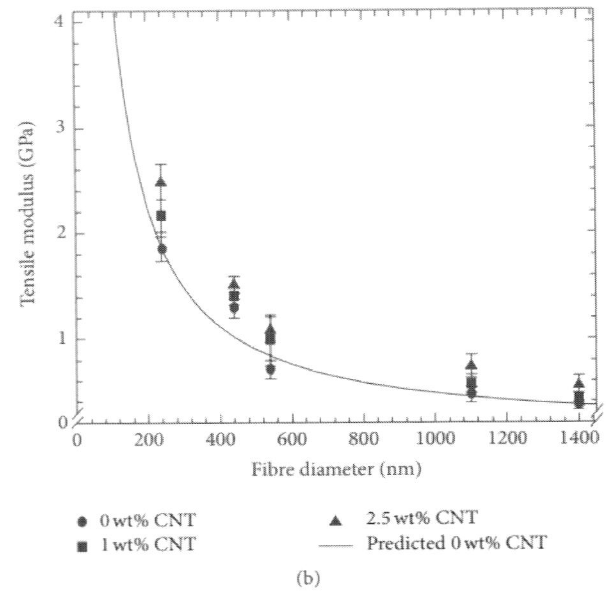

Figure 11: (a) Tensile modulus versus fibre diameter and (b) tensile strength versus fibre diameter. Reprinted from [35], Copyright 2014, with permission from Elsevier.

Functionalisation is also important for the improvement of mechanical properties. Specific tensile strengths of A-MWCNT/PA6 fibres were enhanced from 207 $^{kg_f cm}$/g for PA6 nanofibres to 389 $^{kg_f cm}$/g when using 1 wt.% A-MWCNTs. Functionalisation favoured a better dispersion of the nanotubes in the matrix when compared to P-MWCNT/PA6 nanofibres which resulted in 359 $^{kg_f cm}$/g for the same nanotube content [52]. Breaking strain of A-MWCNT/PA6 nanofibres was improved due to the interfacial bonding between both phases when compared to the P-MWCNT/PA6 nanofibres [50]. Tensile properties of these A-MWCNT/PA6 electrospun fibres, including initial modulus, tensile strength, and breaking strain, were improved after applying a thermal posttreatment [50]. Addition of 1 wt.% O-MWCNTs to PA66 nanofibres increased the tensile modulus and strength by ~70% and ~25%, respectively. Dramatic changes on the storage moduli of the electrospun fibres were also observed; at room temperature, the storage modulus of the 7.5 wt.% O-MWCNTs filled fibres was ~250% higher than the pure PA66 fibres [39]. Storage modulus of PA66 electrospun fibres was enhanced by 118% for A-MWCNTs nanofibres and 69% for AGe based nanofibres at 0.1 wt.% content of the nanofiller. However when the A-MWCNT loading was increased this property decreased and the opposite behaviour was found for AGe based nanofibres, as seen in Figure 12 [31]. These results were attributed to the tendency of A-MWCNTs to aggregate at higher loadings due to the interaction of the functional groups at the graphitic surface.

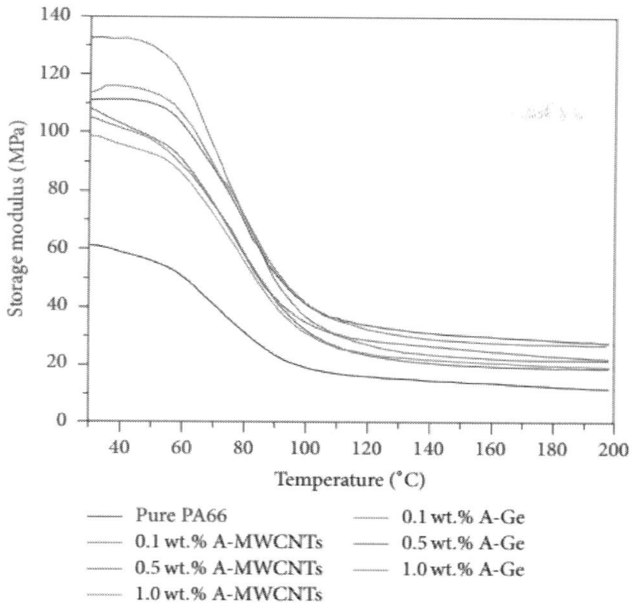

Figure 12: Storage modulus of 1D and 2D amino reinforced carbon/PA66 electrospun fibres. Published under licence in IOP Conference Series: Materials Science and Engineeringby IOP Publishing Ltd.

RGO and GO are shown to be better reinforcement agents when compared to P-MWCNTs and O-MWCNTs in electrospun PA66 nanofibres [32]. This was attributed to the 2D nature of graphene which provides a larger surface area compared to the 1D nanotubes, thus favouring the mechanical properties of the nanofibres. GnPs and GO moderately enhanced the mechanical response at low content of these nanofillers; on the other hand, NGO provided increases of 95%, 73%, and 82% in the tensile strength, Young's modulus, and tensile strain, respectively, compared to PA6 nanofibres. The simultaneous increases in these properties were unusual because the usual effect is that one of these properties increases at the expense of another [41]. The strength and stiffness of PA6 nanofibres and nanoyarns were enhanced by the wrapping of RGO onto these structures, showing improvements on the tensile strength of 74% and 41% for the nanofibre and nanoyarn nanocomposites, respectively [63].

APPLICATIONS

The unique characteristics of electrospun fibres make them candidates for a wide variety of applications. They have been studied in order to develop lightweight, ultra-strong structures for miniaturised and load-bearing applications [35]. Moreover tailoring the nonconducting polymeric matrices with conductive fillers like CNTs and graphene could be used to obtain nanocomposites for sensor applications [54, 75]. Cai et al. developed CNT/LA/PA6 nanofibres for obtaining phase change materials to store and retrieve solar energy [42]. The features obtained in GO/PA6 nanofibres of bimodal fibre diameter promise to have great potential in air/water filter applications [36]. The large surface area and porous nature of carbon based nanofibres offer a very large gas absorptive capacity, making possible greater analyte permeability for sensors [43, 49]. Furthermore these composites could have many advantages such as fast response, long-term stability, high sensitivity, and good reproducibility [43]. Choi et al. also fabricated O-MWCNT/PA66 electrospun fibres for sensing low molecular weight alcohol vapours such as methanol, ethanol, 1-propanol, and 1-butanol [43]. O-MWCNT/PA6 nanofibres served as the nanosized backbone for pyrrole electropolymerisation; the obtained nanocomposite was tested for biosensor applications [54]. A similar approach was used by Uzun et al. to electrochemically synthesise a conducting polymer after a graphite electrode was coated with MWCNT/PA6 in order to obtain biosensors for testing glucose containing beverages [93].

Miao et al. have examined a variety of polymers and nanomaterials for their use in electronic devices [8]. These nanofibres have also attracted a lot of attention for fabricating transparent nanocomposites due to the possibility of obtaining fibre diameters smaller than the wavelength of visible light [94]. Furthermore, the incorporation of 1D and 2D carbon structures studied in this review makes the creation of thin, transparent, and electrically conductive materials possible that are required for numerous applications such as liquid crystal displays, light emitting diodes, transistors, actuators, sensors, organic solar cells, and smart textiles [8, 33, 44,45, 76]. Havel et al. developed thin films of O-MWCNT adsorbed onto PA11 nanofibres in order to obtain a viable candidate to replace ITO, providing sufficient conductivity for most applications, but without the price tag and physical limitations of the current ITO coatings [44]. According to the characteristics found in O-SWCNT/PA6 nanofibres developed by Bak coworkers, these could be applied as transparent electrodes in photoelectronics, such as flexible displays and touch screens, as well as in biological applications, such as actuators [40]. PA66 electrospun fibres were also used as a template to build graphene networks to obtain conductive films while minimising transmittance losses [33]. Ma et al. obtained a patent for producing graphene adsorbed on nanofibres to produce a flexible transparent electrode [57].

Liu and coworkers have provided a review about several electrospun fibres which have found wide applications in biomedical fields [9]. Monty et al. patented a flexible electrode for detecting changes in temperature, humidity, and sodium ion concentration in sweat; this fabric can be made from CNT or graphene adsorbed onto PA electrospun fibres and further functionalised with PPy [74, 75]. CNTs and graphene have been used to reinforce the weak points of existing scaffold materials [13, 18, 62]. Zomer Volpato et al. obtained encouraging results for the use of O-MWCNT/PA6 nanofibres in biomedical applications [53]. In addition, the adjustable morphology and mechanical properties of the electrospun nanofibres are very important in order to mimic specific target tissues that need to be replaced or regenerated [53, 78].

Nonwoven nanofibres can be useful for limited applications such as filtration, tissue scaffolds, implant coating films, and wound dressings [36, 71, 80, 86]. On the other hand, continuous single nanofibres or uniaxial fibre bundles provide a wider variety of applications and could improve the performance of the nanofibres [49, 71]. Teo and Ramakrishna have reviewed these nanofibres and classified different levels of organisation in order to construct nanocomposites using electrospun fibres [4]. Another potential application of these nanofibres could be in the reinforcement of other polymers using these nanostructured carbon based electrospun fibres [58, 79,94].

In summary, polyamide nanofibres reinforced with carbon nanomaterials qualify for a number of applications such transparent electrodes [40, 57], electronic devices [61, 87], solar cells [4, 50, 73], drug delivery systems [17,38], biosensors [54, 62, 93], wearable electronic devices [63, 76, 77], or nanofibrous membranes for filtration [5,59, 80, 82, 86]. The promising applications of these nanocomposite fibres have resulted in the publication of patents by several research groups [57, 61]. It is clear that the study of carbon based/polyamide electrospun will increase the research of these nanocomposites in future.

CONCLUSION

This review showed the different investigations of polyamide electrospun nanocomposites using CNTs or graphene as fillers. The selection of an appropriate solvent for the dispersion of carbon nanomaterials is an important factor for solving one of the main issues of carbon based nanocomposites, the dispersion. The other main challenge which is the alignment of nanomaterials is a feature provided by electrospinning technique. The diverse studies of polyamide electrospun fibres reinforced with these two novel carbon nanomaterials have showed interesting characteristics for the development of a variety of applications in diverse fields of science and technology. However the potential of this technique in polyamides has not been exploited sufficiently when concerning to the reinforcing of the fibres with CNTs and graphene. More efforts need to be made in understanding the behaviour of these different structures of carbon in polyamide matrices in order to achieve the best dispersion and alignment for obtaining nanocomposites with exceptional properties.

REFERENCES

1. D. H. Reneker and I. Chun, "Nanometre diameter fibres of polymer, produced by electrospinning,"Nanotechnolo gy, vol. 7, no. 3, pp. 216–223, 1996. · ·
2. Z.-M.Huang,Y.-Z.Zhang,M.Kotaki,andS.Ramakrishna, "A review on polymer nanofibers by electrospinning and their applications in nanocomposites," Composites Science and Technology, vol. 63, no. 15, pp. 2223–2253, 2003. · ·
3. W.E.TeoandS.Ramakrishna,"Areviewonelectrospinning design and nanofibre assemblies,"Nanotechnology, vol. 17, no. 14, pp. R89–R106, 2006. · ·
4. W.-E. Teo and S. Ramakrishna, "Electrospun nanofibers as a platform for multifunctional, hierarchically organized nanocomposite," Composites Science and Technology, vol. 69, no. 11-12, pp. 1804–1817, 2009. · ·
5. P. Heikkilä, A. Taipale, M. Lehtimäki, and A. Harlin, "Electrospinning of polyamides with different chain compositions for filtration application," Polymer Engineering and Science, vol. 48, no. 6, pp. 1168–1176, 2008. · ·
6. V. Thavasi, G. Singh, and S. Ramakrishna, "Electrospun nanofibers in energy and environmental applications," Energy and Environmental Science, vol. 1, no. 2, pp. 205–221, 2008. ··
7. C. J. Luo, S. D. Stoyanov, E. Stride, E. Pelan, and M. Edirisinghe, "Electrospinning versus fibre production methods: from specifics to technological convergence," Chemical Society Reviews, vol. 41, no. 13, pp. 4708–4735, 2012. · ·
8. J. Miao, M. Miyauchi, T. J. Simmons, J. S. Dordick, and R. J. Linhardt, "Electrospinning of nanomaterials and applications in electronic components and devices," Journal of Nanoscience and Nanotechnology, vol. 10, no. 9, pp. 5507–5519, 2010. · ·
9. H. Liu, X. Ding, G. Zhou, P. Li, X. Wei, and Y. Fan, "Electrospinning of nanofibers for tissue engineering applications," Journal of Nanomaterials, vol. 2013, Article ID 495708, 11 pages, 2013. ··
10. C.-C. Liao, C.-C. Wang, C.-Y. Chen, and W.-J. Lai, "Stretching-induced orientation of polyacrylonitrile nanofibers by an electrically rotating viscoelastic jet for improving the mechanical properties," Polymer, vol. 52, no. 10, pp. 2263–2275, 2011. · ·
11. G. Wang, Z. Tan, X. Liu et al., "Conducting MWNT/ poly(vinyl acetate) composite nanofibres by electrospinning," Nanotechnology, vol. 17, no. 23, pp. 5829–5835, 2006. · ·
12. K. Ketpang and J. S. Park, "Electrospinning PVDF/PPy/ MWCNTs conducting composites," Synthetic Metals, vol. 160, no. 15-16, pp. 1603–1608, 2010. · ·
13. C. Wan and B. Chen, "Poly(ε-caprolactone)/graphene oxide biocomposites: mechanical properties and bioactivity," Biomedical Materials, vol. 6, no. 5, Article ID 055010, 2011. · ·
14. S. Mazinani, A. Ajji, and C. Dubois, "Morphology, structure and properties of conductive PS/CNT nanocomposite electrospun mat," Polymer, vol. 50, no. 14, pp. 3329–3342, 2009. · ·
15. S. R. Givens, K. H. Gardner, J. F. Rabolt, and D. B. Chase, "High-temperature electrospinning of polyethylene microfibers from solution," Macromolecules, vol. 40, no. 3, pp. 608–610, 2007. · ·
16. A. Espíndola-González, A. L. Martínez-Hernández, F. Fernández-Escobar et al., "Natural-synthetic hybrid polymers developed via electrospinning: the effect of PET in chitosan/starch system,"International Journal of Molecular Sciences, vol. 12, no. 3, pp. 1908–1920, 2011. · ·
17. T. J. Sill and H. A. von Recum, "Electrospinning: applications in drug delivery and tissue engineering,"Biomaterials, vol. 29, no. 13, pp. 1989–2006, 2008. · ·
18. H. Haniu, N. Saito, Y. Matsuda et al., "Basic potential of carbon nanotubes in tissue engineering applications," Journal of Nanomaterials, vol. 2012, Article ID 343747, 10 pages, 2012. ··
19. R. Nirmala, R. Navamathavan, H.-S. Kang, M. H. El-Newehy, and H. Y. Kim, "Preparation of polyamide-6/chitosan composite nanofibers by a single solvent system via electrospinning for biomedical applications," Colloids and Surfaces B: Biointerfaces, vol. 83, no. 1, pp. 173–178, 2011. ··
20. L. Yeo and J. R. Friend, "Electrospinning carbon nanotube polymer composite nanofibers," Journal of

Experimental Nanoscience, vol. 1, pp. 177–209, 2006.

21. A. L. Martínez-Hernández, C. Velasco-Santos, and V. M. Castaño, "Carbon nanotubes composites: processing, grafting and mechanical and thermal properties," Current Nanoscience, vol. 6, no. 1, pp. 12–39, 2010. · ·

22. R. Verdejo, M. M. Bernal, L. J. Romasanta, and M. A. Lopez-Manchado, "Graphene filled polymer nanocomposites," Journal of Materials Chemistry, vol. 21, no. 10, pp. 3301–3310, 2011. ··

23. B. Pant, H. R. Pant, D. R. Pandeya et al., "Characterization and antibacterial properties of Ag NPs loaded nylon-6 nanocomposite prepared by one-step electrospinning process," Colloids and Surfaces A: Physicochemical and Engineering Aspects, vol. 395, pp. 94–99, 2012. · ·

24. B. W. Ahn and T. J. Kang, "Preparation and characterization of magnetic nanofibers with iron oxide nanoparticles and poly(ethylene terephthalate)," Journal of Applied Polymer Science, vol. 125, no. 2, pp. 1567–1575, 2012. · ·

25. Y. X. Gan, "Structural assessment of nanocomposites," Micron, vol. 43, no. 7, pp. 782–817, 2012. · ·

26. Y. Lu, Y. Zhang, G. Zhang, M. Yang, S. Yan, and D. Shen, "Influence of thermal processing on the perfection of crystals in polyamide 66 and polyamide 66/clay nanocomposites," Polymer, vol. 45, no. 26, pp. 8999–9009, 2004. · ·

27. G.-M. Kim, G. H. Michler, F. Ania, and F. J. B. Calleja, "Temperature dependence of polymorphism in electrospun nanofibres of PA6 and PA6/clay nanocomposite," Polymer, vol. 48, no. 16, pp. 4814–4823, 2007. · ·

28. G. Nitya, G. T. Nair, U. Mony, K. P. Chennazhi, and S. V. Nair, "In vitro evaluation of electrospun PCL/nanoclay composite scaffold for bone tissue engineering," Journal of Materials Science: Materials in Medicine, vol. 23, no. 7, pp. 1749–1761, 2012. · ·

29. Z. Spitalsky, D. Tasis, K. Papagelis, and C. Galiotis, "Carbon nanotube-polymer composites: chemistry, processing, mechanical and electrical properties," Progress in Polymer Science, vol. 35, no. 3, pp. 357–401, 2010. · ·

30. T. Kuilla, S. Bhadra, D. Yao, N. H. Kim, S. Bose, and J. H. Lee, "Recent advances in graphene based polymer composites," Progress in Polymer Science, vol. 35, no. 11, pp. 1350–1375, 2010. · ·

31. F. Navarro-Pardo, G. Martínez-Barrera, A. L. Martínez-Hernández, et al., "Nylon 6,6 electrospun fibres reinforced by amino functionalised 1D and 2D carbon," IOP Conference Series: Materials Science and Engineering, vol. 40, Article ID 012023, 2012.

32. F. Navarro-Pardo, G. Martínez-Barrera, A. L. Martínez-Hernández et al., "Effects on the thermo-mechanical and crystallinity properties of nylon 6,6 electrospun fibres reinforced with one dimensional (1D) and two dimensional (2D) carbon," Materials, vol. 6, no. 8, pp. 3494–3513, 2013. ··

33. Y.-L. Huang, A. Baji, H.-W. Tien et al., "Self-assembly of graphene onto electrospun polyamide 66 nanofibers as transparent conductive thin films," Nanotechnology, vol. 22, no. 47, Article ID 475603, 2011. · ·

34. J.-S. Jeong, S.-J. Park, Y. H. Shin, Y.-J. Jung, P. S. Alegaonkar, and J.-B. Yoo, "Fabrication of carbon nanotube embedded nylon nanofiber bundles by electrospinning," Solid State Phenomena, vol. 124–126, no. 2, pp. 1125–1128, 2007. · ·

35. A. Baji, Y.-W. Mai, and S.-C. Wong, "Effect of fiber diameter on the deformation behavior of self-assembled carbon nanotube reinforced electrospun polyamide 6,6 fibers," Materials Science and Engineering A, vol. 528, no. 21, pp. 6565–6572, 2011. · ·

36. H. R. Pant, C. H. Park, L. D. Tijing, A. Amarjargal, D.-H. Lee, and C. S. Kim, "Bimodal fiber diameter distributed graphene oxide/nylon-6 composite nanofibrous mats via electrospinning," Colloids and Surfaces A: Physicochemical and Engineering Aspects, vol. 407, pp. 121–125, 2012. ··

37. N. G. Sahoo, H. K. F. Cheng, L. Li, S. H. Chan, Z. Judeh, and J. Zhao, "Specific functionalization of carbon nanotubes for advanced polymer nanocomposites," Advanced Functional Materials, vol. 19, no. 24, pp. 3962–3971, 2009. ·

38. T. Kuila, S. Bose, A. K. Mishra, P. Khanra, N. H. Kim, and J. H. Lee, "Chemical functionalization of graphene and its applications," Progress in Materials Science, vol. 57, no. 7, pp. 1061–1105, 2012. · ·

39. A. Baji, Y.-W. Mai, S.-C. Wong, M. Abtahi, and X. Du, "Mechanical behavior of self-assembled carbon nanotube reinforced nylon 6,6 fibers," Composites Science and Technology, vol. 70, no. 9, pp. 1401–1409, 2010. · ·

40. H. Bak, S. Y. Cho, Y. S. Yun, and H.-J. Jin, "Electrically conductive transparent films based on nylon 6 membranes and single-walled carbon nanotubes," Current Applied Physics, vol. 10, no. 3, pp. S468–S472, 2010. · ·

41. Y. I. Avila-Vega, C. C. Leyva-Porras, M. Mireles, M. Quevedo-López, J. Macossay, and J. Bonilla-Cruz, "Nitroxide-functionalized graphene oxide from graphite oxide," Carbon, vol. 63, pp. 376–389, 2013. · ·

42. Y. Cai, X. Xu, C. Gao et al., "Effects of carbon nanotubes on morphological structure, thermal and flammability properties of electrospun composite fibers consisting of lauric acid and polyamide 6 as thermal energy storage materials," Fibers and Polymers, vol. 13, no. 7, pp. 837–845, 2012. · ·

43. J. Choi, E. J. Park, D. W. Park, and S. E. Shim, "MWCNT-OH adsorbed electrospun nylon 6,6 nanofibers chemiresistor and their application in low molecular weight alcohol vapours sensing," Synthetic Metals, vol. 160, no. 23-24, pp. 2664–2669, 2010. · ·

44. M. Havel, K. Behler, G. Korneva, and Y. Gogotsi, "Transparent thin films of multiwalled carbon nanotubes self-assembled on polyamide 11 nanofibers," Advanced Functional Materials, vol. 18, no. 16, pp. 2322–2327, 2008. · ·

45. J. S. Jeong, S. Y. Jeon, T. Y. Lee et al., "Fabrication of MWNTs/nylon conductive composite nanofibers by electrospinning," Diamond and Related Materials, vol. 15, no. 11-12, pp. 1839–1843, 2006. · ·

46. M. V. Jose, B. W. Steinert, V. Thomas et al., "Morphology and mechanical properties of Nylon 6/MWNT nanofibers," Polymer, vol. 48, no. 4, pp. 1096–1104, 2007. · ·

47. M. Kang and H.-J. Jin, "Electrospun nanofiber of nylon 610/multi-walled carbon nanotube composites,"Key Engineering Materials, vol. 321–323, pp. 934–937, 2006. · ·

48. H. S. Kim, H.-J. Jin, S. J. Myung, M. Kang, and I.-J. Chin, "Carbon nanotube-adsorbed electrospun nanofibrous membranes of nylon 6," Macromolecular Rapid Communications, vol. 27, no. 2, pp. 146–151, 2006. · ·

49. N. L. Lala, V. Thavasi, and S. Ramakrishna, "Preparation of surface adsorbed and impregnated multi-walled carbon nanotube/nylon-6 nanofiber composites and investigation of their gas sensing ability,"Sensors, vol. 9, no. 1, pp. 86–101, 2009. · ·

50. S. Y. Lee, B. Yoo, M. K. Lim, T.-K. Lee, A. R. S. Priya, and K.-J. Kim, "Influence of nylon 6 in I3-/I- redox electrolyte on the photovoltaic performance and stability of dye-sensitized solar cells," Langmuir, vol. 26, no. 9, pp. 6638–6642, 2010. · ·

51. Y. Liu, J. Li, and Z.-J. Pan, "The dispersion of CNT and the conductive property of PA6/MWNTs nanofiber filaments by electrospinning," Advanced Materials Research, vol. 295–297, pp. 1993–1997, 2011. · ·

52. K. Saeed, S.-Y. Park, S. Haider, and J.-B. Baek, "In situ polymerization of multi-walled carbon nanotube/ nylon-6 nanocomposites and their electrospun nanofibers," Nanoscale Research Letters, vol. 4, no. 1, pp. 39–46, 2009. · ·

53. F. Zomer Volpato, S. L. Fernandes Ramos, A. Motta, and C. Migliaresi, "Physical and in vitro biological evaluation of a PA 6/MWCNT electrospun composite for biomedical applications," Journal of Bioactive and Compatible Polymers, vol. 26, no. 1, pp. 35–47, 2011. · ·

54. X. Wang, X. Wang, X. Wang et al., "Novel electrochemical biosensor based on functional composite nanofibers for sensitive detection of p53 tumor suppressor gene," Analytica Chimica Acta, vol. 765, pp. 63–69, 2013. · ·

55. L. Albañil-Sanchez, A. Romo-Uribe, A. Flores, and R. Cruz-Silva, "Electrospun nylon-graphene nanocomposites synthesis and microstructure," MRS Proceedings, vol. 1453, 8 pages, 2012. ·

56. C. Leyva-Porras, C. Ornelas-Gutiérrez, M. Miki-Yoshida, Y. I. Avila-Vega, J. Macossay, and J. Bonilla-Cruz, "EELS analysis of Nylon 6 nanofibers reinforced with nitroxide-functionalized graphene oxide,"Carbon, vol. 70, pp. 164–172, 2014. · ·

57. C.-C. M. Ma, Y.-L. Huang, S.-Y. Yang, and H.-W. Tien, "Transparent electrode with flexibility and method for manufacturing the same," US Patent 0295384, 2013.

58. B. Li, H. Yuan, and Y. Zhang, "Transparent PMMA-based nanocomposite using electrospun graphene-incorporated PA-6 nanofibers as the reinforcement," Composites Science and Technology, vol. 89, pp. 134–141, 2013. · ·

59. H. R. Pant, B. Pant, P. Pokharel et al., "Photocatalytic TiO_2-RGO/nylon-6 spider-wave-like nano-nets via electrospinning and hydrothermal treatment," Journal of Membrane Science, vol. 429, pp. 225–234, 2013. · ·

60. H. R. Pant, B. Pant, C. H. Park et al., "RGO/Nylon-6 composite mat with unique structural features and electrical properties obtained from electrospinning and hydrothermal process," Fibers and Polymers, vol. 14, no. 6, pp. 970–975, 2013. · ·

61. Y. J. Yun and D. Kibong Song, "Graphene electronic device and method of fabricating the same," US Patent 0313523, 2013.

62. Y. J. Yun, W. G. Hong, N.-J. Choi et al., "A 3D scaffold for ultra-sensitive reduced graphene oxide gas sensors," Nanoscale, vol. 6, no. 12, pp. 6511–6514, 2014. · ·

63. Y. J. Yun, W. G. Hong, W.-J. Kim, Y. Jun, and B. H. Kim, "A novel method for applying reduced graphene oxide directly to electronic textiles from yarns to fabrics," Advanced Materials, vol. 25, no. 40, pp. 5701–5705, 2013. · ·

64. J. Shen, M. Shi, B. Yan et al., "Covalent attaching protein to graphene oxide via diimide-activated amidation," Colloids and Surfaces B: Biointerfaces, vol. 81, no. 2, pp. 434–438, 2010. ··

65. T.-W. Chou, L. Gao, E. T. Thostenson, Z. Zhang, and J.-H. Byun, "An assessment of the science and technology of carbon nanotube-based fibers and composites," Composites Science and Technology, vol. 70, no. 1, pp. 1–19, 2010. · ·

66. K. Yang, M. Gu, Y. Guo, X. Pan, and G. Mu, "Effects of carbon nanotube functionalization on the mechanical and thermal properties of epoxy composites," Carbon, vol. 47, no. 7, pp. 1723–1737, 2009. · ·

67. G. Mago, C. Velasco-Santos, A. L. Martinez-Hernandez, D. M. Kalyon, and F. T. Tisher, "Effect of functionalization on the crystallization behavior of MWNT-PBT nanocomposites," Materials Research Society Symposium Proceedings, vol. 1056, pp. 295–300, 2008.

68. Q. Bao, H. Zhang, J.-X. Yang et al., "Graphene-polymer nanofiber membrane for ultrafast photonics,"Advanced Functional Materials, vol. 20, no. 5, pp. 782–791, 2010. · ·

69. M. J. Green, N. Behabtu, M. Pasquali, and W. W. Adams, "Nanotubes as polymers," Polymer, vol. 50, no. 21, pp. 4979–4997, 2009. · ·

70. M. B. Bazbouz and G. K. Stylios, "Novel mechanism for spinning continuous twisted composite nanofiber yarns," European Polymer Journal, vol. 44, no. 1, pp. 1–12, 2008. · ·

71. M. B. Bazbouz and G. K. Stylios, "Alignment and optimization of nylon 6 nanofibers by

electrospinning,"Journal of Applied Polymer Science, vol. 107, no. 5, pp. 3023–3032, 2007. · ·

72. M. B. Bazbouz and G. K. Stylios, "The tensile properties of electrospun nylon 6 single nanofibers,"Journal of Polymer Science, Part B: Polymer Physics, vol. 48, no. 15, pp. 1719–1731, 2010. · ·

73. S. Y. Jan, H. S. Park, S. M. Jo et al., "Graphene composite nanofibre and preparation method thereof," 2013 US Patent 8,519,045, December 2010.

74. C. N. Monty, E. K. Wujcik, and N. J. Blasdel, "Flexible electrode for detecting changes in temperature, humidity and sodium ion concentration in sweat," US Patent 2013/0197319, August 2013.

75. N. J. Blasdel, E. K. Wujcik, J. E. Carletta, K.-S. Lee, and C. N. Monty, "Fabric nanocomposite resistance temperature detector," IEEE Sensors Journal, vol. 15, no. 1, pp. 300–306, 2015. · ·

76. R. Cruz-Silva, A. Morelos-Gomez, H.-I. Kim et al., "Super-stretchable graphene oxide macroscopic fibers with outstanding knotability fabricated by dry film scrolling," ACS Nano, vol. 8, no. 6, pp. 5959–5967, 2014. · ·

77. Y.-S. Wang, S.-M. Li, S.-T. Hsiao et al., "Integration of tailored reduced graphene oxide nanosheets and electrospun polyamide-66 nanofabrics for a flexible supercapacitor with high-volume- and high-area-specific capacitance," Carbon, vol. 73, pp. 87–98, 2014. · ·

78. L. Pan, X. Pei, R. He, Q. Wan, and J. Wang, "Multiwall carbon nanotubes/polycaprolactone composites for bone tissue engineering application," Colloids and Surfaces B: Biointerfaces, vol. 93, pp. 226–234, 2012. · ·

79. A. Kausar, "Mechanical, thermal, and electrical properties of epoxy matrix composites reinforced with polyamide-grafted-MWCNT/poly(azo-pyridine-benzophenone-imide)/ polyaniline nanofibers,"International Journal of Polymeric Materials and Polymeric Biomaterials, vol. 63, no. 16, pp. 831–839, 2014. · ·

80. N. Vitchuli, Q. Shi, J. Nowak et al., "Multifunctional ZnO/Nylon 6 nanofiber mats by an electrospinning–electrospraying hybrid process for use in protective applications," Science and Technology of Advanced Materials, vol. 12, no. 5, Article ID 055004, 2011. · ·

81. R. Zhang and H. Olin, "Carbon nanomaterials as drug carriers: real time drug release investigation,"Materials Science and Engineering C, vol. 32, no. 5, pp. 1247–1252, 2012. · ·

82. S. Zhang, W. S. Shim, and J. Kim, "Design of ultra-fine nonwovens via electrospinning of Nylon 6: spinning parameters and filtration efficiency," Materials and Design, vol. 30, no. 9, pp. 3659–3666, 2009. · ·

83. L. M. Guerrini, M. C. Branciforti, T. Canova, and R. E. S. Bretas, "Electrospinning and characterization of polyamide 66 nanofibers with different molecular weights," Materials Research, vol. 12, no. 2, pp. 181–190, 2009. · ·

84. X. Wang, B. Ding, G. Sun, M. Wang, and J. Yu, "Electro-spinning/netting: a strategy for the fabrication of three-dimensional polymer nano-fiber/nets," Progress in Materials Science, vol. 58, no. 8, pp. 1173–1243, 2013. · ·

85. C. Zhang, Y. Li, W. Wang et al., "A novel two-nozzle electrospinning process for preparing microfiber reinforced pH-sensitive nano-membrane with enhanced mechanical property," European Polymer Journal, vol. 47, no. 12, pp. 2228–2233, 2011. · ·

86. H. R. Pant, M. P. Bajgai, C. Yi et al., "Effect of successive electrospinning and the strength of hydrogen bond on the morphology of electrospun nylon-6 nanofibers," Colloids and Surfaces A: Physicochemical and Engineering Aspects, vol. 370, no. 1–3, pp. 87–94, 2010. · ·

87. R. Nirmala, K. S. Jeon, R. Navamathavan, M. Park, H. Y. Kim, and S.-J. Park, "Enhanced electrical properties of electrospun nylon66 nanofibers containing carbon nanotube fillers and Ag nanoparticles,"Fibers and Polymers, vol. 15, no. 5, pp. 918–923, 2014. · ·

88. S.-Y. Park, Y.-H. Cho, and R. A. Vaia, "Three-dimensional structure of the zone-drawn film of the nylon-6/ layered silicate nanocomposites," Macromolecules, vol. 38, no. 5, pp. 1729–1735, 2005. · ·

89. N. A. Jones, E. D. T. Atkins, and M. J. Hill, "Investigation of solution-grown, chain-folded lamellar crystals of the even-even nylons: 6 6, 8 6, 8 8, 10 6, 10 8, 10 10, 12 6, 12 8, 12 10, and 12 12," Journal of Polymer Science, Part B: Polymer Physics, vol. 38, no. 9, pp. 1209–1221, 2000. · ·

90. J. S. Stephens, D. B. Chase, and J. F. Rabolt, "Effect of the electrospinning process on polymer crystallization chain conformation in nylon-6 and nylon-12," Macromolecules, vol. 37, no. 3, pp. 877–881, 2004. · ·

91. C. T. Lim, E. P. S. Tan, and S. Y. Ng, "Effects of crystalline morphology on the tensile properties of electrospun polymer nanofibers," Applied Physics Letters, vol. 92, no. 14, Article ID 141908, 2008. · ·

92. K. S. Jeon, R. Nirmala, R. Navamathavan, and H. Y. Kim, "Mechanical behavior of electrospun Nylon66 fibers reinforced with pristine and treated multi-walled carbon nanotube fillers," Ceramics International, vol. 39, no. 7, pp. 8199–8206, 2013. · ·

93. S. D. Uzun, F. Kayaci, T. Uyar, S. Timur, and L. Toppare, "Bioactive surface design based on functional composite electrospun nanofibers for biomolecule immobilization and biosensor applications," ACS Applied Materials & Interfaces, vol. 6, no. 7, pp. 5235–5243, 2014. ·

94. M. M. Bergshoef and G. J. Vancso, "Transparent nanocomposites with ultrathin, electrospun nylon-4,6 fiber reinforcement," Advanced Materials, vol. 11, no. 16, pp. 1362–1365, 1999. · ·

CHAPTER 7

RECENT DEVELOPMENT OF GRAPHENE-BASED CATHODE MATERIALS FOR DYE-SENSITIZED SOLAR CELLS

Man-Ning Lu[1,2], Chin-Yu Chang[1], Tzu-Chien Wei[2], and Jeng-Yu Lin[1]

[1]Department of Chemical Engineering, Tatung University, No. 40, Section 3, Chung Shan North Road, Taipei City 104, Taiwan

[2]Department of Chemical Engineering, National Tsing-Hua University, Hsinchu 300, Taiwan

ABSTRACT

Dye-sensitized solar cells (DSSCs) have attracted extensive attention for serving as potential low-cost alternatives to silicon-based solar cells. As a vital role of a typical DSSC, the counter electrode (CE) is generally employed to collect electrons via the external circuit and speed up the reduction reaction of to in the redox electrolyte. The noble Pt is usually deposited on a conductive glass substrate as CE material due to its excellent electrical conductivity, electrocatalytic activity, and electrochemical stability. To achieve cost-efficient DSSCs, reasonable efforts have been made to explore Pt-free alternatives. Recently, the graphene-based CEs have been intensively investigated to replace the high-cost noble Pt CE. In this paper, we provided an overview of studies on the electrochemical and photovoltaic characteristics of graphene-based CEs, including graphene, graphene/Pt, graphene/carbon materials, graphene/conducting polymers, and graphene/inorganic compounds. We also summarize the design and advantages of each graphene-based material and provide the possible directions for designing new graphene-based catalysts in future research for high-performance and low-cost DSSCs.

INTRODUCTION

Nowadays, photovoltaic (PV) technology is considered as one of the widespread and efficient approaches to produce electricity from solar energy due to the increasing attention to low-carbon economy and renewable energy commercialization. On the basis of the basal technology, PV devices are generally classified as first-, second-, and third-generation solar cells. Since dye-sensitized solar cell (DSSC) was first reported in 1991 [1], it has been regarded as one of the promising third-regeneration solar cells due to low manufacturing cost, facile fabrication processes, relatively high power conversion efficiency (PCE), and wide spectral response in visible light region for indoor applications [2].

As illustrated in Figure 1, a typical DSSC basically contains a nanocrystalline semiconductor oxide, a dye sensitizer, an electrolyte redox couple, and a catalyst material as cathode, also called counter electrode (CE). To achieve high cell efficiency, the individual components in DSSCs are necessary to be optimized [3]. As a crucial component in DSSCs, the main functions of a CE is to transport electrons from external circuit and catalyze the reduction reaction of I_3^- to I^- for dye regeneration. The overall I_3^- reduction reaction occurring on the CE surface can be described in (1). Equation (1) can be divided into the following steps (as depicted in (2)–(5)) involved in the reduction mechanism of I_3^- [4]:

Citation: Man-Ning Lu, Chin-Yu Chang, Tzu-Chien Wei, and Jeng-Yu Lin, "Recent Development of Graphene-Based Cathode Materials for Dye-Sensitized Solar Cells," Journal of Nanomaterials, vol. 2016, Article ID 4742724, 21 pages, 2016. doi:10.1155/2016/4742724.

$$I_3^- + 2e^- \longleftrightarrow 3I^- \quad (1)$$

$$I_3^- \longleftrightarrow I_2 + I^- \quad (2)$$

$$I_2 + 2CE \longleftrightarrow I(CE) + I(CE) \quad (3)$$

$$I(CE) + e^- \longleftrightarrow I^-(CE) \quad (4)$$

$$I^-(CE) \longleftrightarrow I^- + CE \quad (5)$$

Equation (2) represents the solution phase reaction, which has been verified to be relatively fast and considered to be in equilibrium [4]. Then I_2 dissociates into two surface I atoms upon adsorption on the CE surface (designated as I(CE)) via (3), and subsequently, I atoms on the CE surface are reduced to I^- ions adsorbed on the CE surface through one-electron transfer as depicted in (4). After that, the adsorbed I^- ions are desorbed from the CE surface to be the solvated I^- ions in electrolyte as illustrated in (5).

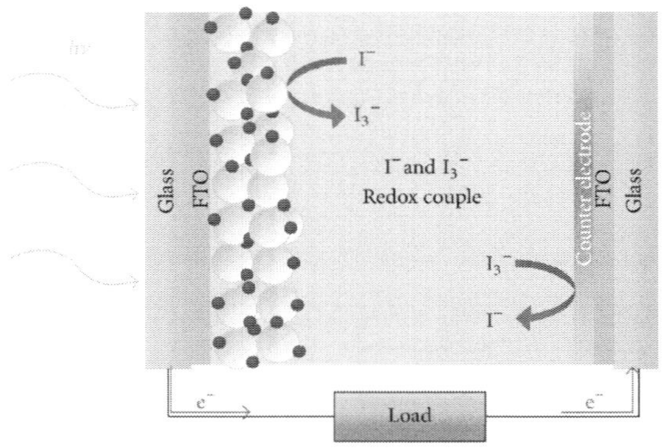

Figure 1: Schematic illustration of the operation principle of a DSSC.

On the basis of the functions of a CE and the reduction mechanism of I_3^-, an efficient CE should exhibit both high electrical conductivity and electrocatalytic activity to keep a low overpotential and to minimize energy loss in DSSCs [5]. Generally, Pt has been proven to be an excellent CE material in DSSCs due to its high electrical conductivity and excellent electrocatalytic activity as well as great electrochemical stability. As a noble and sparse metal, however, the high cost of Pt (ca. $50,000/Kg) restricts its practical applications. Therefore, much effort has been devoted to exploring the efficient Pt-free CEs. To date, a great deal of materials including carbon materials [6], conducting polymers [7–11], and inorganic materials [12–18] have been introduced as alternative CE materials to Pt.

Compared to other conventional carbon materials, graphene, a single-layer structure of two-dimensional graphite, possesses unique features of the strong mechanical strength, high electrical and thermal conductivity, large surface area, and high optical transmittance [19]. Therefore, it has been widely incorporated in the fields of microelectronic and optoelectronic devices, energy storage, and conversion materials [20, 21]. Since Aksay's group revealed the functional graphene sheets (FGSs) as CE material for efficiently catalyzing the redox reaction of I_3^-/I^- in 2010, a variety of graphene-based CE materials have been intensively developed for Pt-free DSSCs [22].

In this work, we comprehensively reviewed the advance of research on graphene-based CE materials for Pt-free DSSCs, with emphasis on composite materials.

GRAPHENE CES

Graphene has been considered as one of the promising Pt-free CE alternatives in DSSCs due to its specific properties of high electrical conductivity, excellent electrocatalytic activity, great anticorrosion resistance, and larger surface area [24, 28]. In general, graphene materials can be synthesized via mechanical exfoliation, epitaxial growth, chemical vapor deposition, thermal exfoliation [29–32], and so forth. Table 1 summarizes the comparison of various graphene CEs in DSSCs. For example, Aksay's group used thermal exfoliation to synthesize FGSs with defects and oxygen-containing function groups (hydroxyl, carbonyl, and epoxide) [24]. They found that the increase in C/O ratio of FGSs is positively associated with their electrocatalytic activity. Nevertheless, the electrical conductivity of FGSs can be reduced by forming more functional groups while the temperature of thermal processing increases over 1500°C. After the optimization of C/O ratio in FGSs, the FGSs-based DSSC reached higher PCE of 5.48% and fill factor (FF) of 0.60 than Pt-based DSSC (4.99% and 0.57). They found that the improved FF value of the FGSs-based DSSC can be attributed to its relatively low charge-transfer resistance (R_{ct}) of $0.64\,\Omega\,cm^2$, which is even lower than that of the Pt CE ($0.79\,\Omega\,cm^2$). Grätzel's group prepared an optically transparent graphene thin film by dropping commercial graphene nanoplatelets on fluorine doped tin oxide (FTO) glass substrates [23]. They found that the graphene CE has a much lower R_{ct} value in an ionic liquid than in the traditional organic solvent by a factor of 5-6, suggesting that the regeneration of I_3^-/I^- in an ionic liquid is superior to that in the traditional organic solvent (as depicted in Figure 2). Choi et al. [33] further prepared graphene CEs using electrophoretic deposition followed by annealing treatment at 200~600°C. They found that the DSSC using the graphene CEs obtained after the annealing treatment at 600°C exhibited the optimized PCE of 5.69%. Kaniyoor and Ramaprabhu [34] synthesized graphene with defect-rich and wrinkled structure through a thermally exfoliated method. Then, the thermally exfoliated graphene (TEG) was suspended in Nafion/ethanol solution and subsequently dropped on FTO glass substrates to form the TEG CEs. The TEG CE demonstrated R_{ct} value of $11.7\,\Omega\,cm^2$, which is close to that of the Pt CE ($6.5\,\Omega\,cm^2$). This indicates that the wrinkled and defect-rich structure of TEG can provide large surface area and high density of defect sides for I_3^-/I^- reduction reaction and thus the electron transfer kinetics at the CE/electrolyte interface is promoted. Zhang et al. synthesized novel 3D structure of graphene nanosheets (GNs) as CE material for DSSCs [28]. The GN was prepared using oxidative exfoliation of graphite followed by hydrazine reduction and annealing process. After the optimization of annealing temperature, it can be found that the DSSC assembled with the GN CE annealed at 400°C exhibited the best cell performance of 6.81%, which is comparable to that of the Pt-based DSSC (7.59%). After being annealed at 400°C, the GN was revealed with a unique 3D structure. This signifies that the annealed GN can afford sufficient surface area for I_3^-/I^- reduction reaction and therefore its R_{ct} value is significantly decreased from 337.2 to $1.2\,\Omega$.

Table 1: Photovoltaic performance of the DSSCs assembled with various graphene CEs

CE	Substrate	Preparation method	Redox couple	Dye	R_{ct} (Ω cm^2)	FF	PCE (%)	Ref
GO-GNP	FTO glass	Drop-casting	I^-/I_3^-	N3	0.64	0.60	4.99	[20]
GNP	FTO glass	Drop-casting	I^-/I_3^-	N719	0.40	0.74	6.89	[29]
RGO	FTO glass	Electrophoretic deposition	I^-/I_3^-	N719	38.00*	0.65	5.69	[46]
TEG	FTO glass	Drop-casting	I^-/I_3^-	N719	11.70	0.54	2.82	[30]
Hemin-RGO	FTO glass	Drop-casting	I^-/I_3^-	N719	6.12	0.31	2.45	[31]
RGO	FTO glass	Screen-printing	I^-/I_3^-	N719	1.20	0.54	6.81	[19]
RGO	ITO glass	Spin-coating	I^-/I_3^-	N719	21.50	0.56	2.13	[47]
3D-NFG	FTO glass	CVD	I^-/I_3^-	N719	45.88	0.60	5.20	[32]
VG	FTO glass	CVD	I^-/I_3^-	N719	0.0073	0.67	5.36	[23]
Honeycomb-like structure graphene	FTO glass	CVD	I^-/I_3^-	N719	20.00*	0.37	7.80	[34]
N-doped graphene	FTO glass	CVD	I^-/I_3^-	N719	—	0.33	3.12	[41]
GNP	FTO glass	Drop-casting	Co^{3+}/Co^{2+}	Y123	3.50*	0.70	9.30	[45]
GNP	FTO glass	Drop-casting	Co^{3+}/Co^{2+}	Y123	3.30	0.72	9.40	[48]
FGS	FTO glass	Screen-printing	I^-/I_3^-	N719	—	0.67	6.79	[20]
FGS	FTO glass	Screen-printing	Co^{3+}/Co^{2+}	D35	—	0.65	4.51	[20]
FGS	FTO glass	Screen-printing	S^{3+}/S^{2+}	D35	—	0.55	3.45	[20]
RGO	FTO glass	Drop-casting	Co^{3+}/Co^{2+}	T123	1.27	0.67	9.30	[49]
N-doped graphene	FTO glass	Drop-casting	Co^{3+}/Co^{2+}	YD-2-o-C8	1.38	0.72	8.3	[50]

*Represents that the unit of the R_{ct} value is Ω.

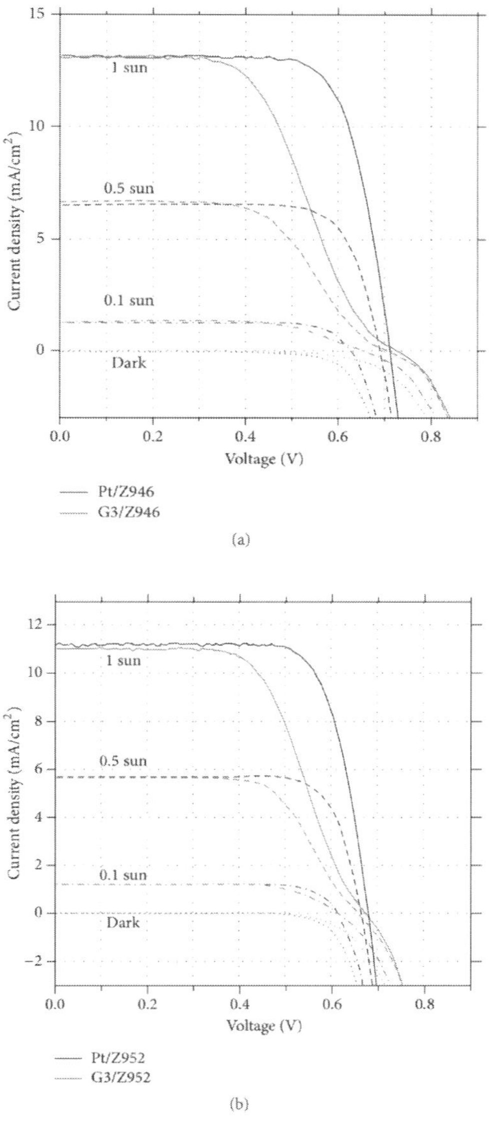

Figure 2: Current-voltage characteristics of a dye-sensitized solar cell with the platinized FTO (black lines) or G3 cathode (red lines) under various light intensities. (a) Z946 electrolyte. (b) Z952 electrolyte [23].

On the other hand, Xu et al. [41] reported a facile and rapid microwave-assisted method to synthesize hemin functionalized reduced graphene oxide (hemin-RGO). It was found that 5-layer hemin-RGO demonstrated R_{ct} value of ca. 7 Ω, comparable to that of the Pt-based CE (7 Ω). Lee et al. fabricated 3D nanofoam of few-layer graphene (3D-NFG) with large-area coverage via the chemical vapor deposition (CVD) technique [42]. The PCE of the DSSC based on the 3D-NFG CE reached 5.2%, which was even close to that of Pt CE (5.7%). Yu et al. prepared the vertically oriented graphene (VG) nanosheets by plasma-enhanced chemical vapor deposition (PECVD), and the as-synthesized VG was found to be with large surface area and abundant oxygen functional groups [43]. According to series of electrochemical characterizations including electrochemical impedance spectroscopy (EIS) and cyclic voltammetry (CV), the electrocatalytic activity of VG was systemically elucidated. The VG CE not only exhibited lower R_{ct} value ($7.3 \times 10^{-3}\,\Omega\,m^2$) than Pt-based CE but also presented higher peak current and smaller peak-to-peak potential separation (ΔE_p) compared to the Pt CE. This signifies the excellent electrocatalytic activity of VG CEs for reduction of $I_3^-$to I^-, therefore resulting in that the DSSC assembled with the VG CE exhibited a relatively higher PCE of 5.36% than that based on the Pt CE (4.36%). Impressively, Wang et al. [44] synthesized the novel graphene sheet with 3D honeycomb-like structure based on a simple reaction between Li_2O and CO. The PCE of the DSSC based on the 3D graphene with honeycomb-like structure reached 7.8%, presenting its great potential as an efficient Pt-free CE in DSSC. Yang et al. reported an N-doped few-layer graphene as CE material in DSSCs. According to the inert nature of the pristine graphene, it is usually with limited defects or edge planes [45]. Compared to the pristine graphene CE, N-doped graphene CE can provide more electrocatalytic sites for charge transfer between CE/electrolyte interfaces and therefore improve the cell conversion performance. Furthermore, they found that N-doped graphene in DSSC demonstrated great long-term stability due to the great adhesion between graphene materials and FTO substrate.

The commonly ruthenium dyes anchored onto the photoelectrode with I_3^-/I^- system are considered to be with slow recombination kinetics and thus this promotes rapid dye regeneration. Nevertheless, the potential of I_3^-/I^- redox couple is 0.35 V (versus the normal hydrogen electrode, NHE) [49]. As for the oxidation potential of the ruthenium dye, it is only ca. 1.1 V versus NHE [46]. In a DSSC, the low operating potential window limits its open-circuit voltage (V_{oc}) (ca. 0.7~0.8 V during the dye generation). To enhance the *V*oc value of DSSCs, redox mediators with more positive potential have been developed, such as $Co(bpy)_3$ $^{3+}$/$Co(bpy)_3$ $^{2+}$ (ca. 0.56 V versus SHE) [47]. Compared to a DSSC with I3 –/I– redox couple, using $Co(bpy)_3$ $^{3+}$/$Co(bpy)_3$ $^{2+}$ redox couple possesses larger *V*oc value up to 0.9~1 V. Kavan's group developed a functionalized graphene CE in $Co(bpy)_3$ $^{3+}$/$Co(bpy)_3$ $^{2+}$ redox mediators. They reported a high-performance DSSC based on a new system including graphene nanoplatelets (GNP) CE and photoanode conjunction with Y123 dye. It should be noted that the GNP-based DSSC in the $Co(bpy)_3$ $^{3+}$/$Co(bpy)_3$ $^{2+}$ redox mediators revealed relatively superior PCE (ca. 9%) to that of the Pt-based DSSC (ca. 8%), signifying that the complex is a promising redox mediator for a novel type of iodine-free DSSCs with *V*oc exceeding 1 V [48]. It was also found that the GNP CE exhibited lower *R*ct value and larger exchange currents in $Co(bpy)_3$ $^{3+}$/$Co(bpy)_3$ $^{2+}$ electrolyte system. This suggests that the GNP CE is one of potential cathode materials in the DSSCs with $Co(bpy)_3$ $^{3+}$/$Co(bpy)_3$ $^{2+}$ electrolyte system. Kavan's group further reported on the improved fabrication protocol toward optically transparent graphene-based layers, yielding wear-resistant films with good adhesion to FTO [50]. They prepared the single-layer graphene oxide (GO) and GO-GNP composite; as a result, the GO-based electrodes showed better mechanical and electrochemical stability. After the aging tests, GO-based electrodes have less losing in electrocatalytic activity. Moreover, the best electrochemical performance and electrocatalytic ability of the CE prepared with 50% of GO and 50% of NGP in $Co(bpy)_3$ $^{3+}$/$Co(bpy)_3$ $^{2+}$ electrolyte system. Roy-Mayhew et al. [24] fabricated the versatile functionalized graphene sheet (FGS) and employed it as CE in the DSSCs with iodine-, cobalt-, and sulfur-based redox mediators. As depicted in Figure 3, the FGS-based DSSCs displayed superior photovoltaic performance to that of the Pt-based DSSCs regardless of the electrolyte systems.

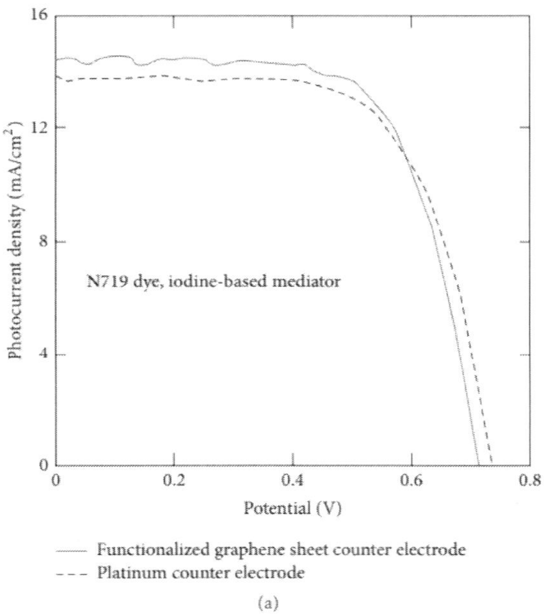

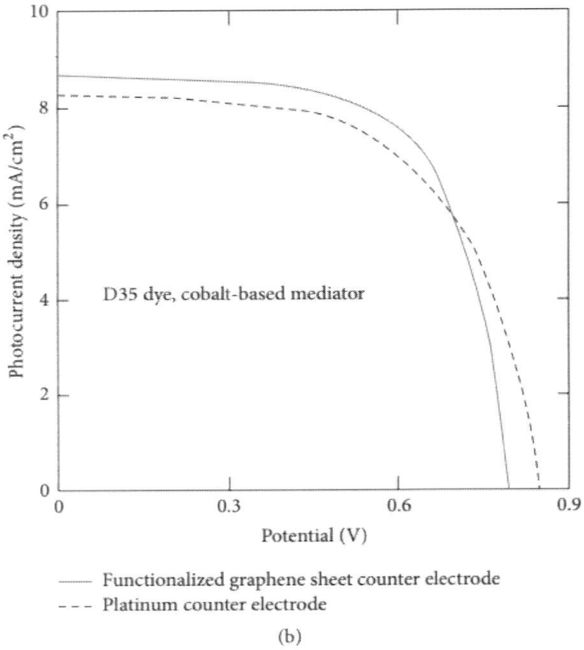

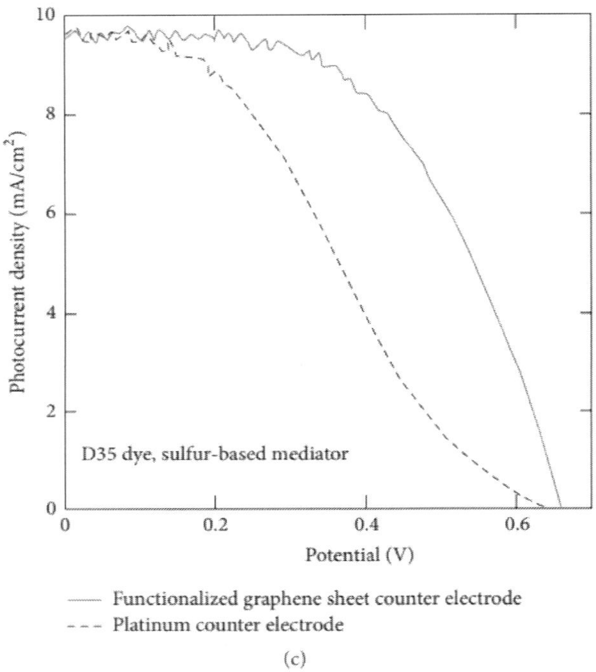

Figure 3: J-V curve characteristics of DSSCs using thermally decomposed chloroplatinic acid (Pt) and FGS counter electrodes. (a) I-based mediator, N719 sensitizer; (b) cobased mediator, D35 sensitizer; (c) S-based mediator, D35 sensitizer. Active area is 0.25 cm² [24].

PT/GRAPHENE COMPOSITES

To enhance the electrocatalytic activity of graphene CEs, a variety of Pt/graphene composite CEs are designed and synthesized, which are summarized in Table 2 for comparison. For example, Bajpai et al.'s [54] synthesized Pt nanoparticles (NP) were deposited directly on to graphene using pulsed laser deposition (PLD) method. Pt-NPs were deposited uniformly over micrometer-sized graphene sheets. They found that the graphene/Pt composite with 27% Pt loading showed higher

PCE, short-circuit current density (J_{sc}) and (V_{oc}) and no loss of the FF as compared to the cells fabricated with standard expansive Pt CE. Wan et al. [55] synthesized graphene/Pt nanocomposites with low Pt loading via a two-step reduction process. The resultant Pt/graphene composite films were then coated on FTO substrates using a simple drop-casting method at room temperature and subsequently used as CEs in DSSCs. The Pt nanoparticles are ca. 4–20 nm in size and uniformly distributed on the surface of the graphene layers. The energy conversion efficiency of the Pt/graphene-based DSSC was found to be 1.9%, close to that of cells with a Pt-based CE. Yen et al. [56] further developed a water-ethylene method to prepare a composite material consisting of Pt nanoparticles and graphene (PtNP/GR). The PCE value (6.35%) of the DSSCs using PtNP/GR CE was higher than Pt CE-based DSSCs (5.27%), which was attributed to the increase in J_{sc} value by 13%. Kim's group [57] prepared aqueous dispersible nanohybrids (NHBs) of graphene nanosheets (GNSs) and Pt nanoparticles (Pt-NPs) via the one-pot reduction of their precursors by using an environmentally friendly chemical, vitamin C. Moreover, the GNS/Pt-NHBs CE was simply fabricated by a facile electrospray approach from the as-prepared stable aqueous colloidal dispersion of GNS/Pt-NHBs. The main advantage of using electrospray to prepare the GNS/Pt-NHBs CE is that the GNS/Pt-NHBs hybrid materials can be directly deposited on the surface of FTO substrates without using any organic binders.

The Pt-NPs were observed to be robustly attached on the surface of the GNSs. The PCE of the DSSC using the GNSs/NHBs CE (7.97%) was approximately two times higher than that of the DSSC with the GNSs (4.44%). When the GNS-NHB CEs were with annealing treatment, the PCE of the DSSC assembled with the annealed GNS-NHBs CE (8.91%) was comparable to that of the DSSC based on Pt CE (8.85%). Dao et al. [58] developed a simple and continuous dry-plasma reduction method to evenly hybrid Pt nanoparticles (Pt-NPs) on reduced graphene oxide (RGO) layer under atmospheric pressure and without using any toxic chemicals. Pt-NPs with a size range of 0.5–4 nm (mostly 2 nm) were found to be dispersed on the surface of RGO. The Pt-NPs/RGO CE displays great electrocatalytic activity as well as excellent long-term stability. The DSSC based on such robust and low-cost CE achieved an impressive PCE of 8.56%. Furthermore, Gong et al. [103] developed a facile electrostatic layer-by-layer self-assembly (ELSA) method to construct ultrathin films composed of graphene nanosheets (GNS) and Pt nanoparticles on the conductive glass as a transparent and high-performance CE. After a series of ELSA treatments, the as-prepared self-assembled film is then sintered and converted to graphene/Pt film on FTO glass substrate. The DSSC with the monolayer GNS/Pt CE achieved a PCE of 7.66%, which was comparable to that using the expensive sputtered Pt CE (8.16%) Yue et al. [104] employed a facile one-step electrochemical deposition method to prepare Pt nanoparticles/graphene nanosheets (PtNP/GN) films, in which the deposition bath was composed of H_2PtCl_6 and GN. After optimizing the amount of GN (ranging from 0–0.25 wt%) in the deposition bath, the PtNP/GN-based DSSC achieved a high PCE of 7.88%, which is increased by 21% compared with a device based on traditional Pt CE. Yeh et al. [25] synthesized graphene/Pt nanoparticles (GN/PtNPs) catalysts with various PtNP loadings (10–60 wt%) using a polyol reduction method. As depicted in Figure 4, the DSSC with the GN/PtNPs-20% CE shows the higher PCE of 8.79% as compared to cells with pristine GN (7.65%) and s-Pt CEs (8.58%). Thus, the efficient and economical GN/PtNPs-20% nanocomposite is a potential candidate for replacing the expensive Pt CE in DSSCs. To reduce the cost and time consumption for production, Saranya et al. [105] tried to employ a microwave-assisted exfoliation method followed by a chemical reduction by chloroplatinic acid for synthesizing graphene nanosheets (GNs)/Pt composites, in which only 160 s reaction time was required to the intercalation and exfoliation of the graphite to form GNs. The device assembled with the as-prepared Pt-decorated GNs achieved a PCE of 5.11%, which is increased by 11% compared to that reported for other similar systems.

Table 2: Photovoltaic performance of the DSSC using various Pt/graphene composite CEs

CE	Substrate	Preparation method	Redox couple	Dye	R_{ct} ($\Omega\,cm^2$)	FF	PCE (%)	Ref
Pt NPs/graphene sheet	FTO glass	Electrodeposition	I^-/I_3^-	N3	2.36	0.59	2.91	[33]
Pt/RGO	FTO glass	Two-step reduction	I^-/I_3^-	N719	34.95	0.60	4.01	[51]
PtNP/GR	FTO glass	Water-ethylene method	I^-/I_3^-	N719	10.45*	0.67	6.35	[52]
GNS/Pt-NHB	FTO glass	Electrospray	I^-/I_3^-	N719	27.00	0.66	7.97	[53]
Pt-NPs/RGO	FTO glass	Dry-plasma reduction	I^-/I_3^-	N719	0.62	0.67	8.56	[54]
GNS/Pt	FTO glass	Electrostatic Layer-by-layer self-assembly	I^-/I_3^-	N719	30.80	0.73	6.09	[55]
PtNP/GN	FTO glass	Electrodeposition	I^-/I_3^-	N719	2.08	0.68	7.88	[56]
GN/PtNPs	FTO glass	Polyol reduction	I^-/I_3^-	N719	2.55	0.70	8.79	[57]
GNs/Pt	ITO glass	Microwave-assisted synthesis/chemical reduction	I^-/I_3^-	N719	2.65*	0.68	5.10	[58]

*Represents that the unit of the R_{ct} value is Ω.

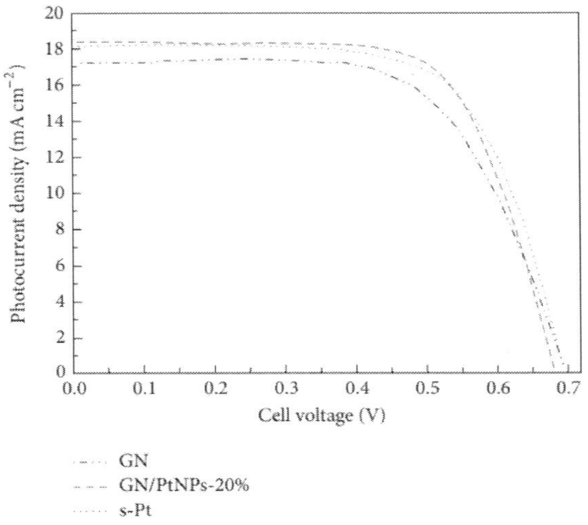

CE	η (%)	V_{oc} (V)	J_{sc} (mA cm^{-2})	FF	R_{ct1} (Ω)
GN	7.65 ± 0.12	0.70 ± 0.02	17.08 ± 0.41	0.64 ± 0.02	9.12
GN/PtNPs-20%	8.79 ± 0.21	0.69 ± 0.01	18.19 ± 0.58	0.70 ± 0.01	5.85
s-Pt	8.58 ± 0.15	0.69 ± 0.01	17.77 ± 0.37	0.70 ± 0.01	6.45

Figure 4: Photocurrent density-voltage curves of DSSCs with CEs containing pristine GN, GN/PtNPs-20%, and s-Pt, obtained at 100 mW cm^{-2} (AM, 1.5 G) [25].

GRAPHENE/CARBON MATERIAL COMPOSITES

To achieve high-performance of graphene-based CEs, lots of research has been reported on hybrid graphene with other kinds of carbon materials including carbon nanotubes (CNTs), carbon black, and mesoporous carbon. For instance, the combination of 1-D CNT with 2-D graphene has been proposed to promote the electron transfer and ionic diffusion and therefore facilitate the charge transfer between CE and electrolyte [106], as shown in Table 3. Choi et al. [59] fabricated a CE composed graphene and multiwalled carbon nanotube (MWCNT) by a CVD method. The incorporation of graphene in the MWCNT matrix can provide another area for I_3^- reduction and thereby the effectively enhanced electrocatalytic activity can be expected for the graphene/MWCNT CE. The cell with the graphene/MWCNT displayed an excellent FF value of 0.7 and exhibited a PCE of 4.46%. Li et al. [60] prepared vertically aligned carbon nanotubes (VACNTs) on a freestanding graphene paper (GP) by CVD. The direct deposition of VACNTs on highly conductive GP can facilitate the ionic diffusion within the composite electrode and electrons transfer at CE/electrolyte interface. As a result, the VACNT/GP CE displayed higher FF and PCE of 0.64 and 6.05%, respectively, compared with pure GP and VACNT/GP CE. Compared to conventional CVD method of growing graphene film to the desired substrates as CEs, electrophoretic deposition (EPD) is a relatively economical and versatile method to fabricate graphene-based CEs since the thickness of graphene film can be controlled by adjusting the deposition parameters such as deposition time and applied voltage. As depicted in Figure 5, Zhu et al. [26] used a facile EPD approach to prepare reduced graphene (RG)/CNT composite CEs. Among the DSSCs with CNT, RG/CNT and Pt CEs, RG/CNT-based CE exhibited the highest FF value and therefore achieved the impressive PCE. The improved FF value can be ascribed to the short pathway of electron transfer within 2-D graphene sheets and the constructed electrical network by connecting graphene sheets with CNTs. Kim et al. [62] also used EPD method to deposit the graphene, single-walled CNT (SWCNT) and graphene-SWCNT composite on FTO glass substrates. Among them, as-fabricated graphene-SWCNT CE not only exhibited excellent electrocatalytic activity but also displayed the optical transmittance of 67% at 550 nm. On the other hand, Ma et al. [70] fabricated different transparent 3-D CNT/graphene CEs by controlling the spraying time and supporting with a mirror to reharvest the reflected light. The 30% transmittance of composite showed the highest conversion efficiency (7.70%) with a mirror for J_{sc} increasing to 0.83 mA cm^{-2}.

Table 3: Photovoltaic performance of the DSSCs based on various graphene/carbon material CEs

CE	Substrate	Preparation method	Redox couple	Dye	R_{ct} (Ω cm^2)	FF	PCE (%)	Ref
GMWNT	FTO glass	CVD	I^-/I_3^-	N719	—	0.70	4.46	[59]
VACNT/GP	GP paper	CVD	I^-/I_3^-	N719	3.40	0.62	6.05	[60]
GPMWNT	FTO glass	CVD	I^-/I_3^-	N719	—	0.70	3.00	[61]
GG-CNT	FTO glass	Electrophoretic deposition	I^-/I_3^-	N719	49.00	0.61	6.17	[26]
Graphene-SWNT	FTO glass	Electrophoretic deposition	I^-/I_3^-	N719	16.20*	0.57	5.17	[62]
OMC-GNS	FTO glass	Doctor-blade	I^-/I_3^-	N719	67.76	0.61	6.82	[63]
G@OMC	FTO glass	Doctor-blade	I^-/I_3^-	N719	3.32	0.63	6.38	[64]
MWNT/GNS	FTO glass	Doctor-blade	I^-/I_3^-	N719	1.10*	0.58	4.00	[65]
Graphene-CB	FTO glass	Doctor-blade	I^-/I_3^-	N719	0.27*	0.57	5.99	[66]
MWNT/Gr-F	FTO glass	Dry spun	I^-/I_3^-	N719	1.70*	0.63	7.55	[67]
CF/GNP	FTO glass	Spin-coating	Co^{3+}/Co^{2+}	Y123	1.12	0.74	9.11	[68]
GMC	FTO glass	Screen-printing	T_2/T^-	N719	1.26	0.69	6.55	[27]
GP/MWNT	FTO glass	Spin-coating	I^-/I_3^-	N719	2.94	0.53	4.66	[69]
GR/MWCNT	FTO glass	Spray	I^-/I_3^-	N719	10.89*	0.49	7.70	[70]
NGC	FTO glass	Screen-printing	I^-/I_3^-	N719	1.78	0.49	6.19	[71]

*Represents that the unit of the R_{ct} value is Ω.

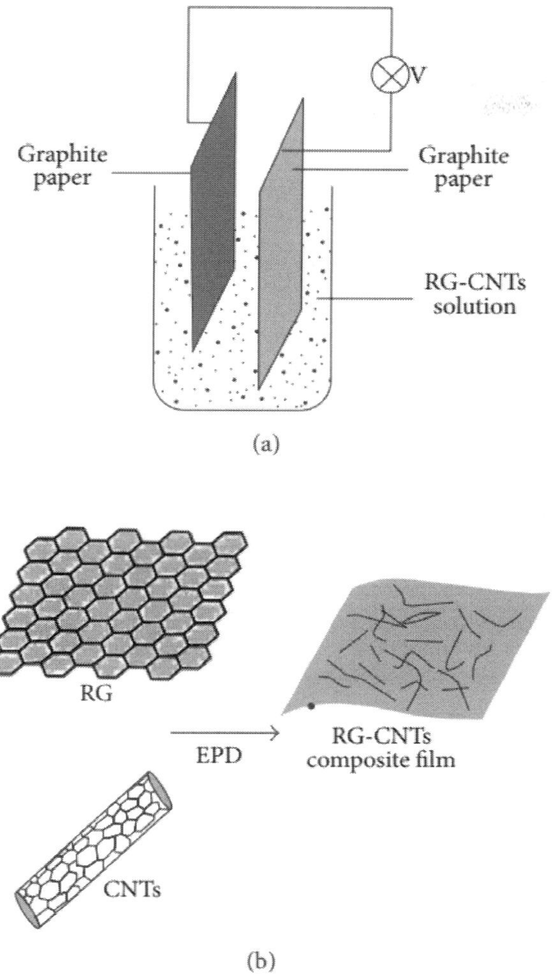

Figure 5: (a) Schematic diagram of EPD process. (b) Structure illustration of RG-CNTs composite films [26].

In addition, Miao et al. [66] revealed highly electrocatalytic composite CEs based on the combination of the rapid electron transport of graphene and high surface area of carbon black. The moderate bundles of graphene homogeneously distributed within carbon black were observed for the composite CE synthesized in the ratio of graphene and carbon black in 1 : 3. This would therefore provide some space for electrolyte diffusion. The improved electron transport and electrolyte of the optimized composite CE would promote the electrons transfer and provide more effective active sites for I_3^- reduction. The DSSC based on the graphene/carbon black showed a PCE of 5.99%. On the other hand, ordered mesoporous carbon (OMC) has attracted extensive attention because of its great chemical stability, fast infiltration of electrolyte, effective catalysis area, and large pore volume [107]. However, OMC with random boundaries usually has low electron mobility and thus its electrocatalytic activity is insufficient as CE material in DSSCs [64]. To address the weakness of OMC, Sun et al. [64] combined OMC with highly conductive reduced graphene oxide (RGO) to form a OMC/RGO composite CE. Therefore, the device with the OMC/RGO composite CE exhibited higher PCE of 6.38% than that based on the OMC CE (5.67%). More recently, Shao et al. [63] prepared OMC/graphene nanosheet (GNS) composite with different weight ratios of GNS ranging from 5 to 50 wt%. The optimized OMC/GNS composite CE with 25 wt% GNS displayed the most excellent electrocatalytic activity and yielded the highest PCE of 6.82%, which rivaled that of the cell with Pt CE (7.08%). The improved photovoltaic performance of the aforementioned OMC/graphene-based composite CEs is mainly ascribed to the increased pathway for electron transfer by the introduction of highly conductive graphene sheets.

In addition to I_3^- /I^- redox couple, several graphene/carbon materials were employed as CEs for iodine-free DSSCs. Recently, Liu's group [27] utilized graphene modified mesoscopic carbon (GMC) as CE in DSSCs with thiolate/disulfide (ET^-/BET) mediator as illustrated in Figure 6 and obtained significantly improved FF value, in which the FF value of mesoscopic carbon (NC) and GMC was 32.5% and 72.5%, respectively. Thus, remarkably enhanced PCE of 6.55% was achieved for the GMC-based device, which was much higher than that of the NC-based one. The EIS measurements confirmed that the R_{ct} value of GMC CE was only 1.26 $\Omega\,cm^2$, which was almost ten times lower than that of NC CE (12.87 $\Omega\,cm^2$). The low R_{ct} value of the GMC CE is accounted for its low FF value. As a result, the GMC-based DSSC with disulfide redox couple system exhibited a superior PCE, which was increased by 35% than that of the GC-based one. As for coupling with $Co(bpy)_3{}^{3+}/Co(bpy)_3{}^{2+}$ redox couple, Stefik et al. [68] developed a novel type of cathode composed of carbon fibers (CF) and graphene nanoplatelets (GNP). The CF/GNP composite CE was fabricated by embedding GNP in a conductive carbon matrix derived from the carbonization of poly(acrylonitrile) (PAN). This approach was found to effectively improve the adhesion of GNP with the conductive glass substrates. After the optimization, the CF/GNP CE containing 20% GNP not only showed the lowest R_{ct} value of 1.12 Ω cm^2 but also displayed its stable mechanical strength against mechanical aging tests. The DSSCs with the optimized CF/GNP CEs have the prominent PCE of 9.11%, which was higher than that of the cell based on Pt CE (8.61%).

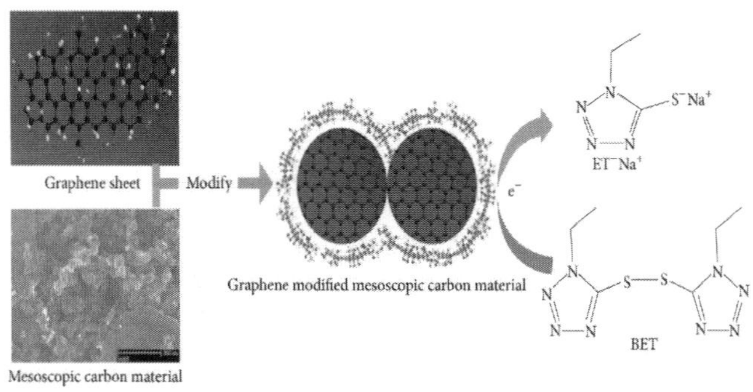

Figure 6: Fabrication of graphene modified mesoscopic carbon (GMC) material toward a new synthesized thiolate/disulfide redox couple (ET^-/BET) [27].

GRAPHENE/CONDUCTING POLYMER COMPOSITES

Organic conducting polymers, such as polyaniline (PANI), poly(3,4-ethylenedioxythiophene) (PEDOT), and polypyrrole (PPy), have also attracted lots of attention to be considered as potential CE materials due to their desirable characteristics of low-cost, environmental stability, high degree of processability, and interesting electrocatalytic properties. As depicted in

Table 4, to improve the electrocatalytic of conducting polymers for I_3^- reduction, highly conductive graphene are generally incorporated into the matrix of conducting polymers to increase their electrical conductivity for rapid electron transfer. For example, Hong et al. [72] fabricated transparent graphene/polystyreneslufonate doped poly(3,4-ethylenedioxythiophene) (PEDOT–PSS) composite films on conductive glass substrates by spin-coating the aqueous mixture of 1-pyrenebutyrate (PB^-) stabilized graphene and PEDOT-PSS. The graphene sheets were observed to be dispersed uniformly in PEDOT-PSS matrix, and the resultant composite film possessed the combined advantages of the excellent electrocatalytic PEDOT-PSS and the high conductive graphene. The PCE of the device was increased from 2.3% to 4.5% with increasing the content of graphene incorporated in the composite film from 0 wt% to 1 wt%. Moreover, Lee et al. [73] used graphene/PEDOT composite film to replace not only the Pt catalyst but also the transparent conducting oxide (TCO) layer in DSSCs. The cell assembled with such Pt- and TCO-free CE achieved a PCE of 6.26%, while the PCEs of DSSCs with Pt/ITO and PEDOT CEs were 6.68% and 5.62%, respectively. Yue et al. [74] prepared graphene/PEDOT-PSS composite film on FTO glass substrates using in situ electropolymerization approach, in which the different contents of graphene flakes were included in the deposition bath. The as-deposited graphene/PEDOT-PSS composite film possessed a lot of clusters for providing the active surface area and facilitating the penetration of the liquid electrolyte. The incorporation of 0.05 wt% graphene within the PEDOT-PSS matrix resulted in the most improvement of the electrocatalytic activity for I_3^- reduction. The DSSC based on such optimized graphene/PEDOT-PSS CE showed a high PC of 7.86%, comparable with the performance of the DSSC using the Pt CE (7.31%). In addition to PEDOT, PANI and PPy conducting polymers have been hybrid with graphene as efficient CEs in DSSCs.

Table 4: Photovoltaic performance of the DSSC using various graphene/conducting polymer CEs

CE	Substrate	Deposition	Redox couple	Dye	R_{ct} (Ω cm^2)	FF	PCE (%)	Ref
Graphene-PEDOT-PSS	ITO glass	Spin-coating	I^-/I_3^-	N719	—	0.48	4.50	[72]
Graphene/PEDOT	TCO glass	Polymerization	I^-/I_3^-	N719	—	—	6.26	[73]
GP/PEDOT-PSS	FTO glass	Electrochemical polymerization	I^-/I_3^-	N719	2.74	0.65	7.86	[74]
PANI/graphene	FTO glass	Electrodeposition	I^-/I_3^-	N719	11.49*	0.67	7.70	[75]
PANI-RGO	FTO glass	Layer-by-layer electrostatic adsorption mechanism	I^-/I_3^-	N719	0.71	0.64	7.84	[76]
Ppy/RGO	ITO glass	electrochemical Oxidative polymerization	I^-/I_3^-	N719	32.95*	0.60	6.45	[77]
PANI/graphene	FTO glass	Polymerization	I^-/I_3^-	N3	1.10	0.67	6.09	[35]

*Represents that the unit of the R_{ct} value is Ω.

He et al. [75] mixed PANI with graphene as PANI/graphene composites via a refluxing process. It was found that PANI was bonded onto graphene without any interfacial separation; the resulting covalent bonding could improve the electron transfer between PANI and graphene. The DSSC employing the PANI/8 wt% graphene composite CE provided an impressive PCE of 7.70% in comparison with 6.40% from the pristine PANI CE-based device. Wang et al. [76] firstly incorporated GO into PANI matrix via layer-by-layer electrostatic adsorption method. To further enhance the electrical conductivity and electrocatalytic activity of the PANI-GO films, they were reduced with hydroiodic acid in the form of PANI-RGO composite films. They also found that the incorporation of RGO into PANI matrix can increase the transparency of PANI and promote the light-harvesting from the rear side of devices. A cell based on the transparent PANI-RGO CE can achieve an impressive PCE of 7.84%, which is comparable to that assembled with Pt CE (8.19%). Liu et al. [77] employed a facile two-step electrochemical process to fabricate PPy/RGO composites. Firstly, the PPy/GO composites were obtained by electrochemical oxidative polymerization. Secondly, the GO incorporated in the PPy/GO composites was effectively reduced to RGO through cyclic voltammetry method to obtain PPy/RGO composites. After optimizing the polymerization period of PPy/GO, the DSSC based on the optimized PPy/RGO CE presented a PCE of 6.45%, which was ca. 90% of that of the device using a thermally deposited Pt CE (7.14%).

In addition to the mixture of graphene and conducting polymers, Wang et al. [35] synthesized a hybrid material of polyaniline (PANI) nanoparticles dispersed on the graphene prepared using an in situ polymerization method (Figure 7). In their work, the graphene sheets function as highly conductive supports for decorating PANI nanoparticles, thus providing rapid electron transfer to highly electrocatalytic PANI nanoparticles and increased electrocatalytic active sites for the reduction of I_3^-. As a result, the DSSC with PANI/graphene counter electrode achieved a PCE of 6.09%, which is comparable to that of the cell with Pt CE (6.88%).

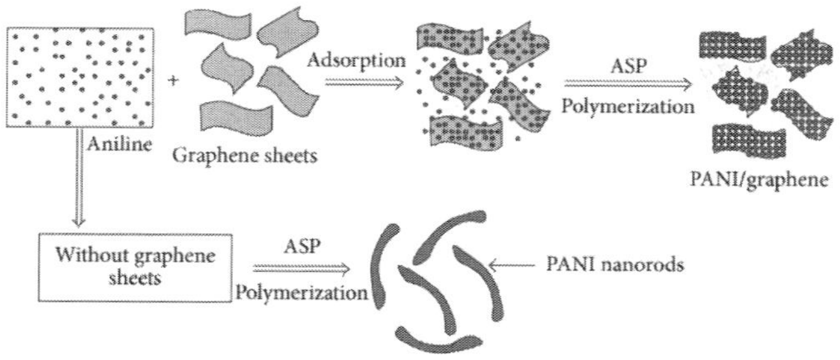

Figure 7: Schematic illumination for the synthesis of PANI/graphene hybrid [35].

GRAPHENE/INORGANIC COMPOUND COMPOSITES

Except for conventional carbon materials and conducting polymers, great deals of studies have been recently reported for exploring low-cost, highly efficient electrocatalytic materials as CEs in DSSCs. Up to date, inorganic compounds, such as transition metal oxides, nitrides, sulfides, and carbides [12–14, 108–110], have demonstrated their promising potential as Pt-free CEs because of their superior electrocatalytic activity. Nevertheless, their electrical conductivity is still insufficient due to numerous defects or grain boundaries in their structures [16, 81, 111]. To address this weakness, the most efficient strategy is to hybridize nanostructured inorganic compounds with highly conductive materials. As mentioned before, graphene is one of the carbon family materials, which can own outstanding electrical conductivity for electron transfer between inorganic compounds nanoparticles and high specific surface area for decorating them. Moreover, the synthesis of nanostructured inorganic compounds on graphene support could provide increased electrocatalytic sites for I_3^- reduction. The synergic effect of the aforementioned advantages of hybridization of graphene and inorganic compounds nanoparticles would promote the charge transfer between CEs and electrolyte. In this section, the recent developments of the composite CEs composed of graphene and inorganic compounds are summarized in Tables 5 and 6 and compared as follows.

Table 5: Photovoltaic performance of the DSSC using various graphene/metal sulfide material CEs

CE	Subtract	Deposition	Redox couple	Dye	R_{ct} (Ω cm^2)	FF	PCE (%)	Ref
G-CoS	FTO glass	CVD + SILAR	I^-/I_3^-	N719	5.05	0.36	3.42	[78]
CoS/graphene	FTO glass	CVD + Dip-coating	I^-/I_3^-	N719	—	0.69	5.04	[78]
NiS/graphene	FTO glass	CVD + Dip-coating	I^-/I_3^-	N719	8.60	0.70	5.25	[79]
CoS_2-G	FTO glass	Doctor-blade	I^-/I_3^-	N719	1.30	0.60	6.55	[80]
NDG/CoS	FTO glass	Spin-coating	I^-/I_3^-	N719	2.58*	0.74	10.71	[81]
CSG	FTO glass	LBL	I^-/I_3^-	N719	5.70*	0.63	5.43	[82]
FGNS	FTO glass	Electrophoretic deposition	I^-/I_3^-	N719	1.79*	0.64	5.54	[83]
CoS/RGO	FTO glass	Electrophoretic deposition + IED	I^-/I_3^-	N719	3.59	0.63	9.39	[84]
NiS_2@RGO	FTO glass	Drop-casting	I^-/I_3^-	N719	2.90	0.69	8.55	[85]
GP/NiS	FTO glass	Doctor-blade	I^-/I_3^-	N719	0.63*	0.68	7.67	[86]
NiS-G	FTO glass	Drop-casting	I^-/I_3^-	N719	8.98*	0.62	8.26	[87]
MoS_2/RGO	FTO glass	Drop-casting	I^-/I_3^-	N719	0.57	0.66	6.04	[36]
MoS_2/graphene	FTO glass	Doctor-blade	I^-/I_3^-	N719	2.17	0.68	5.98	[88]
MoS_2-GNS	FTO glass	Electrophoretic deposition	I^-/I_3^-	N719	2.34	0.59	5.81	[89]
MoS_2/FG	FTO glass	Doctor-blade	I^-/I_3^-	N719	2.67	0.61	6.07	[90]
MoS_2/RGO	FTO glass	Electrophoretic deposition	I^-/I_3^-	N719	5.17	0.67	7.46	[37]
SnS_2@RGO	FTO glass	Doctor-blade	I^-/I_3^-	N719	7.24*	0.67	7.12	[91]
SnS/RGO	FTO glass	Drop-casting	I^-/I_3^-	N719	23.12*	0.49	5.30	[92]
SnS_2/RGO	FTO glass	Drop-casting	I^-/I_3^-	N719	5.29*	0.62	7.47	[92]
Bi_2S_3-reduced graphene oxide	FTO glass	Doctor-blade	I^-/I_3^-	N719	9.20*	0.60	5.50	[93]
RGO/Cu_2S	FTO glass	Doctor-blade	I^-/I_3^-	N719	3.24*	0.69	7.12	[94]
CIS-G	FTO glass	Doctor-blade	I^-/I_3^-	N719	2.30*	0.61	6.40	[95]
$CuInS_2$/RGO	FTO glass	Doctor-blade	I^-/I_3^-	N719	0.65	0.51	6.18	[96]
CZTS/graphene	FTO glass	Doctor-blade	I^-/I_3^-	N719	13.33*	0.66	7.81	[38]

Table 6: Photovoltaic performance of the DSSCs with various graphene/metal oxide material CEs

CE	Subtract	Deposition	Redox couple	Dye	R_{ct} (Ω cm^2)	FF	PCE (%)	Ref
NiO-GP	FTO glass	Drop-casting	I^-/I_3^-	N3	1.72	0.61	3.06	[97]
NiO-NP-RGO	FTO glass	Dry plasma reduction	I^-/I_3^-	N719	13.27	0.62	7.42	[98]
GNS/SiO_2	FTO glass	Dip-coating	I^-/I_3^-	N719	39.80*	0.61	6.82	[99]
GNs@ZnO	FTO glass	Spin-coating	I^-/I_3^-	N719	4	0.67	8.12	[100]
Mn_3O_4/RGO	FTO glass	Doctor-blade	I^-/I_3^-	N3	5.24*	0.61	5.90	[101]
$La_{0.65}Sr_{0.35}MnO_3$/RGO	FTO glass	Spin-coating	I^-/I_3^-	N719	0.71*	0.67	6.57	[102]
FeO_3/GFs	FTO glass	Screen-printing	I^-/I_3^-	N719	5.32*	0.68	7.45	[39]

*Represents that the unit of the R_{ct} value is Ω.

GRAPHENE/METAL SULFIDES

In 2009, Wang et al. [12] first reported an electrochemical deposited cobalt sulfide (CoS) on ITO/PEN film as an efficient CE in DSSCs. Since that, varieties of transition metal sulfides have been investigated as CEs in DSSCs CE. Among them, molybdenum disulfide (MoS_2), a typically two-dimensional layered stricture, exhibits Mo atoms bonding between the three stacked atomic layers (S–Mo–S) by weak van der Waals interplay. MoS_2has two typical surfaces on the crystals, which are terrace sites on the basal planes and edge sites on the side surfaces. Due to the anisotropic bonding and the general tendency to minimize the surface energy, nanoparticles of layer materials usually exhibit platelet-like morphology, in which the basal planes are exposed. In addition, the MoS_2 proposed that the catalytically active sites of MoS_2 lie on the edges of the typical layered material but not on the basal planes [112]. To pursue MoS_2-based CEs with highly efficient performance in DSSCs, as depicted in Figure 8, our group first decorated MoS_2 nanoparticles on reduced graphene oxide (RGO) surface and deposited the composites on FTO glass substrates as efficient CEs using drop-casting method. The extensive cyclic voltammograms (CVs) showed that the cathodic current density of the MoS_2/RGO CE was higher than that of MoS_2, RGO, and sputtered Pt CEs, due to the increased active surface area of the former [36]. As depicted in Figure 9, the enhanced electrocatalytic activity of the MoS_2/RGO CE can be attributed not only to the superior electrical conductivity of RGO but also to the considerable active surface area of the MoS_2 nanoparticles dispersed on the RGO surface. The DSSC assembled with the MoS_2/RGO CE showed a comparable power conversion efficiency of 6.04% to the cell with a Pt CE (6.38%). Subsequently, our group synthesized transparent MoS_2/graphene nanosheet (GNS) composites via one-pot hydrothermal reaction and deposited them on FTO glass substrates using electrophoretic deposition. The resultant transparent MoS_2/GNS composite CE demonstrated both high transmittance (>70%) at visible wavelengths and improved electrocatalytic activity. The DSSC based on the transparent CE still achieved an impassive PCE of 5.81%, which is up to 93% of that obtained with the conventional Pt CE [89]. The use of electrophoretic deposition has the advantages of high deposition rate, controlled thickness, excellent uniformity, large-scale production, and being without any binders. Recently, a hybrid of MoS_2/RGO and CNTs (as depicted in Figure 10) has been directly deposited onto FTO glass substrate by similar electrophoretic deposition and employed as CE materials in DSSCs by our group [37]. Electrochemical studies indicated that the incorporation of CNTs within the MoS_2/RGO matrix can facilitate the electron transport efficiently and therefore enhance the charge-transfer rate for I_3^- reduction. Consequently, the DSSC assembled with the MoS_2/RGO-CNTs CE achieves an impressive PCE of 7.46%, which is higher than that of devices that incorporate MoS_2/RGO CE (6.82%) or Pt CE (7.23%) [37].

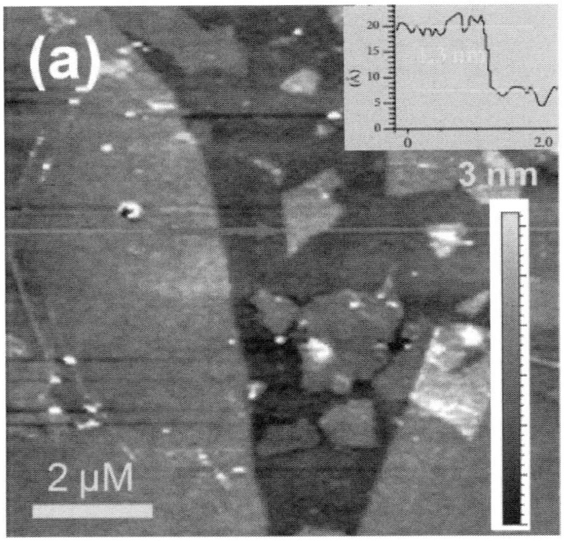

(a)

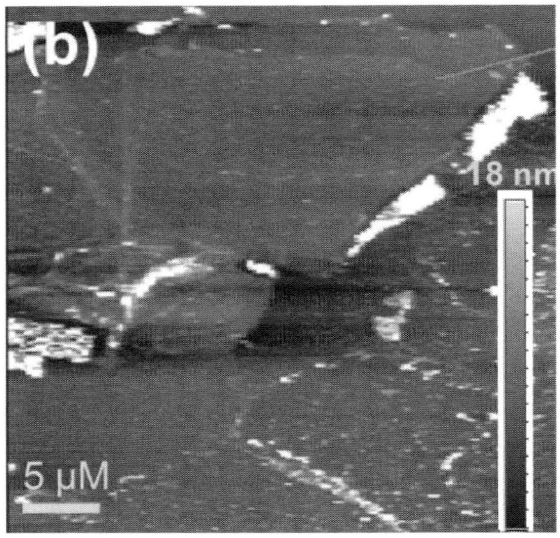

(b)

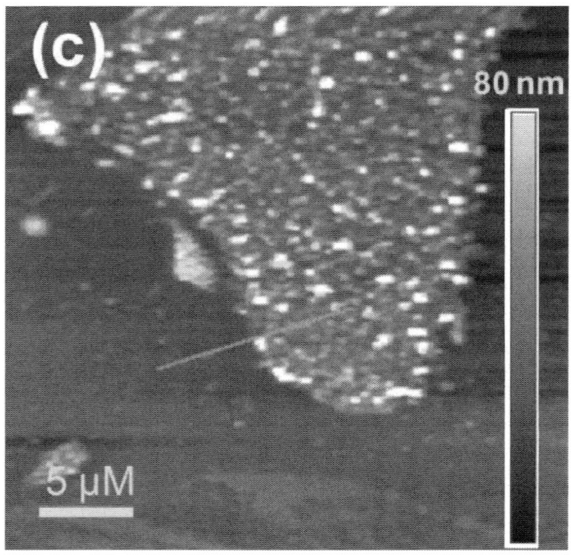

(c)

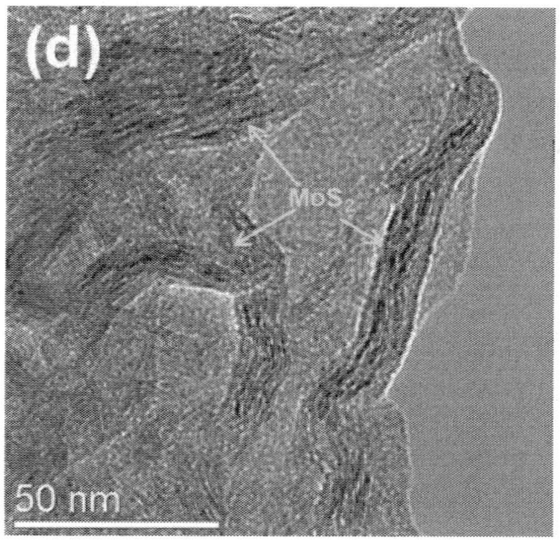

(d)

Figure 8: Typical AFM images of (a) as-synthesized GO, (b) RGO with a low loading of MoS_2 nanoparticles, and (c) RGO with a high loading of MoS_2 nanoparticles. The inset in (a) shows that the height difference between two red arrows is 1.3 nm. (d) TEM image of the MoS_2/RGO nanocomposite [36].

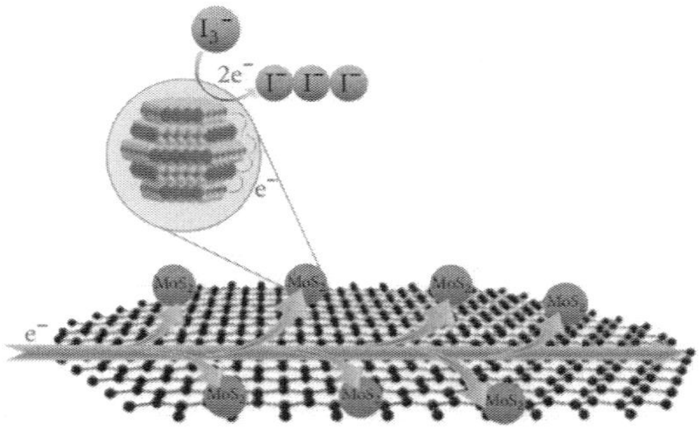

(a)

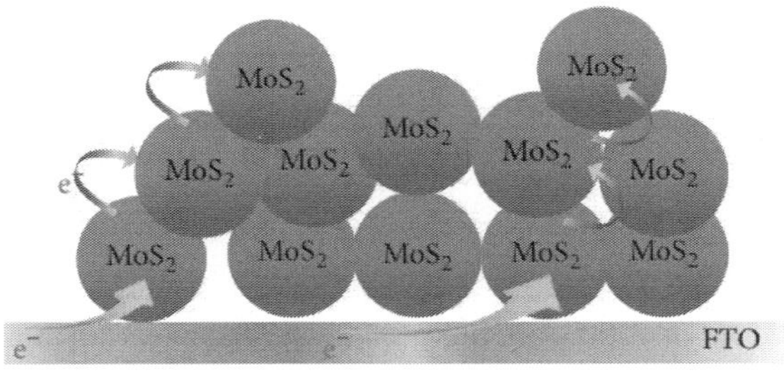

(b)

Figure 9: Schematic of the catalytic mechanisms of (a) MoS_2/RGO and (b) MoS_2 CEs [36].

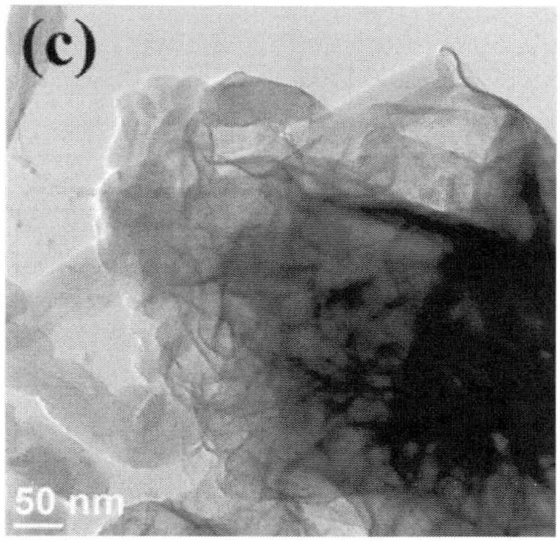

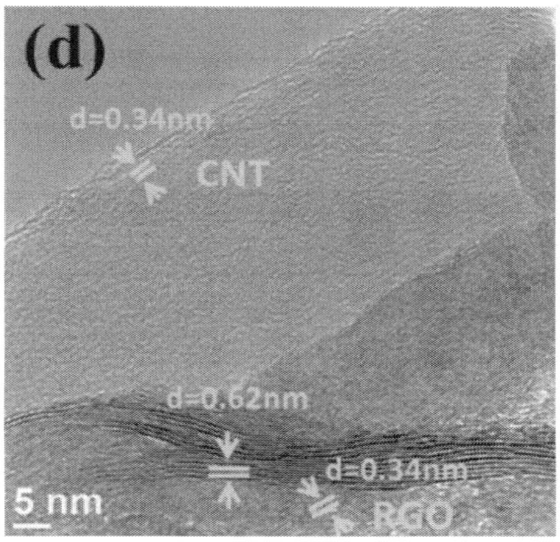

Figure 10: FESEM images of (a) MoS_2/RGO and (b) MoS_2/RGO-CNTs. (c) TEM and (d) HRTEM images of MoS_2/RGO-CNTs [37].

Additionally, Das et al. [78] decorated CoS nanoparticles on graphene film (G-CoS) via successive ionic layer adsorption and reaction (SILAR) process; therefore, the increased catalytic area for I_3^- reduction at CE/electrolyte and CoS nanoparticles was obtained. Consequently, remarkably improved FF and PCE values for the G-CoS based DSSC were achieved, compared to that with pristine CoS CE. Bi et al. [79] synthesized two kinds of metal sulfides (CoS and NiS) and CoS/graphene and NiS/graphene composites as Pt-free CEs. Firstly, 3D graphene was directly deposited on SiO_2 substrate, serving not only as conductivity layer for electron transfer but also as support for decorating CoS or NiS nanoparticles. In recorded CV curves, the cathodic current densities of CoS/graphene and NiS/graphene CEs were obviously higher than those of CoS/FTO, NiS/FTO, and Pt/FTO, indicating that the composites CEs have higher electrocatalytic activity for I_3^- reduction. Moreover, both NiS/graphene and CoS/graphene CEs showed the smaller R_{ct} of 8.6 and 8.8 Ω cm²than that of the NiS/FTO (20.5 Ω cm²), CoS/FTO (26.4 Ω cm²), and Pt/FTO (9.1 Ω cm²) CEs, signifying the significantly improved charge transfer at CE/electrolyte for the NiS/graphene and CoS/graphene CEs. The improved PCE values of 5.25% and 5.04% were found for the DSSCs based on the NiS/graphene and CoS/graphene CEs. This can be ascribed to their improved FF values of 0.72 and 0.69, respectively. More recently, Duan et al. [80] synthesized the CoS_2/graphene composite by a facile hydrothermal reaction and utilized doctor-blade method to prepare CEs. The incorporation of graphene significantly controlled CoS_2particles size and offered large reaction surface at CE/electrolyte. Therefore, CoS_2/graphene composite CE could provide an efficient diffusion channel for electrolyte penetration and enhanced electrocatalytic activity for I_3^- reduction. While the DSSC assembled with the CoS_2/graphene composite CE, it displayed a PCE up to 6.55%, which exceeded that of the Pt-based device (6.20%).

Li et al. [85] prepared the nanocomposites of NiS_2 with reduced graphene oxide (NiS_2@RGO) by a facile hydrothermal reaction. Compared to RGO and NiS_2 CEs, the NiS_2@RGO exhibited superior electrocatalytic activity. Therefore, the device with NiS_2@RGO CE exhibited a higher PCE of 8.55% than that with RGO (3.14%) or NiS_2 (7.02%) CE. This can be explained by the fact that the NiS_2@RGO CE possessed lower R_{ct} value of 2.9 Ω cm² than that of RGO (100.02 Ω cm²) and NiS_2 (8.8 Ω cm²) CEs, therefore revealing the faster charge transfer at CE/electrolyte. Shen et al. [87] used facile and low-temperature hydrothermal method to synthesize the nanocomposites composed of 1-D NiS and 2-D graphene (designated as NiS-G) with the ratio of graphene/NiS ranging from 0.2 to 0.6. Among all devices with NiS-G based CEs, the DSSC based on the NiS-G0.4 CE showed the highest J_{sc} of 17.05 mA cm⁻² and PCE of 8.26%, which was much higher than that with pristine graphene (1.56%) or NiS CE (7.41%). The appropriate proportion of NiS and graphene could be associated with its morphology and the diffusion resistance, confirming that the loading of NiS fine rods can efficiently hinder the agglomeration of adjacent graphene layers and favor the diffusion of the electrolyte species within the NiS-G CE. Yang et al. [91] synthesized the nanocomposite composed of SnS_2 nanoparticles and reduced graphene oxide (designated as SnS_2@RGO). The R_{ct} values for Pt, RGO, SnS_2, and SnS_2@RGO are 24.21, 34.20, 39.73, and 17.96 Ω, respectively. Compared with RGO and SnS_2, SnS_2@RGO composite had the lower R_{ct} value, indicating its higher electrocatalytic activity. Moreover, the DSSCs device with SnS_2@RGO composite CE had a remarkable PCE of 7.12%, which was significantly higher than SnS_2 CE (5.58%) and RGO CE (3.73%) and even comparable to the value of 6.79% obtained with a Pt reference CE. The synergistic effect between RGO and SnS_2 showed both high electrical conductivity and excellent electrocatalytic activity. Zhou et al. also synthesized $CuInS_2$ and RGO by a facile method, and the resultant $CuInS_2$/RGO composite was directly employed as CE material exhibit excellent electrocatalytic activity for the triiodide reduction [95]. The graphene-wrapped $CuInS_2$ composites were exploited as counter electrode for DSSCs and therefore achieved a power conversion efficiency of 6.4%, which is comparable to that of Pt CE (6.9%) [95].

In addition to wrapping transition metal sulfide nanoparticles on graphene surface, the graphene can be incorporated into metal sulfide matrix to serve the conductive network in metal sulfides. Huo et al. [84] developed the sponge-like CoS/reduced graphene oxide (CoS/RGO) by electrophoretic deposition and ion exchange deposition. The as-prepared pristine CoS as CE has a sponge-like morphology with large specific surface area and low charge-transfer resistance at the CE/electrolyte interface. To further enhance the electrocatalytic activity of sponge-like CoS CEs, the various content of RGO was incorporated in the sponge-like CoS. The composite CE with the optimized composition ratio ($CoS/RGO_{0.2}$, RGO: 0.2 mgL⁻¹) revealed the smallest R_{ct} value of 3.59 Ω cm², as well as the highest PCE of 9.39%, which was increased by 27.93% compared with that using Pt CE. Furthermore, Bai et al. [38] synthesized flower-like copper zinc tin sulfide (CZTS) and graphene as composite CE (Figure 11). The CZTS/graphene demonstrated excellent electrocatalytic activity because the incorporation of highly conductive graphene of 2 wt% remarkably reduced its series resistance (R_s) from 22.84 Ω to 13.33 Ω and then enhanced the electrical conductivity of the composite CE. Liu et al. [96] employed 3D $CuInS_2$ microspheres as CE materials and the DSSC based on the CE showed a PCE of only 3.31%. To improve the cell efficiency of the DSSC, the $CuInS_2$ nanomaterial was hybridized with highly conductive RGO and its cell performance was increased to 6.18%.

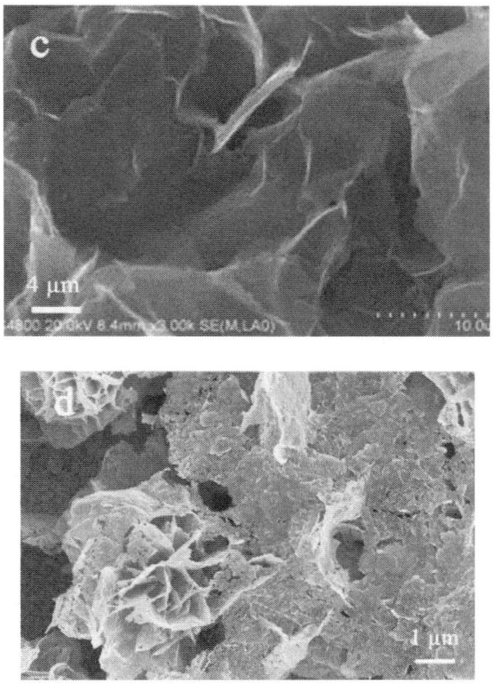

Figure 11: (a) and (b) SEM images of CZTS microspheres. The inset in (a) is HRTEM image of CZTS nanoparticle, (c) SEM image of graphene, and (d) SEM image of annealed CZTS/graphene composites films [38].

Graphene/Metal Oxide Composites

Bajpai et al. [97] synthesized NiO nanoparticles homogeneously deposited over few-layered graphene platelets (GPs) by pulsed laser ablation. The device with NiO-GP CE yielded a PCE of 3.06%, which outperformed the cells using unsupported NiO nanoparticles (2.03%) or pristine GPs (2.46%) and was even comparable to a conventional Pt-based DSSC (3.57%). Furthermore, Dao et al. [98] employed dry-plasma reduction to hybridize the NiO nanoparticles (NiO-NPs) on the surface of RGO. The resultant NiO-NP-RGO CE displayed R_{ct} value of 1.93 Ωcm^2, much lower than that of NiO-NP-immobilized CE (44.39 Ωcm^2) and a GO-coated CE (12.19 Ωcm^2). In addition, the shunt resistant (R_h) value of the NiO-NP-RGO CE measured at a high frequency range was found to be 2.44 Ωcm^2, which was slightly lower than the values found for RGO CE (2.45 Ωcm^2) and NiO-NP CE (2.53 Ωcm^2). This phenomenon can be associated with the decoration of NiO-NPs on the RGO surface, which constructs lots of bridges for facilitating electron transfer between NiO-CPs and RGO. Gong et al. [99] synthesized a porous graphene (GNS)/SiO_2 nanocomposite converted from graphene oxide mixed with SiO_2 nanoparticles through facile hydrazine hydrate reduction. To substantiate the formation of pores within the composite film, the GNS/SiO_2 nanocomposite exhibited a narrow pore size distribution centered at 4.0 nm; thus, SiO_2 nanoparticles played a significant role in the GNS layers to build up porous nanostructured architecture. Compared to SiO_2 and GNS CEs, the GNS/SiO_2 composite CE demonstrated the lower R_s and R_{ct} values, displaying its fast electron and charge transfer. Additionally, the porous structure of the GNS/SiO_2 composite film could provide larger surface area than the GNS film, thus enhancing the accessibility of the electrolyte to the inside of CE and being favorable for the reduction of I_3^- to I^-. As a consequence, DSSC assembled with GNS/SiO_2 CE achieved high cell efficiency of 6.82%, considered as a promising potential candidate to replace conventional Pt CE. Recently, Chang et al. [100] demonstrated that the combination of graphene nanosheets (GNS) and ZnO nanorods can be a highly efficient 3-D CE in DSSCs. The use of the ZnO nanorods as a 3D framework nanostructure could prevent the aggregations of GNs. Unlike conventional chemical functionalization of graphene, the electrocatalytic active sites are created by damaging the conjugated structure in the graphitic basal plan with a concomitant decrease in the electrical conductivity. This novel hybrid system has proved that the GNs were efficiently isolated from each other to prevent the aggregation and restacking of GNs, which increased the active defective sites for the redox reaction of I^-/I_3^- to improve the electrocatalytic performance. Therefore, the novel hybrid nanoarchitecture exhibited improved performance as a promising candidate CE for DSSCs due to the fast electron transport network and more active sites for catalyzing the I_3^- reduction. The PCE of the DSSC based on the GNs@ZnO nanorods CE was reached 8.12%, which was comparable to the device using Pt CE (8.82%). Interestingly, the inorganic transition metal oxides, Fe_2O_3, exhibited approximately identical adsorption energy of iodine compared to that of Pt in earlier research [113, 114]. Nevertheless, the cells with Fe_2O_3 still obtained low overall conversion efficiency due to the poor electron transport efficiency across Fe_2O_3 particles. Yang et al. [39] explored Fe_2O_3 nanoparticles (NPs) anchored onto 3D graphene frameworks (GFs), where Fe_2O_3 NPs act as highly active sites for reduction and the 3D graphene frameworks form an interconnected electron transfer highway system. As can be seen in Figure 12, The DSSCs fabricated with the Fe_2O_3/GFs CEs showed a higher PCE of 7.45% in comparison to 7.29% for the DSSCs with Pt CEs [39]. By EIS, CV and Tafel polarization measurements (also summarized in Figure 12), 3D Fe_2O_3/GFs not only obtained high efficiency but also provided multidimensional pathways to facilitate the transport of electrons in the bulk electrode [39]. Due to the fast electron transfer in the interpenetrating graphene frameworks, the shuttle electrons easily cross the graphene sheets to the catalytic Fe_2O_3 NPs sites where the electrons are used to reduce I_3^-. Consequently, Fe_2O_3 NPs have lower R_s and R_{ct} to reduce the interface loss of charge transfer and enhance charge collection efficiency, thereby boosting the photovoltaic performance of DSSCs.

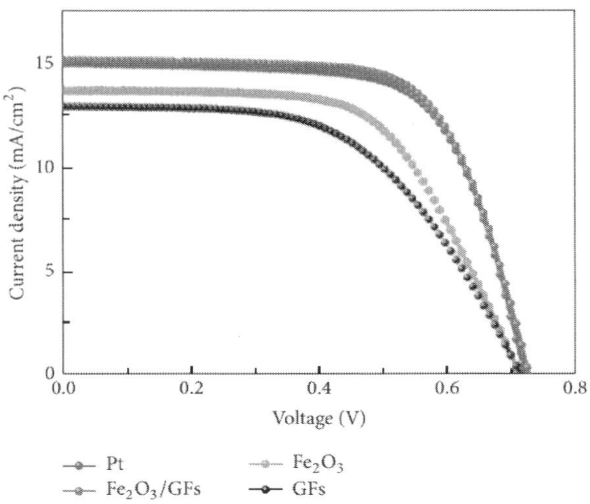

(a)

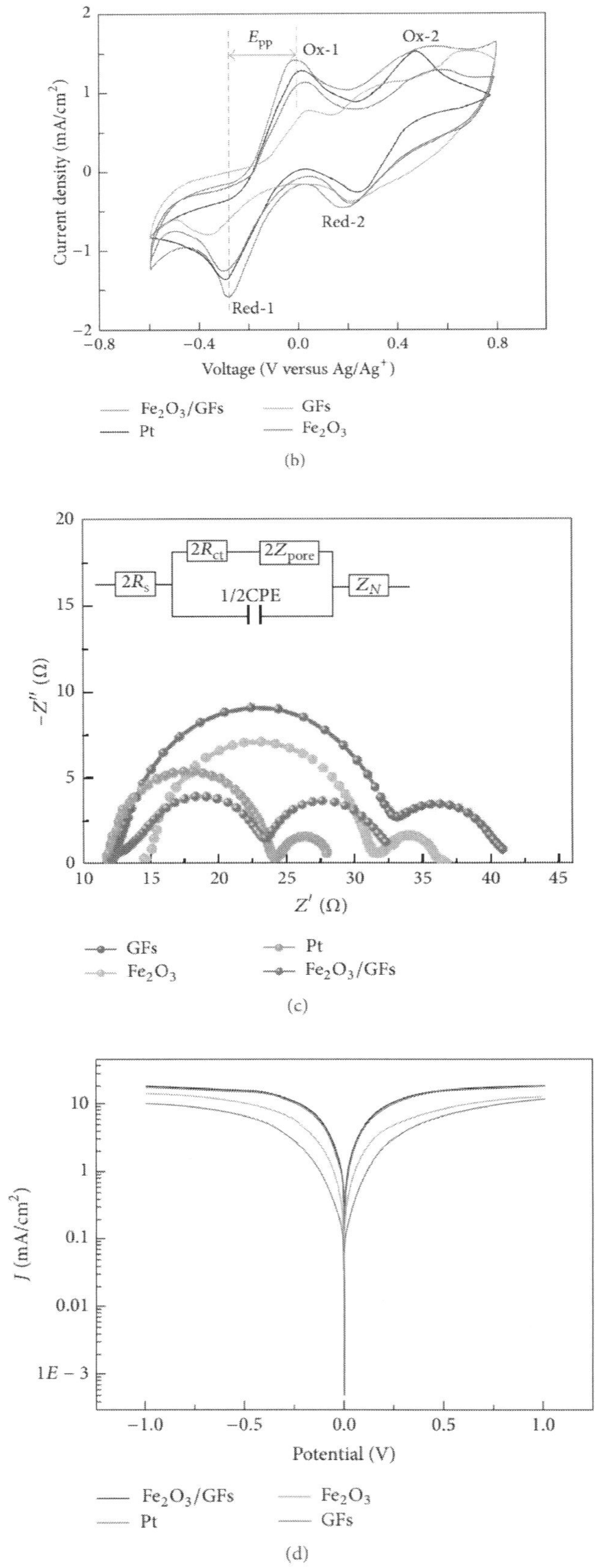

Figure 12: (a) Current-voltage characteristics of the DSSCs with different CEs under one sun illumination (AM 1.5 G). (b) Cyclic voltammograms of various CEs for I^-/I_3^- redox species recorded at a scan rate of $50\,mV\,s^{-1}$. Electrochemical impedance spectra (c) and (d) Tafel polarization curves of the symmetric dummy cells fabricated with different CEs. (A colored version of this figure can be viewed online [39].)

Other Inorganic Compound/Graphene Composites

Combination $Ni_{12}P_5$ and graphene (graphene-$Ni_{12}P_5$) as a unique composite CE also displayed interesting characteristics of improved electrocatalytic activity, electrical conductivity, and electrolyte penetration [40]. After evaluating the intrinsic electrochemical features of the $Ni_{12}P_5$, graphene and graphene-$Ni_{12}P_5$ as CEs in DSSCs, the graphene-$Ni_{12}P_5$ composite shows optimized electrochemical features, including lower charge-transfer resistance and diffusion impedance by the combination of both contributions of the high electrocatalytic activity of $Ni_{12}P_5$ nanocrystallites and fast diffusion process of the electrolyte species in the graphene. The main key could relative that the graphene-$Ni_{12}P_5$ composite as CE had more active sites and spacers among the graphene layers to accelerate the diffusion of electrolyte species.

On the other hand, some studies have demonstrated that $Ni_{0.85}Se$ presented excellent electrocatalytic performance for the I_3^- reduction and can be a potential material as CEs [115]. As can be seen in Figure 13, mesoporous $Ni_{0.85}Se$ nanospheres on reduced graphene oxide ($Ni_{0.85}Se$@RGO) were further prepared because the mesoporous structure of $Ni_{0.85}Se$ nanospheres and RGO both possessed large specific surface area and fast electron transfer channels. Consequently, the $Ni_{0.85}Se$@RGO CE could provide increased catalytic active sites to reduce I_3^- and relatively fast electron transfer channels, thus significantly enhancing the electrocatalytic activity of $Ni_{0.85}Se$@RGO CE. The PCE of the device with the $Ni_{0.85}Se$@RGO CE reached 7.82%, which was slightly higher than that of Pt CE (7.54%).

Figure 13: Typical SEM ((a) and (b)), TEM ((c) and (d)), HRTEM (e), and EDS (f) images of mesoporous $Ni_{0.85}Se$@RGO nanospheres [40].

In order to achieve the high-performance of DSSCs with CEs composed of inorganic compounds and carbon materials, graphene sheets doped with heteroatoms such as nitrogen (N), boron (B), and phosphorous (P) can improve their electrocatalytic active sites [116]. Therefore, Balamurugan et al. [117] hybridized iron nitride (FeN) core-shell nanoparticles grown on nitrogen-doped graphene (NG) as CE material for application in DSSCs. In particular, the superior performance of DSSCs with newly developed core-shell FeN/NG nanohybrids was attributed to the high electrical conductivity, large surface area, good surface hydrophilicity, enhanced electrolyte/CE interaction, and excellent electrocatalytic activity toward the reduction of iodine redox species. The PCE of the DSSC with core-shell FeN/NG CE reached 10.86%, which was also superior to that of a cell with a conventional Pt CE (9.93%) under the same experimental conditions.

SUMMARY

DSSCs have been widely regarded as promising photovoltaic devices to convert incident light to electricity. In accordance with the working mechanism of DSSCs, CEs have to possess both high electrical conductivity and excellent electrocatalytic activity for collecting electron back from external circuit, constructing rapid electron transfer network, and catalyzing I_3^- reduction reaction. Moreover, low-cost alternative CEs should be explored to replace high-cost noble Pt CE for large-scale applications. As a result, we review a great deal of graphene-based CE materials for low-cost Pt-free DSSCs since graphene has the following specific properties of high electrical conductivity, great anticorrosion resistance, and large surface area. In this study, the electrochemical properties of various graphene-based CEs and their effects on cell performance were systemically discussed. It is found that the pristine graphene does not display sufficient electrocatalytic activity due to its limited electroactive sites. To improve this issue, three approaches have been conducted: (i) the decoration of excellent electrocatalytic nanomaterials on the surface of highly conductive graphene, (ii) the incorporation of graphene in the matrix of electrocatalytic materials to improve its electrical conductivity, and (iii) the introduction of highly conductive materials to connect graphene layers for promoting electron transfer between them. Although lots of studies have been carried out to improve the various aspects of graphene-based CEs in DSSCs, the device performance of most DSSCs is still relatively lower compared with that using a conventional Pt CE.

On the basis of strategies (i) and (iii), although the graphene decorated with highly electrocatalytic nanoparticles has high electrocatalytic activity, it still requires some conductive materials to connect graphene layers for promoting electron transfer. Therefore, the fabrication of the novel hybrids composed of conductive materials/graphene decorated with highly electrocatalytic nanoparticles would be a promising approach to design highly efficient Pt-free alternatives for low-cost DSSCs in the near future.

ACKNOWLEDGMENTS

This research was supported by Ministry of Science and Technology, Taiwan (NSC 101-2221-M-036-035 and MOST 103-2221-E-036-014-MY3).

REFERENCES

1. B. O'Regan and M. Grätzel, "A low-cost, high-efficiency solar cell based on dye-sensitized colloidal TiO_2 films," Nature, vol. 353, no. 6346, pp. 737–740, 1991.
2. N. Sridhar and D. Freeman, "A study of dye sensitized solar cells under indoor and low level outdoor lighting: comparison to organic and inorganic thin film solar cells and methods to address maximum power point tracking,"

in Proceedings of the 26th European Photovoltaic Solar Energy Conference and Exhibition, Hamburg, Germany, September 2011.

3. G. Calogero, A. Bartolotta, G. Di Marco, A. Di Carlo, and F. Bonaccorso, "Vegetable-based dye-sensitized solar cells," Chemical Society Reviews, vol. 44, no. 10, pp. 3244–3294, 2015.

4. G. Syrrokostas, A. Siokou, G. Leftheriotis, and P. Yianoulis, "Degradation mechanisms of Pt counter electrodes for dye sensitized solar cells," Solar Energy Materials and Solar Cells, vol. 103, pp. 119–127, 2012.

5. G. Boschloo and A. Hagfeldt, "Characteristics of the iodide/triiodide redox mediator in dye-sensitized solar cells," Accounts of Chemical Research, vol. 42, no. 11, pp. 1819–1826, 2009.

6. M. Wu and T. Ma, "Platinum-free catalysts as counter electrodes in dye-sensitized solar cells,"ChemSusChem, vol. 5, no. 8, pp. 1343–1357, 2012.

7. L. Yang, L. Wu, M. Wu, G. Xin, H. Lin, and T. Ma, "High-efficiency flexible dye-sensitized solar cells fabricated by a novel friction-transfer technique," Electrochemistry Communications, vol. 12, no. 7, pp. 1000–1003, 2010.

8. A. Kanciurzewska, E. Dobruchowska, A. Baranzahi, E. Carlegrim, M. Fahlman, and M. A. Gîrţu, "Study on poly(3,4-ethylene dioxythiophene)-poly(styrenesulfonate) as a plastic counter electrode in dye sensitized solar cells," Journal of Optoelectronics and Advanced Materials, vol. 9, no. 4, pp. 1052–1059, 2007.

9. Y. Saito, W. Kubo, T. Kitamura, Y. Wada, and S. Yanagida, "I-/I3- redox reaction behavior on poly (3,4-ethylenedioxythiophene) counter electrode in dye-sensitized solar cells," Journal of Photochemistry and Photobiology A: Chemistry, vol. 164, no. 1–3, pp. 153–157, 2004.

10. J. Xia, N. Masaki, K. Jiang, and S. Yanagida, "The influence of doping ions on poly(3,4-ethylenedioxythiophene) as a counter electrode of a dye-sensitized solar cell," Journal of Materials Chemistry, vol. 17, no. 27, pp. 2845–2850, 2007.

11. Y. Chen, K. S. Kang, K. J. Han, K. H. Yoo, and J. Kim, "Enhanced optical and electrical properties of PEDOT: PSS films by the addition of MWCNT-sorbitol," Synthetic Metals, vol. 159, no. 17-18, pp. 1701–1704, 2009.

12. M. Wang, A. M. Anghel, B. Marsan et al., "CoS supersedes Pt as efficient electrocatalyst for triiodide reduction in dye-sensitized solar cells," Journal of the American Chemical Society, vol. 131, no. 44, pp. 15976–15977, 2009.

13. Q. W. Jiang, G. R. Li, and X. P. Gao, "Highly ordered TiN nanotube arrays as counter electrodes for dye-sensitized solar cells," Chemical Communications, no. 44, pp. 6720–6722, 2009.

14. M. Wu, Q. Zhang, J. Xiao et al., "Two flexible counter electrodes based on molybdenum and tungsten nitrides for dye-sensitized solar cells," Journal of Materials Chemistry, vol. 21, no. 29, pp. 10761–10766, 2011.

15. G. R. Li, J. Song, G. L. Pan, and X. P. Gao, "Highly Pt-like electrocatalytic activity of transition metal nitrides for dye-sensitized solar cells," Energy and Environmental Science, vol. 4, no. 5, pp. 1680–1683, 2011.

16. Q. W. Jiang, G. R. Li, S. Liu, and X. P. Gao, "Surface-nitrided nickel with bifunctional structure as low-cost counter electrode for dye-sensitized solar cells," The Journal of Physical Chemistry C, vol. 114, no. 31, pp. 13397–13401, 2010.

17. X. Lin, M. Wu, Y. Wang, A. Hagfeldt, and T. Ma, "Novel counter electrode catalysts of niobium oxides supersede Pt for dye-sensitized solar cells," Chemical Communications, vol. 47, no. 41, pp. 11489–11491, 2011.

18. M. Wu, Y. Wang, X. Lin et al., "Economical and effective sulfide catalysts for dye-sensitized solar cells as counter electrodes," Physical Chemistry Chemical Physics, vol. 13, no. 43, pp. 19298–19301, 2011.

19. F. Bonaccorso, L. Colombo, G. Yu et al., "Graphene, related two-dimensional crystals, and hybrid systems for energy conversion and storage," Science, vol. 347, no. 6217, Article ID 1246501, 2015.

20. A. H. Castro Neto, F. Guinea, N. M. R. Peres, K. S. Novoselov, and A. K. Geim, "The electronic properties of graphene," Reviews of Modern Physics, vol. 81, no. 1, pp. 109–162, 2009.

21. F. Bonaccorso, A. Lombardo, T. Hasan, Z. Sun, L. Colombo, and A. C. Ferrari, "Production and processing of graphene and 2d crystals," Materials Today, vol. 15, no. 12, pp. 564–589, 2012.

22. J. D. Roy-Mayhew and I. A. Aksay, "Graphene materials and their use in dye-sensitized solar cells,"Chemical Reviews, vol. 114, no. 12, pp. 6323–6348, 2014.

23. L. Kavan, J. H. Yum, and M. Grätzel, "Optically transparent cathode for dye-sensitized solar cells based on graphene nanoplatelets," ACS Nano, vol. 5, no. 1, pp. 165–172, 2011.

24. J. D. Roy-Mayhew, D. J. Bozym, C. Punckt, and I. A. Aksay, "Functionalized graphene as a catalytic counter electrode in dye-sensitized solar cells," ACS Nano, vol. 4, no. 10, pp. 6203–6211, 2010.

25. M. H. Yeh, L. Y. Lin, J. S. Su et al., "Nanocomposite graphene/Pt electrocatalyst as economical counter electrode for dye-sensitized solar cells," ChemElectroChem, vol. 1, no. 2, pp. 416–425, 2014.

26. G. Zhu, L. Pan, T. Lu, T. Xu, and Z. Sun, "Electrophoretic deposition of reduced graphene-carbon nanotubes composite films as counter electrodes of dye-sensitized solar cells," Journal of Materials Chemistry, vol. 21, no. 38, pp. 14869–14875, 2011.

27. G. Liu, X. Li, H. Wang et al., "An efficient thiolate/disulfide redox couple based dye-sensitized solar cell with a graphene modified mesoscopic carbon counter electrode," Carbon, vol. 53, pp. 11–18, 2013.

28. D. W. Zhang, X. D. Li, H. B. Li et al., "Graphene-based counter electrode for dye-sensitized solar cells,"Carbon, vol. 49, no. 15, pp. 5382–5388, 2011.

29. Y. Sun, Q. Wu, and G. Shi, "Graphene based new energy materials," Energy and Environmental Science, vol. 4, no. 4, pp. 1113–1132, 2011.

30. D. R. Dreyer, S. Park, C. W. Bielawski, and R. S. Ruoff, "The chemistry of graphene oxide," Chemical Society Reviews, vol. 39, no. 1, pp. 228–240, 2010.

31. J. Hass, W. A. De Heer, and E. H. Conrad, "The growth and morphology of epitaxial multilayer graphene," Journal of Physics Condensed Matter, vol. 20, no. 32, Article ID 323202, 2008.

32. A. Reina, X. Jia, J. Ho et al., "Large area, few-layer graphene films on arbitrary substrates by chemical vapor deposition," Nano Letters, vol. 9, no. 1, pp. 30–35, 2009.

33. H. Choi, H. Kim, S. Hwang, Y. Han, and M. Jeon, "Graphene counter electrodes for dye-sensitized solar cells prepared by electrophoretic deposition," Journal of Materials Chemistry, vol. 21, no. 21, pp. 7548–7551, 2011.

34. A. Kaniyoor and S. Ramaprabhu, "Thermally exfoliated graphene based counter electrode for low cost dye sensitized solar cells," Journal of Applied Physics, vol. 109, no. 12, Article ID 124308, 2011.

35. G. Wang, W. Xing, and S. Zhuo, "The production of polyaniline/graphene hybrids for use as a counter electrode in dye-sensitized solar cells," Electrochimica Acta, vol. 66, pp. 151–157, 2012.

36. C.-J. Liu, S.-Y. Tai, S.-W. Chou et al., "Facile synthesis of MoS_2/graphene nanocomposite with high catalytic activity toward triiodide reduction in dye-sensitized solar cells," Journal of Materials Chemistry, vol. 22, no. 39, pp. 21057–21064, 2012.

37. J.-Y. Lin, A.-L. Su, C.-Y. Chang, K.-C. Hung, and T.-W. Lin, "Molybdenum disulfide/reduced graphene oxide-carbon nanotube hybrids as efficient catalytic materials in dye-sensitized solar cells,"ChemElectroChem, vol. 2, no. 5, pp. 720–725, 2015.

38. L. Bai, J. N. Ding, N. Y. Yuan, H. W. Hu, Y. Li, and X. Fang, "Cu_2ZnSnS_4/graphene composites as low-cost counter electrode materials for dye-sensitized solar cells," Materials Letters, vol. 112, pp. 219–222, 2013.

39. W. Yang, X. Xu, Z. Li et al., "Construction of efficient counter electrodes for dye-sensitized solar cells: Fe_2O_3 nanoparticles anchored onto graphene frameworks," Carbon, vol. 96, pp. 947–954, 2016.

40. Y. Y. Dou, G. R. Li, J. Song, and X. P. Gao, "Nickel phosphide-embedded graphene as counter electrode for dye-sensitized solar cells," Physical Chemistry Chemical Physics, vol. 14, no. 4, pp. 1339–1342, 2012.

41. C. Xu, J. Li, X. Wang et al., "Synthesis of hemin functionalized graphene and its application as a counter electrode in dye-sensitized solar cells," Materials Chemistry and Physics, vol. 132, no. 2-3, pp. 858–864, 2012.

42. J.-S. Lee, H.-J. Ahn, J.-C. Yoon, and J.-H. Jang, "Three-dimensional nano-foam of few-layer graphene grown by CVD for DSSC," Physical Chemistry Chemical Physics, vol. 14, no. 22, pp. 7938–7943, 2012.

43. K. Yu, Z. Wen, H. Pu et al., "Hierarchical vertically oriented graphene as a catalytic counter electrode in dye-sensitized solar cells," Journal of Materials Chemistry A, vol. 1, no. 2, pp. 188–193, 2013.

44. H. Wang, K. Sun, F. Tao, D. J. Stacchiola, and Y. H. Hu, "3D honeycomb-like structured graphene and its high efficiency as a counter-electrode catalyst for dye-sensitized solar cells," Angewandte Chemie—International Edition, vol. 125, no. 35, pp. 9380–9384, 2013.

45. W. Yang, X. Xu, Z. Tu et al., "Nitrogen plasma modified CVD grown graphene as counter electrodes for bifacial dye-sensitized solar cells," Electrochimica Acta, vol. 173, pp. 715–720, 2015.

46. A. K. Chandiran, N. Tetreault, R. Humphry-Baker et al., "Subnanometer Ga_2O_3 tunnelling layer by atomic layer deposition to achieve 1.1 V open-circuit potential in dye-sensitized solar cells," Nano Letters, vol. 12, no. 8, pp. 3941–3947, 2012.

47. S. M. Feldt, E. A. Gibson, E. Gabrielsson, L. Sun, G. Boschloo, and A. Hagfeldt, "Design of organic dyes and cobalt polypyridine redox mediators for high-efficiency dye-sensitized solar cells," Journal of the American Chemical Society, vol. 132, no. 46, pp. 16714–16724, 2010.

48. L. Kavan, J.-H. Yum, M. K. Nazeeruddin, and M. Grätzel, "Graphene nanoplatelet cathode for Co(III)/(II) mediated dye-sensitized solar cells," ACS Nano, vol. 5, no. 11, pp. 9171–9178, 2011.

49. T. Daeneke, A. J. Mozer, T.-H. Kwon et al., "Dye regeneration and charge recombination in dye-sensitized solar cells with ferrocene derivatives as redox mediators," Energy & Environmental Science, vol. 5, no. 5, pp. 7090–7099, 2012.

50. L. Kavan, J.-H. Yum, and M. Graetzel, "Optically transparent cathode for Co(III/II) mediated dye-sensitized solar cells based on graphene oxide," ACS Applied Materials and Interfaces, vol. 4, no. 12, pp. 6999–7006, 2012.

51. C.-T. Hsieh, B.-H. Yang, and Y.-F. Chen, "Dye-sensitized solar cells equipped with graphene-based counter electrodes with different oxidation levels," Diamond and Related Materials, vol. 27-28, pp. 68–75, 2012.

52. L. Kavan, J.-H. Yum, and M. Grätzel, "Graphene nanoplatelets outperforming platinum as the electrocatalyst in co-bipyridine-mediated dye-sensitized solar cells," Nano Letters, vol. 11, no. 12, pp. 5501–5506, 2011.

53. P. Zhai, C.-C. Lee, Y.-H. Chang, C. Liu, T.-C. Wei, and S.-P. Feng, "A significant improvement in the electrocatalytic stability of N-doped graphene nanosheets used as a counter electrode for $[Co(bpy)_3]^{3+/2+}$ based porphyrin-sensitized solar cells," ACS Applied Materials and Interfaces, vol. 7, no. 3, pp. 2116–2123, 2015.

54. R. Bajpai, S. Roy, P. Kumar et al., "Graphene supported platinum nanoparticle counter-electrode for enhanced performance of dye-sensitized solar cells," ACS Applied Materials & Interfaces, vol. 3, no. 10, pp. 3884–3889, 2011.

55. L. Wan, Q. Zhang, S. Wang et al., "A two-step reduction method for synthesizing graphene nanocomposites with a low loading of well-dispersed platinum nanoparticles for use as counter electrodes in dye-sensitized solar cells," Journal of Materials Science, vol. 50, no. 12, pp. 4412–4421, 2015.

56. M.-Y. Yen, C.-C. Teng, M.-C. Hsiao et al., "Platinum nanoparticles/graphene composite catalyst as a novel composite counter electrode for high performance dye-sensitized solar cells," Journal of Materials Chemistry, vol. 21, no. 34, pp. 12880–12888, 2011.

57. Y.-G. Kim, Z. A. Akbar, D. Y. Kim, S. M. Jo, and S.-Y. Jang, "Aqueous dispersible graphene/Pt nanohybrids by green chemistry: application as cathodes for dye-sensitized solar cells," ACS Applied Materials & Interfaces, vol. 5, no. 6, pp. 2053–2061, 2013.

58. V.-D. Dao, N. T. Q. Hoa, L. L. Larina, J.-K. Lee, and H.-S. Choi, "Graphene–platinum nanohybrid as a robust and low-cost counter electrode for dye-sensitized solar cells," Nanoscale, vol. 5, no. 24, pp. 12237–12244, 2013.

59. H. Choi, H. Kim, S. Hwang, M. Kang, D.-W. Jung, and M. Jeon, "Electrochemical electrodes of graphene-based carbon nanotubes grown by chemical vapor deposition," Scripta Materialia, vol. 64, no. 7, pp. 601–604, 2011.

60. S. Li, Y. Luo, W. Lv et al., "Vertically aligned carbon nanotubes grown on graphene paper as electrodes in lithium-ion batteries and dye-sensitized solar cells," Advanced Energy Materials, vol. 1, no. 4, pp. 486–490, 2011.

61. H. Choi, H. Kim, S. Hwang, W. Choi, and M. Jeon, "Dye-sensitized solar cells using graphene-based carbon nano composite as counter electrode," Solar Energy Materials and Solar Cells, vol. 95, pp. 323–325, 2011.

62. H. Kim, H. Choi, S. Hwang, Y. Kim, and M. Jeon, "Fabrication and characterization of carbon-based counter electrodes prepared by electrophoretic deposition for dye-sensitized solar cells," Nanoscale Research Letters, vol. 7, no. 1, article 53, 2012.

63. L.-L. Shao, M. Chen, T.-Z. Ren, and Z.-Y. Yuan, "Ordered mesoporous carbon/graphene nano-sheets composites as counter electrodes in dye-sensitized solar cells," Journal of Power Sources, vol. 274, pp. 791–798, 2015.

64. W. Sun, T. Peng, Y. Liu, N. Huang, S. Guo, and X. Zhao, "Ordered mesoporous carbon-decorated reduced graphene oxide as efficient counter electrode for dye-sensitized solar cells," Carbon, vol. 77, pp. 18–24, 2014.

65. T. Battumur, S. H. Mujawar, Q. T. Truong et al., "Graphene/carbon nanotubes composites as a counter electrode for dye-sensitized solar cells," Current Applied Physics, vol. 12, no. 1, pp. e49–e53, 2012.

66. X. Miao, K. Pan, Q. Pan et al., "Highly crystalline graphene/carbon black composite counter electrodes with controllable content: synthesis, characterization and application in dye-sensitized solar cells,"Electrochimica Acta, vol. 96, pp. 155–163, 2013.

67. J. Velten, A. J. Mozer, D. Li et al., "Carbon nanotube/graphene nanocomposite as efficient counter electrodes in dye-sensitized solar cells," Nanotechnology, vol. 23, no. 8, Article ID 085201, 2012.

68. M. Stefik, J.-H. Yum, Y. Hu, and M. Grätzel, "Carbon-graphene nanocomposite cathodes for improved Co (II/III) mediated dye-sensitized solar cells," Journal of Materials Chemistry A, vol. 1, pp. 4982–4987, 2013.

69. L.-H. Chang, C.-K. Hsieh, M.-C. Hsiao et al., "A graphene-multi-walled carbon nanotube hybrid supported on fluorinated tin oxide as a counter electrode of dye-sensitized solar cells," Journal of Power Sources, vol. 222, pp. 518–525, 2013.

70. J. Ma, W. Shen, C. Li, and F. Yu, "Light reharvesting and enhanced efficiency of dye-sensitized solar cells based 3D-CNT/graphene counter electrodes," Journal of Materials Chemistry A, vol. 3, no. 23, pp. 12307–12313, 2015.

71. J. Ma, C. Li, F. Yu, and J. Chen, "'Brick-like' N-doped graphene/carbon nanotube structure forming three-dimensional films as high performance metal-free counter electrodes in dye-sensitized solar cells,"Journal of Power Sources, vol. 273, pp. 1048–1055, 2015.

72. W. Hong, Y. Xu, G. Lu, C. Li, and G. Shi, "Transparent graphene/PEDOT-PSS composite films as counter electrodes of dye-sensitized solar cells," Electrochemistry Communications, vol. 10, no. 10, pp. 1555–1558, 2008.

73. K. S. Lee, Y. Lee, J. Y. Lee, J.-H. Ahn, and J. H. Park, "Flexible and platinum-free dye-sensitized solar cells with conducting-polymer-coated graphene counter electrodes," ChemSusChem, vol. 5, no. 2, pp. 379–382, 2012.

74. G. Yue, J. Wu, Y. Xiao et al., "Functionalized graphene/poly(3,4-ethylenedioxythiophene):polystyrenesulfonate as counter electrode catalyst for dye-sensitized solar cells,"Energy, vol. 54, pp. 315–321, 2013.

75. B. He, Q. Tang, M. Wang, C. Ma, and S. Yuan, "Complexation of polyaniline and graphene for efficient counter electrodes in dye-sensitized solar cells: enhanced charge transfer ability," Journal of Power Sources, vol. 256, pp. 8–13, 2014.

76. Y.-S. Wang, S.-M. Li, S.-T. Hsiao et al., "Thickness-self-controlled synthesis of porous transparent polyaniline-reduced graphene oxide composites towards advanced bifacial dye-sensitized solar cells,"Journal of Power Sources, vol. 260, pp. 326–337, 2014.

77. W. Liu, Y. Fang, P. Xu et al., "Two-step electrochemical synthesis of polypyrrole/reduced graphene oxide composites as efficient pt-free counter electrode for plastic dye-sensitized solar cells," ACS Applied Materials and Interfaces, vol. 6, no. 18, pp. 16249–16256, 2014.

78. S. Das, P. Sudhagar, S. Nagarajan et al., "Synthesis of graphene-CoS electro-catalytic electrodes for dye sensitized solar cells," Carbon, vol. 50, no. 13, pp. 4815–4821, 2012.

79. H. Bi, W. Zhao, S. Sun et al., "Graphene films decorated with metal sulfide nanoparticles for use as counter electrodes of dye-sensitized solar cells," Carbon, vol. 61, pp. 116–123, 2013.

80. X. Duan, Z. Gao, J. Chang et al., "CoS_2-graphene

composite as efficient catalytic counter electrode for dye-sensitized solar cell," Electrochimica Acta, vol. 114, pp. 173–179, 2013.

81. E. Bi, H. Chen, X. Yang, W. Peng, M. Grätzel, and L. Han, "A quasi core–shell nitrogen-doped graphene/cobalt sulfide conductive catalyst for highly efficient dye-sensitized solar cells," Energy & Environmental Science, vol. 7, no. 8, pp. 2637–2641, 2014.

82. L. Sun, Y. Bai, N. Zhang, and K. Sun, "The facile preparation of a cobalt disulfide-reduced graphene oxide composite film as an efficient counter electrode for dye-sensitized solar cells," Chemical Communications, vol. 51, no. 10, pp. 1846–1849, 2015.

83. X. Miao, K. Pan, G. Wang et al., "Well-dispersed cos nanoparticles on a functionalized graphene nanosheet surface: a counter electrode of dye-sensitized solar cells," Chemistry—A European Journal, vol. 20, no. 2, pp. 474–482, 2014.

84. J. Huo, J. Wu, M. Zheng, Y. Tu, and Z. Lan, "High performance sponge-like cobalt sulfide/reduced graphene oxide hybrid counter electrode for dye-sensitized solar cells," Journal of Power Sources, vol. 293, pp. 570–576, 2015.

85. Z. Li, F. Gong, G. Zhou, and Z.-S. Wang, "NiS_2/reduced graphene oxide nanocomposites for efficient dye-sensitized solar cells," The Journal of Physical Chemistry C, vol. 117, no. 13, pp. 6561–6566, 2013.

86. X. Zuo, R. Zhang, B. Yang et al., "NiS nanoparticles anchored on reduced graphene oxide to enhance the performance of dye-sensitized solar cells," Journal of Materials Science: Materials in Electronics, vol. 26, no. 10, pp. 8176–8181, 2015.

87. J. Shen, R. Cheng, Y. Luo et al., "Growth of NiS/graphene nanocomposites for enhanced performance of dye sensitized solar cells," Journal of Solid State Electrochemistry, vol. 19, no. 4, pp. 1045–1052, 2015.

88. G. Yue, J.-Y. Lin, S.-Y. Tai, Y. Xiao, and J. Wu, "A catalytic composite film of MoS_2/graphene flake as a counter electrode for Pt-free dye-sensitized solar cells," Electrochimica Acta, vol. 85, pp. 162–168, 2012.

89. J.-Y. Lin, C.-Y. Chan, and S.-W. Chou, "Electrophoretic deposition of transparent MoS_2-graphene nanosheet composite films as counter electrodes in dye-sensitized solar cells," Chemical Communications, vol. 49, no. 14, pp. 1440–1442, 2013.

90. J.-Y. Lin, G. Yue, S.-Y. Tai et al., "Hydrothermal synthesis of graphene flake embedded nanosheet-like molybdenum sulfide hybrids as counter electrode catalysts for dye-sensitized solar cells," Materials Chemistry and Physics, vol. 143, no. 1, pp. 53–59, 2013.

91. B. Yang, X. Zuo, P. Chen et al., "Nanocomposite of tin sulfide nanoparticles with reduced graphene oxide in high-efficiency dye-sensitized solar cells," ACS Applied Materials & Interfaces, vol. 7, no. 1, pp. 137–143, 2015.

92. B. Yang, P. Chen, X. Zuo et al., "Structural phase transition from tin (IV) sulfide to tin (II) sulfide and the enhanced performance by introducing graphene in dye-sensitized solar cells," Electrochimica Acta, vol. 176, pp. 797–803, 2015.

93. G. Li, X. Chen, and G. Gao, "Bi_2S_3 microspheres grown on graphene sheets as low-cost counter-electrode materials for dye-sensitized solar cells," Nanoscale, vol. 6, no. 6, pp. 3283–3288, 2014.

94. E. Bi, Y. Su, H. Chen et al., "A hybrid catalyst composed of reduced graphene oxide/Cu_2S quantum dots as a transparent counter electrode for dye sensitized solar cells," RSC Advances, vol. 5, no. 12, pp. 9075–9078, 2015.

95. L. Zhou, J. Luo, X. Yang et al., "Graphene-wrapped $CuInS_2$ composites for efficient dye-sensitized solar cells," Functional Materials Letters, vol. 8, no. 1, Article ID 1550011, 2015.

96. M. Liu, G. Li, and X. Chen, "One-pot controlled synthesis of spongelike $CuInS_2$ microspheres for efficient counter electrode with graphene assistance in dye-sensitized solar cells," ACS Applied Materials & Interfaces, vol. 6, no. 4, pp. 2604–2610, 2014.

97. R. Bajpai, S. Roy, N. Koratkar, and D. S. Misra, "NiO nanoparticles deposited on graphene platelets as a cost-effective counter electrode in a dye sensitized solar cell," Carbon, vol. 56, pp. 56–63, 2013.

98. V.-D. Dao, L. L. Larina, K.-D. Jung, J.-K. Lee, and H.-S. Choi, "Graphene–NiO nanohybrid prepared by dry plasma reduction as a low-cost counter electrode material for dye-sensitized solar cells," Nanoscale, vol. 6, no. 1, pp. 477–482, 2014.

99. F. Gong, Z. Li, H. Wang, and Z.-S. Wang, "Enhanced electrocatalytic performance of graphene via incorporation of SiO_2 nanoparticles for dye-sensitized solar cells," Journal of Materials Chemistry, vol. 22, no. 33, pp. 17321–17327, 2012.

100. Q. Chang, Z. Ma, J. Wang et al., "Graphene nanosheets@ZnO nanorods as three-dimensional high efficient counter electrodes for dye sensitized solar cells," Electrochimica Acta, vol. 151, pp. 459–466, 2015.

101. Q. Zhang, Y. Liu, Y. Duan et al., "Mn_3O_4/graphene composite as counter electrode in dye-sensitized solar cells," RSC Advances, vol. 4, no. 29, pp. 15091–15097, 2014.

102. K. Xiong, G. Li, C. Jin, and S. Jin, "$La_{0.65}Sr_{0.35}MnO_3$@RGO nanocomposites as an effective counter electrode for dye-sensitized solar cells," Materials Letters, vol. 164, pp. 609–612, 2016.

103. F. Gong, H. Wang, and Z.-S. Wang, "Self-assembled monolayer of graphene/Pt as counter electrode for efficient dye-sensitized solar cell," Physical Chemistry Chemical Physics, vol. 13, no. 39, pp. 17676–17682, 2011.

104. G. Yue, J. Wu, Y. Xiao et al., "Platinum/graphene hybrid film as a counter electrode for dye-sensitized solar cells," Electrochimica Acta, vol. 92, pp. 64–70, 2013.

105. K. Saranya, N. Sivasankar, and A. Subramania, "Microwave-assisted exfoliation method to develop platinum-decorated graphene nanosheets as a low cost counter electrode for dye-sensitized solar cells,"RSC Advances, vol. 4, no. 68, pp. 36226–36233, 2014.

106. C.-T. Hsieh, B.-H. Yang, and J.-Y. Lin, "One- and two-dimensional carbon nanomaterials as counter electrodes for dye-sensitized solar cells," Carbon, vol. 49, no. 9, pp. 3092–3097, 2011.

107. G. Wang, W. Xing, and S. Zhuo, "Application of mesoporous carbon to counter electrode for dye-sensitized solar cells," Journal of Power Sources, vol. 194, no. 1, pp. 568–573, 2009.

108. H. Sun, D. Qin, S. Huang et al., "Dye-sensitized solar cells with NiS counter electrodes electrodeposited by a potential reversal technique," Energy & Environmental Science, vol. 4, no. 8, pp. 2630–2637, 2011.

109. F. Gong, H. Wang, X. Xu, G. Zhou, and Z.-S. Wang, "In situ growth of $Co_{0.85}Se$ and $Ni_{0.85}Se$ on conductive substrates as high-performance counter electrodes for dye-sensitized solar cells," Journal of the American Chemical Society, vol. 134, no. 26, pp. 10953–10958, 2012.

110. J. S. Jang, D. J. Ham, E. Ramasamy, J. Lee, and J. S. Lee, "Platinum-free tungsten carbides as an efficient counter electrode for dye sensitized solar cells," Chemical Communications, vol. 46, no. 45, pp. 8600–8602, 2010.

111. G. R. Li, F. Wang, J. Song, F. Y. Xiong, and X. P. Gao, "TiN-conductive carbon black composite as counter electrode for dye-sensitized solar cells," Electrochimica Acta, vol. 65, pp. 216–220, 2012.

112. B. Lei, G. R. Li, and X. P. Gao, "Morphology dependence of molybdenum disulfide transparent counter electrode in dye-sensitized solar cells," Journal of Materials Chemistry A, vol. 2, no. 11, pp. 3919–3925, 2014.

113. L. J. Brennan, M. T. Byrne, M. Bari, and Y. K. Gun'ko, "Carbon nanomaterials for dye-sensitized solar cell applications: a bright future," Advanced Energy Materials, vol. 1, no. 4, pp. 472–485, 2011.

114. S. Das, P. Sudhagar, V. Verma et al., "Amplifying charge-transfer characteristics of graphene for triiodide reduction in dye-sensitized solar cells," Advanced Functional Materials, vol. 21, no. 19, pp. 3729–3736, 2011.

115. X. Zhang, Y. Yang, S. Guo, F. Hu, and L. Liu, "Mesoporous $Ni_{0.85}Se$ nanospheres grown in situ on graphene with high performance in dye-sensitized solar cells," ACS Applied Materials & Interfaces, vol. 7, no. 16, pp. 8457–8464, 2015.

116. D. Yu, E. Nagelli, F. Du, and L. Dai, "Metal-free carbon nanomaterials become more active than metal catalysts and last longer," The Journal of Physical Chemistry Letters, vol. 1, no. 14, pp. 2165–2173, 2010.

117. J. Balamurugan, T. D. Thanh, N. H. Kim, and J. H. Lee, "Nitrogen-doped graphene nanosheets with FeN core-shell nanoparticles as high-performance counter electrode materials for dye-sensitized solar cells,"Advanced Materials Interfaces, vol. 3, no. 1, Article ID 1500348, 2016.

CHAPTER 8

HYDROTHERMAL FABRICATION OF SILVER NANOWIRES-SILVER NANOPARTICLES-GRAPHENE NANOSHEETS COMPOSITES IN ENHANCING ELECTRICAL CONDUCTIVE PERFORMANCE OF ELECTRICALLY CONDUCTIVE ADHESIVES

Hongru Ma[1], Jinfeng Zeng[1], Steven Harrington[2], Lei Ma[2], Mingze Ma[1], Xuhong Guo[1,3,4,5] and Yanqing Ma[1,4,5]

[1]School of Chemistry and Chemical Engineering, Shihezi University, Shihezi 832003, China

[2]Tianjin International Center of Nanoparticle and Nanosystems, Tianjin University, Tianjin 300072, China

[3]State Key Laboratory of Chemical Engineering, East China University of Science and Technology, Shanghai 200237, China

[4]Key Laboratory for Green Processing of Chemical Engineering of Xinjiang Bingtuan, Shihezi University, Shihezi 832003, China

[5]Engineering Research Center of Materials–Oriented Chemical Engineering of Xinjiang Bingtuan, Shihezi University, Shihezi 832003, China

ABSTRACT

Silver nanowires-silver nanoparticles-graphene nanosheets (AgNWs-AgNPs-GN) hybrid nanomaterials were fabricated through a hydrothermal method by using glucose as a green reducing agent. The charge carriers of AgNWs-AgNPs-GN passed through defect regions in the GNs rapidly with the aid of the AgNW and AgNP building blocks, leading to high electrical conductivity of electrically conductive adhesives (ECA) filled with AgNWs-AgNPs-GN. The morphologies of synthesized AgNWs-AgNPs-GN hybrid nanomaterials were characterized by field emission scanning electron microscope (FESEM), and high resolution transmission electron microscopy (HRTEM). X-ray diffraction (XRD) and laser confocal micro-Raman spectroscopy were used to investigate the structure of AgNWs-AgNPs-GN. The resistance of cured ECAs was investigated by the four-probe method. The results indicated AgNWs-AgNPs-GN hybrid nanomaterials exhibited excellent electrical properties for decreasing the resistivity of electrically conductive adhesives (ECA). The resistivity of ECA was 3.01 × 10^{-4} Ω·cm when the content of the AgNWs-AgNPs-GN hybrid nanomaterial was 0.8 wt %.

INTRODUCTION

Metal nanomaterials have attracted a great deal of research attention and have tremendous potential in a variety of high-performance antibacterial applications [1], catalysts [2,3,4,5], sensors [6,7,8], optoelectronic devices and electronics [9,10,11]. Metal-based nanomaterials usually have excellent thermal, mechanical and electrical properties [12] compared to conventional materials. Particularly, one-dimensional (1D) nanomaterials such as nanotubes [13,14], nanowires [15,16] and nanorods [17,18] fulfill an important role in the development of conductive nanomaterials. The higher aspect ratio and smaller dimensions of 1D nanomaterials allow them to effectively transport electrical carriers along one controllable path and they have demonstrated that an increase in their aspect ratios will ultimately increase the electrical conductivity of composites [19,20]. It is expected that distinct properties from the 1D nanomaterials will be exhibited in the metal-based nanomaterials' hybrid nanomaterial. Many kinds of 1D

Citation: Hongru Ma, Jinfeng Zeng, Steven Harrington, Lei Ma, Mingze Ma, Xuhong Guo and Yanqing Ma, Hydrothermal Fabrication of Silver Nanowires-Silver Nanoparticles-Graphene Nanosheets Composites in Enhancing Electrical Conductive Performance of Electrically Conductive Adhesives, doi:10.3390/nano6060119.

nanomaterials have been investigated to produce conductive composites, especially AgNWs, and they have drawn prodigious interest for several reasons, such as their stable chemical properties and high electrical conductivity [21,22]. Furthermore, random networks of AgNWs are promising candidates for conductive devices. As a result, AgNWs have been examined for potential use in many applications including chemical and biological sensors and electrically conductive adhesives (ECAs).

For this paper, silver nanoparticles (AgNPs) were introduced into ECAs. This is because AgNPs can fill gaps in two possible ways. In the first, some of the conductive paths are formed by filling gaps between linked AgNWs. In the second, more conductive paths can be formed by filling the gaps between unattached silver flakes. In addition, AgNWs may improve the conductivity of ECAs in yet another way. They are excellent materials to supply junctions for wire-built networks, and so the number of conductive pathways can be increased and the effective contact area can be enlarged. The contact resistance of AgNWs is less than that of AgNPs in ECA samples. Therefore, the conductivity of ECAs is improved in the presence of both AgNWs and AgNPs. Taking this information into account, the conductivity of ECAs with a mixture of AgNWs-AgNPs-GN is better than that of ECAs containing only AgNPs-GN.

Regardless of whether AgNWs or AgNPs are used, both tend to coalesce during applications due to their high surface energies and van der Waals forces [23]. However, silver-graphene hybrid nanomaterials are very fascinating materials that combine the key features of silver and graphene. In this paper, AgNWs-AgNPs-graphene nanosheets hybrid nanomaterial is used as conductive filler in ECA instead of partial silver flakes. The electrical conductivity of the ECA can be increased by the inclusion of conductive fillers into the insulating resin matrix. Graphene has drawn interest as well because of its large surface area and high electrical conductivity [24,25,26,27]. However, graphene synthesis from graphene oxide introduces a considerable amount of defects, reducing electrical conductivity [28,29]. This work demonstrates that a system containing AgNWs and AgNPs is an efficient way to provide the electronic transmission channels of Ag-Ag junctions on the surface of graphene [30], while simultaneously decreasing the resistivity of the ECA containing AgNWs-AgNPs-GN hybrid nanomaterials.

EXPERIMENTAL SECTION

Materials

Natural graphite flakes (325 mesh) were obtained from Alfa Aesar. Hydrochloric (HCl, 38%) and hydrogen peroxide (H_2O_2, 30%) were purchased from Tianjin Fuyu Fine Chemical Co. Ltd. (Tianjin, China). Potassium permanganate ($KMnO_4$, ≥99.5%) and glucose were supplied by Tianjin Shengyu Chemical Co. Ltd. Sodium nitrate ($NaNO_3$, ≥99.0%) was supplied by Aladdin industrial Co. (Shanghai, China). Concentrated sulfuric acid H_2SO_4 (>98%) was obtained from Chengdu Area Kelong Chemical Co. Ltd. (Chengdu, China). Silver nitrate ($AgNO_3$, >99.8%) was purchased from Xi'an Chemical Co. (Xi'an, China). Polyvinylpyrrolidone (PVP, K30) and sodium chloride (NaCl, AR) were obtained from Tianjin Guangfu Fine Chemical Research Institute. Glycerolwas obtained from Tianjin Zhiyuan Chemical Reagent Co. Ltd. (Tianjin, China). Epoxy resin (862), used as epoxy binder was purchased from Guangzhou Topyon Trade Co. Ltd. (Guangzhou, China). The catalysts 1-(2-Cyanoethyl)-2-ethyl-4-methylimidazde (2E4MZ-CN) and 4-Methylcyclohexane-1, 2-dicarboxylic Anhydride (MHHPA) were obtained from TCI (Shanghai) Development Co. Ltd. The micro silver flakes were purchased from Strem Chemicals, Inc. (Newburyport, MA, USA). All reagents and solvents were used without further purification.

Synthesis of AgNWs

The AgNWs were synthesized in ways similar to what has already been described in literature [31]. Briefly, 5.86 g PVP was slowly added into 190 mL glycerol in a three-necked round bottle flask until all PVP was dissolved with gentle mechanical stirring (50 rpm) and heating. After cooling, $AgNO_3$ (1.58 g) was added into the mixed solution. The flask was then immersed into oil bath. The glycerol solution (10 mL) of NaCl (59 mg) and H_2O (0.5 mL) was added into the three-necked round bottle flask. The mixture was heated to 160 °C. When the temperature reached 160 °C, heating ceased and the mixture was allowed to cool to room temperature. Gentle mechanical stirring (50 rpm) was constantly applied during the entire process of AgNWs growth. After, the obtained product was washed with water through centrifugation, and subsequently re-dispersed in water.

Preparation of Graphene Oxide and AgNWs-AgNPs-GN

Graphene oxide (GO) was prepared using a modified Hummers method [32]. In a typical procedure, the glucose (0.08 g) was added to graphene oxide (60 mL, 1 mg/mL) under stirring. Then 90 mg AgNWs were added to above solution with ultra-sonication for 1 h, and the mixture solution was heated in the stainless steel autoclave with a Teflon liner of 100 mL capacity under the condition of 180 °C for 18 h. After the autoclave was air-cooled to room temperature unaided, the product was isolated by centrifugation and washed with distilled water, then dried by lyophilization for characterization and experiments.

Preparation of the ECA Based on AgNWs-Ag-NPs-GN

The polymer matrix contains epoxy resin, 2E4MZ-CN and HMMPA in a weight ratio of 1:0.0185:0.85. The content of AgNWs-AgNPs-GN hybrid nanomaterial varies from 0.0%, 0.2%, 0.5%, 0.8% to 1.1%, and 30% polymer matrix. Two strips of polyimide adhesive tape were affixed to the surface of a cleaned glass slide separated by a distance of 3 mm. The uncured ECA mixture was bladed uniformly into the gap between the two strips. The polyimide tapes were removed from the glass slide when the ECA was cured at a temperature of 150 °C for 2 h.

CHARACTERIZATION OF AGNWS-AGNPS-GN

The dispersion and morphology of the AgNWs-AgNPs-GN were characterized by field emission scanning electron microscopy (FE-SEM) on a Hitachi Limited (Tokyo, Japan) and high resolution transmission electron microscopy (HR-TEM) on a FEI Tecnai G20 (FEI, Hillsboro, OR, USA). The X-ray diffraction spectra (XRD) measurements were performed using a Bruker D8 advanced X-ray diffractometer (Bruker, Germany) with Cu Kα irradiation (λ = 1.5406 Å) in the scanning angle range of 10°–90° at a scanning rate of 10°/min at 40 mA and 40 kV, which was used to investigate the crystalline structure of the AgNWs-AgNPs-GN. The Raman spectra of AgNWs-AgNPs-GN were obtained using laser confocal micro-Raman spectroscopy (LabRAM HR800, Horiba Jobin Yvon, Paris, France) equipped with a 532 nm laser. X-ray photoelectron spectroscopy (XPS) was completed with a monochromatic MgKαX-ray source (*hv* = 1486.6 eV) on an AMICUS/ESCA 3400 (Shimadzu, Japan), which was used to analyze the elements and the functional groups on the surface of AgNWs-AgNPs-GN. The thickness of the cured ECA was measured by a thickness gauge (Shanghai Chuanlu Measuring Tool Limited Liability Co., Shanghai, China). Resistance was measured by the RTS-8 four point probe Resistivity Measurement system (Guangzhou, China).

RESULTS AND DISCUSSION

The synthesis of the AgNWs-AgNPs-GN hybrid nanomaterial is illustrated in Figure 1. GO was synthesized according to a modified Hummers method and dispersed in water by ultrasonic treatment to form a brown suspension. The AgNWs were prepared via reduction by glycerol. Then an appropriate amount of AgNWs was added into the GO aqueous dispersion and sonicated for 2 h to achieve a homogeneous mixture. Next, the mixture containing a certain amount of glucose was added to a stainless steel autoclave with a capacity of 100 mL and Teflon liner. The color of the mixture changed from brown to black, indicating that GN was obtained. The AgNWs were fused and produced some AgNPs due to the high reaction temperature, but as a result the lengths of the AgNWs were reduced. Fortunately, the AgNPs filled the gaps among the AgNWs and formed the conductive network.

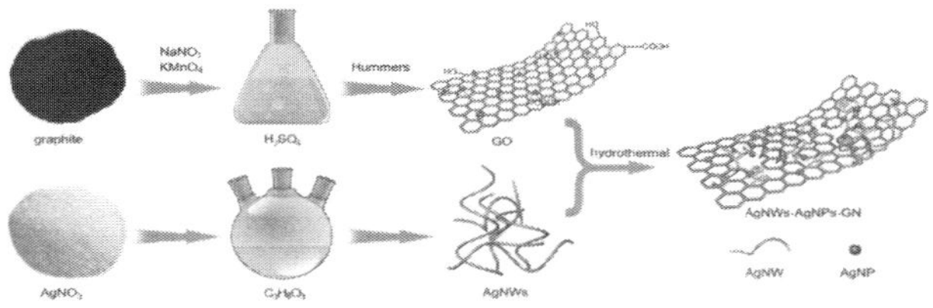

Figure 1: Schematic of preparing the silver nanowires-silver nanoparticles-graphene nanosheets (AgNWs-AgNPs-GN).

The AgNWs as shown in Figure 2A possessed an average length of about 4 μm and a diameter of about 50 nm. A large number of AgNWs agglomerated due to high surface energy. There was an irregularly large number of AgNWs and AgNPs that coexisted on the GN layers observed from Figure 2B, and the agglomeration of AgNWs decreased distinctly.

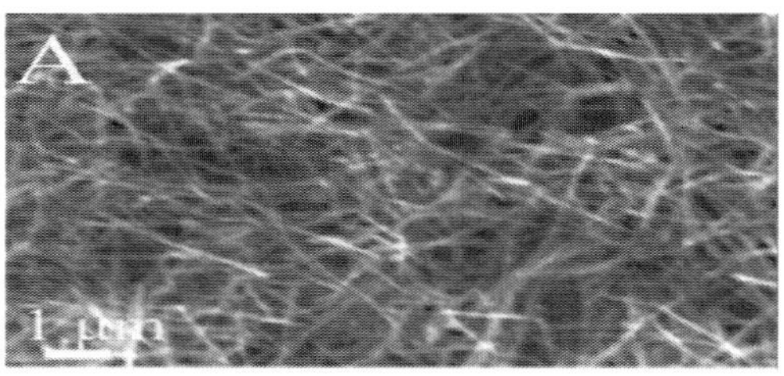

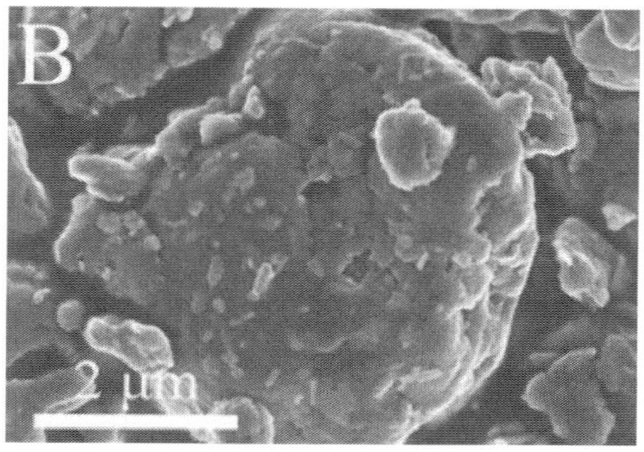

Figure 2: Field emission scanning electron microscopy (FE-SEM) images of the AgNWs (A) and AgNWs-AgNPs-GN (B).

Figure 3 present the high resolution transmission electron microscopy (HRTEM) images of the synthesized AgNWs-AgNPs-GN hybrid nanomaterial with different magnifications. It can be seen that some AgNWs and AgNPs were present on the surface of the GN. The average diameter of the AgNWs was 55 nm and their lengths ranged from 100 nm to 500 nm. The average diameter of the AgNPs was measured to be 90 nm. In addition, it can be observed that AgNWs and AgNPs are in the same plane and form conductivity networks on the surface of the GN (denoted by the red lines). These conductivity networks may improve the electrical properties of AgNWs-AgNPs-GN hybrid nanomaterial and further enhance the electrical properties of ECAs containing AgNWs-AgNPs-GN hybrid nanomaterial.

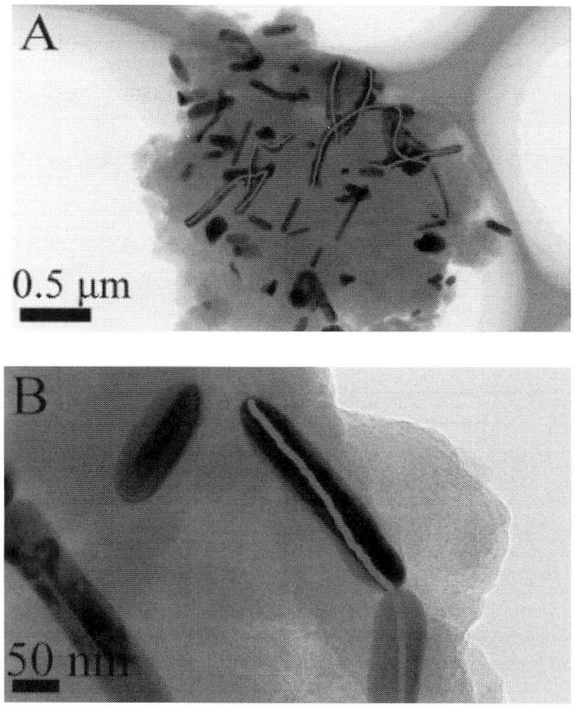

Figure 3: High resolution transmission electron microscopy (HR-TEM) images of the AgNWs-AgNPs-GN (A) and magnification of one segment of the AgNWs-AgNPs-GN (B).

The crystal structures of the AgNWs-AgNPs-GN hybrid nanomaterial were characterized using an X-ray powder diffractometer (XRD) (Figure 4). It can be seen that there is an intense and sharp diffraction peak in Figure 4b at 2θ = 11.09°, which is attributed to the (001) lattice plane of GO.

An interlayer d-spacing between the GO nanosheets can be calculated at 0.80 nm. This indicated the oxygen-containing functional groups were introduced into the GO layers [33]. For AgNWs-AgNPs-GN hybrid nanomaterials (Figure 4a), the characteristic diffraction peak of GO disappeared, indicating that the GO was reduced through a hydrothermal method by using glucose as a reducing agent. The peaks at 2θ = 38.17°, 44.24° and 64.57° can be assigned to the (111), (200) and (220) crystalline planes of silver, respectively [34]. These peaks match well with the data from the PDF card (PDF2-2004) for silver. The XRD results show that the silver was present on the surface of the GN, which corresponded to what was observed via SEM and TEM.

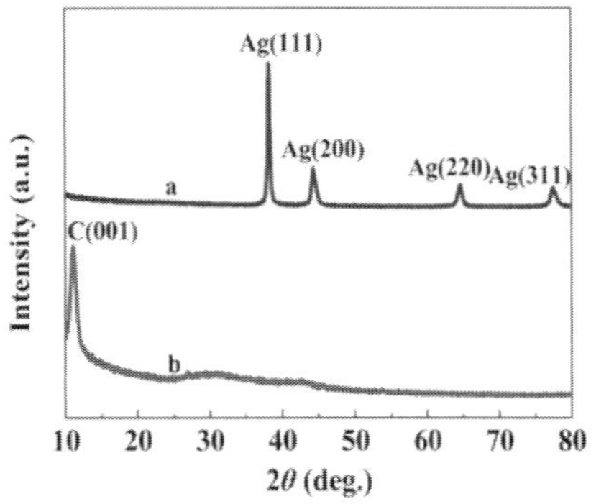

Figure 4: X-ray powder diffractometer (XRD) patterns of graphene oxide (GO) and AgNWs-AgNPs-GN (a: AgNWs-GNs, b: GO).

Raman spectroscopy is very sensitive to the microstructure of nanomaterials, and so it was used to illustrate the GN crystalline structure. As shown in Figure 5, the D band at 1351.3 cm^{-1} represents structural defects related to the size of the sp^2 domains [5,35]. The G band at 1589.6 cm^{-1} corresponds to the ordered sp^2 hybridized graphene structure [36]. The relative intensity ratio of the D and G band elucidates the disorder degree of GN. It was calculated that the I_D/I_G ratio was 1.11 for AgNWs-AgNPs-GN (Figure 5a), 1.14 for GO (Figure 5b) and 0.86 for GNs (Figure 5c). The increase of the I_D/I_G ratio of AgNWs-AgNPs-GN is due to the incorporation of AgNWs-AgNPs, in which the AgNWs-AgNPs graft on the surface of the GNs and replace some of the sp^2 carbon sites and generate more sp^3 carbon forms. The intensity ratio of I_D/I_G of both the AgNWs-AgNPs-GN and the GN decreases compared to that of GO, suggesting there is a decrease of oxygen-functional groups on GO.

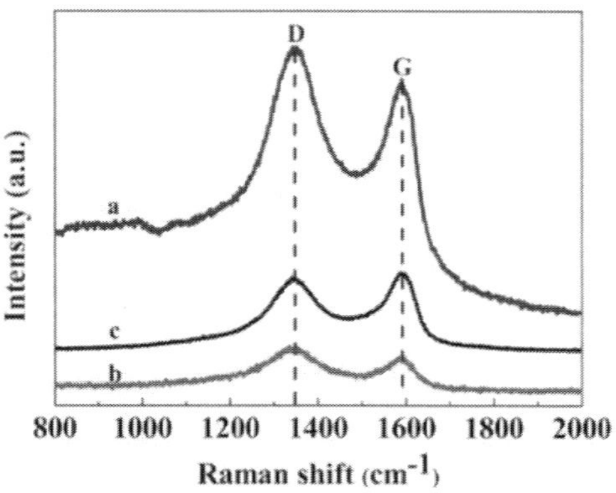

Figure 5: Raman spectra of GO, GN and AgNWs-AgNPs-GN (a: AgNWs-GNs, b: GO, c: GN).

XPS was used to provide further insight into the nanostructure of AgNWs-AgNPs-GN and GO, the results of which are shown in Figure 6. The peaks of the elements C and O appeared at 287.6 and 533.1 eV, respectively. The C, O atomic mass ratio of AgNWs-AgNPs-GN was 5.46, which was a significant increase compared to that of GO (1.78). This ratio indicated that the oxygen-containing functional groups on the surface of the GO were effectively removed via a hydrothermal method. The insert is the XPS signature of the Ag 3d doublet ($3d_{5/2}$ and $3d_{3/2}$) for the AgNWs-AgNPs deposited on the rGO. The Ag $3d_{5/2}$ and $3d_{3/2}$ peaks of AgNWs-AgNPs-GN nanocomposites appeared at 368.3 and 374.3 eV, respectively. The spin-orbit splitting of doublet components for Ag $3d_{5/2}$ and $3d_{3/2}$ was calculated to be 6 eV, indicating the presence of Ag in the hybrid nanomaterial [37].

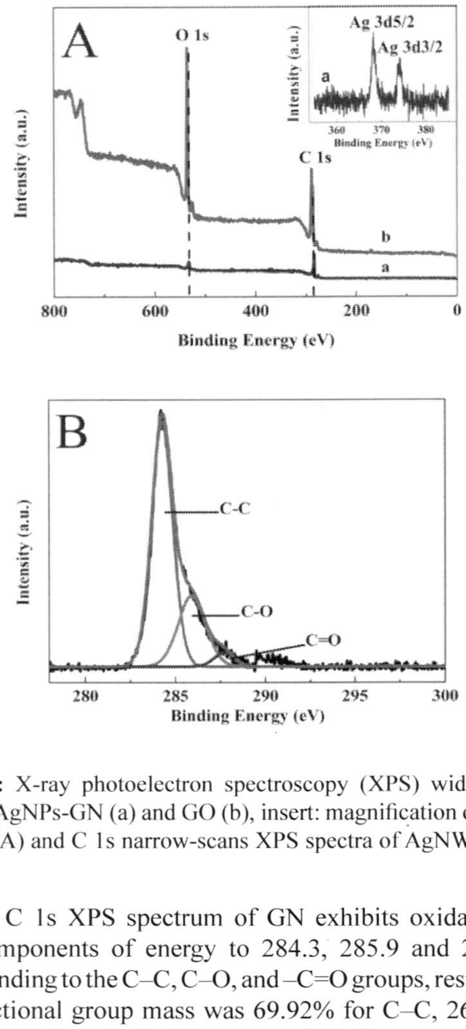

Figure 6: X-ray photoelectron spectroscopy (XPS) wide-scans of AgNWs-AgNPs-GN (a) and GO (b), insert: magnification of segment of a line (A) and C 1s narrow-scans XPS spectra of AgNWs-AgNPs-GN (B).

The C 1s XPS spectrum of GN exhibits oxidation with three components of energy to 284.3, 285.9 and 287.7 eV, corresponding to the C–C, C–O, and –C=O groups, respectively. The functional group mass was 69.92% for C–C, 26.28% for C–O, 3.80% for –C=O. The mass of C–C was much higher than that of GO [38], providing further evidence that oxygen-containing functional groups were effectively removed.

MEASUREMENTS OF ELECTRICAL PROPERTIES

The uncured ECAs were bladed into five identical specimens. Five points for each specimen were selected after the ECA was cured at 150 °C for 2 h. The resistance of these five points was measured by the four-point probe resistivity measurement system. Then the resistivity of ECAs was calculated using Equation (1).

$$\rho=R_L\times T \quad (1)$$

where ρ is the calculated resistivity value of cured ECA (Ω·cm); R_L issquare resistance (Ω); and T is the thickness of cured ECA (cm).

The ECAs were prepared via mixing matrix resin and silver flakes. AgNWs-AgNPs-GN was introduced into the mixture in order to fill the gaps between the unattached silver flakes, as well as to build a new conductivity network in the matrix resin. The resistivity of ECAs containing AgNWs-AgNPs-GN decreased with the increase of the AgNWs-AgNPs-GN content, which was observed from Figure 7. The resistivity of the ECAs dipped as low as 3.01×10^{-4} Ω·cm when the content of AgNWs-AgNPs-GN reached 0.8%. However, the resistivity of the ECAs significantly increased after the content of AgNWs-AgNPs-GN surpassed 0.8%. One cause of this phenomenon is that AgNPs filled the gaps between the AgNWs, and the AgNWs-AgNPs formed a random conductive network on the surface of graphene which improved the number of conductive paths. Therefore, the resistivity of the AgNWs-AgNPs-GN hybrid nanomaterial was decreased (as shown inFigure 8). Another cause is that the conductive networks are formed in matrix resin due to AgNWs-AgNPs-GN linking to unattached silver flakes, or they connect each other. However, excess AgNWs-AgNPs-GN led to a significant increase of resistivity of ECAs and resulted in electrical performance deterioration. When the content of AgNWs-AgNPs-GN filler increased to 1.1 wt %, the resistivity of the ECAs increases and the conductivity performance goes down [39]. The optimal content of AgNWs-AgNPs-GN is 0.8 wt %. The attractive interactions between conductive fillers contributed to generating an agglomeration-free conductive network and a more conductive channel [40].

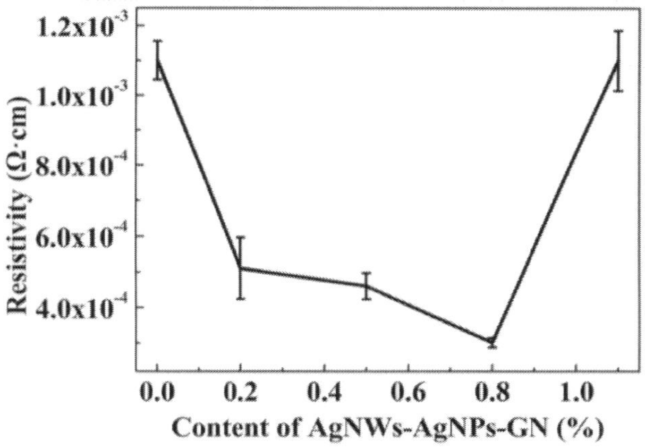

Figure 7: Resistivity of electrically conductive adhesives (ECA)-changing trend image under different contents of AgNWs-AgNPs-GN.

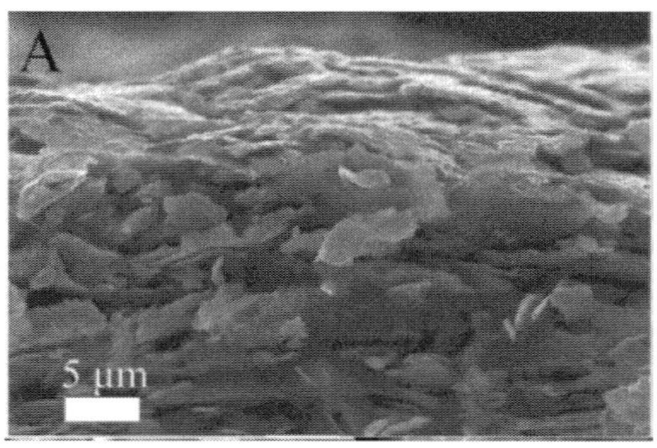

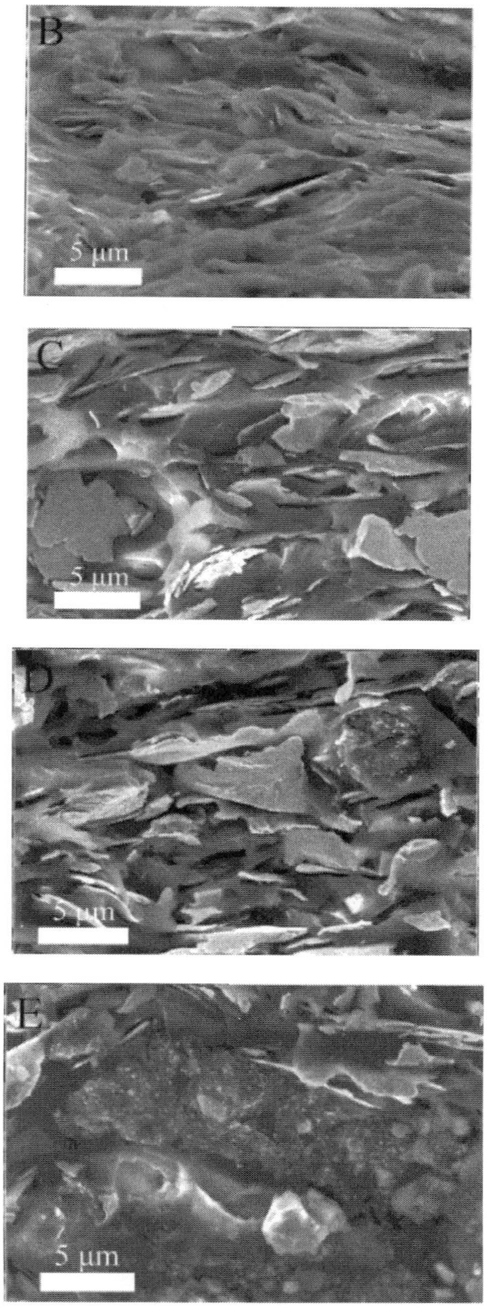

Figure 8: SEM images of cross-section morphology of electrically conductive adhesives (ECA) filled with 0.0 wt % AgNWs-AgNPs-GN (A), 0.2 wt % AgNWs-AgNPs-GN (B), 0.5 wt % AgNWs-Ag-NPs-GN (C), 0.8 wt % AgNWs-AgNPs-GN (D), and 1.1 wt % Ag-NWs-AgNPs-GN (E).

As shown in the SEM images of the cross-section of cured ECAs in Figure 8, for the ECA filled with silver flakes only, there were many gaps among the silver flakes providing more electrical pathways. The AgNWs-AgNPs-GN can be inserted between the gaps of silver flakes when a small amount of AgNWs-AgNPs-GN is added. It can also be observed that the conductive fillers were enwrapped by the epoxy resin without agglomerating and adhering into the blocks seen in Figure 8B,C. The aggregation began to form when the content of AgNWs-AgNPs-GN reached 0.8%. This provided a desirable environment for the uniform distribution of the conductive fillers, and a favorable conductive network was formed. However, when a great quantity of AgNWs-AgNPs-GN was added, the AgNWs-AgNPs-GN at a high content easily agglomerated as shown in Figure 8D,E, which created a large number of contact points between the silver flakes, increasing the tunneling resistance and further decreasing the electric properties of the ECAs.

CONCLUSIONS

Electrically conductive adhesive incorporated with AgNWs-AgNPs-GN can be prepared with high electrical conductivity. In this study, AgNWs-AgNPs-GN nanomaterial was chosen as a nanofiller and exhibited a well-dispersed morphology and decreased the electrical resistivity of the electrically conductive adhesive. A physical network formation of AgNWs-AgNPs was indicative of when the AgNWs-AgNPs-GN content was increased. The electrically conductive adhesive prepared by 69.2% silver flakes and 0.8% AgNWs-AgNPs-GN reached an electrical resistivity of 3.01×10^{-4} Ω·cm. It is expected that this electrically conductive adhesive with AgNWs-AgNPs-GN will be a promising material for the development of electronic packaging in the near future.

ACKNOWLEDGMENTS

This work was financially supported by the Program for Changjiang Scholars and Innovative Research Team in University (PCSIRT, No.IRT1161) and the Program of Science and Technology Innovation Team in Bingtuan (No. 2011CC001).

AUTHOR CONTRIBUTIONS

X.H.G. and Y.Q.M. designed and administered the experiments. H.R. Ma, J.F. Zeng and Y.Q. Ma performed experiments. M.Z. Ma collected data. X.H. Guo and Y.Q. Ma gave technical support and conceptual advice. S. Harrington and L. Ma modified the whole manuscript in English sentences. All authors contributed to the analysis and discussion of the data and writing the manuscript.

REFERENCES

1. Yuan, W.; Gu, Y.; Li, L. Green synthesis of graphene/Ag nanocomposites. *Appl. Surf. Sci.* 2012, *261*, 753–758.
2. Hui, K.S.; Hui, K.N.; Dinh, D.A.; Tsang, C.H.; Cho, Y.R.; Zhou, W.; Hong, X.; Chun, H.H. Green synthesis of dimension-controlled silver nanoparticle-graphene oxide with *in situ* ultrasonication. *Acta Mater.* 2014, *64*, 326–332.
3. Zhang, C.; Li, C.; Chen, Y.; Zhang, Y. Synthesis and catalysis of Ag nanoparticles trapped into temperature-sensitive and conductive polymers. *J. Mater. Sci.* 2014, *49*, 6872–6882.
4. Davis, D.J.; Raji, A.R.O.; Lambert, T.N.J.; Vigil, A.; Li, L.; Nan, K.; Tour, J.M. Silver-Graphene Nanoribbon Composite Catalyst for the Oxygen Reduction Reaction in Alkaline Electrolyte. *Electroanalysis* 2014, *26*, 164–170.
5. Zahed, B.; Monfared, H.H. A comparative study of silver-graphene oxide nanocomposites as a recyclable catalyst for the aerobic oxidation of benzyl alcohol: Support effect. *Appl. Surf. Sci.* 2015, *328*, 536–547.
6. Deng, X.; Tang, H.; Jiang, J. Recent progress in graphene-material-based optical sensors. *Anal. Bioanal. Chem.* 2014, *406*, 6903–6916.
7. Yang, B.; Liu, Z.; Guo, Z.; Zhang, W.; Wan, M.; Qin, X.; Zhong, H. *In situ* green synthesis of silver-graphene oxide nanocomposites by using tryptophan as a reducing and stabilizing agent and their application in SERS. *Appl. Surf. Sci.*2014, *316*, 22–27.
8. Nia, P.M.; Lorestani, F.; Meng, W.P.; Alias, Y. A novel non-enzymatic H_2O_2 sensor based on polypyrrole nanofibers–silver nanoparticles decorated reduced graphene oxide nano composites. *Appl. Surf. Sci.* 2015, *332*, 648–656.
9. Huang, S.Y.; Zhang, K.; Yuen, M.M.F.; Fu, X.Z.; Sun, R.; Wong, C.P. Facile synthesis of flexible graphene-silver composite papers with promising electrical and thermal conductivity performances. *RSC Adv.* 2014, *4*, 34156–34161.
10. Cho, H.; Somu, S.; Lee, J.Y.; Jeong, H.; Busnaina, A. High-rate nanoscale offset printing process using directed assembly and transfer of nanomaterials. *Adv. Mater.* 2015, *27*, 1759–1766.
11. Tsai, S.L.; Liu, Y.K.; Pan, H.; Liu, C.H.; Lee, M.T. The Coupled Photothermal Reaction and Transport in a Laser Additive Metal Nanolayer Simultaneous Synthesis and Pattering for Flexible Electronics. *Nanomaterials* 2016, *6*.
12. Ho, P.H.; Liou, Y.T.; Chuang, C.H.; Lin, S.W.; Tseng, C.Y.; Wang, D.Y.; Chen, C.C.; Hung, W.Y.; Wen, C.Y.; Chen, C.W. Self-crack-filled graphene films by metallic nanoparticles for high-performance graphene heterojunction solar cells.*Adv. Mater.* 2015, *27*, 1724–1729.
13. Kwon, Y.B.S.; Yim, J.; Kim, M.; Kim, J. Dispersion, hybrid interconnection and heat dissipation properties of functionalized carbon nanotubes in epoxy composites for electrically conductive adhesives (ECAs). *Microelectron. Reliab.* 2011, *51*, 812–818.
14. Cui, H.W.; Li, D.S.; Fan, Q. Using a functional epoxy, micron silver flakes, nano silver spheres, and treated single-wall carbon nanotubes to prepare high performance electrically conductive adhesives. *Electron. Mater. Lett.* 2013, *9*, 299–307.
15. Li, L.; Jia, J.L.; Tong, K.; Ren, Z.; Hu, W.; Niu, X.F.; Chen, Y.S.; Pei, Q.B. Silver Nanowire Percolation Network Soldered with Graphene Oxide at Room Temperature and Its Application for Fully Stretchable Polymer Light-Emitting Diodes. *ACS Nano* 2014, *8*, 1590–1600.
16. Tran, Q.T.; Hoa, H.T.M.; Yoo, D.H.; Cuong, T.V.; Hur, S.H.; Chung, J.S.; Kim, E.J.; Kohl, P.A. Reduced graphene oxide as an over-coating layer on silver nanostructures for detecting NH_3 gas at room temperature. *Sens. Actuators B Chem.*2014, *194*, 45–50.
17. Si, G.; Zhao, Y.; Lv, J.; Lu, M.; Wang, F.; Liu, H.; Xiang, N.; Huang, T.J.; Danner, A.J.; Teng, J.; Liu, Y.J. Reflective plasmonic color filters based on lithographically patterned silver nanorod arrays. *Nanoscale* 2013, *5*, 6243–6248.

18. He, Y.; Lu, G.; Shen, H.; Cheng, Y.; Gong, Q. Strongly enhanced Raman scattering of graphene by a single gold nanorod. *Appl. Phys. Lett.* 2015, *107*.
19. Liu, S.; Weng, B.; Tang, Z.R.; Xu, Y.J. Constructing one-dimensional silver nanowire-doped reduced graphene oxide integrated with CdS nanowire network hybrid structures toward artificial photosynthesis. *Nanoscale* 2015, *7*, 861–866.
20. Zhi, J.; Zhao, W.; Liu, X.; Chen, A.; Liu, Z.; Huang, F. Highly Conductive Ordered Mesoporous Carbon Based Electrodes Decorated by 3D Graphene and 1D Silver Nanowire for Flexible Supercapacitor. *Adv. Funct. Mater.* 2014,*24*, 2013–2019.
21. Peng, P.; Hu, A.; Huang, H.; Gerlich, A.P.; Zhao, B.; Zhou, Y.N. Room-temperature pressureless bonding with silver nanowire paste: Towards organic electronic and heat-sensitive functional devices packaging. *J. Mater. Chem.* 2012, *22*, 12997–13001.
22. Wei, Y.; Chen, S.; Li, F.; Lin, Y.; Zhang, Y.; Liu, L. Highly stable and sensitive paper-based bending sensor using silver nanowires/layered double hydroxides hybrids. *ACS Appl. Mater. Interfaces* 2015, *7*, 14182–14191.
23. Dar, R.A.; Khare, N.G.; Cole, D.P.; Karna, S.P.; Srivastava, A.K. Green synthesis of a silver nanoparticle-graphene oxide composite and its application for As(III) detection. *RSC Adv.* 2014, *4*, 14432–14440.
24. Song, B.; Li, L.; Lin, Z.; Wu, Z.K.; Moon, K.S.; Wong, C.P. Water-dispersible graphene/polyaniline composites for flexible micro-supercapacitors with high energy densities. *Nano Energy* 2015, *16*, 470–478.
25. Li, Y.; Tang, J.; Huang, L.; Wang, Y.; Liu, J.; Ge, X.; Tjong, S.C.; Li, R.K.Y.; Belfiore, L.A. Facile preparation, characterization and performance of noncovalently functionalized graphene/epoxy nanocomposites with poly(sodium 4-styrenesulfonate). *Compos. Part A Appl. Sci. Manuf.* 2015, *68*, 1–9.
26. Joshi, A.C.; Markad, G.B.; Haram, S.K. Rudimentary simple method for the decoration of graphene oxide with silver nanoparticles: Their application for the amperometric detection of glucose in the human blood samples. *Electrochim. Acta* 2015, *161*, 108–114.
27. Choi, H.J.; Jung, S.M.; Seo, J.M.; Chang, D.W.; Dai, L.; Baek, J.B. Graphene for energy conversion and storage in fuel cells and supercapacitors. *Nano Energy* 2012, *1*, 534–551.
28. Naito, K.; Yoshinaga, N.; Tsutsumi, E.; Akasaka, Y. Transparent conducting film composed of graphene and silver nanowire stacked layers. *Synth. Metals* 2013, *175*, 42–46.
29. Lee, M.S.; Lee, K.S.; Kim, Y.; Lee, H.; Park, J.; Choi, K.H.; Kim, H.K.; Kim, D.G.; Lee, D.Y.; Nam, S.; *et al.* High-performance, transparent, and stretchable electrodes using graphene-metal nanowire hybrid structures. *Nano Lett.*2013, *13*, 2814–2821.
30. Yang, S.B.; Choi, H.; Leeda, S.; Choi, C.G.; Choi, S.Y.; Kim, I.D. Improved optical sintering efficiency at the contacts of silver nanowires encapsulated by a graphene layer. *Small* 2015, *11*, 1293–1300.
31. Zeng, J.F.; Tian, X.; Song, J.S.; Wei, Z.; Steven, H.; Yao, Y.G.; Ma, L.; Ma, Y.Q. Green Synthesis of AgNPs/ Reduced Graphene Oxide Nanocomposites and Effect on the Electrical Performance of Electrically Conductive Adhesives. *J. Mater. Sci. Mater. Electron.* 2016, *27*, 3540–3548.
32. Willian, S.; Hummers, J.R.; Richard, E.O. Preparation of Graphitic Oxide. *J. Am. Chem. Soc.* 1958, *80*.
33. Li, J.; Xie, H.; Li, Y.; Liu, J.; Li, Z. Electrochemical properties of graphene nanosheets/polyaniline nanofibers composites as electrode for supercapacitors. *J. Power Sources* 2011, *196*, 10775–10781.
34. Hsu, K.C.; Chen, D.H. Microwave-assisted green synthesis of Ag/reduced graphene oxide nanocomposite as a surface-enhanced Raman scattering substrate with high uniformity. *Nanoscale Res. Lett.* 2014, *9*, 1–9.
35. Song, B.; Sizemore, C.; Li, L.; Huang, X.; Lin, Z.; Moon, K.S.; Wong, C.P. Triethanolamine functionalized graphene-based composites for high performance supercapacitors. *J. Mater. Chem. A* 2015, *3*, 21789–21796.
36. Silva, L.C.O.; Silva, G.G.; Ajayan, P.M.; Soares, B.G. Long-term behavior of epoxy/graphene-based composites determined by dynamic mechanical analysis. *J. Mater. Sci.* 2015, *50*, 6407–6419.
37. Huang, Q.; Wang, J.; Wei, W.; Yan, Q.; Wu, C.; Zhu, X. A facile and green method for synthesis of reduced graphene oxide/Ag hybrids as efficient surface enhanced Raman scattering platforms. *J. Hazard. Mater.* 2015, *283*, 123–130.
38. Yang, C.; Gu, H.; Lin, W.; Yuen, M.M.; Wong, C.P.; Xiong, M.; Gao, B. Silver nanowires: From scalable synthesis to recyclable foldable electronics. *Adv. Mater.* 2011, *23*, 3052–3056.
39. Cui, H.W.; Fan, Q.; Li, D.S. Surface functionalization of micro silver flakes and their application in electrically conductive adhesives for electronic package. *Int. J. Adhes. Adhes.* 2014, *48*, 177–182.
40. Amoli, B.M.; Hu, A.; Zhou, N.Y.; Zhao, B. Recent progresses on hybrid micro-nano filler systems for electrically conductive adhesives (ECAs) applications. *J. Mater. Sci. Mater. Electron.* 2015, *26*, 4730–4745.

CHAPTER 9

INVESTIGATION OF THE STRUCTURAL, ELECTRICAL, AND OPTICAL PROPERTIES OF THE NANO-SCALE GZO THIN FILMS ON GLASS AND FLEXIBLE POLYIMIDE SUBSTRATES

Fang-Hsing Wang[1], Kun-Neng Chen[2], Chao-Ming Hsu[3], Min-Chu Liu[1] and Cheng-Fu Yang[4]

[1]Department of Electrical Engineering and Graduate Institute of Optoelectronic Engineering, National Chung Hsing University, Taichung 402, Taiwan

[2]Department of Electrical Engineering, Kun-Shan University, Tainan 710, Taiwan

[3]Department of Mechanical Engineering, National Kaohsiung University of Applied Science, Kaohsiung 807, Taiwan

[4]Department of Chemical and Materials Engineering, National University of Kaohsiung, Kaohsiung 81141, Taiwan

ABSTRACT

In this study, Ga_2O_3-doped ZnO (GZO) thin films were deposited on glass and flexible polyimide (PI) substrates at room temperature (300 K), 373 K, and 473 K by the radio frequency (RF) magnetron sputtering method. After finding the deposition rate, all the GZO thin films with a nano-scale thickness of about 150 ± 10 nm were controlled by the deposition time. X-ray diffraction patterns indicated that the GZO thin films were not amorphous and all exhibited the (002) peak, and field emission scanning electron microscopy showed that only nano-scale particles were observed. The dependences of the structural, electrical, and optical properties of the GZO thin films on different deposition temperatures and substrates were investigated. X-ray photoemission spectroscopy (XPS) was used to measure the elemental composition at the chemical and electronic states of the GZO thin films deposited on different substrates, which could be used to clarify the mechanism of difference in electrical properties of the GZO thin films. In this study, the XPS binding energy spectra of $Ga2p_{3/2}$ and $Ga2p_{1/2}$ peaks, $Zn2p_{3/2}$ and $Zn2p_{1/2}$ peaks, the Ga3d peak, and O_{1s} peaks for GZO thin films on glass and PI substrates were well compared.

INTRODUCTION

Recent advances in nanotechnology have contributed to the development of photon upconversion materials as promising new-generation candidates of fluorescent bioprobes and spectral converters for biomedical and optoelectronic applications [1]. Also, dimension-reduced process to make them approaching two-dimensional nanostructures into precisely controlled lamellar or called layer nanomaterials are currently achievable. In particular, layer-by-layer (LbL) assembly is known as a highly versatile method for fabrication of controlled layered structures from various kinds of component materials [2]. Zhuang *et al.* published a review focusing on networks‘ growth processes of metal-organic frameworks and crystalline coordination from the view of surface chemistry [3]. They mainly focused on the methods involving liquid phases, and those methods can be summarized under the term of “liquid phase epitaxy” (LPE). The LPE method can deposit thin films having crystalline, high quality coatings and with predefined orientations as the thin films are deposited on the different substrates. Indium-tin oxide (ITO)-based transparent conduction oxide (TCO) thin films have been widely used in the applications of solar cells, flat panel displays [4], and other optoelectronic products due to their excellent conductivity and transparency. Much more interest has been given to ZnO-based thin films such as undoped ZnO thin films [5], Al-doped

Citation: Fang-Hsing Wang, Kun-Neng Chen, Chao-Ming Hsu, Min-Chu Liu and Cheng-Fu Yang, Investigation of the Structural, Electrical, and Optical Properties of the Nano-Scale GZO Thin Films on Glass and Flexible Polyimide Substrates, doi:10.3390/nano6050088.

ZnO (AZO) thin films [6], and F-doped ZnO (FZO) thin films [7] because they have the stability properties even under hydrogen plasma. For that, they are potential candidates for the applications of solar cell technology based on thin-film silicon.

Al-doped ZnO (AZO) are the major materials for ZnO-based thin films, because they present favorable electrical properties and can be investigated as indium-free TCO thin films. In spite of studies on the preparation, characterization, and effect of doping on the properties of ZnO-based thin films, certain effects of either some dopants or preparation procedures still remain unclear. Because Ga is less receptive to oxidation, Ga_2O_3-doped ZnO TCO materials have been reported to have a better stable property [8,9]. When Ga_2O_3 is served as a dopant, the resistivity of Ga_2O_3-doped ZnO thin films is lower than that using other group-III elements [10,11], because Ga has a more compatible covalent bond length with Zn than other group-III elements [12]. TCO thin films deposited on different substrates [13], for example glass and polyimide (PI), are widely used throughout the semiconductor and electronics industries. The polymer substrates are cheaper, lighter, and more flexible as compared with the conventional glass substrates; they could be effectively used in applications such as flexible display and flexible solar cells. For that, the necessity of studying the deposition process of TCO thin films on polymer substrates has been increased, as the polymer substrates are suitable for flat-panel displays (FPDs) and optoelectronics [14,15,16].

Many researchers have reported on the Ga_2O_3-doped ZnO thin films in regard to the different doping concentrations of Ga_2O_3. In the past, the highly conductive and transparent Ga_2O_3-doped ZnO thin films had been deposited at high growth rates by radio frequency magnetron sputtering. In the present paper, a ceramic target of ZnO with 3 wt % Ga_2O_3($ZnO:Ga_2O_3$ = 97:3 in wt %, abbreviated as GZO) was used to deposit the GZO thin films on the Eagle 173 glass and flexible polyimide (PI) substrates by using the radio frequency (RF) magnetron sputtering process. Then the detail analysis of defects for nano-scale thin films on different substrtaes could be achieved by X-ray photoelectron spectroscopy (XPS) [3]. In this study, the dependences of the structural, electrical, and optical properties of the GZO thin films on different deposition temperatures and substrates were investigated. Also, XPS is a surface-sensitive and a quantitative spectroscopic technique that can be used to measure the variation of electron or x-ray excited from one lower energy layer to one higher energy layer. In order to clarify the mechanism of difference in electrical properties, we would investigate the change of the chemical structures of the GZO thin films deposited on different substrates by XPS.

EXPERIMENTAL PROCEDURES

In this work, the radio frequency (RF) (13.56 MHz) magnetron sputtering process was used to deposit the GZO thin films. ZnO (97 wt %, 5 N, Admat Inc., Norristown, PA, USA) and Ga_2O_3 (3 wt %, 5 N, Admat Inc., Norristown, PA, USA) were mixed, ground, calcined at 1273 K for 2 h, and sintered at 1673 K for 2 h to form the ceramic target with a 2 in diameter. The area of Eagle 173 glass (Corning Inc., New York, NY, USA) and polyimide (abbreviated as PI, Taimide Tech. Inc., Hsinchu County, Taiwan) were used as substrates and their areas were about 33 mm × 33 mm. Before the deposition process was started, the base chamber pressure of the sputtering system was pumped to less than 1×10^{-6} Torr, then the deposition parameters were controlled at different pressures and powers. The optimal deposition parameters were a RF power of 50 W and a working pressure of 5 mTorr because the deposited GZO thin films had the flattest surface and the acceptable deposition rate. During the deposition process, only pure argon (99.999%) was introduced in the chamber, the flow rate of argon was 20 sccm. Deposition temperature was also used as a parameter, where GZO thin films were deposited at room temperature (RT, 300 K), 373 K, and 473 K, respectively. The RT was characterized as that the chamber was unheated and the desired temperatures of 373 K and 473 K were achieved by heating of the sample stage using a neon lamp and monitoring the temperature using a thermocouple. Thickness of the GZO thin films was one of the most important parameters to influence the characteristics of the GZO thin films. For this reason, thicknesses of the GZO thin films were measured using field emission scanning electron microscopy (FESEM, JSM-6700, JEOL, Tokyo, Japan). Deposition rate and thickness of the GZO thin films were determined by averaging five data sets obtained by FESEM. After calculating the deposition rate, the thicknesses of all GZO thin films were about 150 ± 10 nm by controlling the deposition time. The roughness (or flatness) was measured using atomic force microscopy (AFM), and the crystalline structure was measured using X-ray diffraction (XRD) patterns with Cu Kα radiation. The electrical properties of the GZO thin films were determined by Hall effect measurement (Ecopia, HMS-3000, Bridge Technol., Chandler Heights, AZ, USA) at room temperature using the Van der Pauw method with four pressed indium balls onto the corners of the samples under a 0.55 T magnetic field. While the optical transmittances of the GZO thin films on glass and PI substrates were measured by using an ultraviolet-visible spectroscopy (UV-Vis) spectrophotometer (Hitachi U3300, Kenichi Sato, Osaka, Japan). The surface chemical composition and bonding of the GZO thin films were analyzed using the X-ray photoemission spectroscopy (XPS, ULVAC-PHI, 5000 Versaprobe, Physical (PHI) Electronics, Chanhassen, MN, USA) after the thin films' surfaces were pre-cleaning for 3 min by sputtering. The parameters for XPS were monochromatized Al Kα 187.85 eV, presputter setting: 2 kV, sputter area: 2 mm × 2 mm, sputter rate: 12.5 nm/min, sputter time: 0.3 min, and the Ar ions with energy of ~2 keV, respectively.

RESULTS AND DISCUSSION

Because the GZO thin films were deposited using a sputtering method in a pure Ar atmosphere, we believe that substrate would be the important factor that would influence the properties of the deposited thin films. We used FESEM to observe the surface morphologies of the GZO thin films deposited on different substrates under different deposition temperatures. The morphologies of the GZO thin films as a function of the deposition substrate and temperature are shown in Figure 1, which indicates that as those deposition parameters are changed, the surface morphologies apparently changed as well. AsFigure 1a–e show, only nano-scale particles were observed, and the surface morphologies had no apparent change, regardless of the variations of substrates and deposition temperatures. As RT (Figure 1a,b) and 373 K (as Figure 1c,d show) were used as the deposition temperatures, the GZO thin films deposited on glass substrates had less pores than those deposited on PI substrates. As different deposition temperatures were used, the morphology of the GZO thin films exhibited a roughness surface. However, the variations of average crystallization sizes are dependent on the substrate and deposition temperature and they have not easily been calculated from the surface observation. We will illustrate the variations of grain sizes from the XRD patterns and the following equation [17], and the results are also shown in Table 1:

$$D = (0.9\lambda)/(\beta\cos\theta) \quad (1)$$

where D is the size of nano-scale particles; λ is 1.54 Å; β is the full width at half-maximum (FWHM); and θ is the diffraction angle, respectively. As Table 1 shows, the crystallization size increased with the increase of the deposition temperature. This is caused by the fact that as the deposition temperature increases, the surface energy of the GZO thin films during the deposition process increases, and the chance for particles' growth of the GZO thin films increases, and then their crystallization sizes increase.

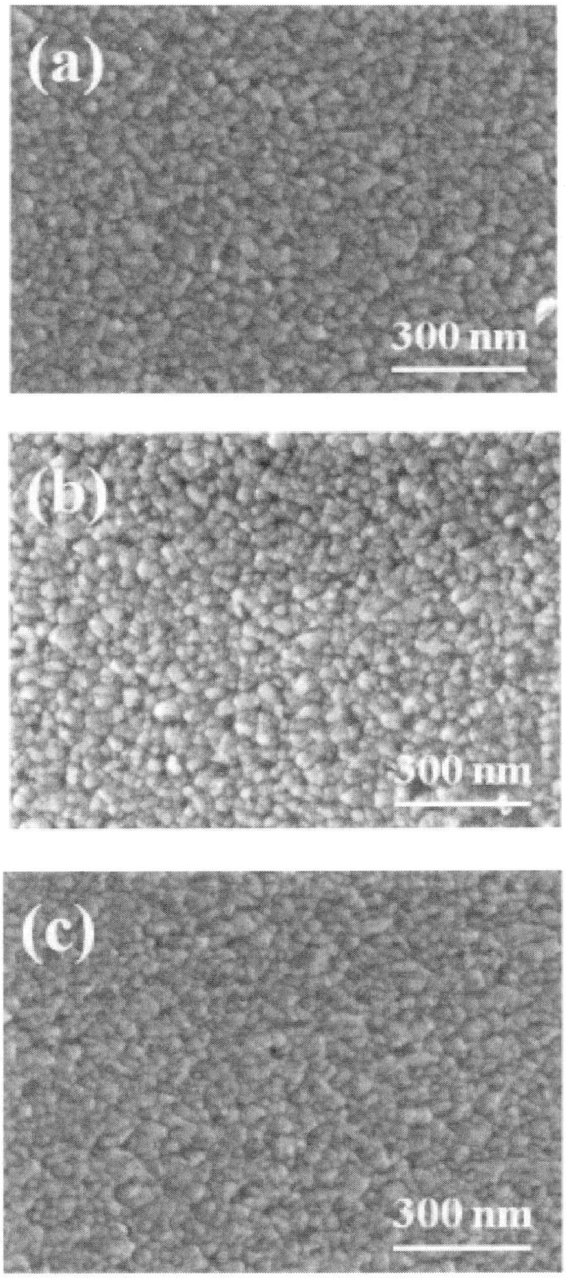

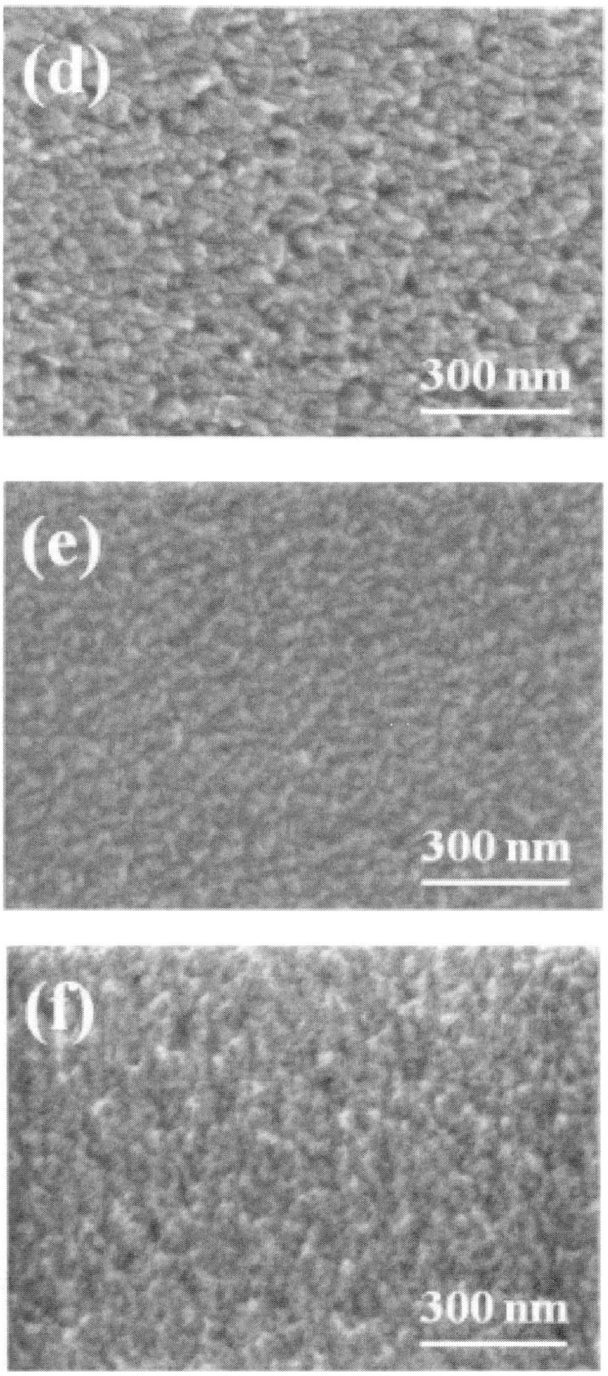

Figure 1: Surface morphologies of the Ga_2O_3-doped ZnO (GZO) thin films as a function of substrate and deposition temperature. (**a**) Glass; (**b**) polyimide (PI); (**c**) 373 K-Glass; (**d**) 373 K-PI; (**e**) 473 K-Glass; and (**f**) 473 K-PI, respectively.

Table 1: The 2θ value of the (002) peak, crystallite size (*D*), lattice constant (c), the full width at half-maximum (FWHM) value, and roughness of the Ga_2O_3-doped ZnO (GZO) thin films deposited on different substrates and at different temperatures and ZnO deposited on different substrates. RT: room temperature; PI: polyimide

Substrates	2θ (°)	Crystallite Size (nm)	c (nm)	FWHM (°)	Roughness (nm)
RT glass	34.20	31.34	0.5240	0.407	1.46
RT PI	34.14	28.12	0.5249	0.412	1.44
373 K glass	34.24	24.71	0.5234	0.336	1.25
373 K PI	34.18	23.08	0.5243	0.360	1.22
473 K glass	34.36	20.42	0.5216	0.265	1.12
473 K PI	34.36	20.18	0.5216	0.296	1.10
RT ZnO glass	34.12	23.42	0.5252	0.382	1.24
RT ZnO PI	34.10	24.45	0.5255	0.406	1.26

The surface roughness was also calculated as a function of substrate and deposition temperature; the results of the RT-deposited GZO thin films are shown in Figure 2 and the results of all the GZO thin films are also compared in Table 1. The root mean square (RMS) surface roughness of the GZO thin films on glass and PI substrates was measured to be 0.30 nm by AFM. As the deposition temperature was RT, 373 K, and 473 K, the measured values were 1.46 nm, 1.25 nm, and 1.12 nm when the substrate was glass and the measured values were 1.44 nm, 1.22 nm, and 1.10 nm when the substrate was PI, respectively. Apparently, the flatness was improved as the deposition temperature was raised.

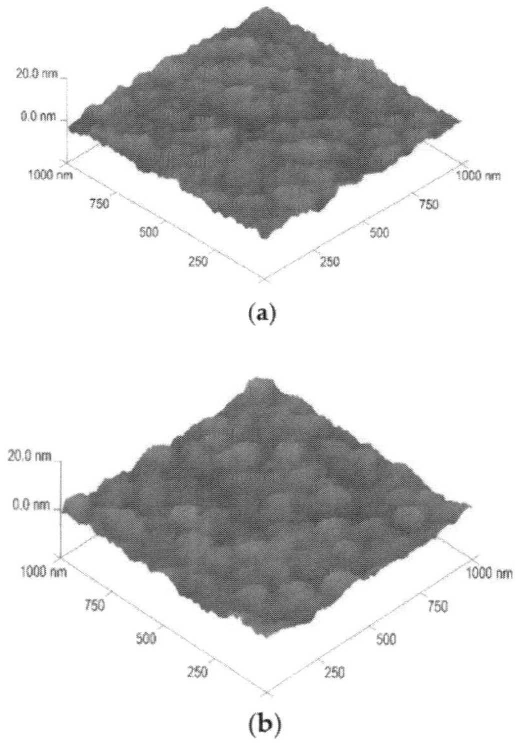

Figure 2: Atomic force microscopy (AFM) analysis of the GZO thin films as a function of substrate at room temperature. (**a**) Glass and (**b**) PI, respectively.

XRD patterns of the GZO thin films as a function of deposition temperature and substrate were shown in Figure 3, and all the GZO thin films exhibited the (002) peak. XRD patterns of the RT-deposited ZnO thin films are also added in Figure 3 as a reference for the GZO thin films. When glass was used as the substrate, the (002) peaks of the GZO thin films prepared with deposition temperatures of RT, 373 K, and 473 K were situated at 2θ = 34.20°, 34.24°, and 34.36°; when PI was used as the substrate, the (002) peaks of the GZO thin films prepared with deposition temperatures of RT, 373 K, and 473 K were situated at 2θ = 34.14°, 34.18°, and 34.36°, respectively. The (002) peaks of the GZO thin films prepared with a deposition temperature of RT on glass and PI substrates were situated at 2θ = 34.12° and 34.10°. The lattice constant (c) was calculated by using the 2θ values shown in Figure 3 and the results were also compared in Table 1. When glass was used as the substrate, the calculated lattice constant was 0.5240 nm, 0.5234 nm, and 0.5215 nm for deposition temperatures of RT, 373 K, and 473 K; when PI was used as the substrate, the calculated lattice constant was 0.5249 nm, 0.5243 nm, and 0.5215 nm for deposition temperatures of RT, 373 K, and 473 K, respectively. The calculated lattice constants of the ZnO thin films were 0.5252 nm and 0.5255 nm when glass and PI were used as substrates. All the calculated lattice constants c of the GZO thin films in Table 1 being smaller than those of the ZnO thin films is significant, because the radius of Ga^{3+} ions (62 pm) is smaller than that of Zn^{2+} ions (72 pm). Table 1 also shows an important result: the lattice constant of the GZO thin films decreases with the increase of the deposition temperature, independent of the substrate. The lattice parameters of TCO-based thin films usually depend on the concentration of foreign atoms, defects, external strain, and the difference of their ionic radii with respect to the substituted matrix ions. Those results suggest that as the deposition temperature increases, more Ga^{3+} ions will substitute the sites of Zn^{2+} ions and the lattice constant decreases. The study of Li *et al.* suggested that the oxygen vacancies might reduce the lattice parameters, at least the c value [18].

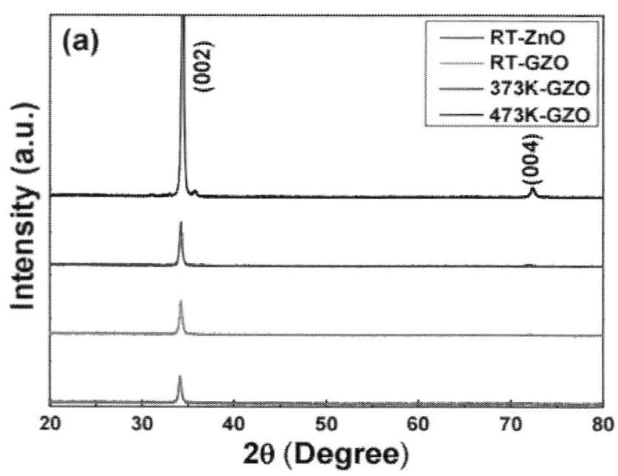

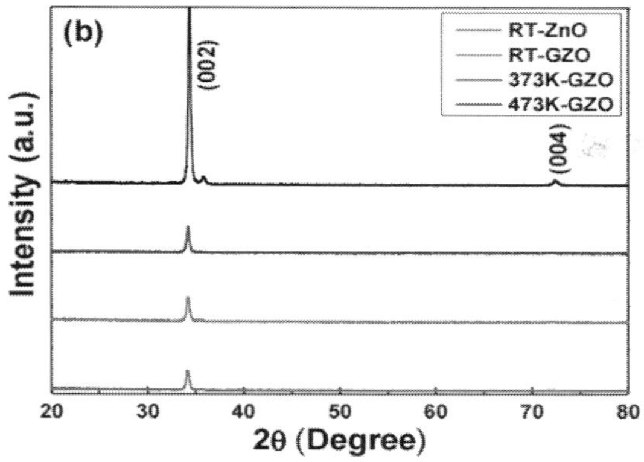

Figure 3: X-ray diffraction (XRD) patterns of the undoped ZnO thin films and GZO thin films as a function of deposition temperature and on different substrates: (**a**) glass and (**b**) PI. a.u.: arbitrary unit.

As Figure 3 shows, the full width at half maximum (FWHM) values of the (002) peak of the GZO thin films deposited on different substrates were calculated, and the results are compared in Table 1. The FWHM values of the (002) peak for GZO thin films deposited on glass substrates were 0.407, 0.336, and 0.265 for substrate temperatures of RT, 373 K, and 473 K; the FWHM values of the (002) peak for GZO thin films deposited on PI substrates were 0.412, 0.360, and 0.292 for substrate temperatures of RT, 373 K, and 473 K, respectively. These results suggest that the GZO thin films deposited at higher temperatures have a better crystalline structure.

Figure 4 shows the transmission spectra of the GZO thin films plotted against the different deposition temperature and on different substrates; the measured wavelength was in the region of 300–1000 nm. The transmission spectrum of the RT-deposited ZnO thin films is also added in Figure 4 as a reference for the GZO thin films. For the GZO thin films, the absorption edges were blue-shifted and the transmittance ratios were higher as compared with that of the RT-deposited ZnO thin films. As Figure 4 shows, the maximum transmittance ratios of all the GZO thin films in the range of visible light (400–700 nm) were more than 95%, regardless of deposition temperature and substrate. We know that many factors will affect the transmission spectrum of the GZO thin films. The increase of the transmission ratio in the optical band is caused by the increase in carrier density, because a higher deposition temperature can cause the GZO thin films to have fewer defects. The higher deposition temperature could even cause the GZO thin films to have better crystallization; the average and maximum transmittance ratios of all the GZO thin films in the region of visible light had no apparent variation as the deposition temperature increased. The reason is that the optical band gap (*Eg*) of the GZO thin films is in the region of ultraviolet light, thus the optical energy in the region of visible light cannot excite the electrons, and then the GZO thin films in the range of visible light will have high transmittance ratios. As Figure 4 shows, the average transmittance ratios of all the GZO thin films in the range of visible light were about 90.9%–92.8% and in the range of near-infrared light (700–1500 nm) they were about 84.8%–88.9%, respectively. The detailed optical properties of the GZO thin films deposited on different substrates and at different temperatures and ZnO thin films deposited on different substrates are compared in Table 2.

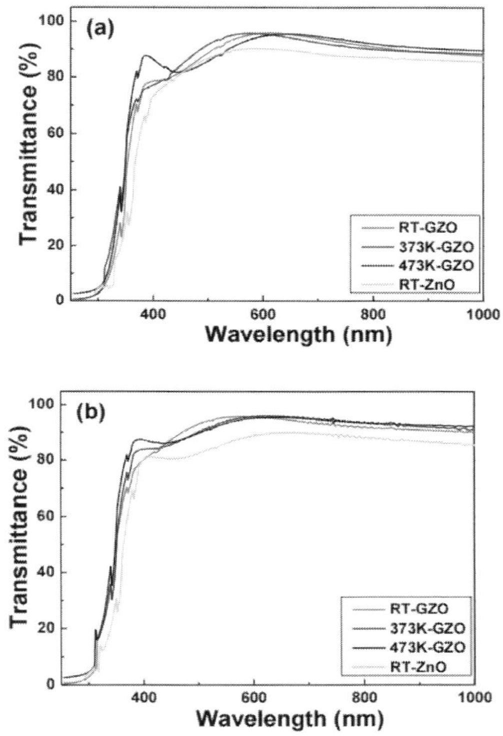

Figure 4: Transmittance spectra of the undoped ZnO thin films and GZO thin films deposited as a function of deposition temperature and different substrates: (**a**) glass and (**b**) PI.

Table 2: Optical properties and the optical band gap (*Eg*) values of the GZO thin films deposited on different substrates and at different temperatures and ZnO thin films deposited on different substrates

Substrates	Transmittance Ratio-Visible	Transmittance Ratio-Infrared	Maximum Transmittance Ratio	Absorption Edge (nm)	*Eg* Value (eV)
RT glass	90.9%	85.0%	95.9%	360	3.612
373 K glass	91.6%	84.8%	95.8%	354	3.674
473 K glass	90.5%	87.2%	95.6%	352	3.677
RT PI	92.8%	86.8%	95.8%	356	3.551
373 K PI	92.4%	87.0%	96.1%	354	3.580
473 K PI	92.4%	88.9%	95.5%	350	3.622
RT ZnO glass	87.1%	82.6%	90.2%	370	3.320
RT ZnO PI	86.1%	82.1%	89.9%	368	3.317

In the transmission spectra of all the GZO thin films, a greater sharpness was noticeable in the curves of the absorption edges, and the optical band edges had no apparent shift as the deposition temperature increased. In the past, determination of the optical band gap (Eg) was often necessary to develop the electronic band structure of a thin-film material. However, using extrapolation methods, the Eg values of the GZO thin films can be determined from the absorption edge for direct interband transition, which can be calculated using the relation in Equation (3) as follows:

$$\alpha h\nu = C \times (h\nu - Eg)^{1/2} \tag{2}$$

where C is the constant for the direct transition, h is Planck's constant, and ν is the frequency of the incident photon [19]; α is the optical absorption coefficient, which is calculated using Lambert's law as follows:

$$\alpha = \ln((1/T)/h) \tag{3}$$

where T and h are the thin film's transmittance ratio and thickness.

Figure 5 plots (αhν) against (hν) (energy) in accordance with Equation (3), and the Eg values can be found by extrapolating a straight line at (αhν) = 0. The calculated Eg values of the GZO thin films as a function of the deposition temperature and on different substrates are also shown in Figure 5. The linear dependence of (αhν) on (hν) indicates that the GZO thin films are direct transition-type semiconductors. As the deposition temperature increased from RT to 473 K and glass was used as the substrate, the Eg values increased from 3.612 eV to 3.677 eV and the wavelength of the absorption edge decreased from about 360 nm (transmittance ratio over 60%) to about 352 nm, where the wavelengths of the absorption edge were blue-shifted as compared with those investigated by Roth *et al.* [20]. As the deposition temperature increased from RT to 373 K and PI was used as the substrate, the Eg values increased from 3.551 eV to 3.622 eV and the wavelength of the absorption edge decreased from 356 nm to 350 nm. As compared with the results deposited at room temperature by Gong *et al.*, the deposited GZO thin films had Eg values changed from 3.42 eV to 3.50 eV [21] and had an absorption edge at the wavelength of about 400 nm. Apparently, for all the GZO thin films investigated in this study the Eg values were larger than those investigated by Gong *et al.* The Eg values are 3.320 eV (glass) and 3.317 eV (PI) in this study (Eg is about 3.24 eV in [22]). As compared with those of the RT-deposited ZnO thin films, the reason for the blue-shift in the absorption edge of the GZO thin films (Figure 3) is caused by the increase of the Eg values. The detailed Eg values of the GZO thin films deposited on different substrates and at different temperatures and ZnO thin films deposited on different substrates are also compared in Table 2.

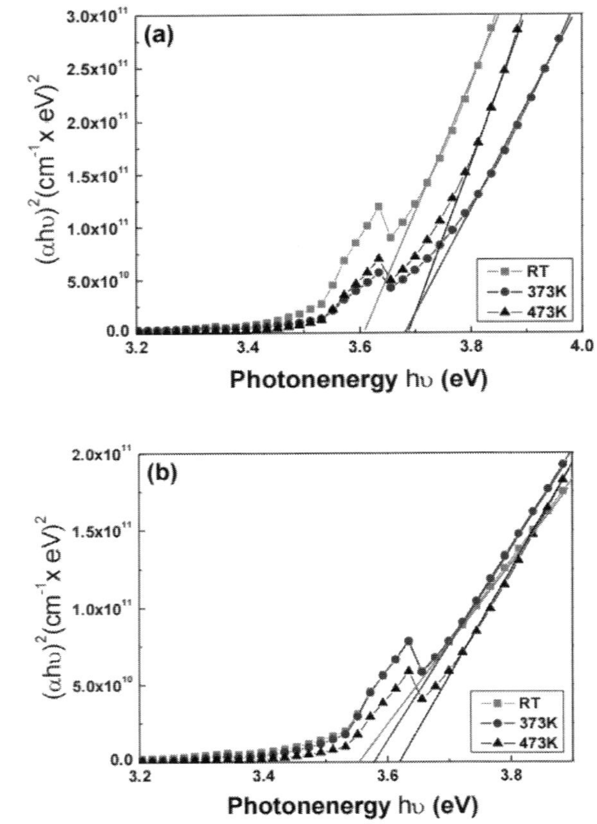

Figure 5: The $(\alpha h\nu)^2$ *vs.* hν-Eg plots of the GZO thin films deposited as a function of deposition temperature and on different substrates: (**a**) glass and (**b**) PI.

Two competing phenomena are generally agreed to be dominant in affecting the absorption edge in heavily doped semiconductors. The first is the well-known Burstein-Moss band-filling effect which negatively shifts the measured band-edge energy with the decrease of the carrier concentration; the second phenomenon that affects the optical absorption edge is the increase of donor density, which will cause a change in the nature and strength of the interaction potentials between donors and the host crystal. The carrier concentration, which will be shown in Figure 5a, is higher than that investigated by Roth *et al.* [20], and the wavelength of the absorption edge can be shifted to a lower value of about 350 nm. The Eg value of Ga_2O_3 is about 4.9 eV [23] and the Eg value of ZnO is about 3.24 eV. We believe that the Eg values of all the deposited GZO thin films higher than those of the ZnO thin films are caused by the addition of Ga_2O_3 into the ZnO.

The variations of the carrier concentration, carrier mobility, and resistivity of the GZO thin films at different deposition temperatures on glass and PI substrates are compared Figure 6. Figure 6a shows that the carrier concentration of the GZO thin films increased with the raising deposition temperature. When the GZO thin films are deposited on glass or PI substrates by using the RF sputtering process, many defects and pores will exist and inhibit the electrons' movement. Using a higher deposition temperature during the deposition process can lead to an enhancement of the thin films' densification and crystallization; the increase in the diffraction intensity of the (002) peak, as shown in Figure 3, has proven those results. That can decrease the numbers of defects and pores in the GZO thin films and can decrease the inhibition of the barrier electron transportation [24]. Also, as the deposition temperature was raised from RT to 473 K, the carrier (electron) concentration increased from 8.92 × 10^{20} to 11.1 × 10^{20} cm^{-3} for glass substrates and from 8.74 × 10^{20} to 11.6 × 10^{20} cm^{-3} for PI substrates. Figure 6b shows that the carrier mobility of the GZO thin films increased with the raising deposition temperature and was almost independent of the used substrates. Figure 6b also shows that as the same deposition temperature was used, the carrier mobility of the GZO thin films on glass substrates was higher than that of the GZO thin films on PI substrates. As the deposition temperature was raised from RT to 473 K, as Figure 6b shows, the carrier mobility increased from 3.58 to 8.25 cm^2/V-s for glass substrates and increased from 2.92 to 8.28 cm^2/V-s for PI substrates. The improvement in electron mobility could be attributed to the re-crystallization of the GZO thin films that takes place during the deposition process with the increase of the deposition temperature.

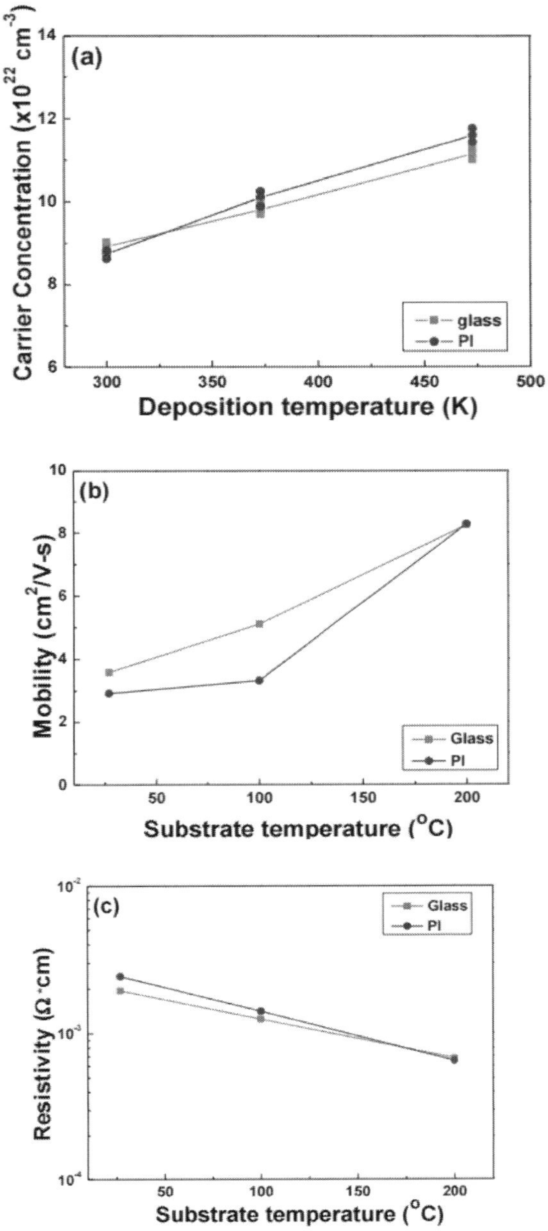

Figure 6: Variations of the (**a**) Hall mobility; (**b**) carrier concentration; and (**c**) resistivity of the GZO thin films as a function of used substrates and deposition temperature.

Figure 6c shows the dependence of the resistivity of the GZO thin films on deposition temperature and the used substrate. Resistivity of the GZO thin films is proportional to the reciprocal value of the product of the carrier concentration N and the mobility μ:

$$\rho = 1/(N \times e \times \mu) \tag{4}$$

As Equation (4) shows, both the carrier concentration and the carrier mobility contribute to the resistivity. The results inFigure 6c indicate that the resistivity decreased with the increase of the deposition temperature. Thus, the increases in carrier concentration and electron mobility should be taken into account when we analyze the decrease in resistivity of the GZO thin films. Those results in Figure 6a,b suggest that reduction in resistivity also comes from an increase in electron mobility in addition to the increase of the electron concentration. In this study, the minimum resistivity of the GZO thin films at a deposition temperature of 473 K on different substrates is mainly influenced by both the carrier concentration and the carrier mobility being at their maximum. As the deposition temperature was raised from RT to 473 K, the resistivity decreased from 1.95 × 10^{-3} to 0.680 × 10^{-3} Ω·cm for glass substrates and from 2.44 × 10^{-3} to 0.651 × 10^{-3} Ω·cm for PI substrates, respectively.

Figure 7 shows a typical widescan spectrum of the GZO thin films as a function of the used substrates. The photoelectron peaks of the main elements, Zn, O, and Ga, and Auger Zn and Ga *LMM* (electron excited from L layer to M layer) and O *KLL*(electron excited from K layer to L layer) peaks were obtained. As Figure 7 shows, some small peaks with intensities lower than each main photoelectron peak were also observed.

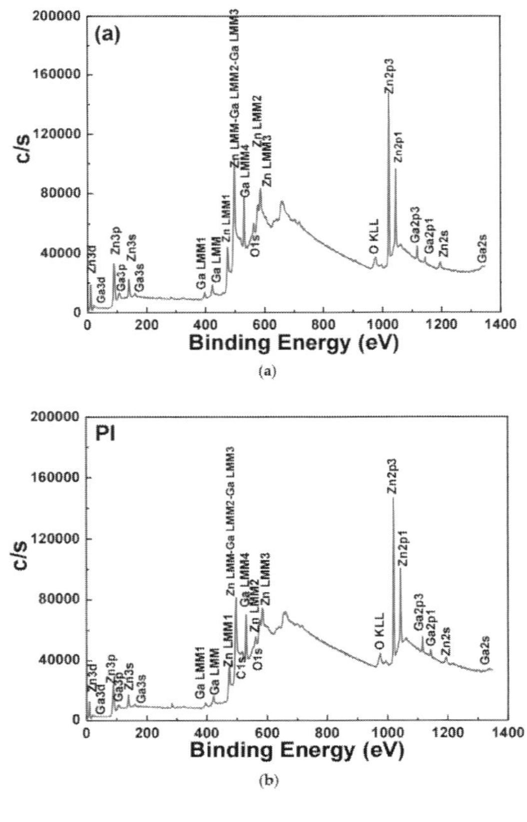

Figure 7: Typical widescan spectrum of the GZO thin films as a function of used substrates. (**a**): glass; (**b**): PI.

To clarify the mechanism of improvement in electrical properties, the chemical structures of the GZO thin films were investigated by XPS. The XPS spectra for the $Ga2p_{1/2}$, $Ga2p_{3/2}$, $Zn2p_{1/2}$, $Zn2p_{3/2}$, and Ga3d peaks of the GZO thin films are shown in Figure 8 as a function of the used substrates, and the values are compared in Table 3; those thin films were deposited at RT. The XPS spectra for the $Ga2p_{1/2}$, $Ga2p_{3/2}$, $Zn2p_{1/2}$, $Zn2p_{3/2}$, and Ga3d peaks of the ZnO thin films are also shown in Figure 8 for comparison. When glass was used as the substrate, the binding energy of each constituent element was positioned at 1145.22 eV ($Ga2p_{1/2}$), 1118.36 eV ($Ga2p_{3/2}$), 1045.23 eV ($Zn2p_{1/2}$), 1022.10 eV ($Zn2p_{3/2}$), and 20.72 eV (Ga3d), respectively, as calibrated to 285.43 eV (C1s); When PI was used as the substrate, the binding energy of each constituent element was positioned at 1143.35 eV ($Ga2p_{1/2}$), 1116.52 eV ($Ga2p_{3/2}$), 1043.64 eV ($Zn2p_{1/2}$), 1020.47 eV ($Zn2p_{3/2}$), and 19.06 eV (Ga3d), respectively. We know the binding energies of Ga_2O_3 and Ga metal are 1117.4 ± 0.5 eV and 1116.5 ± 0.2 eV. As Figure 8a shows, no apparent peak could be observed in the XPS spectra of the Ga3d peak of the ZnO thin films. For that, the results in Figure 8a suggest that the binding state of Ga_2O_3 will dominate the GZO thin films on glass and the binding state of Ga will dominate the GZO thin films on PI.

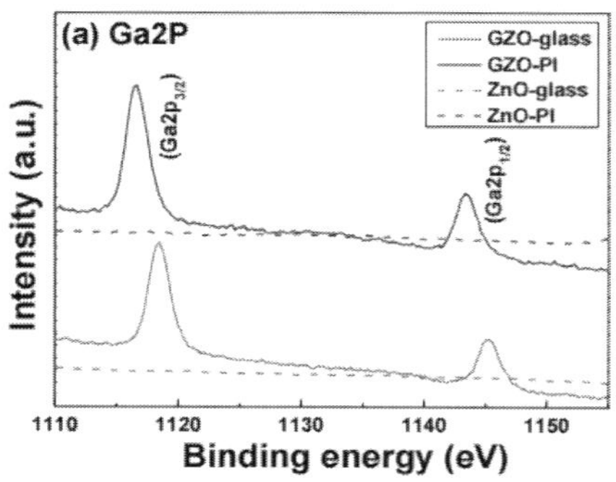

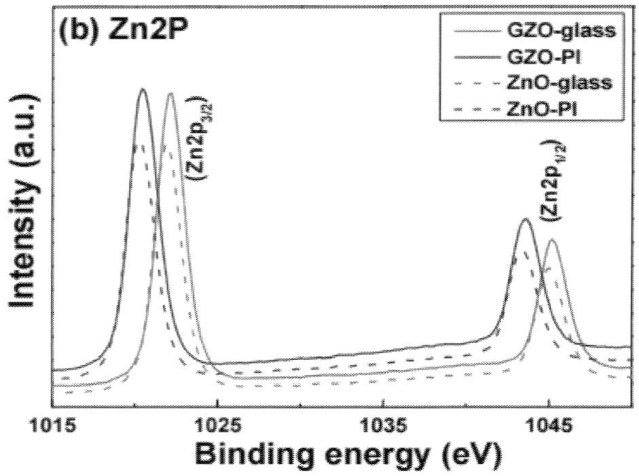

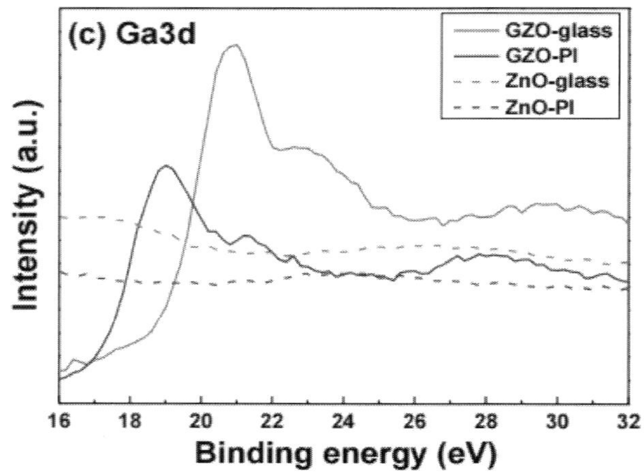

Figure 8: Binding energy spectra of (**a**) $Ga2p_{3/2}$ and $Ga2p_{1/2}$ peaks; (**b**) $Zn2p_{3/2}$ and $Zn2p_{1/2}$ peaks; and (**c**) Ga3d peak for the undoped ZnO thin films and GZO thin films on different substrates.

Table 3: Comparison of binding energies of $Ga2p_{3/2}$, $Ga2p_{1/2}$, $Zn2p_{3/2}$, $Zn2p_{1/2}$, and Ga3d peaks for GZO thin films on different substrates

Title	Glass	PI
$Ga2P_{1/2}$ Binding energy (eV)	1145.22	1143.35
$Ga2P_{3/2}$ Binding energy (eV)	1118.36	1116.52
$Zn2p_{1/2}$ Binding energy (eV)	1045.23	1043.64
$Zn2p_{3/2}$ Binding energy (eV)	1022.10	1020.47
Ga3d Binding energy (eV)	20.72	19.06

As Figure 8b shows, two similar peaks were observed in the XPS spectra of the ZnO and GZO thin films. The binding energy peak at 1022.40 ± 0.10 eV could be due to $Zn2p_{3/2}$ in the ZnO_{1-x} structure and the binding energy peak at 1021.1 ± 0.40 eV was due to metallic zinc [25]. The results in Figure 8b suggest that the binding state of ZnO_{1-x} will dominate the GZO thin films on glass and the binding state of Zn will dominate the GZO thin films on PI. The binding energy of the Ga3d peak for Ga_2O_3 (Ga^{3+}) is 20.7 eV; the binding energies of the Ga3d peak for Ga_2O (Ga^{1+}) or GaO (Ga^{2+}) are about 20.7 eV; and the binding energy of the Ga3d peak for Ga metal is 18.6 ± 0.3 eV, respectively. Figure 8c shows the binding energy of the Ga3d peak of the GZO thin films on different substrates. When glass was used as the substrate, the binding energy of the Ga3d peak was about 20.7 eV, which suggests the binding energy of the Ga3d peak is Ga_2O_3 (Ga^{3+}); when PI was used as the substrate, the binding energy of the Ga3d peak was about 18.6 ± 0.3 eV, which suggests the binding energy of the Ga3d peak is the Ga metal. No apparent peaks could be observed in the XPS spectra of the Ga3d peak of the ZnO thin films. For that, the binding energies in Figure 8c reveal the same results as those of Figure 8a. The results in Figure 8 showed significant variations in the core binding energies (~1.55 eV) as we varied the substrates. The change can be related to the difference between the intrinsic properties of the deposition materials as well as to the differences in the interaction with the substrates [26].

Figure 9a shows the XPS spectra for the Gaussian-resolved components of O_{1s} of the GZO thin films on glass substrates;

those thin films were deposited at RT. The bonding energy components were centered at 531.17 eV and 529.35 eV for glass and PI substrates, respectively. As shown in Figure 9a, two shoulders were visible on the low-energy side and high-energy side of the GZO thin films, indicating that at least three different bonds existed. For GZO thin films deposited on glass substrates, the bonding state of the O_{1s} spectrum was resolved into three components centered at 530.21 eV, 531.18 eV, and 532.39 eV for the O_I, O_{II}, and O_{III} peaks, respectively. As shown in Figure 9b, only one shoulder was visible on the high-energy side of the GZO thin films, and we also found that three different bonds existed. For GZO thin films deposited on PI substrates, the bonding state of the O_{1s} spectrum was resolved into three components centered at 529.03 eV, 529.83 eV, and 531.02 eV for the O_I, O_{II}, and O_{III} peaks, respectively. Table 4 shows the area ratios of the O_I, O_{II}, and O_{III} peaks; the area beneath the peaks is clearly different for these two substrates.

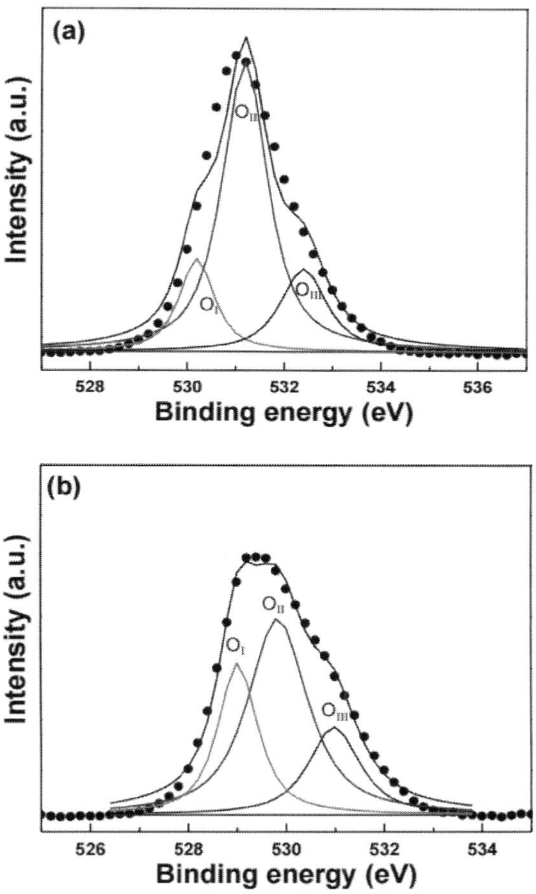

Figure 9: Binding energy spectra of O_{1s} peaks for GZO thin films on different substrates: (**a**) glass and (**b**) PI.

Table 4: Comparison of area ratios of O_I, O_{II}, and O_{III} peaks for GZO thin films on different substrates

Title	Glass-RT	PI-RT
O_I peak	14.1%	20.3%
O_{II} peak	81.1%	52.6%
O_{III} peak	4.8%	27.0%

For GZO thin films deposited on glass (PI) substrates, the low bonding energy component of the O_I peak located at 530.21 ± 0.15 eV (529.03 ± 0.15 eV) is attributed to O^{2-} ions on the wurtzite structure of the hexagonal Zn^{2+} ion array, surrounded by Zn atoms with their full complement of nearest-neighbor O^{2-} ions [25,27]. The second binding energy component of the O_{II}peak located at 531.18 ± 0.15 eV (529.83 ± 0.15 eV) is associated with O^{2-} ions in the oxygen vacancies within the matrix of ZnO_{1-x} [28], which is also known as V_O-like bonding. The highest bonding energy component of the O_{III} peak located at 532.39 ± 0.15 eV (531.02 ± 0.15 eV) implies the presence of hydrated oxide species on the film surface [29]. As Figure 9a shows, when glass was used as the substrate and the deposition was RT, the area ratio of the O_{II} peak was larger than those of the O_I and O_{III} peaks. These results suggest that there are more oxygen vacancies existing in GZO thin films to cause the increase in the carrier concentration. When PI was used as the substrate, because the PI substrates had less surface energy, that would cause the GZO thin films to have poor adhesive force, and the GZO thin films on PI substrates would have poor crystallization, as Figure 3 shows. The area ratio of the O_{II} peak apparently decreased and the area ratios of the O_I and O_{III}peaks increased. These results are caused by the decrease of oxygen vacancies and the increase of oxygen absorption, which will degenerate the electrical properties of the GZO thin films, as Figure 6 shows. These results suggest that even with the same deposition parameters, the GZO thin films deposited on different substrates will have different results.

CONCLUSIONS

In this study, the high-transmittance nano-scale GZO thin films were successfully deposited on glass and PI substrates by using the RF sputtering process. For all GZO thin films, the transmittance ratios of the GZO thin films at 400–700 nm were more than 94% regardless of deposition temperature and substrate. As the deposition temperature increased from RT to 473 K and glass (PI) was used as the substrate, the FWHM values of the (002) peak for GZO thin films deposited on glass (PI) substrates were 0.407 (0.412), 0.336 (0.360), and 0.265 (0.292) for substrate temperatures of RT, 373 K, and 473 K; the *Eg*values increased from 3.612 (3.551) eV to 3.677 (3.622) eV and the wavelength of the absorption edge decreased from about 360 (356) nm to about 352 (350) nm, respectively. From the binding energy spectra of $Ga2p_{3/2}$ and $Ga2p_{1/2}$ peaks, $Zn2p_{3/2}$and $Zn2p_{1/2}$ peaks, the Ga3d peak, and O_{1s} peaks, we had proved that GZO thin films on different substrates had different chemical and electronic states, and thus different electrical properties.

ACKNOWLEDGMENTS

The authors acknowledge financial supports of MOST 104-2221-E-390-013-MY2, MOST 104-2622-E-390-004-CC3, and MOST 103-2221-E-005-040-MY2.

AUTHOR CONTRIBUTIONS

Kun-Neng Chen, Chao-Ming Hsu, and Min-Chu Liu helped carrying out the deposition and measurement processes of GZO thin films and data analysis; Fang-Hsing Wang and Cheng-Fu Yang organized the paper, encouraged in paper writing, and helped proceeding the experimental processes and measurements.

REFERENCES

1. Tsang, M.K.; Baia, G.; Hao, J. Stimuli responsive upconversion luminescence nanomaterials and films for various applications. *Chem. Soc. Rev.* 2015, *44*, 1585–1607.
2. Ariga, K.; Yamauchi, Y.; Rydzek, G.; Ji, Q.; Yonamine, Y.; Wu, K.C.-W.; Hill, J.P. Layer-by-layer Nanoarchitectonics: Invention, Innovation, and Evolution. *Chem. Lett.* 2014, *43*, 36–68.
3. Zhuang, J.L.; Terfort, A.; Wöll, C. Formation of oriented and patterned films of metal–organic frameworks by liquid phase epitaxy: A review. *Coord. Chem. Rev.* 2016, *307*, 391–424.
4. Stadler, A. Transparent conducting oxides—An up-to-date overview. *Materials* 2012, *5*, 661–683.
5. Fay, S.; Kroll, U.; Bucher, C.; Vallat-Sauvain, E.; Shah, A. Low pressure chemical vapour deposition of ZnO layers for thin-film solar cells: Temperature induced morphological changes. *Sol. Energy Mater. Sol. Cell* 2005, *86*, 385–397.
6. Huang, C.C.; Wang, F.H.; Yang, C.F. Effects of deposition temperature and hydrogen flow rate on the properties of the Al-doped ZnO thin films and amorphous silicon thin-film solar cells. *Appl. Phys. A* 2013, *112*, 877–883.
7. Wang, F.H.; Yang, C.F.; Lee, Y.H. Deposition of F-doped ZnO transparent thin films using ZnF_2-doped ZnO target under different sputtering substrate temperatures. *Nanoscale Res. Lett.* 2014, *9*.
8. Assuncao, V.; Fortunato, E.; Marques, A.; Aguas, H.; Ferreira, I.; Costa, M.E.V.; Martins, R. Influence of the deposition pressure on the properties of transparent and conductive ZnO:Ga thin-film produced by r.f. sputtering at room temperature. *Thin Solid Films* 2003, *427*, 401–405.
9. Kim, B.G.; Kim, J.Y.; Lee, S.J.; Park, J.H.; Lima, D.G.; Park, M.G. Structural, electrical and optical properties of Ga-doped ZnO films on PET substrate. *Appl. Surf. Sci.* 2010, *257*, 1063–1067.
10. Ohkubo, I.; Ohtomo, A.; Ohnishi, T.; Mastumoto, Y.; Koinuma, H. In-plane and polar orientations of ZnO thin films grown on atomically flat sapphire. *Surf. Sci.* 1999, *433*, L1043–L1048.
11. Hartnagel, H.L.; Dawar, A.L.; Jain, A.K.; Jagadish, C. *Semiconducting Transparent Thin Films*; Institute of Physics Publishing: Philadelphia, PA, USA, 1995.
12. Park, H.C.; Byun, D.; Angadi, B.; Park, D.H.; Choi, W.K.; Choi, J.W.; Jung, Y.S. Photoluminescence of Ga-doped ZnO film grown on *c*-Al_2O_3 (0001) by plasma-assisted molecular beam epitaxy. *J. Appl. Phys.* 2007, *102*.
13. Lu, S.K.; Huang, J.T.; Lee, T.H.; Wang, J.J.; Liu, D.S. Flexibility of the Indium Tin Oxide Transparent Conductive Film Deposited onto the Plastic Substrate. *Smart Sci.* 2014, *2*, 7–12.
14. Fortunato, E.; Goncalves, A.; Assuncao, V.; Marques, A.; Aguas, H.; Pereira, L.; Ferreira, I.; Martins, R. Growth of ZnO:Ga thin films at room temperature on polymeric substrates: Thickness dependence. *Thin Solid Films* 2003, *442*, 121–126.
15. Yu, Z.N.; Zhao, J.J.; Xia, F.; Lin, Z.J.; Zhang, D.P.; Leng, J.; Xue, W. Enhanced electrical stability of flexible indium tin oxide films prepared on stripe SiO_2 buffer layer-coated polymer substrates by magnetron sputtering. *Appl. Surf. Sci.*2011, *257*, 4807–4810.
16. Wang, F.H.; Kuo, H.H.; Yang, C.F.; Liu, M.C. Role of SiN_x Barrier Layer on the Performances of Polyimide Ga_2O_3-doped ZnO p-i-n Hydrogenated Amorphous Silicon Thin Film Solar Cells. *Materials* 2014, *7*, 948–962.
17. Azaroff, L.V. *Element of X-ray Crystallography*; McGraw-Hill: New York, NY, USA, 1968.
18. Li, X.; Wang, Y.; Liu, W.; Jiang, G.; Zhu, C. Study of oxygen vacancies' influence on the lattice parameter in ZnO thin film. *Mater. Lett.* 2012, *85*, 25–28.
19. Wang, F.H.; Huang, C.C.; Yang, C.F.; Tzeng, H.Z. Optical and Electrical Properties of the Different Magnetron Sputter Power 300 °C-Deposited Ga_2O_3-ZnO Thin Films and Applications in p-i-n α-Si:H Thin-Film Solar Cells. *Int. J. Photoenergy* 2013, *2013*.
20. Roth, A.P.; Webb, J.B.; Williams, D.F. Band-gap narrowing in heavily defect-doped ZnO. *Phys. Rev. B* 1982, *25*, 7836–7839.
21. Gong, L.; Lu, J.; Ye, Z. Transparent and conductive Ga-doped ZnO films grown by RF magnetron sputtering on polycarbonate substrates. *Sol. Energy Mater. Sol. Cell* 2010, *94*, 937–941.
22. Khan, Z.R.; Khan, M.S.; Zulfequar, M.; Khan, M.S. Optical and Structural Properties of ZnO Thin Films Fabricated by Sol-Gel Method. *Mater. Sci. Appl.* 2011, *2*, 340–345.
23. Orita, M.; Ohta, H.; Hirano, M.; Hosono, H. Deep-ultraviolet transparent conductive β-Ga_2O_3 thin films. *Appl. Phys. Lett.* 2000, *77*, 4166–4168.
24. Igasaki, Y.; Saito, H. Substrate temperature dependence of electrical properties of ZnO:Al epitaxial films on sapphire (1210). *J. Appl. Phys.* 1991, *69*, 2190–2195.
25. Choi, B.H.; Im, H.B.; Song, J.S.; Yoon, K.H. Optical and electrical properties of Ga_2O_3-doped ZnO films prepared by r.f. sputtering. *Thin Solid Films* 1990, *193–194*, 712–720.
26. Burkstrand, J.M. Substrate effects on the electronic structure of metal overlayers—An XPS study of polymer-metal interfaces. *Phys. Rev.* 1979, *20*, 2853–4858.
27. Yu, C.F.; Chen, S.H.; Sun, S.J.; Chou, H. Influence of the grain boundary barrier height on the electrical properties of Gallium doped ZnO thin films. *Appl. Surf. Sci.* 2011, *257*, 6498–6502.
28. Rao, L.K.; Vinni, V. Novel mechanism for high speed growth of transparent and conducting tin oxide thin films by spray pyrolysis. *Appl. Phys. Lett.* 1993, *63*, 608–610.
29. Fan, J.C.C.; Goodenough, J.B. X-ray photoemission spectroscopy studies of Sn-doped indium-oxide films. *J. Appl. Phys.* 1997, *48*, 3524–3531.

CHAPTER 10

EFFECT OF NA DOPING ON THE NANOSTRUCTURES AND ELECTRICAL PROPERTIES OF ZNO NANOROD ARRAYS

Lu Yue[1,2], Zhiqiang Zhang[1,2], Yanyan Ma[1,2], and Wenhui Zhang[1,2]

[1]Jiangsu Collaborative Innovation Center for Ecological Building Materials and Environmental Protection Equipment, Yancheng Institute of Technology, Yancheng 224051, China

[2]Key Laboratory for Advanced Technology in Environmental Protection of Jiangsu Province, Yancheng Institute of Technology, Yancheng 224051, China

ABSTRACT

The p-type ZnO nanorod arrays were prepared by doping Na with hydrothermal method. The structural, electrical, and optical properties were explored by XRD, Hall-effect, PL, and Raman spectra. The carrier concentrations and the mobility of Na-doped ZnO nanorod arrays are arranged from 1.4×10^{16} cm^{-3} to 1.7×10^{17} cm^{-3} and 0.45 cm^2 v^{-1} s^{-1} to 10^6 cm^2 v^{-1} s^{-1}, respectively.

INTRODUCTION

Zinc oxide (ZnO) is a promising material candidate for photoelectronic devices due to its direct wide electron energy bandgap of 3.37 eV at room temperature and large exciton binding energy of 60 meV. ZnO is conventionally used as a substrate for GaN and has advantages over GaN such as larger exciton binding energy, lower refraction index, and shallower acceptor levels. Despite its advantages over GaN, the application of ZnO in optoelectronics is hampered by the lack of stable p-type doping due to self-compensation by donor-like native defects, low solubility of p-type dopants, and formation of deep acceptor levels [1]. The n-type ZnO is available even without any intentional doping, but it is very difficult to dope ZnO p-type. Many groups have realized p-type ZnO conversion by doping group V elements (N [2], P [3], As [4], and Sb [5]) and groups I and IB (Li [6], Na [7–9], K [10], Ag [11], and Cu [12]), but the stability was not good. Among the group V elements, N is the most suitable dopant because it has about the same ionic radius as O and thus should readily be substituted on O sites. Indeed, it remains difficult to achieve good quality p-type conduction in N-doped ZnO due to either a low dopant solubility or a high defect ionization energy. Theoretically, group I element substituted on Zn sites is shallow acceptors [13], but instead these dopants tend to occupy the interstitial sites, partly due to their small atomic radius, and therefore act mainly as donors. On the other hand, reports of Na-doped p-type ZnO films are rather poor [14]. Up to now, various methods have been employed to prepare p-type ZnO including molecular beam epitaxy (MBE) [15], chemical vapor deposition [16], and pulsed laser deposition [17], in which complicated, high temperature deposition and expensive vacuum systems have been employed. In contrast, the hydrothermal method was proven to be an ideal technique for the development of p-type ZnO films owing to its simplicity and easy control of the composition at low cost [18–20]. To date, there is no report of Na doping of p-type ZnO nanorods by a hydrothermal method. Here, we report a simple, reliable, and low-cost method for growing and doping p-type ZnO nanorod arrays in aqueous solution, with Na as the p-type dopant.

Citation: Lu Yue, Zhiqiang Zhang, Yanyan Ma, and Wenhui Zhang, "Effect of Na Doping on the Nanostructures and Electrical Properties of ZnO Nanorod Arrays," Journal of Nanomaterials, vol. 2016, Article ID 3040536, 5 pages, 2016. doi:10.1155/2016/3040536.

EXPERIMENTAL SECTION

A high-resistance Si (100) wafer, the resistivity of which is above 500 Ω cm, was used as the substrate. Precleaned silicon substrate was coated with a ZnO seed layer by sol-gel spin coating. $Zn(AC)_2 \cdot 2H_2O$ were first dissolved in the mixture of 2-methoxyethanol and monoethanolamine. The molar ratio of monoethanolamine to $Zn(AC)_2 \cdot 2H_2O$ was maintained at 1 : 1 and the concentration of $Zn(AC)_2 \cdot 2H_2O$ was 75 mM. The sol was aged for 24 h at room temperature. The above solution was then spin-coated at 500 rpm for 15 s and then at 5000 rpm for 60 s, followed by drying at 120°C for 15 min. The procedure from coating to drying was repeated five times before the substrates were placed into a 360°C tube furnace in air for 20 min to form ZnO seed layer, evaporate the solvent, and remove any organic residue. Subsequently, ZnO nanorod arrays were grown directly by placing the substrate at the bottom of the beaker in 20 mL of aqueous solution of 12.5 mM $Zn(AC)_2 \cdot 2H_2O$, 12.5 mM hexamethylenetetramine (HMTA, $C_6H_{12}N_4$), and different amounts of NaCl in a 50 mL autoclave. Different Na doping levels were obtained by varying the molar ratio of Na/Zn (n_{Na}/n_{Zn}) (0, 2.8, 5.6, 8.4, and 11.2) in the precursor solution. Finally the autoclave was sealed and heated at 95°C for 2 h. Subsequently, the samples are washed with deionized water for several times to remove the residual salt and amino complex and dried in a flow of nitrogen gas.

Powder X-ray diffraction (XRD) analyses were performed on a Philips PW-1830 X-ray diffractometer with Cu Kα irradiation (λ = 1.5406Å) at a scanning speed of 0.014°/sec over the 2θ range of 30–40°. The electronic morphology of the samples was examined by Hitachi S-4800 scanning electron microscope (SEM). The electrical properties were characterized by a Hall effect measurement system. Photoluminescence and Raman spectra of samples on Si substrate were recorded using Jobin-Yvon Lab-Ram high-resolution spectrometer with He-Cd laser (λ = 325 nm).

RESULTS AND DISCUSSION

XRD patterns of the hydrothermally grown samples were shown in Figure 1. The strong peak corresponding to ZnO (002) is observed in each film, which indicates high c-axis oriented in all samples. It can be seen that the diffraction peak shows an obvious shift to a smaller angle of about 34.56, 34.52, 34.50, 34.38, and 34.52 for the sample with n_{Na}/n_{Zn} of 0, 2.8, 5.6, 8.4, and 11.2, respectively. This demonstrates that Na doping increases the lattice constant as a result of larger ionic radius of Na^+ (0.089 nm) compared with that of Zn^{2+} (0.070 nm) [8].

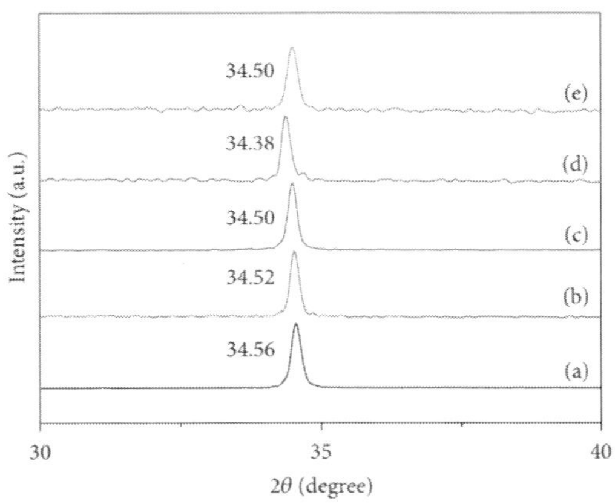

Figure 1: XRD patterns of the samples with n_{Na}/n_{Zn} of (a) 0, (b) 2.8, (c) 5.6, (d) 8.4, and (e) 11.2.

Figure 2 shows the surface morphologies of samples grown by simple aqueous solution approach on Si substrates. The nanorods are vertically well-aligned on the substrate surface. The diameter and length of pure ZnO are in the range of around 30 nm and 754 nm, respectively. Figure 3 shows a plot of diameter and length of ZnO nanorod arrays as functions of n_{Na}/n_{Zn}. With the increase of n_{Na}/n_{Zn}, the diameter of ZnO nanorod increased to 50 nm, 60 nm, 70 nm, and 70 nm, and the length decreased to 150 nm, 112 nm, 98 nm, and 71 nm, respectively.

(a)

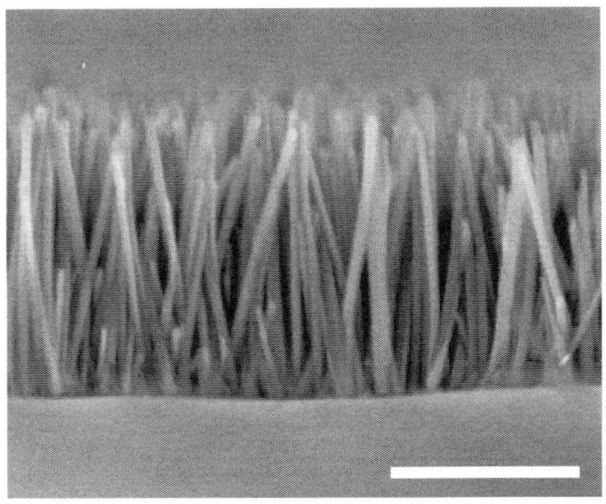

(b)

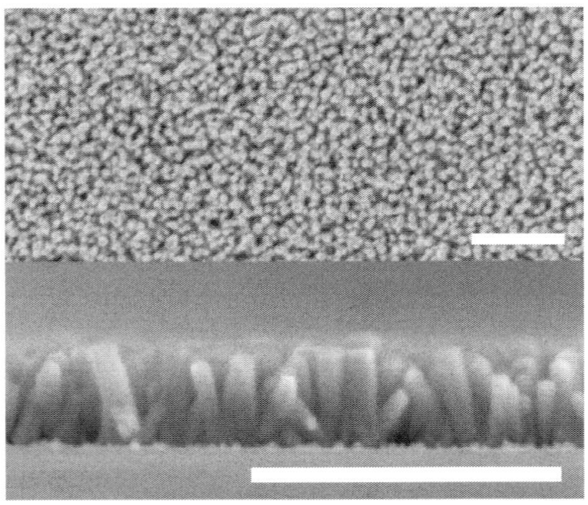

(c)

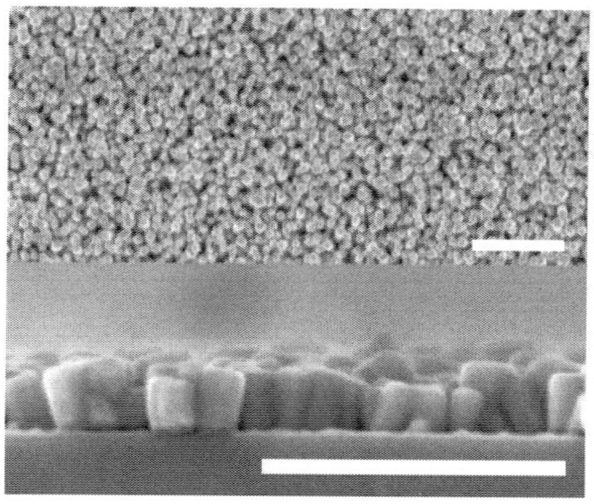

(d)

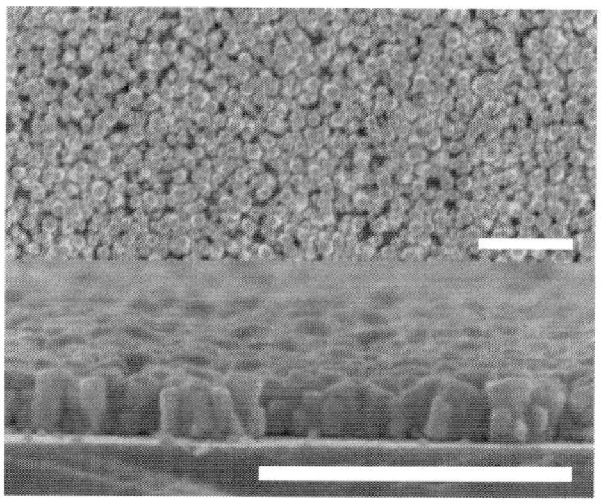

(e)

(f)

Figure 2: SEM images of the samples with n_{Na}/n_{Zn} of (a, b) 0, (c) 2.8, (d) 5.6, (e) 8.4, and (f) 11.2. All the scale bars are 500 nm.

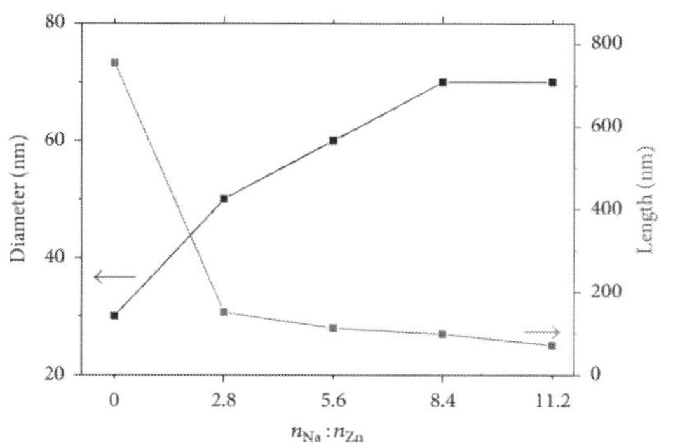

Figure 3: Diameter and length of ZnO nanorod arrays as functions of n_{Na}/n_{Zn}.

Table 1 summarizes the electrical properties of ZnO films. Under Hall effect measurements, the undoped sample showed n-type conductivity, while the doped sample with n_{Na}/n_{Zn} of 2.8–8.4 showed p-type conductivity, but when n_{Na}/n_{Zn} was increased to 11.2, the film showed n-type conductivity again. In contrast, the electrical properties of samples with n_{Na}/n_{Zn} of 2.8–8.4 showed higher mobility and carrier concentration. Mobility was up to 106 $cm^2 v^{-1} s^{-1}$ when n_{Na}/n_{Zn} was 2.8, and the carrier concentration was 1.7×10^{17} cm^{-3} $cm^{-3}$when n_{Na}/n_{Zn} was 5.6. Mobility and carrier concentration were significantly improved by doping. We think this is due to the right amount of Na ions in the Zn lattice sites acting as acceptors, compensation Zn_i, and V_o intrinsic donor defects in the lattice. Most of the Na^+ occupies the lattice position, and the lattice expansion is small. The above XRD analysis further illustrates the rationality of our analysis. Electrical properties of Na-doped ZnO films by different processes are shown in Table 2. It can be seen that Na-doped nanorod arrays in this study showed excellent electrical properties.

Table 1: Electrical properties of ZnO films at room temperature

n_{Na}/n_{Zn}	Carrier type	Mobility ($cm^2 v^{-1} s^{-1}$)	Carrier concentration (cm^{-3})
0	n	40	1.4×10^{16}
2.8	p	106	3.1×10^{16}
5.6	p	41.9	1.7×10^{17}
8.4	p	99.4	6.5×10^{16}
11.2	n	10.4	4.3×10^{16}

Table 2: Electrical properties of Na-doped ZnO films by different processes

Material	Mobility ($cm^2 v^{-1} s^{-1}$)	Carrier concentration (cm^{-3})	References
ZnO:Na	1.41	5.19×10^{16}	[9]
ZnO:Na	0.138–0.676	6.90×10^{16}–2.16×10^{17}	[21]
ZnO:Na	0.22	5.3×10^{16}	[22]
ZnO:Na	0.43	2.5×10^{17}	[23]
ZnO:Na	0.402	1.81×10^{15}	[24]
ZnO:Na nanowires	2.1	1.3×10^{16}	[7]
ZnO:Na nanorod arrays	41.9–106	3.1×10^{16}–1.7×10^{17}	Our work

Usually, PL spectra of ZnO at room temperature were composed of a near band edge UV emission peak at around 375 nm and a deep level green emission peak at 450–550 nm. Researchers generally believe that the near band edge UV emission peak was caused by the recombination luminescence of the band edge exciton, while the deep level green emission peak was caused by the recombination of the defects such as O vacancy and Zn interstitial. Strong excitonic emission at 375 nm was observed, as shown in Figure 4. The UV emission was so strong, which indicates that all the 1D ZnO arrays were of high UV emission efficiency and that the green emission could hardly be observed in the PL spectrum, indicating that the crystalline quality of ZnO was high and lattice oxygen vacancy concentration is very low.

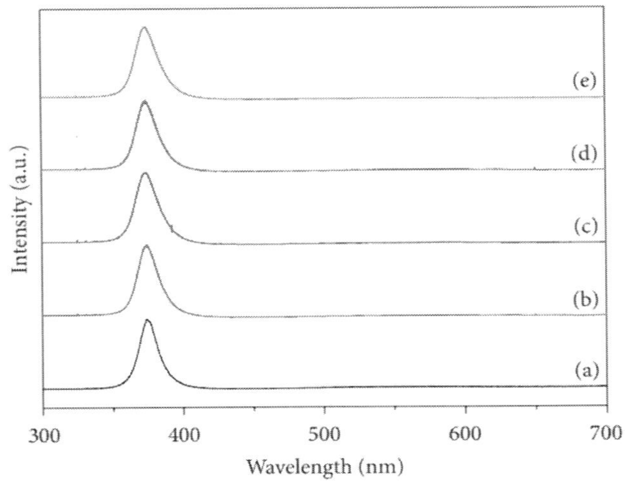

Figure 4: PL spectrum of the samples with n_{Na}/n_{Zn} of (a) 0, (b) 2.8, (c) 5.6, (d) 8.4, and (e) 11.2.

Figure 5 shows the doped samples before and after 325 nm excitation of Raman spectra. Raman spectra can be sensitive to detect the crystallization of the material and the defect state. Undoped Al (LO) mode peak at 578.1 cm^{-1} corresponds to the El longitudinal optical phonon (LO) mode. When the molar ratio of Na and Zn was 2.8, 5.6, 7.8, and 11.2, the peak position was at 577.3 cm^{-1}, 574.9 cm^{-1}, 574.2 cm^{-1}, and 575.8 cm^{-1}, which moved 0.8 cm^{-1}, 3.2 cm^{-1}, 3.9 cm^{-1}, and 2.3 cm^{-1}, respectively. We observed that the peaks move to lower frequency, because this peak and the oxygen atom vacancies and zinc atoms interstitial related to its move reflect the changes in the concentration of free carriers in the ZnO crystal. So the concentration of free carriers in the ZnO crystals doped Na changed, and when the doping concentration of Al (LO) peak of the amount of movement is also different, this also shows the degree of change of free carrier concentration to be different.

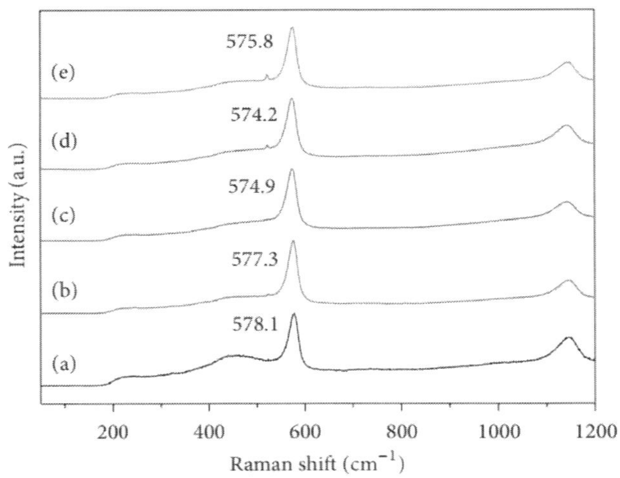

Figure 5: Raman spectra of the samples with n_{Na}/n_{Zn} of (a) 0, (b) 2.8, (c) 5.6, (d) 8.4, and (e) 11.2.

CONCLUSION

In summary, Na-doped ZnO nanorod arrays with high crystalline quality were successfully fabricated on Si substrate by a simple hydrothermal method for the first time. With the increase of n_{Na}/n_{Zn}, the diameter of ZnO nanorod gradually increased from 30 nm to 70 nm, and the length decreased from 754 nm to 71 nm. Structural, electrical, and optical properties were investigated. The p-type carrier concentrations and mobility of Na-doped ZnO nanorod arrays arranged from 3.1×10^{16} cm^{-3} to 1.7×10^{17} cm^{-3} and 41.9 $cm^2\ v^{-1}\ s^{-1}$ to 10^6 $cm^2\ v^{-1}\ s^{-1}$, respectively.

ACKNOWLEDGMENTS

This study is supported by the National Natural Science Foundation of China (no. 51402252) and the Natural Science Foundation of Jiangsu Province (no. BK20140463).

REFERENCES

1. Y. X. Liu, H. L. Zhang, Z. X. Zhang, Y. Z. Xie, and E. Q. Xie, "Conversion of p-type to n-type conductivity in undoped ZnO films by increasing operating temperature," Applied Surface Science, vol. 257, no. 4, pp. 1236–1238, 2010.
2. M. Ding, D. X. Zhao, B. Yao, B. H. Li, Z. Zhang, and D. Shen, "The p-type ZnO film realized by a hydrothermal treatment method," Applied Physics Letters, vol. 98, no. 6, Article ID 062102, 2011.
3. Y. T. Shih, J. F. Chien, M. J. Chen, J. R. Yang, and M. Shiojiri, "P-type ZnO:P films fabricated by atomic layer deposition and thermal processing," Journal of the Electrochemical Society, vol. 158, no. 5, pp. H516–H520, 2011.
4. V. Vaithianathan, B.-T. Lee, and S. S. Kim, "Preparation of As-doped p-type ZnO films using a Zn_3As_2ZnO target with pulsed laser deposition," Applied Physics Letters, vol. 86, no. 6, Article ID 062101, 2005.
5. F. Friedrich, I. Sieber, C. Klimm, M. Klaus, C. Genzel, and N. H. Nickel, "Sb-doping of ZnO: phase segregation and its impact on p-type doping," Applied Physics Letters, vol. 98, no. 13, Article ID 131902, 2011.
6. Y. J. Zeng, Z. Z. Ye, W. Z. Xu et al., "Dopant source choice for formation of p-type ZnO: Li acceptor,"Applied Physics Letters, vol. 88, no. 6, Article ID 062107, 2006.
7. W. Liu, F. X. Xiu, K. H. Sun et al., "Na-doped p-type ZnO microwires," Journal of the American Chemical Society, vol. 132, no. 8, pp. 2498–2499, 2010.
8. L. Q. Zhang, Y. Z. Zhang, Z. Z. Ye et al., "The fabrication of Na doped p-type Zn1–xMgxO films by pulsed laser deposition," Applied Physics A: Materials Science and Processing, vol. 106, no. 1, pp. 191–196, 2012.
9. Y. Li, Y. Z. Zhang, H. P. He, Z. Z. Ye, J. Jiang, and L. Cao, "Growth of Na doped p-type non-polar a-plane ZnO films by pulsed laser deposition," Materials Letters, vol. 76, pp. 81–83, 2012.
10. M. K. Gupta, N. Sinha, and B. Kumar, "p-type K-doped ZnO nanorods for optoelectronic applications,"Journal of Applied Physics, vol. 109, no. 8, Article ID 083532, 2011.
11. M. A. Myers, J. H. Lee, Z. X. Bi, and H. Y. Wang, "High quality p-type Ag-doped ZnO thin films achieved under elevated growth temperatures," Journal of Physics-Condensed Matter, vol. 24, no. 22, Article ID 229501,

2012.

12. B. Marí, M. Sahal, M. A. Mollar, F. M. Cerqueira, and A. P. Samantilleke, "p-Type behaviour of electrodeposited ZnO:Cu films," Journal of Solid State Electrochemistry, vol. 16, no. 6, pp. 2261–2265, 2012.
13. C. H. Park, S. B. Zhang, and S.-H. Wei, "Origin of p-type doping difficulty in ZnO: the impurity perspective," Physical Review B—Condensed Matter and Materials Physics, vol. 66, no. 7, Article ID 073202, 2002.
14. B. Y. Zhang, B. Yao, Y. F. Li et al., "Investigation on the formation mechanism of p-type Li-N dual-doped ZnO," Applied Physics Letters, vol. 97, no. 22, Article ID 222101, 2010.
15. M. Suja, S. B. Bashar, M. M. Morshed, and J. L. Liu, "Realization of Cu-doped p-type ZnO thin films by molecular beam epitaxy," ACS Applied Materials & Interfaces, vol. 7, no. 16, pp. 8894–8899, 2015.
16. S. B. Bashar, M. Suja, M. Morshed, F. Gao, and J. Liu, "An Sb-doped p-type ZnO nanowire based random laser diode," Nanotechnology, vol. 27, no. 6, Article ID 065204, 2016.
17. L. Cao, L. P. Zhu, J. Jiang, Y. Li, Y. Z. Zhang, and Z. Z. Ye, "Preparation and properties of p-type Ag-doped ZnMgO thin films by pulsed laser deposition," Journal of Alloys and Compounds, vol. 516, pp. 157–160, 2012.
18. X. Fang, J. H. Li, D. Zhao, D. Shen, B. Li, and X. Wang, "Phosphorus-doped p-Type ZnO nanorods and ZnO nanorod p-n homojunction LED fabricated by hydrothermal method," Journal of Physical Chemistry C, vol. 113, no. 50, pp. 21208–21212, 2009.
19. S. N. Das, J.-H. Choi, J. P. Kar, T. I. Lee, and J.-M. Myoung, "Fabrication of p-type ZnO nanowires based heterojunction diode," Materials Chemistry and Physics, vol. 121, no. 3, pp. 472–476, 2010.
20. Q. X. Xia, K. S. Hui, K. N. Hui et al., "High quality p-type N-doped AZO nanorod arrays by an ammonia-assisted hydrothermal method," Materials Letters, vol. 78, pp. 180–183, 2012.
21. Z. Zheng, Y. F. Lu, Z. Z. Ye, H. P. He, and B. H. Zhao, "Carrier type- and concentration-dependent absorption and photoluminescence of ZnO films doped with different Na contents," Materials Science in Semiconductor Processing, vol. 16, no. 3, pp. 647–651, 2013.
22. X. H. Pan, Y. S. Zhou, S. S. Chen et al., "Effect of Na contents on fabrication of p-type non-polar m-plane ZnO films," Journal of Crystal Growth, vol. 404, pp. 54–58, 2014.
23. W. Chen, X. H. Pan, S. S. Chen et al., "Investigation on non-polar m-plane ZnO and Na-doped p-type ZnO films grown by plasma-assisted molecular beam epitaxy," Applied Physics A: Materials Science and Processing, vol. 121, no. 1, pp. 77–82, 2015. View at Publisher · View at Google Scholar · View at Scopus
24. P. Ding, X. H. Pan, Z. Z. Ye et al., "Realization of p-type non-polar a-plane ZnO films via doping of Na acceptor," Solid State Communications, vol. 156, pp. 8–11, 2013.

CHAPTER 11 EFFECT OF TIO_2 RUTILE ADDITIVE ON ELECTRICAL PROPERTIES OF PPY/TIO_2 NANOCOMPOSITE

Trong Tung Nguyen[1,2] and Ngoc Huyen Duong[2]

[1]College of Television Vietnam, Thuong Tin, Hanoi 158500, Vietnam

[2]School of Engineering Physics, Hanoi University of Science and Technology, No.1 Daicoviet, Hanoi 112400, Vietnam

ABSTRACT

Polypyrrole/titanium dioxide nanocomposite (PPy/TiO_2) was synthesized by in situ chemical polymerization of pyrrole (Py) monomer in colloidal suspension of TiO_2 rutile. TEM and SEM images show that the TiO_2 is covered by PPy forming a core-shell structure. The PPy/TiO_2 core-shell will create n-p junction and bring in an inversion layer on the PPy-shell surface. The feature is accounted for the modification in electrical properties of the PPy/TiO_2 nanocomposite. On the exposure to oxygen the conductivity of the nanocomposite exhibits an increase in 16–18-folds that are accounted for the interaction between oxygen (an electron acceptor) and the inversion layer. The cyclic voltammetry diagrams have shown that at around 15% TiO_2 and scan rate 100 mV/s the nanocomposites can reach a specific capacitance about 176 F/g.

INTRODUCTION

Conducting polymer (CP) is typical conjugated class that possesses the unique structure of linear chain consisting of alternative single and double bonds [1]. Due to the susceptivity of pi-electron on the double bonds to a variety of counterions as well as to surrounding environment, the conductivity of conjugated polymer can be modulated in various ways. The feature offers effective approaches to modify the electrical properties of the material for possible applications [2]. For example, the coupling of CPs with active nanostructure can be used as a mean to tailor specific properties to design new materials with desirable characteristics. Beyond a simple combination of its property, the interaction between molecules component in nanocomposite can bring in new features or synergistic effects. Common CPs, polypyrrole (PPy) and its derivatives, have attracted a great deal of attention because of their high electrical conductivity, good environmental stability, and simple synthesis and processing [3]. As a p-type semiconductor with bandgap $E_g \sim 3.2$ eV, PPy has been shown to be a promising material for a wide range of application such as conducting wire, sensors, electrochemical capacitor, and electronics [2, 4–6].

On the other hand, titanium dioxide (TiO_2) is a typical oxide metal and n-type semiconductor that exhibits strong electrochemical and photochemical activity [7]. Apart from major application in vanishes, paint, and so on relying on the brightness and high refractive index of TiO_2, the electrochemical and photochemical features have enlarged the application of TiO_2 in photocatalyst, solar cells, and so on [8, 9]. The combination of $TiO_2$with SnO_2 and $ZnSnO_3$ in heterostructures has been found to boost the adsorption abilities for water molecules via a water-collection process [10, 11]. The coupling of TiO_2 with PPy in a nanocomposite is expected to improve their unique features, namely, sensitivity,

Citation: Trong Tung Nguyen and Ngoc Huyen Duong, "Effect of TiO2 Rutile Additive on Electrical Properties of PPy/TiO_2 Nanocomposite," Journal of Nanomaterials, vol. 2016, Article ID 4283696, 6 pages, 2016. doi:10.1155/2016/4283696.

electrical conductivity, and electrochemical properties [12–17]. In general, TiO_2 crystallines have three main structural morphologies under normal conditions: rutile, anatase, and brookite. Among those three structural morphologies rutile is more stable while anatase and brookite are metastable at room temperature. The difference in morphology, crystalline, and electronic structure results in different overall properties. In terms of facial interaction and charge exchange, the rutile TiO_2 with respect to band gap E_g ~ 3.1 eV could be more effective in the coupling with PPy [18].

It is well known that rutile is the favored polymorph under acidic conditions in hydrothermal synthesis. In this study, pure-phase rutile nanoparticles were synthesized via a low-temperature hydrothermal method using $TiCl_4$ as the Ti precursor in HCl aqueous solution which was then refluxed to facilitate the formation of pure-phase rutile [19–21].

Based on these assumptions, an attempt was carried out to investigate effect of TiO_2 nanocrystallites in rutile polymorphs on TiO_2/PPy nanocomposite synthesized by in situ chemical polymerization.

EXPERIMENTAL

Pyrrole (Py) 99.5% and $TiCl_4$ (Aldrich Co.), ammonium persulphate (APS, Kanto Chemical Co. Inc.), and titanium chloride $TiCl_4$ (Aldrich Co.) were used as precursors in the experiment. The PPy/TiO_2nanocomposite was synthesized in two steps as follows.

Firstly, TiO_2 is prepared by the thermal hydrolysis of a titanium precursor ($TiCl_4$) at relative low temperature (<100°C) in HCl acidic aqueous solution to enhance the rutile formation. The procedure was carried out as follows: A solution of 40 mM $TiCl_4$ and 1.0 M HCl in a flask was heated up to 80°C for 2 hours to make TiO_2colloidal dispersions. After cooling down to room temperature, the colloidal dispersion was kept in refrigerator to have TiO_2 sedimentation at the bottom of the flask.

Secondly, the resulting TiO_2 sedimentation was extracted, weighted, and mixed with an aqueous solution of 0.1 M Py and 1.0 M HCl with a given TiO_2/Py ratio. Then, the polymerization was carried out by adding drop-by-drop a solution of 0.1 M APS and 1.0 M HCl to the Py solution. After 1.5 hours of continuously stirring the polymerization was terminated by pouring ethanol into the mixture. The resulting black nanocomposites formed in the solution were filtered, then cleaned by distilled water, and kept in 1.0 M HCl solution.

The morphology, structure, particle size, and size distribution of the TiO_2, PPy, and PPy/TiO_2 nanocomposite were characterized by FESEM (Hitachi S4800) and TEM (Jeol JEM1010) and X-ray diffraction (XRD D8 Advance Bruker). The change in electronic structure of the materials was analyzed by Raman spectra using LabRAM HR800 (Horiba) with a 632.8 nm excitation laser ranging from 100 to 2339 cm^{-1} at a resolution of 4 cm^{-1}.

The change in conductivity of the PPy/TiO_2 composite was determined as its electrical resistor variation exposure to oxygen O_2 by reducing pressure. The energy storage is determined by cyclic voltammetry spectra with scanning rate of 100 mV/s, within 0.5 V to 0.5 V. The test PPy/TiO_2 layers were drop-coated on interdigital Pt electrodes and the layer resistor changes were acquired and analyzed by Keithley Digital Multimeters connecting with a personal computer.

RESULTS AND DISCUSSION

The XRD made on the resulting TiO_2 shows the positions of all diffraction peaks correspond to rutile (JCP2.2CA number 00-021-1276). As can be seen from Figure 1, a strong peak is observed at 2θ of 27.37° assigned to (110) plane refraction and the other peaks are observed at 2θ of 36.10°, 39.16°, 41.26°, 44.01°, 54.36°, 56.59°, 62.92°, 64.10°, and 68.91° standing for the refraction at (101), (200), (111), (210), (211), (220), (002), (310), and (301) planes, respectively. However, in comparison to the normal powder XRD pattern, the intensity of the (101) peak is relatively heightened and its full width at half maximum (FWHM) is shrunk. The feature likely indicates that the resulting TiO_2 rutile crystallites are major nanorods growth preferably in [001] direction and the (101) planes are the side wall of TiO_2 nanorods which are preferentially exposed to X-ray [9,22].

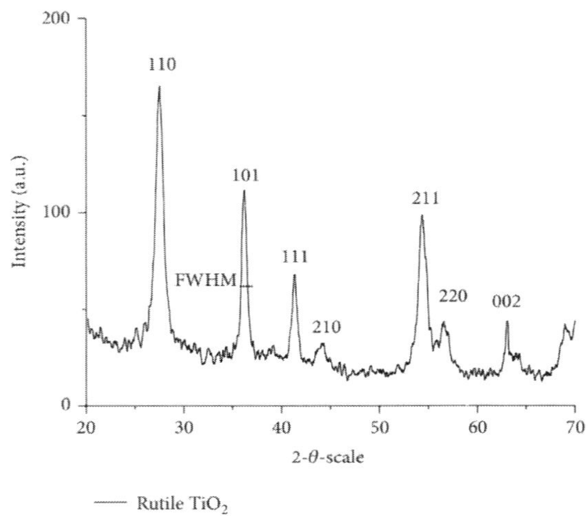

Figure 1: XRD spectrum of TiO_2 synthesized by pyrolysis of $TiCl_4$ in 1.0 M HCl aqueous solution.

As can be seen from TEM images (Figure 2), the neat PPy has a granular structure with mean size around 30–40 nm and the TiO_2 crystallizes in the needle-like form with mean diameter around 9-10 nm and length of around 200–250 nm that is expected growth along [001] direction as discussed in XRD. The PPy granular and TiO_2 needle-like structures are commonly observed with materials synthesized in HCl medium [20]. The resulting PPy/TiO_2 nanocomposite shows a coupling core-shell structure of granular PPy covering TiO_2needles. From chemical view of consideration, at the first oxidization stage the TiO_2 surface (donor sources) in the reaction medium acts as nucleation sites for the Py radical monomers (acceptor sources) adsorbed and anchored on. In the following stage, PPy chains grow up on the TiO_2 surface forming core-shell structures. The structures show the evidence of strong bonds between TiO_2 and PPy via donor-acceptor interaction [13, 14]. As a result, the core-shell structure will form p-n junction between PPy and TiO_2. The thickness of PPy layer growth on surface of TiO_2 then depends on the TiO_2 concentration in the reaction medium. The interfacial interaction between TiO_2 and PPy in the nanocomposites is expected to modify their chemical and electronic structures.

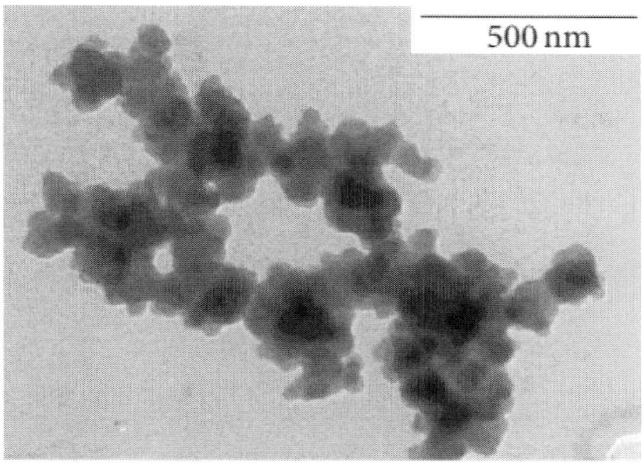

(a) PPy

(b) TiO_2

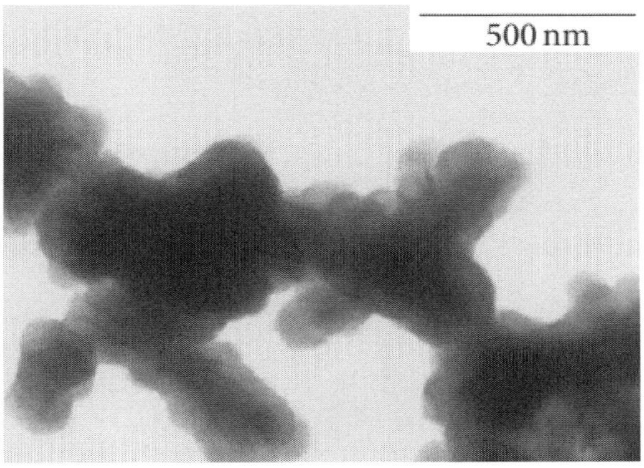

(c) PPy/TiO_2

Figure 2: TEM images of (a) PPy and (b) TiO_2 nanoparticles and (c) PPy/TiO_2nanocomposites.

As shown in Figure 3 the Raman spectrum of a pure TiO_2 sample reveals three strong Raman bands 232, 446, and 609 cm^{-1} and these three bands are in agreement with data observed in the spectra of a rutile phase. The predominance of rutile structure in the resulting TiO_2 sedimentation has accounted for the effect of HCl in the formation of TiO_2 during pyrolysis.

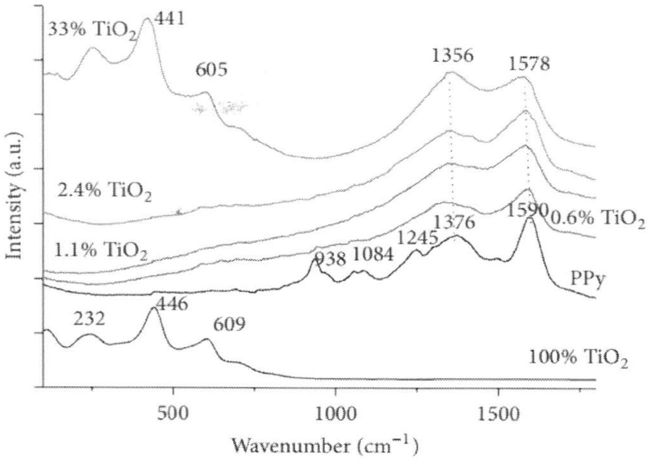

Figure 3: Raman spectra of TiO_2, PPy, and PPy/TiO_2 nanocomposite with different TiO_2concentration.

The Raman spectrum of neat PPy shows the instinct vibrational modes at 1590 cm^{-1} (G band) and 1340 cm^{-1}(D band), and 980 cm^{-1} band indicates that certain sites of the PPy chain are still in the benzoid form (reduced state), where the vibrational mode at 938 cm^{-1}, assigned to C-H out of the plane deformation of the quinoid form, is related to the oxidized sites [23, 24]. The vibrational modes assigning a band at 1082 cm^{-1} was correlated to the bipolaron state. The presence of TiO_2 in the nanocomposite causes a change in relative intensity and a red-shift of the vibrational bands standing for C=C bond from 1376 cm^{-1} to 1356 cm^{-1} and 1590 cm^{-1} to 1578 cm^{-1} indicating the fact that PPy shifts to reduction structure. The effect indicates the degree of reduction of PPy as the result of coupling with TiO_2 in the nanocomposites.

It is well known that the conductivity of both components (PPy, TiO_2) as well as resulting nanocomposites is sensitive to oxygen (oxidizing agent) [25]. Upon exposure to open air, oxygen is adsorbed onto the surface and abstracts electrons from the material causing ionization of the atom and yielding a positive hole; hence, the material conductance increases with oxygen concentration. Changing the air pressure means to change the oxygen concentration in the environment; the conductivity of the materials is varied. The O_2 sensing profiles of PPy, PPy/TiO_2 in Figure 4 show the materials resistance rapidly increasing as air pressure reduces (the vacuum pump on) and fast returning back to the origin value as air flows in (the vacuum pump off). Typically, O_2sensitivity of PPy/TiO_2 is 16–18 times higher than those of PPy. From the physical view of consideration, the change in conductivity involves the extraction of electrons from the valence band of semiconducting PPy as a result of oxygen interaction (physical adsorption) thus converting the materials to be p-type semiconductor. The adsorption and desorption of oxygen molecule on PPy/TiO_2 surface then cause the variation in conductivity of the materials. The more oxygen concentration is in the environment, the more electrons are extracted; as a result, the material conductivity is increased and vice versa. Higher oxygen sensitivity indicates the fact that more oxygen adsorption sites are created on the PPy/TiO_2 surface. The fast response and recovery and the stability in base line indicate that the physical adsorptions with weak bonding are dominant in the oxygen-nanocomposites interaction.

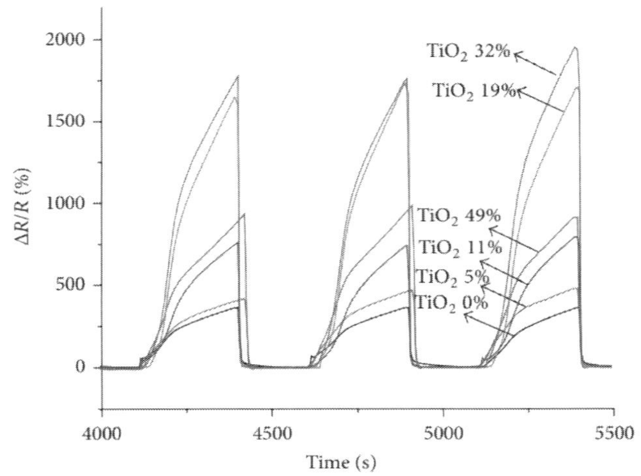

(a)

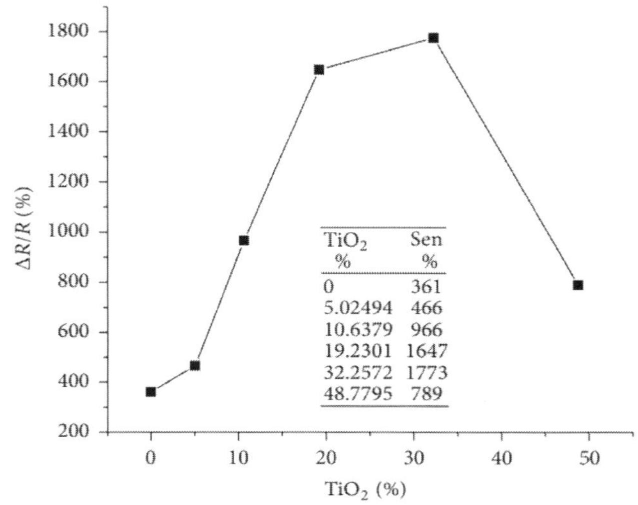

(b)

Figure 4: (a) The sensitivity of of PPy/TiO_2 upon exposure to oxygen, (b) the oxygen sensitivity of the materials as function of TiO_2 concentration.

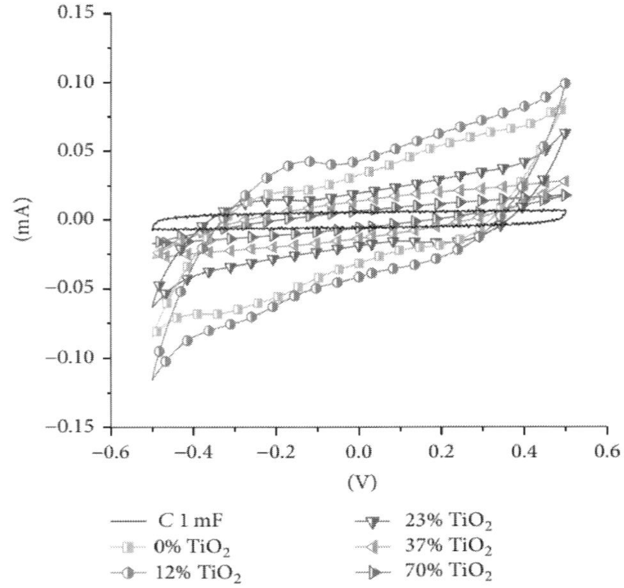

(a)

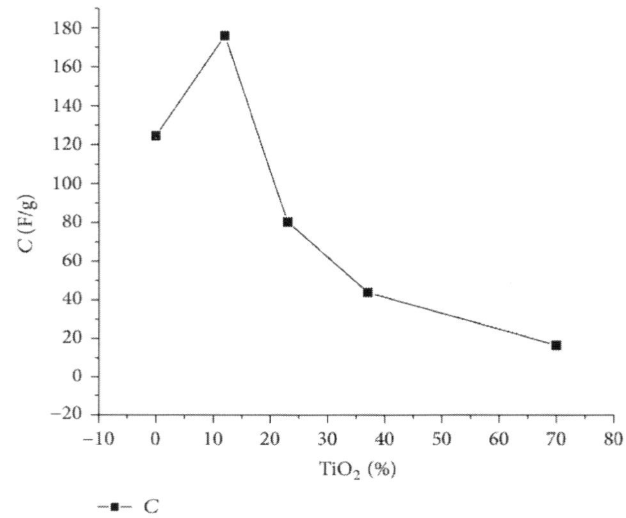

(b)

Figure 5: (a) CV diagram of PPy/TiO_2 nanocomposites with scanning rate of 100 mV/s, (b) calculated specific capacitance of the materials as function of TiO_2 concentration.

Cyclic voltammetry spectra with scanning rate of 100 mV/s, within −0.5 V to 0.5 V, are used to determine the specific capacitance of PPy/TiO_2 in comparison with a capacitor 1000 μF as reference. The CV spectra of samples are shown in Figure 5. The equivalent capacitance of samples is calculated by $C = dQ/dV = i/S$, where $S = dV/dt$ and current i. The specific capacitance of PPy/TiO_2 is calculated by C/m, where m is the nanocomposite weight used in the sample. Experiments show that the specifics capacitance of neat PPy is 135 F/g and changing when TiO_2 is added. The capacitance achieves the peak as TiO_2 content is about 15% and reduces when TiO_2 is increased 12 F/g with 70% TiO_2.

The change in conductivity, sensitivity, and specific capacitance of PPy/TiO_2 nanocomposite can be explained by the impact of core-shell structure as shown in Figure 6. The PPy/TiO_2 core-shell structure will form n-p junction and depletion region with a width of around 10–20 nm that depends on doping condition [26, 27]. The PPy shell with thickness of 20–30 nm actually locates the space charge region. Similar to CMOS capacitor, the change in the depletion width can be used to explain the change in specific capacitance of the PPy/TiO_2. As a result of the diffusion, electrons from n-type TiO_2 diffuse into p-type PPy and form an inversion region of rich electron in a very thin layer near the PPy surface. Upon exposure to the air, the interaction between oxygen (an electron acceptor) and the rich electron layer results in an additive electron extraction from PPy valence band and then changes the PPy conductivity.

The adsorption and desorption of oxygen molecule on PPy surface then cause the variation in conductivity of the materials. The more oxygen concentration is in the environment, the more electrons are extracted; as a result, the material conductivity is increased and vice versa. Higher oxygen sensitivity indicates the fact that more oxygen adsorption sites are created on the PPy surface. The fast response and fast recovery in the oxygen sensitivity profile indicate that the physical adsorptions with weak bonding are dominant in the oxygen-nanocomposites interaction.

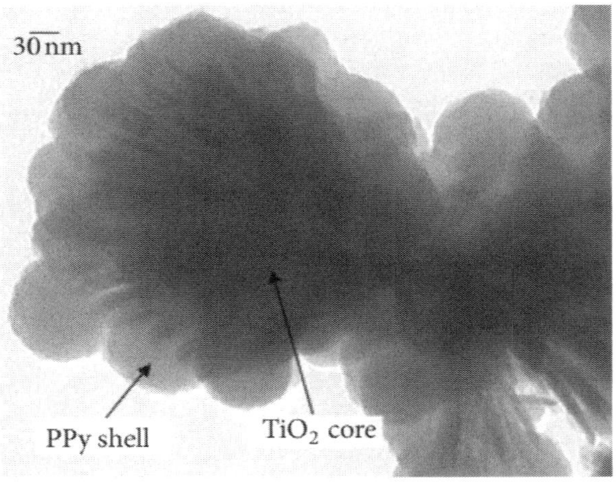

Figure 6: TEM image of the core-shell structure of PPy/TiO_2 nanocomposite.

Thus, the PPy/TiO_2 coupling result in the formation of the inversion layer then creates more additive oxygen deficiency on the PPy-shell surface. The feature is accounted for the modification in conductance, oxygen sensitivity, and specific capacitance of the PPy/TiO_2 nanocomposite.

CONCLUSIONS

PPy/TiO_2 nanocomposite synthesized by in situ chemical polymerization was created in the form of granular PPy and embedded TiO_2 nanoparticles forming core-shell structure. The PPy/TiO_2 core-shell results in thin n-p junctions in which the TiO_2 core effectively modifies the PPy band gap as well as overall the electrical properties of the coupling. Consequently, the conductance and specific capacitance of the nanocomposite change as a function of TiO_2 concentration. The n-p contact between TiO_2 and PPy brings in an additive oxygen deficiency to the PPy shell surface and then increases the oxygen sensitivity. The process is revertible as a result from the physical O_2 adsorption and desorption which act as doping and dedoping on semiconducting PPy.

ACKNOWLEDGMENT

The authors gratefully acknowledge financial support from Basic Research Project Grant no. 103.02-2012.32 in Aid by National Foundation for Science and Technology Development (NAFOSTED), Vietnam.

REFERENCES

1. T. A. Skotheim and J. Reynolds, Conjugated Polymers: Theory, Synthesis, Properties, and Characterization, CRC Press, New York, NY, USA, 2006.
2. T. A. Skotheim and J. Reynolds, Handbook of Conducting Polymers, 2 Volume Set, CRC Press, 2007.
3. A. F. Diaz, K. K. Kanazawa, and G. P. Gardini, "Electrochemical polymerization of pyrrole," Journal of the Chemical Society, Chemical Communications, no. 14, pp. 635–636, 1979.
4. R. K. Sharma, A. C. Rastogi, and S. B. Desu, "Pulse polymerized polypyrrole electrodes for high energy density electrochemical supercapacitor," Electrochemistry Communications, vol. 10, no. 2, pp. 268–272, 2008.
5. G. A. Snook, P. Kao, and A. S. Best, "Conducting-polymer-based supercapacitor devices and electrodes,"Journal of Power Sources, vol. 196, no. 1, pp. 1–12, 2011.
6. W. Sun and X. Chen, "Preparation and characterization of polypyrrole films for three-dimensional micro supercapacitor," Journal of Power Sources, vol. 193, no. 2, pp. 924–929, 2009.
7. K. Hashimoto, H. Irie, and A. Fujishima, "TiO_2 photocatalysis: a historical overview and future prospects," Japanese Journal of Applied Physics, vol. 44, no. 12, pp. 8269–8285, 2005.
8. Z. Zhang, Y. Huang, K. Liu, L. Guo, Q. Yuan, and B. Dong, "Multichannel-improved charge-carrier dynamics in well-designed hetero-nanostructural plasmonic photocatalysts toward highly efficient solar-to-fuels conversion," Advanced Materials, vol. 27, no. 39, pp. 5906–5914, 2015.
9. B. Liu and E. S. Aydil, "Growth of oriented single-crystalline rutile TiO_2 nanorods on transparent conducting substrates for dye-sensitized solar cells," Journal of the American Chemical Society, vol. 131, no. 11, pp. 3985–3990, 2009.
10. Z. Yang, Z. Zhang, K. Liu, Q. Yuan, and B. Dong, "Controllable assembly of SnO_2 nanocubes onto TiO_2electrospun nanofibers toward humidity sensing applications," Journal of Materials Chemistry C, vol. 3, no. 26, pp. 6701–6708, 2015.
11. Z. Zhang, J. Huang, B. Dong, Q. Yuan, Y. He, and O. S. Wolfbeis, "Rational tailoring of $ZnSnO_3$/TiO_2heterojunctions with bioinspired surface wettability for high-performance humidity nanosensors,"Nanoscale, vol. 7, no. 9, pp. 4149–4155, 2015.
12. C. A. Ferreira, S. C. Domenech, and P. C. Lacaze, "Synthesis and characterization of polypyrrole/TiO_2composites on mild steel," Journal of Applied Electrochemistry, vol. 31, no. 1, pp. 49–56, 2001.
13. D. M. Lenz, M. Delamar, and C. A. Ferreira, "Application of polypyrrole/TiO_2 composite films as corrosion protection of mild steel," Journal of Electroanalytical Chemistry, vol. 540, pp. 35–44, 2003.
14. G. K. R. Senadeera, T. Kitamura, Y. Wada, and S. Yanagida, "Enhanced photoresponses of polypyrrole on surface modified TiO_2 with self-assembled monolayers," Journal of Photochemistry and Photobiology A: Chemistry, vol. 184, no. 1-2, pp. 234–239, 2006.
15. K. Jurewicz, S. Delpeux, V. Bertagna, F. Béguin, and E. Frackowiak, "Supercapacitors from nanotubes/

polypyrrole composites," Chemical Physics Letters, vol. 347, no. 1–3, pp. 36–40, 2001.

16. D. Zhang, X. Zhang, Y. Chen, P. Yu, C. Wang, and Y. Ma, "Enhanced capacitance and rate capability of graphene/polypyrrole composite as electrode material for supercapacitors," Journal of Power Sources, vol. 196, no. 14, pp. 5990–5996, 2011.
17. Y. Gao, K. Ding, X. Xu, Y. Wang, and D. Yu, "PPy film/ TiO_2 nanotubes composite with enhanced supercapacitive properties," RSC Advances, vol. 4, no. 52, pp. 27130–27134, 2014.
18. C. Yu, G. Li, S. Kumar, H. Kawasaki, and R. Jin, "Stable $Au_{25}(SR)_{18}/TiO_2$ composite nanostructure with enhanced visible light photocatalytic activity," The Journal of Physical Chemistry Letters, vol. 4, no. 17, pp. 2847–2852, 2013.
19. G. Li and K. A. Gray, "Preparation of mixed-phase titanium dioxide nanocomposites via solvothermal processing," Chemistry of Materials, vol. 19, no. 5, pp. 1143–1146, 2007.
20. Z. Wang, D. Xia, G. Chen, T. Yang, and Y. Chen, "The effects of different acids on the preparation of TiO_2nanostructure in liquid media at low temperature," Materials Chemistry and Physics, vol. 111, no. 2-3, pp. 313–316, 2008.
21. G. Li, N. M. Dimitrijevic, L. Chen, J. M. Nichols, T. Rajh, and K. A. Gray, "The important role of tetrahedral Ti^{4+} sites in the phase transformation and photocatalytic activity of TiO_2 nanocomposites,"Journal of the American Chemical Society, vol. 130, no. 16, pp. 5402–5403, 2008.
22. J. Wu, S. Lo, K. Song et al., "Growth of rutile TiO_2 nanorods on anatase TiO_2 thin films on Si-based substrates," Journal of Materials Research, vol. 26, no. 13, pp. 1646–1652, 2011.
23. Y. Furukawa, S. Tazawa, Y. Fujii, and I. Harada, "Raman spectra of polypyrrole and its 2,5-13C-substituted and C-deuterated analogues in doped and undoped states," Synthetic Metals, vol. 24, no. 4, pp. 329–341, 1988.
24. M. J. L. Santos, A. G. Brolo, and E. M. Girotto, "Study of polaron and bipolaron states in polypyrrole by in situ Raman spectroelectrochemistry," Electrochimica Acta, vol. 52, no. 20, pp. 6141–6145, 2007.
25. X. Li, R. Ramasamy, and P. K. Dutta, "Study of the resistance behavior of anatase and rutile thick films towards carbon monoxide and oxygen at high temperatures and possibilities for sensing applications,"Sensors and Actuators B: Chemical, vol. 143, no. 1, pp. 308–315, 2009.
26. O. Inganäs, T. Skotheim, and I. Lundström, "Polypyrrole-semiconductor Schottky barriers," Journal of Applied Physics, vol. 54, no. 6, pp. 3636–3639, 1983. V
27. P. Syed Abthagir and R. Saraswathi, "Junction properties of metal/polypyrrole Schottky barriers," Journal of Applied Polymer Science, vol. 81, no. 9, pp. 2127–2135, 2001.

CHAPTER 12

STRUCTURAL AND ELECTRICAL STUDIES ON ZNO-BASED THIN FILMS BY LASER IRRADIATION

Shanyue Zhao,[1] Yinqun Hua,[1,2] Ruifang Chen,[2] Jian Zhang,[1] and Ping Ji[1]

[1]School of Material and Science Engineering, Jiangsu University, Zhenjiang 212013, China

[2]School of Mechanics Engineering, Jiangsu University, Zhenjiang 212013, China

ABSTRACT

The effects of laser irradiation on the structural and electrical properties of ZnO-based thin films were investigated. The XRD pattern shows that the thin films were highly textured along the -axis and perpendicular to the surface of the substrate. Raman spectra reveal that Bi_2O_3 segregates mainly at ZnO-ZnO grain boundaries. After laser irradiation processing, the grain size of the film was reduced significantly, and the intrinsic atomic defects of grain boundaries and Bi element segregated at the grain boundary were interacted frequently and formed the composite defects of acceptor state. The nonlinear coefficient increased to 24.31 and the breakdown voltage reduced to 5.34 V.

INTRODUCTION

ZnO varistors exhibit high nonlinear coefficient, low leakage current, and high surge absorbing ability. They have been extensively applied on high-voltage protection of electrical equipment since the 1970s, such as inhibiting lightning overvoltage of electrical power system [1–5]. However, modern-day electronic components, mainly based on silicon, are more susceptible not only to electrical overvoltage transients, but also to damage by electromagnetic pulses or electromagnetic interference (EMI) noise that is now strictly regulated by law throughout the world [6, 7]. Therefore, the industry has high demands for low-voltage varistors. For a low-voltage varistor, it is required to be of few microns in thickness to reduce nonlinear voltage. ZnO films with good quality were grown through different methods, such as hydrothermal method [8], sol-gel technique [9], electron-beam evaporation [10], pulse laser deposition [11], and magnetron sputtering [12]. Among all, magnetrons puttering shows several advantages such as good adhesion between film and substrate, low substrate temperature, scalability to large areas, great thickness uniformity with a deposition rate range, and high film density [12–15].

Recently, laser irradiation technique has been used on thin films as an efficient method to improve the crystalline quality and increase the electron donors [16–19]. Laser irradiation has several advantages, including fast crystallization at room temperature, possibility of local crystallization, crystallization of thin films on low melting point substrates, and increasing charge carriers through a photoconductivity effect. Tsay and Wang [18] performed ZnO thin films on glass by sol-gel and used KrF excimer laser for annealing. They reported that the crystallinity levels and average crystallite size of thin films with excimer laser irradiation were greater than those of thermally annealed thin films. Tsang et al. [19] fabricated transparent conducting AZO thin films using KrF excimer laser irradiation. They reported that the electrical and optical

Citation: Shanyue Zhao, Yinqun Hua, Ruifang Chen, Jian Zhang, and Ping Ji, "Structural and Electrical Studies on ZnO-Based Thin Films by Laser Irradiation," Journal of Nanotechnology, vol. 2016, Article ID 9385725, 6 pages, 2016. doi:10.1155/2016/9385725.

characteristics of AZO thin films irradiated by excimer laser improved significantly.

To the best of our knowledge, no report on ZnO-based thin film varistors with laser irradiation for low-voltage application exists. In our previous work [20], we studied the effect of laser shock on electrical property of ZnO varistor ceramics. In the present work, ZnO-based thin films of high quality were prepared by RF magnetron sputtering technique on Si (111) substrates. After examinations, the films were subsequently treated by laser irradiation. The paper has studied the structural and electrical properties of the ZnO-based thin film varistors for low-voltage application.

EXPERIMENT

ZnO-based thin films were deposited on Si (111) substrates by RF magnetron sputtering technique and followed by annealing at 700°C for 4 h in the air. The target, ZnO-based ceramic with 60 mm in diameter and 5 mm in thickness, was formed by dry pressing and subsequently sintered in the air. The raw materials of the target were mixed powders and the matching composition was 96.5 mol% ZnO + 2 mol% Bi_2O_3 + 1.5 mol% (MnO_2, Co_2O_3, Cr_2O_3, and Sb_2O_3). It is the optimized composition investigated by the authors in our previous works. Si (111) which was used as the substrate for ZnO thin films deposition were cleaned in an ultrasonic bath with ethanol for 15 min than dried for use. The substrates were sprayed gold in the metal-spraying equipment for 5 min, to prepare a layer of bottom electrode on the Si substrate. Using the sky-0826138 RF magnetron sputtering instrument, the background vacuum degree of the chamber was less than 1×10^{-4} Pa. The sputtering pressure was 3.0 Pa, and the Argon flow was 30 sccm. The sputtering time was 60 min, and the sputtering power was 150 W. The substrate temperature was 300°C. The film thickness was controlled within 2 microns. After the laser irradiation processing, the laser irradiation area was covered with gold to prepare the top electrode. Figure 1 shows the structure of the ZnO-based thin film varistors.

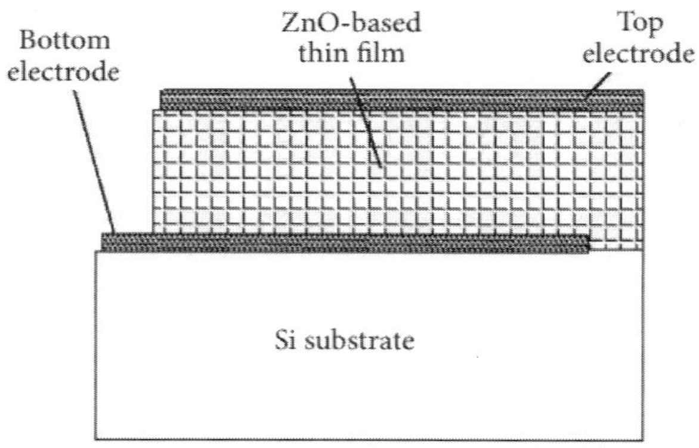

Figure 1: The structure of the ZnO-based thin film varistors.

The 0.2 mm aluminum foil adhered to the film surface as the absorption layer. The homemade liquids were coated on the aluminum foil surface as the constraint layer. The thickness was 0.5 mm. The laser irradiation parameters are as followings: light spot diameter of 6 mm, wavelength of 1064 nm, pulse width of 5 ns, pulse energy of 0.8 J. Finally, the samples were set in the mold to conduct the laser irradiation experiment. After the laser irradiation processing, the films' surfaces were fine, which showed that as long as the parameters of laser irradiation were appropriate, laser irradiation could be applied on the films completely.

The structural properties were characterized by X-ray diffractometer (XRD) with CuKα radiation. The surface morphology and root mean square (RMS) roughness level of the films were examined by tapping atomic force microscope (AFM). Plane-view micrographs of ZnO-based thin films were taken with a field-emission scanning electron microscope (FE-SEM). Raman spectra were excited with the 532 nm line of a laser at an incident power of 10 mW and obtained in the range 0–3000 cm^{-1}. The J-V curves are measured by transistor characteristic meter. The breakdown voltage U_{1mA} (V) and leakage current J_L (μA/mm²) are measured by piezoresistor general measuring instrument.

RESULTS AND DISCUSSIONS

The X-ray diffraction patterns of the ZnO-based thin films before and after irradiation are shown in Figure 2. Only the (002) peaks are observed in the XRD patterns, which shows that the films are highly textured along the c-axis and perpendicular to the surface of the substrate. This diffraction peak is indexed to those of hexagonal wurtzite zinc oxide and no diffraction peaks of any other phases or impurities are detected [8, 9, 11]. After irradiation, the intensity of (002) ZnO peak increases and there is a shift to larger angles. This trend with the shift of 2θ in XRD patterns suggests that the Bi^{3+} ions occupied the Zn^{2+} substitutional sites [15, 21]. This result also shows that laser irradiation can facilitate the substitution of Bi^{3+} ions to Zn^{2+} ions [18].

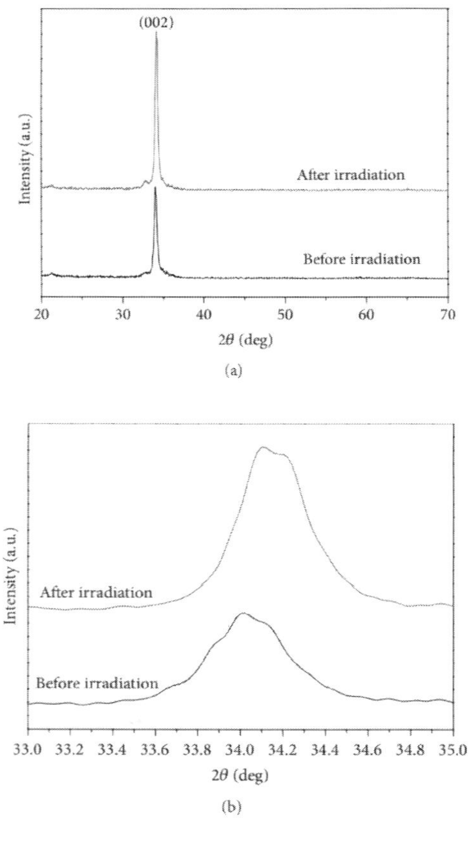

Figure 2: XRD patterns of ZnO-based thin films.

Figure 3 shows the AFM morphologies for the films, which illustrate the surface morphology of the thin films. (a) is the sample before irradiation and (b) is the sample after irradiation. After laser irradiation, the film surface morphology changes significantly. Corresponding to the above two samples, the RMS surface roughnesses are 31.4 and 2.8 nm. The film treated by laser irradiation has a relative smooth and dense surface morphology. At the initial state, the laser shock wave passes through the rigid aluminum foil adhering to the film surface and reaches the bottom of the aluminum foil. For the impedance mismatch of the aluminum foil and the film, strong contact pressure appears between the contact point of the aluminum foil and the film. The sharp protrusions are compressed at the strong contact pressure and this is characterized by the great reduction of surface roughness. Figure 4 shows the morphology and particle sizes of prepared film by SEM. The histograms reveal the distribution of average grain size, which are reasonably described by the Gaussian function, showing tight size distribution with average sizes of approximately 191 and 139 nm in diameter, respectively.

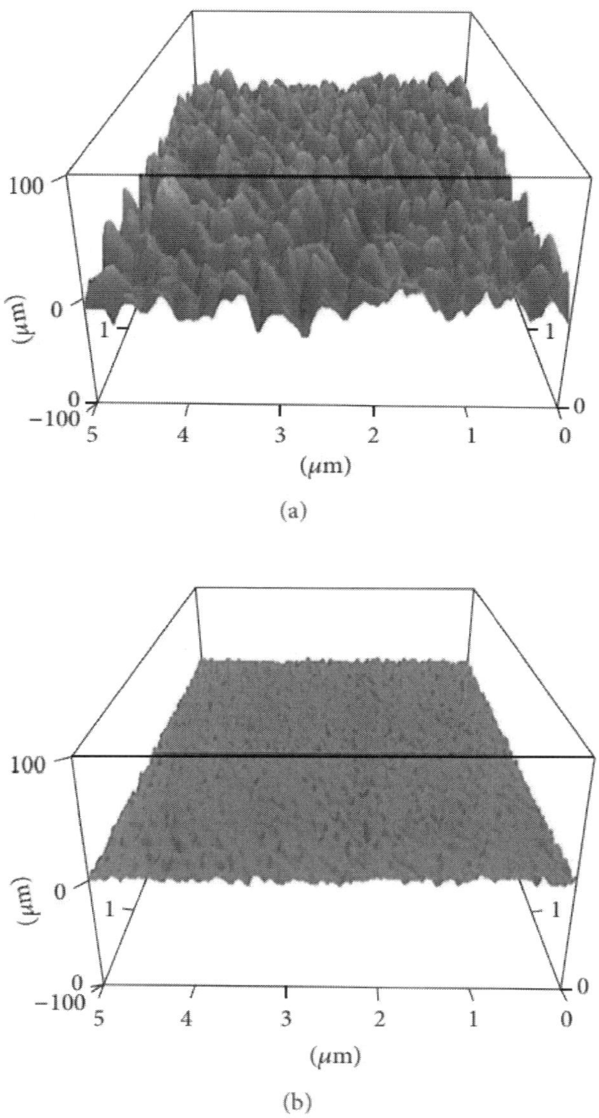

Figure 3: AFM micrographs of the ZnO-based thin films.

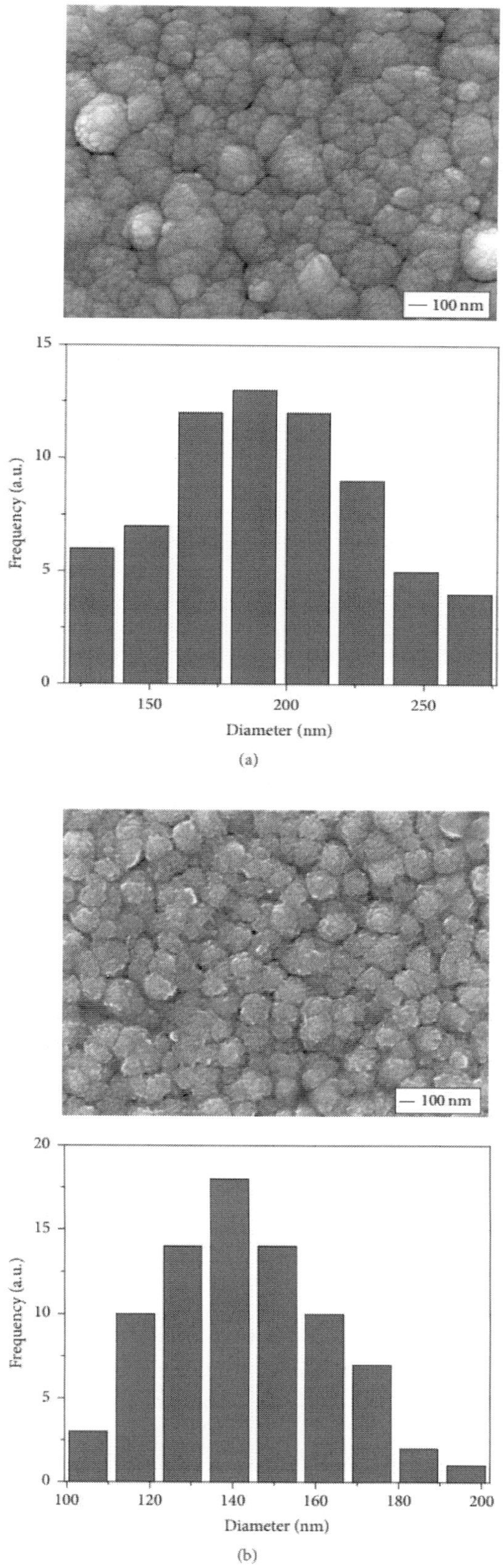

Figure 4: SEM micrographs of the ZnO-based thin films.

Figure 5 shows the Raman spectrum of ZnO-based thin film. A larger peak shift at 437 cm^{-1} is observed, which is almost the same position as the standard ZnO. It indicates that the thin film is almost free of stress. Theoretically, it is difficult to substitute Zn^{2+} by Bi^{3+} in ZnO crystal, for the radius of Bi^{3+} is much larger than Zn^{2+}. The former results indicate that Bi_2O_3 segregates mainly at ZnO-ZnO grain boundaries as Bi-rich phases during sintering in the ZnO-Bi_2O_3 system. The Bi elements distribute evenly throughout the thin film, and the distribution of Bi appears in certain regularity, which are almost distributed at the grain boundaries. After laser irradiation processing, grain boundaries increase with the grain refinement. In this condition, the intrinsic atomic defects of grain boundaries and Bi element segregated at the grain boundary interact frequently, forming the composite defects of acceptor state [22, 23]. So the interface state density N_S of grain boundaries increase greatly, which promotes the increasing of barrier height Φ_B and the nonlinear coefficient. Therefore, the pressure-sensitive electrical properties of ZnO-based ceramic film of laser irradiation state are excellent.

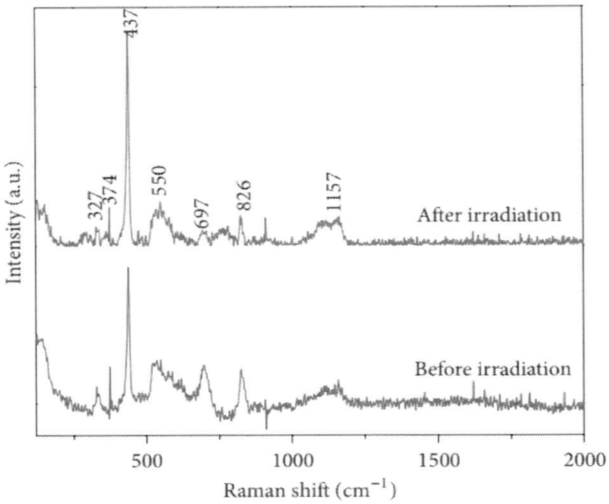

Figure 5: Raman spectroscopy of ZnO-based thin films.

Table 1 is the result of electrical properties. The nonlinear coefficient is α, breakdown voltage is U_{1mA} (V), and leakage current is J_L (μA/mm²). After laser irradiation processing, the nonlinear coefficient is 24.31, which increases by 40.8%, and breakdown voltage is 5.34 V, which decreases by 20.7%. Leakage current density is as low as 2.03 μA/mm², which decreases by 12.1%. Similar investigations have been done by other researchers. Horio et al. [13] prepared the ZnO/Pr_6O_{11} films on the glass substrates by the RF magnetron sputtering technique, and the film thickness was 600 nm/400 nm. The breakdown voltage was 20 V, and the nonlinear coefficient was 10. Miśta et al. [14] prepared the Bi-doped films on the ITO/Glass using Zn-Bi target by the RF magnetron sputtering technique. The results displayed that the breakdown voltage was from a few volts to dozens of volts, and the nonlinear coefficient was 15. Jeong [24] prepared low-voltage ZnO varistors and investigated the field failure; after polishing, the breakdown voltage was 9.71 V, and the nonlinear coefficient was 20.35. Therefore, the electrical properties improve significantly by the laser irradiation processing.

Table 1: Electrical properties of films before and after laser-irradiation processing

Parameters	Before irradiation (I)	After irradiation (II)	Rate (%)
α	17.26	24.31	40.8
$U_{1\,mA}$ (V)	6.73	5.34	−20.7
J_L (μA/mm^2)	2.31	2.03	−12.1

The grain boundary of ZnO-based ceramic film is a thin disorder area, where a large number of interface states and electron traps exist. They are able to capture the free electrons in the ZnO grains. The electrons of grain boundaries are depleted, and depletion layers are formed. Consequently, electronic barriers with certain depth have been formed from grain surfaces into body. The barrier height is Φ_B, which is called Double Schottky barrier [25]. It is the source of voltage-sensitive electrical property.

Nonlinear characteristics of ZnO-based film originate from the grain boundary barrier. To study the impact of laser irradiation processing on the electric properties of ZnO-based ceramic film, we adopted the double Schottky barrier model proposed by Gupta and Carlson [26] to explain the change of electrical property. At preswitch region, there is a relationship between the current density and the extra electric field obtained from the model, which is shown in [27]

$$J = AT^2 \exp\left[\frac{\left(\beta E^{0.5} - \Phi_B\right)}{(kT)}\right]. \quad (1)$$

J is current density, A is Richardson constant, T is absolute temperature, E is electric field intensity, Φ_B is barrier height, and K is Boltzmann constant. $\beta = \{1/(\gamma\omega)[2e^3/(4\pi\varepsilon 0\varepsilon r)]\}^{1/2}$ [17], γ is grain number of unit length, ω is Barrier width, e is electron charge (1.602×10^{-19} C), ε_0 is vacuum dielectric constant (8.85×10^{-14} F/cm), and εr is relative dielectric constant (8.5). Setting T as 25∘ C, ln J-$E^{0.5}$ fitting diagram of film and film of laser irradiation state are shown in Figure 5. The boundary characteristic parameters were calculated by Figure 6, which were listed in Table 2, including the barrier height Φ_B, barrier width ω, interface-state density N_D, and donor concentration N_S.

Table 2: The characteristic parameters of grain-boundary of ZnO-based ceramic films before and after laser-irradiation processing

State	Barrier height Φ_B (eV)	Barrier width ω (Å)	Donor concentration N_D (10^{22} cm^{-3})	Density of interface states N_S (10^{14} cm^{-2})
Before irradiation (I)	1.86	1.13	13.5	15.3
After irradiation (II)	2.41	0.31	227	71.6

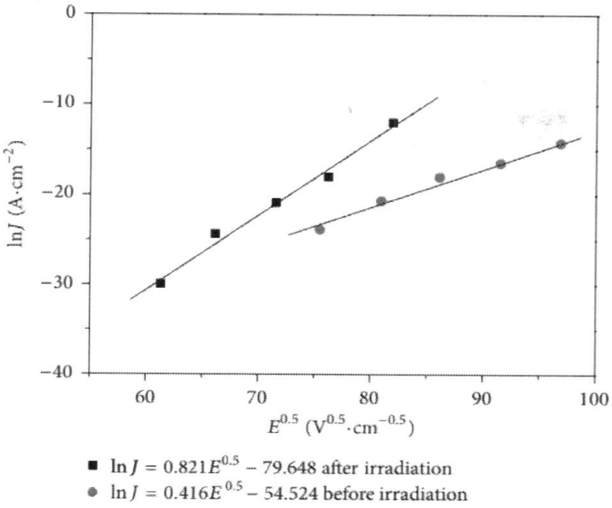

Figure 6: The relation patterns of ln J-$E^{0.5}$ of the ZnO-based thin films before and after irradiation.

Table 2 shows that after laser irradiation processing, barrier height Φ_B increased, barrier width decreased, donor concentration N_D and interface-state density N_S also increased greatly. According to [27]

$$\alpha \approx \left(\frac{\nu}{E}\right)\Phi_B{}^{3/2}. \quad (2)$$

Nonlinear coefficient α is proportional to potential barrier height Φ_B. Φ_B rose from 1.86 to 2.41, which prompted the increasing of α from 12.94 to 21.38. So the increasing of barrier height can effectively improve the nonlinear coefficient. According to formula $\Phi_B = e^2 N^2_S / 2\varepsilon_0\varepsilon_r N_D$ [18], barrier height Φ_B is closely related to the interface-state density N_S and donor concentration N_D. After laser irradiation processing, the thin film produced the lattice distortion, which increased the amount of intrinsic atomic defects of ZnO. There is no second phase diffraction peak in the Xray diffraction patterns of samples, but which cannot exclude the doping elements as indeterminate type segregating at the grain boundaries.

CONCLUSIONS

After laser irradiation processing, electrical properties of the films have been improved in different degrees. The nonlinear coefficient was 24.31, which was increased by 40.8%. The breakdown voltage was 5.34 V, which was reduced by 20.7%. Leakage current density was as low as 2.03 μA/mm^2, which was reduced by 12.7%. The grain size was reduced and the clearance between the adjacent particles also decreased. The film produced the lattice distortion, prompting the increaseing of ZnO intrinsic atomic defects. With the refinement of the grain, grain boundaries were increasing. In this condition, the intrinsic atomic defects of grain boundaries and Bi element segregated at the grain boundaries interact frequently, forming the composite defects of acceptor state. The interface-state density N_S of the grain boundaries increased greatly, increasing the barrier height Φ_B. Therefore, the pressure-sensitive electrical properties of ZnO-based ceramic film of laser irradiation state are excellent.

REFERENCES

1. M. Matsuoka, "Nonohmic properties of zinc oxide ceramics," Japanese Journal of Applied Physics, vol. 10, no. 6, pp. 736–746, 1971.
2. T. K. Gupta, "Application of zinc oxide varistors," Journal of the American Ceramic Society, vol. 73, no. 7, pp. 1817–1840, 1990.
3. D. R. Clarke, "Varistor ceramics," Journal of the American Ceramic Society, vol. 82, no. 3, pp. 485–502, 1999.
4. S. Anas, R. V. Mangalaraja, M. Poothayal, S. K. Shukla, and S. Ananthakumar, "Direct synthesis of varistor-grade doped nanocrystalline ZnO and its densification through a step-sintering technique," Acta Materialia, vol. 55, no. 17, pp. 5792–5801, 2007.
5. P. R. Bueno, J. A. Varela, and E. Longo, "SnO_2, ZnO and related polycrystalline compound semiconductors: an overview and review on the voltage-dependent resistance (non-ohmic) feature,"Journal of the European Ceramic Society, vol. 28, no. 3, pp. 505–529, 2008.
6. R. Puyané, "Applications and product development in varistor technology," Journal of Materials Processing Technology, vol. 55, no. 3-4, pp. 268–277, 1995.
7. A. Rafferty, Y. Gun'ko, and R. Raghavendra, "An investigation of co-fired varistor-ferrite materials,"Journal of the European Ceramic Society, vol. 24, no. 7, pp. 2005–2013, 2004.
8. Z. N. Urgessa, O. S. Oluwafemi, and J. R. Botha, "Hydrothermal synthesis of ZnO thin films and its electrical characterization," Materials Letters, vol. 79, no. 15, pp. 266–269, 2012. Y. M. Li, L. H. Xu, X. G. Li, X. G. Shen, and A. L. Wang, "Effect of aging time of ZnO sol on the structural and optical properties of ZnO thin films prepared by sol-gel method," Applied Surface Science, vol. 256, no. 14, pp. 4543–4547, 2010.
9. H. Z. Wu, K. M. He, D. J. Qiu, and D. M. Huang, "Low-temperature epitaxy of ZnO films on Si(001) and silica by reactive e-beam evaporation," Journal of Crystal Growth, vol. 217, no. 1, pp. 131–137, 2000.
10. X. Q. Wei, J. Z. Huang, M. Y. Zhang, Y. Du, and B. Y. Man, "Effects of substrate parameters on structure and optical properties of ZnO thin films fabricated by pulsed laser deposition," Materials Science and Engineering B, vol. 166, no. 2, pp. 141–146, 2010.
11. D. R. Hernández-Socorro, Z. Montiel-González, S. E. Rodil-Posada et al., "Effect of 8 MeV Si ions irradiation and thermal annealing in ZnO thin films," Journal of Crystal Growth, vol. 354, no. 1, pp. 169–173, 2012.
12. N. Horio, M. Hiramatsu, M. Nawata, K. Imaeda, and T. Torii, "Preparation of zinc oxide/metal oxide multilayered thin films for low-voltage varistors," Vacuum, vol. 51, no. 4, pp. 719–722, 1998.
13. W. Mišta, J. Ziaja, and A. Gubański, "Varistor performance of nanocrystalline Zn-Bi-O thin films prepared by reactive RF magnetron sputtering at room temperature," Vacuum, vol. 74, no. 2, pp. 293–296, 2004.
14. M. Jiang, X. Liu, and H. Wang, "Conductive and transparent Bi-doped ZnO thin films prepared by

rf magnetron sputtering," Surface and Coatings Technology, vol. 203, no. 24, pp. 3750–3753, 2009.

15. Y. Zhao and Y. Jiang, "Effect of KrF excimer laser irradiation on the properties of ZnO thin films,"Journal of Applied Physics, vol. 103, no. 11, Article ID 114903, 2008.

16. Y. Hou and A. H. Jayatissa, "Effect of laser irradiation on gas sensing properties of sol-gel derived nanocrystalline Al-doped ZnO thin films," Thin Solid Films, vol. 562, no. 26, pp. 585–591, 2014.

17. C.-Y. Tsay and M.-C. Wang, "Structural and optical studies on sol-gel derived ZnO thin films by excimer laser annealing," Ceramics International, vol. 39, no. 1, pp. 469–474, 2013.

18. W. M. Tsang, F. L. Wong, M. K. Fung, J. C. Chang, C. S. Lee, and S. T. Lee, "Transparent conducting aluminum-doped zinc oxide thin film prepared by sol-gel process followed by laser irradiation treatment," Thin Solid Films, vol. 517, no. 2, pp. 891–895, 2008.

19. Y. Q. Hua, G. Ji, and Z. Z. Sun, "Effect of laser shock wave on electrical property of ZnO varistor ceramics," Laser Technology, vol. 31, no. 1, pp. 133–136, 2011 (Chinese).

20. I. Lorite, J. Wasik, T. Michalsky, R. Schmidt-Grund, and P. Esquinazi, "Hydrogen influence on the electrical and optical properties of ZnO thin films grown under different atmospheres," Thin Solid Films, vol. 556, no. 1, pp. 18–22, 2014.

21. F. Oba and I. Tanaka, "Effect on oxidation of chemical bonding around 3d transition-metal impurities in ZnO," Japanese Journal of Applied Physics, vol. 38, no. 6, pp. 3570–3574, 1999.

22. F. Oba, H. Adachi, and I. Tanaka, "Energetics and electronic structure of point defects associated with oxygen excess at a tilt boundary of ZnO," Journal of Materials Research, vol. 15, no. 10, pp. 2167–2175, 2000.

23. J.-S. Jeong, "Field failure mechanism and reproduction due to moisture for low-voltage ZnO varistors,"Microelectronics Reliability, vol. 53, no. 9–11, pp. 1632–1637, 2013.

24. L. J. Brillson, H. L. Mosbacker, M. J. Hetzer et al., "Dominant effect of near-interface native point defects on ZnO Schottky barriers," Applied Physics Letters, vol. 90, no. 10, Article ID 102116, 2007.

25. T. K. Gupta and W. G. Carlson, "A grain-boundary defect model for instability/stability of a ZnO varistor," Journal of Materials Science, vol. 20, no. 10, pp. 3487–3500, 1985.

26. S. A. Pianaro, P. R. Bueno, P. Olivi, E. Longo, and J. A. Vareia, "Effect of Bi_2O_3 addition on the microstructure and electrical properties of the SnO_2·CoO·Nb_2O_5 varistor system," Journal of Materials Science Letters, vol. 16, no. 8, pp. 634–638, 1997.

CHAPTER 13

AN EASY-TO-FABRICATE LOW-TEMPERATURE TiO_2 ELECTRON COLLECTION LAYER FOR HIGH EFFICIENCY PLANAR HETEROJUNCTION PEROVSKITE SOLAR CELLS

B. Conings[1], L. Baeten[1], T. Jacobs[1], R. Dera[1], J. D'Haen[1], J. Manca[1] andH.-G. Boyen[1]

[1]Instituut voor Materiaalonderzoek, Universiteit Hasselt, Wetenschapspark 1, 3590 Diepenbeek, Belgium

ABSTRACT

Organometal trihalide perovskite solar cells arguably represent the most auspicious new photovoltaic technology so far, as they possess an astonishing combination of properties. The impressive and brisk advances achieved so far bring forth highly efficient and solution processable solar cells, holding great promise to grow into a mature technology that is ready to be embedded on a large scale. However, the vast majority of state-of-the-art perovskite solar cells contains a dense TiO_2 electron collection layer that requires a high temperature treatment (>450 °C), which obstructs the road towards roll-to-roll processing on flexible foils that can withstand no more than ~150 °C. Furthermore, this high temperature treatment leads to an overall increased energy payback time and cumulative energy demand for this emerging photovoltaic technology. Here we present the implementation of an alternative TiO_2 layer formed from an easily prepared nanoparticle dispersion, with annealing needs well within reach of roll-to-roll processing, making this technology also appealing from the energy payback aspect. Chemical and morphological analysis allows to understand and optimize the processing conditions of the TiO_2 layer, finally resulting in a maximum obtained efficiency of 13.6% for a planar heterojunction solar cell within an ITO/TiO_2/$CH_3NH_3PbI_{3-x}Cl_x$poly(3-hexylthiophene)/Ag architecture.

INTRODUCTION

Since the first reports on hybrid solar cells based on organolead trihalide perovskites, these materials have thoroughly engrossed the photovoltaic community.[1,2] Already from the birth of this new class of solar cells, high photon-to-electron conversion efficiencies have been reported for cells containing either $CH_3NH_3PbI_3$ or $CH_3NH_3PbI_{3-x}Cl_x$, thereby almost immediately overtaking the record efficiencies of organic solar cells and dye-sensitized solar cells, and leading up to a current world-record of 17.9% in a matter of only a few years.[3] This fast-paced progress to such high efficiencies announces this new technology as a true potential rival for Si-based photovoltaics. In first instance, the perovskite was regarded as a new kind of dye that could boost the efficiency of dye-sensitized solar cells.[4,5] The dye is grafted onto a porous metal-oxide structure to optimize light absorption and simultaneously guarantee favorable charge transport. Therefore (and in unawareness of its true capabilities), the perovskite solar cell concept initially often inherited the porous metal-oxide structure, which at the same time would be ostensibly required to guarantee sufficiently large area for exciton dissociation.[6] However, not much later it was discovered that the typical three-dimensional organolead trihalide perovskites exhibit very high charge

Citation: B. Conings, L. Baeten, T. Jacobs, R. Dera, J. D'Haen, J. Manca and H.-G. Boyen, An easy-to-fabricate low-temperature TiO2 electron collection layer for high efficiency planar heterojunction perovskite solar cells, http://dx.doi.org/10.1063/1.4890245.

carrier diffusion lengths (especially for the mixed halide with Cl, up to >1 μm) and low exciton binding energy (<50 meV), while having a very high absorption coefficient and charge carrier mobility.[7–10] Moreover, it has been shown that the perovskite actually functions as the ambipolar layer in a p-i-n junction.[11] The combination of these observations raises serious doubts regarding the necessity of the well-established porous metal-oxide matrix in perovskite-based solar cells, even though there are many high efficiencies reported based on mesoporous TiO_2.[6,12–15] Indeed, several groups pointed out that a simpler device architecture, where a perovskite layer is sandwiched between selective contacts, is able to deliver comparable or even better performance than a mesoporous TiO_2/perovskite compound: Efficiencies of around 11% were obtained for solution-processed planar heterojunction $CH_3NH_3PbI_{3-x}Cl_x$-based solar cells, and even up to 15.4% by vapor deposition.[16–18] In the latter cases the compact TiO_2 layer was deposited by heating a precursor gel at high temperature (~500 °C), lacking compatibility with flexible substrates and thus presenting a drawback in view of industrial scale roll-to-roll fabrication. This can in principle be remedied by using poly(3,4-ethylenedioxythiophene) polystyrene sulfonate (PEDOT:PSS) as bottom selective contact (thereby inverting the device structure),[19,20] though the hygroscopic nature of PEDOT:PSS would likely hamper the long-term stability of the moisture-sensitive perovskite.[21] It was then shown that the high temperature treatment for TiO_2 can be circumvented by using a layer consisting of TiO_2nanoparticles, rather than one that is *in situ* formed by annealing a precursor gel. Two strategies have been demonstrated so far, either the deposition of a nanoparticle dispersion[22] or the direct precipitation of TiO_2 nanoparticles from solution onto the substrate.[23] Both methods use $TiCl_4$ as the titanium source, which produces HCl fumes when in contact with humidity and thus has to be handled with great care. Another method does use titanium isopropoxide (TIP) as titanium source, but requires the addition of graphene flakes to the resulting TiO_2 dispersion to obtain optimal performance.[24] Here we present the chemical and morphological characterization of a TIP-based sub-150 °C processed TiO_2 electron acceptor layer consisting of nanoparticles, and we investigate the influence of this layer on the morphology of the subsequently deposited perovskite layer. The nanoparticulate TiO_2 layer is deposited from an alcoholic dispersion that is very easy to synthesize and the resulting planar heterojunction solar cells outperform previously reported analogs containing high-temperature processed TiO_2 layers. The resulting maximum obtained conversion efficiency of 13.6% is to the authors' knowledge the highest reported value for a *solution processed*planar heterojunction solar cell based on mixed halide perovskite $CH_3NH_3PbI_{3-x}Cl_x$.

Our TiO_2 nanoparticle dispersion was prepared using a slight modification of a procedure reported by Katoch *et al.*[25] A 470 μl of TIP was added dropwise to a solution of 123 μl nitric acid in 2.5 ml ethanol under mild stirring. After stirring for 2 h, 114 μl of demineralized water was added for the hydrolysis to proceed and the solution was left stirring for another hour. The result was then diluted with 1-propanol. This transparent dispersion is ready for use, with no need for centrifugation, washing, and redispersing, and is stable for months (Figure 1(a): a 0.127 M dispersion). Fig. 1(b) shows a Dynamic Light Scattering (DLS) measurement of the dispersion, revealing a narrow size distribution with an average particle size of 6 nm. An X-ray diffraction (XRD) measurement of a dried powder sample is shown in Figure 1(c), confirming the anatase phase of the material, and confirming the crystal diameter measured with DLS (via Scherrer equation). Recently, it was shown by Wojciechowski *et al.* that the addition of a small amount of titanium diisopropoxide bis(acetylacetonate) (TiAcAc) to a TiO_2 dispersion can serve as "electronic glue" between the nanoparticles in the eventual nanoparticulate thin film, enhancing its conductivity and thus improving the resulting solar cell built on it.[22] In addition, the TiAcAc can also assist in the film formation itself. Figures 1(d)–1(f) show Atomic Force Microscopy (AFM) micrographs of 25–30 nm thin TiO_2 films spincoated on indium-tin-oxide (ITO) (3000 rpm; 1000 rpm/s; 30 s) from a 0.127 M dispersion with the addition of 8 mol.% of TiAcAc (0.08 mole of TiAcAc per mole of TiO_2), annealed for different times at 135 °C in air. Smooth films are obtained with a low root-mean-square roughness of 1 nm or less on ITO as base layer.

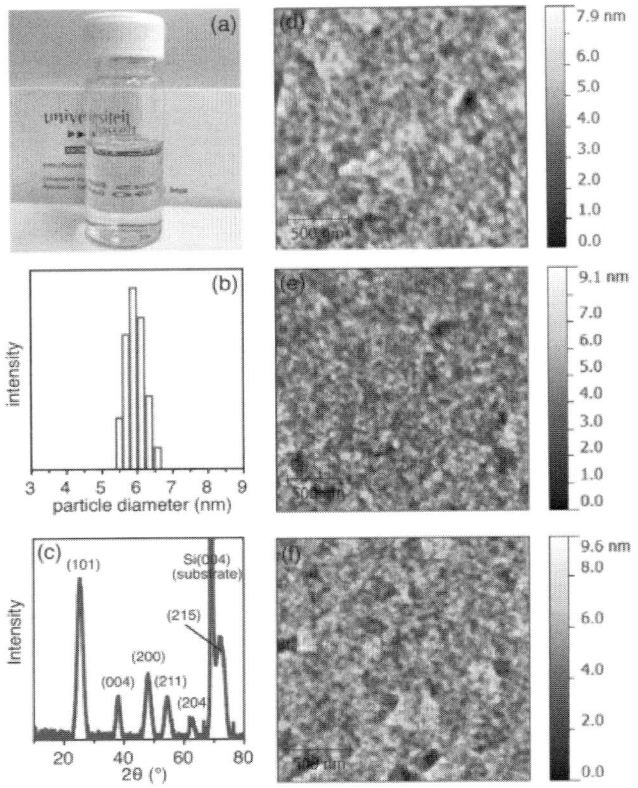

Figure 1: (a) Photographic image of a bottle of TiO_2 dispersion with a concentration of 0.127 M. (b) DLS measurement of the TiO_2 dispersion. (c) XRD spectrum of a TiO_2 powder sample. (d)–(f) AFM images of a TiO_2 nanoparticulate layer on ITO with the addition of 8 mol.% TiAcAc prior to spincoating (2 × 2 μm), annealed in air at 135 °C for 10 min (d), 45 min (e), and 90 min (f).

The thermally treated TiO_2-covered ITO substrates were then used to build planar heterojunction $CH_3NH_3PbI_{3-x}Cl_x$ perovskite solar cells according to a previously published procedure.[17] In short, a 3:1 methylammonium iodide:$PbCl_2$ solution in dimethylsulfoxide (DMSO) was spincoated (3000 rpm; 800 rpm/s; 540 s) on the TiO_2 in a nitrogen-filled glovebox, followed by a thermal treatment at 100 °C for 60 min. A layer of poly(3-hexylthiophene) (P3HT) (nominally 60 nm) was then spincoated on top, followed by the evaporation of 200 nm of silver at a pressure of 1×10^{6} mbar to complete the devices. A key observation is the influence of the thermal treatment of the TiO_2 on the performance of otherwise identical solar cells. Devices made from unannealed TiO_2 exhibit a reasonable open-circuit voltage (V_{oc}) but a negligible short-circuit current (J_{sc}) and a very low fill factor (FF), resulting in close to zero photovoltaic output (Fig. 2(a)). The extremely low J_{sc} and FF indicate the presence of a very high series resistance, likely induced by the presence of precursor products. Indeed, when the TiO_2 is annealed at 135 °C for only 10 min, we already find a reasonable conversion efficiency of around 9%. J_{sc} keeps increasing with a longer TiO_2thermal treatment, finding a maximum at 45 min, leading to an average efficiency of 12.6%. Efficiencies of even up to 13.6% (Fig. 2(b)) were obtained using an adapted annealing treatment as recently described by Saliba *et al.*, where the perovskite is slowly heated to 100 °C and then fast to 130 °C.[26] Longer TiO_2 treatments tend to have a vastly detrimental effect, initially resulting in highly fluctuating performance, and in the case of 90 min annealing even in a breakdown of all cells during the first measurement. It must not be overlooked that the best efficiencies obtained with our new low temperature processed TiO_2 layer are considerably higher than those obtained by using a sintered TiO_2 layer (500 °C) we reported before (J_{sc} = 20.8 ± 0.4 mA/cm²; V_{oc} = 0.919 ± 0.006 V; FF = 0.54 ± 0.01; η = 10.4 ± 0.4%).[17]

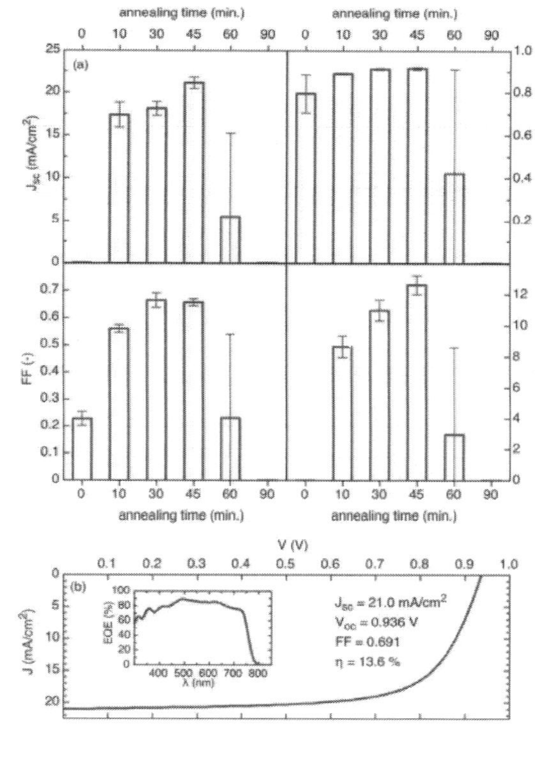

Figure 2: (a) Photovoltaic parameters of a series of solar cells built on TiO_2 layers that were annealed for different times. Error bars represent the standard deviation of at least 4 cells. The large error bars for the case of 60 min of annealing reflect extreme fluctuations between individual cells, with occurrences of both shunting and decently performing cells. All cells were measured in the forward bias to J_{sc}- direction after 10 min of AM1.5 illumination. (b) JV-characteristic of the best cell, using a 45 min annealed TiO_2 layer and a flash-anneal of the perovskite. Inset: External quantum efficiency spectrum.

To find the origin of the performance dependence on the thermal history of the TiO_2 layer, chemical analysis of the layers was performed using X-ray photoelectron Spectroscopy (XPS). No other elements than the ones expected from the starting products (C, N, O, Ti) were found in the TiO_2 layers, confirming their chemical purity. To probe more into detail, high-resolution core level spectra were taken of relevant elements. Fig. 3(a) shows N1s spectra of TiO_2 layers that were annealed at 135 °C for times from 0 to 90 min. In the pristine case there are two clear components: a higher binding energy component that represents nitrates (407.5 eV) and a low binding energy component that corresponds with nitrogen in a carbonated environment (400.5 eV), resulting from nitric acid, and both TIP and TiAcAc, respectively.[27] As the TiO_2 layer is annealed, the nitrate component shrinks, which is logical since the annealing temperature of 135 °C is well above the boiling point of nitric acid (83 °C). However, the nitrite component vanishes in favor of the component at lower binding energy, indicating that the nitric acid is disintegrated and its nitrogen being incorporated in the organic matrix of TiAcAc and TIP residue. This is seconded by the corresponding C1s core level spectra (Fig. 3(b)), as will be clarified in the following. The C1s comprises three components: one at 284.6 eV (C–C bonds), one at 286.8 eV (C–N and C–O bonds), and one at 289 eV (carboxyl).[27] We observe that the C–N/C–O component at 286.8 eV decreases upon annealing, while the carboxylic component at 289 eV increases; these are the organic moieties from the TIP and TiAcAc that oxidize due to the annealing being executed in air. The small remainder of the peak at 286.8 eV must therefore be a C–N contribution, corresponding with the nitrogen peak at 400.5 eV. Notice that the complete disappearance of nitric acid at 45 min of annealing coincides with the maximum obtained efficiency of the corresponding solar cells (Fig. 2(a)), indicating that the presence of residual nitric acid induces a parasitic series resistance in the device and could possibly cause interfacial recombination.

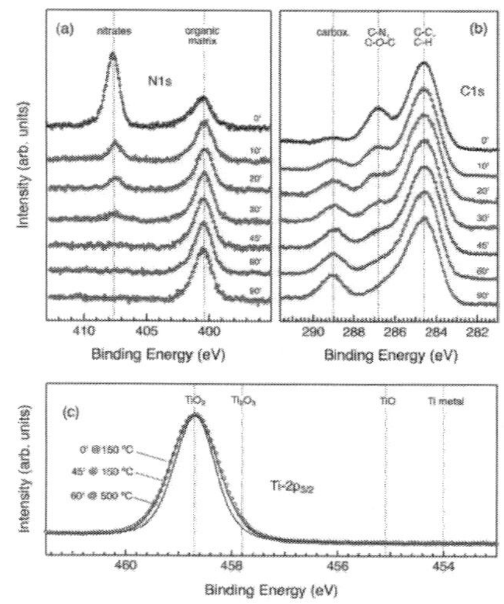

Figure 3: (a) High-resolution N1s core level spectra of TiO_2 layers annealed for 0–90 min. (b) High-resolution C1s core level spectra of TiO_2layers annealed for 0–90 min. (c) High resolution Ti$2p_{3/2}$ core level spectra of a TiO_2 layer before (diamonds) and after (blue line) annealing for 45 min at 135 °C together with a spectrum acquired on a TiO_2 layer sintered for 60 min at 500 °C (red line). For comparison, characteristic line positions of different oxidation states have been added.

We turn our attention now to the Ti2p spectra in Fig. 3(c), comparing a pristine versus a 45 min annealed layer. It is immediately obvious that there is no elemental Ti present in either case (454 eV); all Ti is in the 4+ oxidation state (459 eV).[27] It has previously been speculated that TiAcAc might decompose upon heating, forming TiOx that acts as mortar between TiO_2 particles, thus increasing the resulting layer's conductivity.[22] While annealing does induce decomposition of the carbonated part of TiAcAc (as shown above), the presence of Ti in different oxidation states (e.g., 3+ or 2+) should be corroborated by additional peaks appearing at the low binding energy side of the main line. In contrast, our measured $Ti2p_{3/2}$ signal comprising only a Ti^{4+} component that is unchanged upon annealing at 135 °C, reveals the absence of any Ti^{3+}[28] or Ti^{2+}[29] contributions to the spectrum. In fact, when analyzing the $Ti2p_{3/2}$ photoemission peak using a physically motivated line shape,[30] a single symmetric line turns out to be sufficient to describe the experimental data, thus ruling out the presence of any additional components with oxidation state other than 4+.

At this point one might speculate that a degraded form of TiAcAc with 4+ oxidation state for Ti is capable of bridging the voids between TiO_2nanoparticles such that the resulting layer has a better conductivity. The source of an improved conductivity can indeed be traced back by comparing the $Ti2p_{3/2}$ core level region for a layer after annealing at 135 °C with that of a conventional TiO_2 sol-gel film sintered at 500 °C,[31]both measured under identical experimental conditions (Fig. 3(c)). Clearly, the low-temperature deposition induces a significant line broadening as compared to the sintered film. As the line shape of the latter represents the combined effects of (i) line broadening due to the finite life time of the photoionized states of the probed Ti ions and (ii) the broadening induced by the finite energy resolution of the spectrometer, the increased line width in case of the low-temperature sample clearly points to an additional broadening effect. This additional line broadening most likely arises from small fluctuations/deviations in stoichiometry around the probed Ti ions, which can best be expressed as $TiO_{2\pm}$, reflecting a distribution of different "chemically shifted" components that superimpose to form the measured photoemission line. Consequently, a significant density of, e.g., defects acting as dopants must be present in the low-temperature deposited titania films, which can most probably be assigned to the "mortar" material, easily explaining their improved conductivity.

Based on the analysis of the N1s and C1s spectra it is clear why the TiO_2 layer needs a thermal treatment of 45 min to obtain optimally functioning solar cells (decomposition of NO_3), though it is still obscure why a longer annealing step results in much less performance and in worst-case even shunting cells. To shed more light onto the negative effect of prolonged annealing, we now examine the perovskite layer. Scanning Electron Microscopy (SEM) images were recorded to investigate whether the annealing time could have an influence on the consecutively deposited perovskite. The series of images in Fig. 4 reveal that there is indeed a correlation between the thermal history of the TiO_2 and the morphology of the eventual perovskite layer. The smoothest layer is obtained for a pristine TiO_2 layer, however the photovoltaic properties are, as mentioned, overshadowed in this case by the detrimental presence of HNO_3. An intermediate annealing time of the TiO_2(10–45 min) returns grainy to needle-like perovskite morphologies that closely resemble those we previously reported, accompanied by decent photovoltaic performance. Those films do show a number of pinholes but, contrary to foregoing hypotheses, this does not need to degrade the resulting performance significantly.[17] A treatment beyond 45 min appears to induce a considerably coarser perovskite morphology that shows much larger holes of several square micrometers. This is the likely cause why corresponding solar cells shunt easily; at such holes there is only a thin layer of P3HT (nominally 60 nm) that separates the TiO_2 from the top silver contact, so the occurrence of occasional short circuits is conceivable. It also deserves to be mentioned that the gradual change in morphology seen in the SEM results is paralleled by a decrease in peak height of the (hh0) peaks in XRD (not shown), indicating a reduced preferential orientation of the perovskite crystals. This might lead to inferior charge transport and could be an additional reason for the low performance of solar cells containing overly annealed TiO_2 layers.

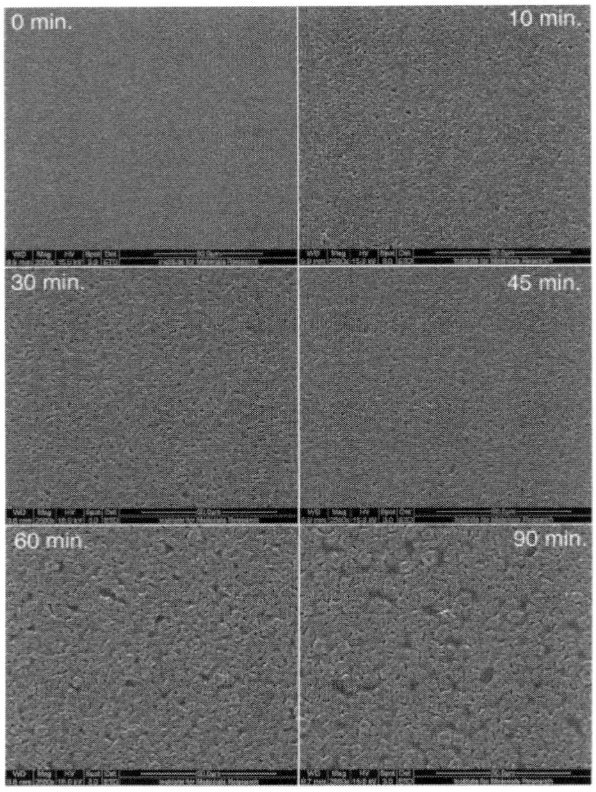

Figure 4: SEM images of perovskite layers deposited on TiO_2 base layer that were annealed at 135 °C in air for different times from 0 to 90 min.

The origin of the perovskite morphology dependence on the thermal history of the TiO_2 layer can be explained by the presented chemical analysis. The observed increased presence of carboxyl moieties on the TiO_2 surface as function of annealing time points to an enhanced wettability of the surface for the DMSO-based perovskite precursor, intuitively implying better film formation. In spite of that, the eventual perovskite layers show an opposite trend, becoming rougher when deposited on longer annealed TiO_2. This can be rationalized by considering that the actual perovskite is formed only when the precursor film is heated, i.e., the initial solvent wetting properties are subordinate to the subsequent crystallization mechanism of the dried film. As has been shown before, a high quality perovskite film is often more easily formed when a preformed matrix or a textured surface is available, functioning as a kind of seed layer.[11,32] Therefore, we propose that the organic residue from the starting products (at 286.8 eV in the C1s spectra in Fig. 3(b)) assists in the formation of smoother perovskite films. Upon the conversion into carboxylic compounds (at 289 eV), there are fewer seeding options for the perovskite crystals, causing them to form more rough and discrete structures, rather than an uninterrupted film.

Finally, it is interesting to remark that our extremely smooth anatase TiO_2 layers, having "minimal" contact area with the perovskite (i.e., a rather sharp bilayer interface), still allow for very high photocurrents. This is in contrast with a recent report from Yella *et al.*, who observe that rather rough rutile TiO_2 layers, having a more intimate contact with the perovskite, are more efficient at extracting electrons. An explanation for this apparent inconsistency might be offered by their use of the triiodide-based perovskite (that generally has lower charge carrier diffusion lengths) as opposed to the mixed halide variety in our case (typically with higher diffusion length).[7] In the former case a rougher TiO_2 film may assist in compensating for the slightly inferior charge carrier diffusion length of the triiode perovskite by creating an ersatz bulk heterojunction, whereas in the latter case any additional roughness would be redundant.

In conclusion, we introduced a low-temperature processable TiO_2 electron collection layer, customized for planar heterojunction perovskite solar cells. The layer is deposited from an easy one-pot synthesis nanoparticle dispersion and requires a thermal treatment at only 135 °C, in contrast with the typical 450 °C–500 °C baking steps for conventional TiO_2 base layers, making it compatible with roll-to-roll fabrication on plastic foils. Even slightly lower temperatures could deliver the same outcome, provided longer annealing times and/or lower pressure. In addition, the resulting planar heterojunction devices significantly outperform their otherwise identical counterparts containing a high temperature processed TiO_2 layer. From this initial set of experiments, an efficiency of 13.6% is achieved, which is at the time of writing the highest value for solution-processed planar heterojunction mixed halide perovskite solar cells so far, and the highest value for a perovskite solar cell using P3HT as hole transporting layer. The results also show that the chemical composition and thermal history of the nanoparticulate TiO_2 film have a consequential impact on the device performance. Finally, it cannot go unnoticed that the observed high photocurrents and fill factors are obtained even with films exhibiting a significant amount of pinholes, illustrating that the interplay between the selective contact and the perovskite is probably an even more important consideration than the smoothness of the perovskite itself.

ACKNOWLEDGMENTS

This work was financially supported by the Interreg project Organext and the Research Foundation Flanders (FWO) within the Odysseus program. B.C. is a postdoctoral fellow of FWO. The authors would like to acknowledge B. Ruttens for XRD measurement, T. Vandenreijt for technical assistance, and Aslihan Babayigit for extraordinary logistics.

REFERENCES

1. N.-G. Park, *J. Phys. Chem. Lett.* **4**, 2423–2429 (2013). http://dx.doi.org/10.1021/jz400892a
2. H. J. Snaith, *J. Phys. Chem. Lett.* **4**(21), 3623–3630 (2013).http://dx.doi.org/10.1021/jz4020162
3. NREL, see http://www.nrel.gov/ncpv/images/efficiency_chart.jpg, 2014.
4. A. Kojima, K. Teshima, Y. Shirai, and T. Miyasaka, *J. Am. Chem. Soc.* **131**(17), 6050–6051 (2009).http://dx.doi.org/10.1021/ja809598r
5. J.-H. Im, C.-R. Lee, J.-W. Lee, S.-W. Park, and N.-G. Park, *Nanoscale* **3**(10), 4088–4093 (2011).http://dx.doi.org/10.1039/c1nr10867k
6. J. H. Heo, S. H. Im, J. H. Noh, T. N. Mandal, C.-S. Lim, J. A. Chang, Y. H. Lee, H.-J. Kim, A. Sarkar, Md. K. Nazeeruddin, M. Gratzel, and S. I. Seok, *Nat. Photonics* **7**(6), 486–491 (2013).http://dx.doi.org/10.1038/nphoton.2013.80
7. S. D. Stranks, G. E. Eperon, G. Grancini, C. Menelaou, M. J. P. Alcocer, T. Leijtens, L. M. Herz, A. Petrozza, and H. J. Snaith, *Science* **342**(6156), 341–344 (2013).http://dx.doi.org/10.1126/science.1243982
8. G. Xing, N. Mathews, S. Sun, S. S. Lim, Y. M. Lam, M. Grätzel, S. Mhaisalkar, and T. C. Sum, *Science* **342**(6156), 344–347 (2013).http://dx.doi.org/10.1126/science.1243167
9. C. Wehrenfennig, G. E. Eperon, M. B. Johnston, H. J. Snaith, and L. M. Herz, *Adv. Mater.* **26**(10), 1584–1589 (2014).http://dx.doi.org/10.1002/adma.201305172
10. V. D'Innocenzo, G. Grancini, M. J. P. Alcocer, A. R. S. Kandada, S. D. Stranks, M. M. Lee, G. Lanzani, H. J. Snaith, and A. Petrozza, *Nat. Commun.* **5**, 3586 (2014). http://dx.doi.org/10.1038/ncomms4586
11. J. M. Ball, M. M. Lee, A. Hey, and H. J. Snaith, *Energy Environ. Sci.* **6**(6), 1739–1743 (2013).http://dx.doi.org/10.1039/c3ee40810h
12. J. Burschka, N. Pellet, S.-J. Moon, R. Humphry-Baker, P. Gao, M. K. Nazeeruddin, and M. Gratzel, *Nature (London)* **499**(7458), 316–319 (2013).http://dx.doi.org/10.1038/nature12340
13. A. Abrusci, S. D. Stranks, P. Docampo, H.-L. Yip, A. K. Y. Jen, and H. J. Snaith, *Nano Lett.* **13**(7), 3124–3128 (2013).http://dx.doi.org/10.1021/nl401044q
14. J. H. Noh, S. H. Im, J. H. Heo, T. N. Mandal, and S. I. Seok, *Nano Lett.* **13**(4), 1764–1769 (2013).http://dx.doi.org/10.1021/nl400349b
15. N. J. Jeon, J. Lee, J. H. Noh, M. K. Nazeeruddin, M. Grätzel, and S. I. Seok, *J. Am. Chem. Soc.* **135**(51), 19087–19090 (2013).http://dx.doi.org/10.1021/ja410659k
16. G. E. Eperon, V. M. Burlakov, P. Docampo, A. Goriely, and H. J. Snaith, *Adv. Funct. Mater.* **24**(1), 151–157 (2014).http://dx.doi.org/10.1002/adfm.201302090
17. B. Conings, L. Baeten, C. De Dobbelaere, J. D'Haen, J. Manca, and H.-G. Boyen, *Adv. Mater.* **26**(13), 2041–2046 (2014).http://dx.doi.org/10.1002/adma.201304803

18. M. Liu, M. B. Johnston, and H. J. Snaith, *Nature (London)* **501**(7467), 395–398 (2013).http://dx.doi.org/10.1038/nature12509

19. P. Docampo, J. M. Ball, M. Darwich, G. E. Eperon, and H. J. Snaith, *Nat. Commun.* **4**, 2761 (2013).http://dx.doi.org/10.1038/ncomms3761

20. J. You, Z. Hong, Y. Yang, Q. Chen, M. Cai, T.-B. Song, C.-C. Chen, S. Lu, Y. Liu, H. Zhou, and Y. Yang, *ACS Nano* **8**(2), 1674–1680 (2014).http://dx.doi.org/10.1021/nn406020d

21. K. Kawano, R. Pacios, D. Poplavskyy, J. Nelson, D. D. C. Bradley, and J. R. Durrant, *Sol. Energy. Mater. Sol. Cells* **90**(20), 3520–3530 (2006).http://dx.doi.org/10.1016/j.solmat.2006.06.041

22. K. Wojciechowski, M. Saliba, T. Leijtens, A. Abate, and H. J. Snaith, *Energy Environ. Sci.* **7**(3), 1142–1147 (2014).http://dx.doi.org/10.1039/c3ee43707h

23. A. Yella, L.-P. Heiniger, P. Gao, M. K. Nazeeruddin, and M. Grätzel, *Nano Lett.* **14**(5), 2591–2596 (2014).http://dx.doi.org/10.1021/nl500399m

24. J. T.-W. Wang, J. M. Ball, E. M. Barea, A. Abate, J. A. Alexander-Webber, J. Huang, M. Saliba, I. Mora-Sero, J. Bisquert, H. J. Snaith, and R. J. Nicholas, *Nano Lett.* **14**(2), 724–730 (2014).http://dx.doi.org/10.1021/nl403997a

25. A. Katoch, H. Kim, T. Hwang, and S. Kim, *J. Sol-Gel Sci. Technol.* **61**(1), 77–82 (2012).http://dx.doi.org/10.1007/s10971-011-2593-6

26. M. Saliba, K. W. Tan, H. Sai, D. T. Moore, T. Scott, W. Zhang, L. A. Estroff, U. Wiesner, and H. J. Snaith, "Influence of Thermal Processing Protocol upon the Crystallization and Photovoltaic Performance of Organic-Inorganic Lead Trihalide Perovskites," *J. Phys. Chem. C*(published online).http://dx.doi.org/10.1021/jp500717w

27. J. F. Moulder, W. F. Stickle, P. E. Sobol, and K. D. Bomben, *Handbook of X-ray Photoelectron Spectroscopy* (Physical Electronics, Inc., Minnesota, USA, 1995).

28. F. Werfel and O. Brummer, *Phys. Scr.* **28**(1), 92–96 (1983).http://dx.doi.org/10.1088/0031-8949/28/1/013

29. NIST X-ray Photoelectron Spectroscopy Database, Version 4.1, http://srdata.nist.gov/xps/

30. G. K. Wertheim and S. B. Dicenzo, *J. Electron Spectrosc.* **37**(1), 57–67 (1985).http://dx.doi.org/10.1016/0368-2048(85)80082-7

31. C. De Dobbelaere, J. Mullens, A. Hardy, and M. K. Van Bael, *Thermochim. Acta* **520**(1–2), 121–133 (2011).http://dx.doi.org/10.1016/j.tca.2011.03.028

32. D. Liu and T. L. Kelly, *Nat. Photon.* **8**(2), 133–138 (2014).http://dx.doi.org/10.1038/nphoton.2013.342

CHAPTER 14

THE INFLUENCE OF ATMOSPHERE ON ELECTRICAL PROPERTY OF COPPER OXIDE NANOPARTICLES

Minghan Chen, Huichao Zhu, Xiaogan Li, Jun Yu, Haitao Cai, Xiaotong Quan, Kaili Wang, and Jiaqi Zhang

School of Electronic Science and Technology, Faculty of Electronic Information and Electrical Engineering, Dalian University of Technology, Dalian 116024, China

ABSTRACT

In the present work the authors have fabricated copper oxide nanoparticles through a water bath method and investigated their composition and microstructure using XRD and TEM methods. Their electrical properties were investigated under different atmospheres and temperatures after painting them onto the insulating substrates with interdigital Pt and Au electrodes. The electric currents are found dramatically dependent on the variation of humidity, temperature, and ratio of oxygen to nitrogen. In the end, the HCHO gas sensitivity of these nanoparticles has been investigated.

INTRODUCTION

Today, more and more attention has been paid to gas sensors based on semiconducting metal oxides for detection of hazardous, humid, flammable, or toxic gases, due to their rich resource, low cost, and easy fabrication. Plenty of n-type semiconducting metal oxides such as ZnO [1], Fe_2O_3 [2], SnO_2 [3], WO_3 [4], and MoO_3 [5] are widely investigated due to their extensive sensing performance. Recently, increasing interest has been focused on p-type semiconducting metal oxides such as CuO [6] and NiO [7]. Copper oxides, including CuO and Cu_2O, are p-type semiconductors because of presence of acceptors levels attributed to copper vacancies. There have been some reports on CuO nanoparticles [8], nanowires [9], nanorods [10], nanoribbons [11], nanosheets [12], and microspheres [13] towards various target gases such as NO_2, H_2S, C_2H_5OH, CO, and NH_3 in the working temperature ranging from 200°C to 400°C. However, all synthesis methods of these copper oxide nanomaterials need harsh conditions such as high temperature or high pressure, no matter chemical vapor deposition, or hydrothermal process. There are few reports on copper oxide nanomaterials synthesized through low temperature conditions. In this study, we report the fabrication of copper oxide nanoparticles through a water bath process at relatively low temperature and normal pressure. We investigated influence of different atmospheres and temperatures on the electrical property of copper oxide nanoparticles.

EXPERIMENTAL

The copper oxide nanoparticles were synthesized using a water bath method under relatively low temperature and normal pressure [14]. The following chemical materials are employed in the synthesis process: copper (II) nitrate hemi-pentahydrate ($Cu(NO_3)_2 2.5H_2O$, analytical reagent), methenamine ($(CH_2)_6N_4$, analytical reagent), and deionized water. The synthesis was conducted by the thermal decomposition of a

Citation: Minghan Chen, Huichao Zhu, Xiaogan Li, et al., "The Influence of Atmosphere on Electrical Property of Copper Oxide Nanoparticles," Journal of Nanomaterials, vol. 2014, Article ID 461269, 6 pages, 2014. doi:10.1155/2014/461269.

Cu^{2+} amino complex with reagent-grade chemicals. Methenamine, $(CH_2)_6N_4$, a nontoxic water-soluble and nonionic tetradentate cyclic tertiary amine was chosen to comply simultaneously with the precipitation of the divalent transition metal ion Cu^{2+} and the nucleation growth of its stable oxide form, copper (II) oxide. An equimolar (0.1 M) aqueous solution (deionized water) of copper (II) nitrate hemipentahydrate and methenamine was prepared in a bottle with a screw cap. The bottle is then heated at a constant temperature of 100°C for 6 hours in a regular laboratory water bath kettle. Subsequently, the homogeneous nanoparticles are thoroughly washed with deionized water to remove any contamination from residual salts or amino complex. It is noted that the water bath fabrication method is highly reliable and several kinds of metal oxide nanoparticles are fabricated under different conditions.

Characterizations of the as grown copper oxide nanoparticles were performed using the following techniques. The XRD patterns were collected with a Bruker D8 advance X-ray powder diffractometer (Bruker, Germany) equipped with a scintillation counter using Cu Kα radiation (λ = 1.5406A). The goniometer scanning rate was of 0.4° min^{-1}. The as grown nanoparticles were pressed onto a Si substrate before scanning. The morphology and microstructure information of as grown nanoparticles were investigated through a high-resolution transmission electron microscope (HRTEM, Tecnai G2 F20, Netherland) with a super twin objective lens and working at 200 kV with a LaB_6 filament. The influences of different temperatures and atmospheres on the electrical transport properties of annealed copper oxide nanoparticles are investigated with multiple gas mix and transport controller and a data acquisition/processing instrument (Keithley 2000 multimeter) and Agilent B1500A semiconductor device analyzer.

RESULTS AND DISCUSSION

Figure 1 shows the XRD pattern of the as grown copper oxide nanoparticles, which indicated that the nanoparticles include CuO, Cu_2O, $Cu(OH)_2$, $Cu(OH)_2H_2O$, and $Cu(OH)_23H_2O$ according to references PDF#42-0638, PDF#42-0746, PDF#65-3288, PDF#36-0545, and PDF#48-1548, respectively. Among these nanoparticles, $Cu(OH)_2$, $Cu(OH)_2H_2O$, and $Cu(OH)_23H_2O$ are thermolabile materials, which will transform into CuO under relatively low heating environment. It is well known that only when the temperature is higher than 900°C, all the CuO will transform into Cu_2O because of the instability under high temperature.

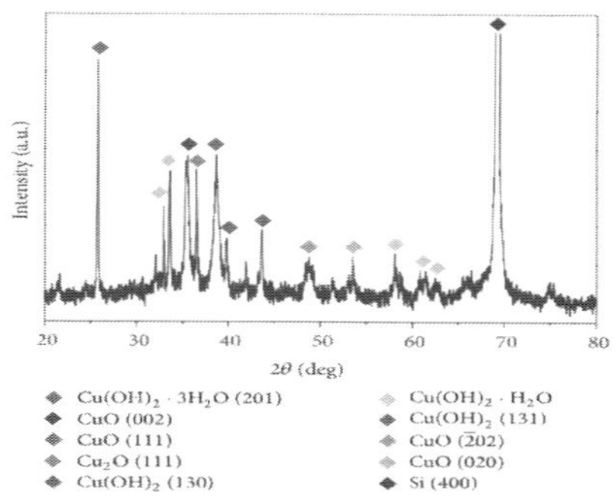

Figure 1: The XRD pattern of as grown copper oxide nanoparticles.

Figure 2 shows the HRTEM analysis of as grown copper oxide nanoparticles. An irregular size of nanoparticle distributed from hundreds nanometers to micron in length and tens to hundreds nanometers in width. Under high-resolution observation, the individual nanoparticle has polycrystalline structure, the size of each crystalline grain distributed in tens nanometers scale. These nanoparticles undoubtedly have active chemical properties because of their large specific surface area, such as high surface reaction activity and efficient transmission channel for analyte molecules to reach the active sites.

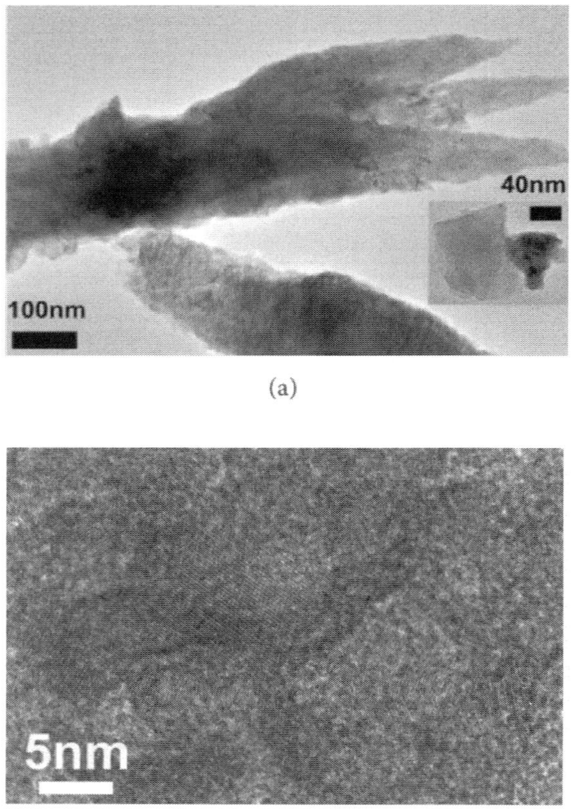

(b)

Figure 2: The HRTEM images of as grown copper oxide nanoparticles.

Figure 3 shows the schematic diagrams of the devices based on copper oxide nanoparticles for electrical property investigation. One device was fabricated on a SiO_2/Si substrate with Pt interdigital electrodes; the other was fabricated on an alumina substrate with Au interdigital electrodes. The device fabrication is as described in the following process. The copper oxide nanoparticles were initially dispersed in ethanol and then were uniformly coated onto a 300 nm-SiO_2/Si (or alumina) substrate provided with interdigitated Pt (or Au) electrodes with trace and gap size of 20 μm (or 250 μm) on the insulating side. In order to remove the residual water and ethanol and to ensure a good adherence of copper oxide nanoparticles to substrates and electrodes, a subsequent heating process was performed in a tube furnace. The devices were first heated at 200°C for 6 hours, then slowly heated up to 500°C for 2 hours, and finally cooled down to room temperature naturally. After the heating process, the water, crystal water, and hydroxyl inside of copper oxide nanoparticles were driven out.

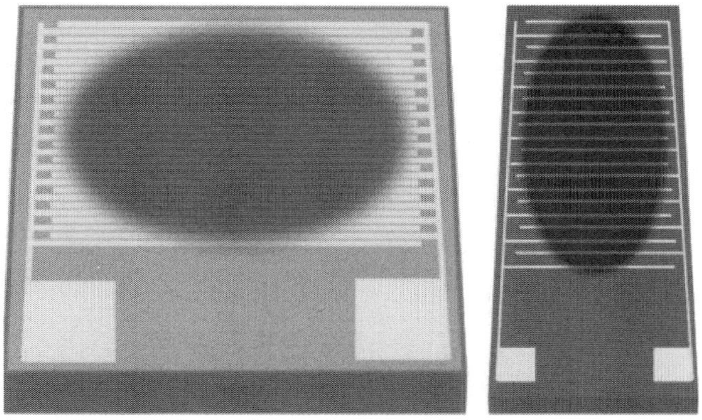

Figure 3: The schematic diagrams of the devices based on copper oxide nanoparticles.

The I-V sweep measurements were performed under dc voltage sweep at room temperature in room air (humid), commercial air (dry), nitrogen (dry), and oxygen (dry), respectively. The voltage was scanned as −10 V → 0 V → 10 V → 0 V → −10 V with a step of 20 mV, as shown in Figure 4(a). The I-V curves in three kinds of dry gas atmosphere are similar symmetrical rectifying curves, which indicates the existence of small Schottky barriers between copper oxide nanoparticles and metal electrodes. When the voltage varied along the opposite direction, the two curves kept good coincidence with each other. However, the situation in room air atmosphere with 85% RH was dramatically different. First, the current values were much larger than those measured in dry gas atmosphere. Second, when the voltage varied along the opposite direction, the two curves are entirely not coincident and show an interesting loop-like feature. The numbers 1, 2, 3, and 4 indicate the voltage variation −10 V → 0 V, 0 V → 10 V, 10 V → 0 V, and 0 V → −10 V, respectively. The curve segments 1 and 3 indicate that when the voltage got smaller, the current also got smaller in the meantime, which is apparently reasonable. Yet the curve segments 2 and 4 are very different; in detail, the current got smaller when voltage got larger than −5 V and 5 V.

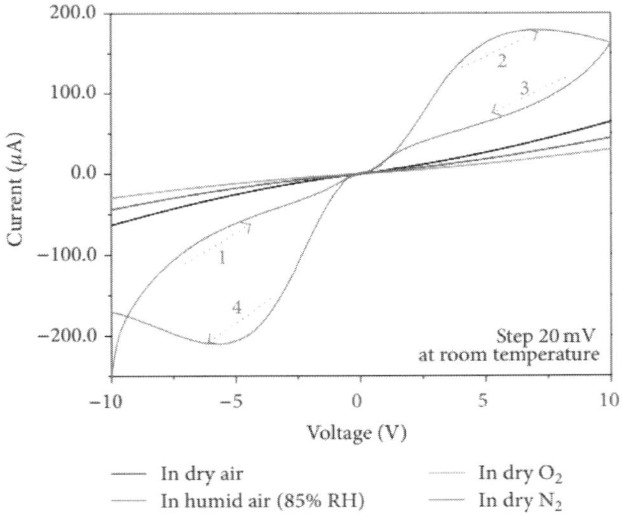

(a)

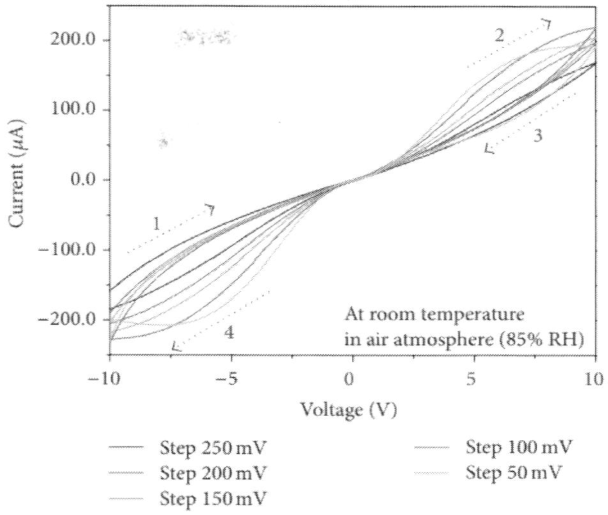

(b)

Figure 4: (a) Current variation of copper oxide nanoparticles under voltage from −10 V to 10 V to −10 V under different atmospheres; (b) current variation of copper oxide nanoparticles under voltage from −10 V to 10 V to −10 V with different voltage variation step values in humid air atmosphere.

The only difference between the humid room air atmosphere and commercial dry air atmosphere is in the water vapor. The contrast experiments show the dramatic influence of water vapor on the electrical conductance of copper oxide nanoparticles. It is well known that in a water molecule, the oxygen atom and two hydrogen atoms share electrons in covalent bonds, but the sharing is not equal. In the covalent bond between oxygen and hydrogen, the oxygen atom attracts electrons stronger than the hydrogen atoms. The unequal sharing of electrons gives the water molecule a slight negative charge near its oxygen atom and a slight positive charge near its hydrogen atoms [15]. As a result, the polar water molecules in humid air and the electrons on the surface of copper oxide nanoparticles will adsorb each other. More and more water molecules were attracted to form an ultrathin water layer in the surfaces of copper oxide nanoparticles. In this water layer, the following hydrolysis reaction and ionization process might happen:

$$CuO + H_2O \longrightarrow Cu(OH)_2 \quad (1)$$

$$Cu(OH)_2 \longrightarrow Cu^{2+} + 2OH^- \quad (2)$$

Both positive and negative ions will enhance the conductance. When the speed of water molecules aggregation is faster than speed of hydrolysis reaction and ionization process, the current may become saturated and even smaller because the concentration variation of ions may reach a bottleneck, as Figure 4(a) shows.

As shown in Figure 4(b), the dc sweep voltage was scanned as −10 V → 0 V → 10 V → 0 V → −10 V with different voltage variation step value of 50 mV, 100 mV, 150 mV, 200 mV, and 250 mV in room air atmosphere with 85% RH, respectively. With increasement of voltage variation step value, the loop-like curve became thinner and thinner. Because the voltage variation step value decided the entire sweeping time, the larger variation step value resulted in the shorter entire sweeping time and the less adsorbed water molecules; the smaller variation step value resulted in the longer entire sweep time and the more adsorbed water molecules. This theory can well explain the different shapes of I-V curves.

In order to investigate the influence of oxygen adsorption on the electrical conductivity of copper oxide nanoparticles at high temperature, a gas sensor on alumina substrate provided with interdigitated Au electrodes (trace and gap size: 250 μm) on its surface through aforementioned process is fabricated. Before the coating process, fine gold wires (100 μm in diameter) were connected to the interdigitated Au electrodes in assistance of gold glue, with a subsequent thermal solidification treatment (800°C for 60 min).

Figure 5(a) shows the I-V character of the copper oxide nanoparticles from −40 V to 40V with a step of 50 mV in commercial dry air atmosphere at different temperatures of 100°C, 150°C, 200°C, and 250°C, respectively. The figure shows that the resistance of the copper oxide nanoparticles is highly sensitive to temperature. Figure5(b) shows the I-V character of the copper oxide nanoparticles from −40 V to 40 V with a step of 50 mV at 200°C in different mixture gas atmospheres with different oxygen concentrations of 5%, 10%, 15%, and 20%, respectively. At the same voltage, the current increased along with the increase of oxygen concentration. Figure5(c) shows the time-resolved current (or resistance) variation of copper oxide nanoparticles in different mixture gas atmospheres with oxygen concentration of 5%, 10%, 15%s and 20%, respectively. The voltage is 10 V and the temperature is 200°C. One can observe that each current curve slowly increases along with the time due to an increasing oxygen adsorption. It is also obvious that the current level of four curves got a monotonous increase along with the increasing density of oxygen, which is also due to the oxygen adsorption. Because copper oxide is well known as a p-type semiconductor, the majority carrier is hole and the minority carrier is electron. The adsorption of oxygen will decrease the electron concentration and increase the hole concentration in copper oxide, which makes the resistance decrease.

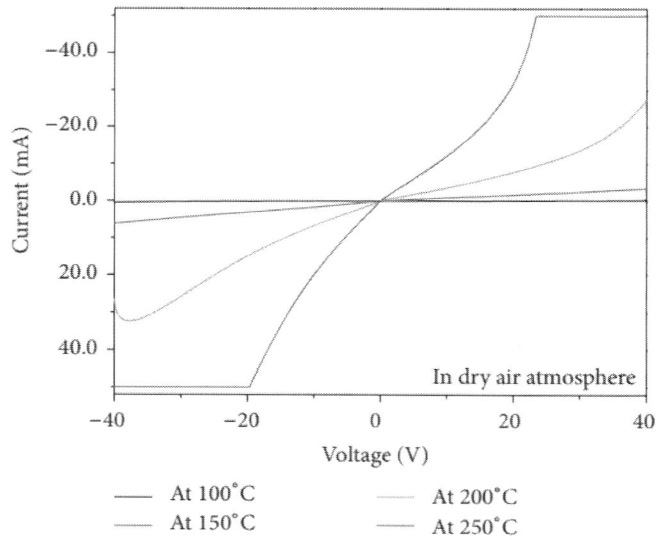

(a)

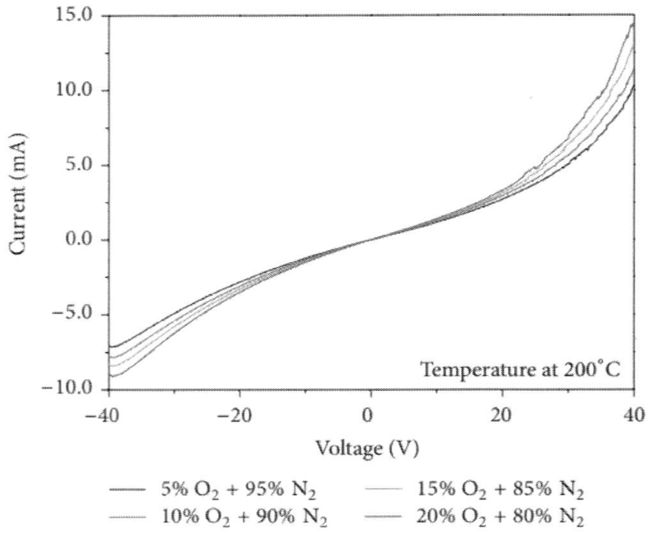

(b)

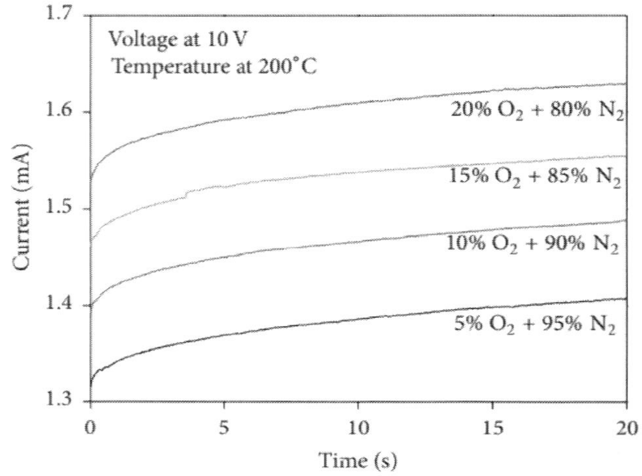

(c)

Figure 5: (a) I-V character of the copper oxide nanoparticles under voltage from −40 V to 40 V with a step of 50 mV in dry air atmosphere at different temperatures; (b) I-Vcharacter of the copper oxide nanoparticles at 200°C in different atmospheres; (c) time-resolved current variation of copper oxide nanoparticles in different atmospheres.

The HCHO sensing properties were investigated at atmospheric pressure through a dynamic test system which included a horizontal tube furnace, a multigas mix and transport controller and a data acquisition/processing instrument (Keithley 2000 multimeter), using a constant synthetic air flow (100 sccm) as both reference gas and carrier gas for the analyte dispersion. Time dependencies of the resistance variation at different temperatures (300°C and 400°C) of copper oxide nanoparticles upon exposure towards 4 pulses of HCHO (10, 20, 30, and 40 ppm) in dry air are shown in Figure 6. The results presented are highly reproducible on similar devices. The exposure time in HCHO is 4 minutes and the intermission time in air is 6 minutes. The applied voltage between the double Au electrodes is kept at 15 V during the entire process.

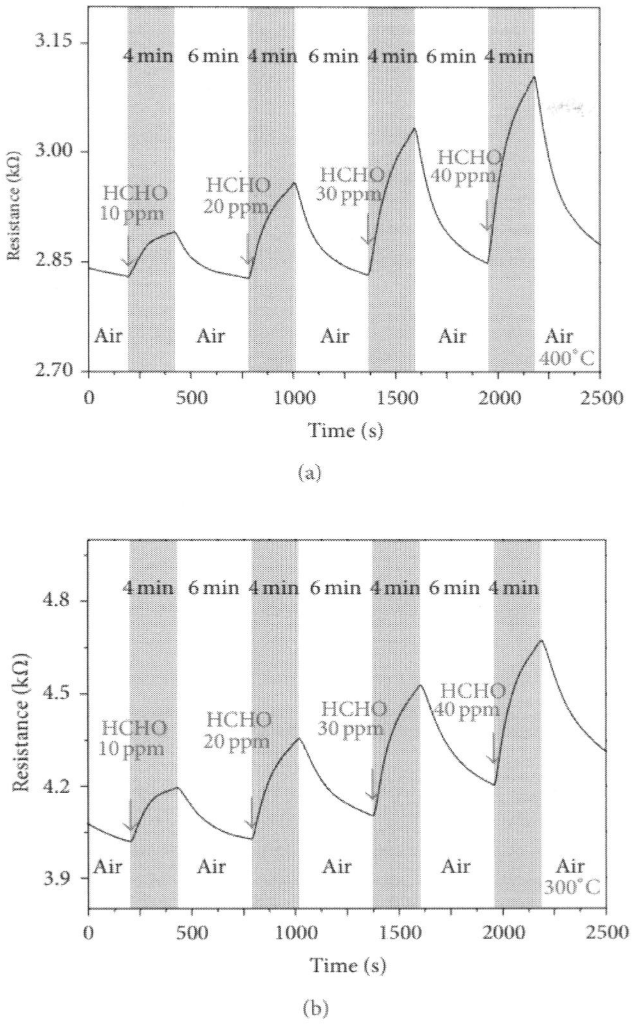

Figure 6: The HCHO sensitivity of copper oxide nanoparticles at different temperatures.

At both temperatures the resistance of copper oxide nanoparticles gradually increased during sensing of HCHO and gradually decreased since HCHO stopped. The copper oxide nanoparticles showed stable p-type conductivity at both 300°C and 400°C according to the decrease of resistance when the environment atmosphere is air. The sensing mechanism of resistive gas sensors based on metal oxide semiconductors is ascribed to variations of majority carrier surface concentration under the influence of adsorbed substances [16]. Particularly, the surface depletion model and grain boundary model are used to describe the sensing mechanism of metal oxide semiconductors [17]. For p-type metal oxide semiconductors, the adsorbed oxygen will form acceptor levels near the valence band at the surface, resulting in the upward band bending and formation of an accumulation layer of holes. The upward band bending between nanoparticles will enhance the transport of holes, the majority carriers in p-type metal oxide semiconductors. During exposing to reductive gas, the reductive gas molecules would react with the preadsorbed oxygen species, resulting in the cancellation of holes and an increase of the resistance. This theory can very well explain the variation of resistance of copper oxide nanoparticles during sensing HCHO at 300°C and 400°C.

CONCLUSIONS

In conclusion, we have synthesized copper oxide nanoparticles and investigated the influence of different atmospheres and temperatures on their electrical conductivity. We found that the I-V character of copper oxide nanoparticles is very sensitive to humid air atmosphere with 85% RH at room temperature and also sensitive to the variation of temperature and oxygen concentration. Further, at high temperature they are very sensitive to HCHO concentration variation.

ACKNOWLEDGMENTS

This work was supported by the National Natural Science Foundation of China (Grant nos. 61107028, 61131004, 61001054, and 61274076) and the Fundamental Research Funds for the Central Universities (Grant no. DUT12JB11).

REFERENCES

1. J. Xu, Q. Pan, Y. Shun, and Z. Tian, "Grain size control and gas sensing properties of ZnO gas sensor,"Sensors and Actuators B, vol. 66, no. 1, pp. 277–279, 2000.
2. F. Zhang, H. Yang, X. Xie et al., "Controlled synthesis and gas-sensing properties of hollow sea urchin-like α-Fe_2O_3 nanostructures and α-Fe_2O_3 nanocubes," Sensors and Actuators B, vol. 141, no. 2, pp. 381–389, 2009.
3. Y.-J. Choi, I.-S. Hwang, J.-G. Park, K. J. Choi, J.-H. Park, and J.-H. Lee, "Novel fabrication of an SnO_2nanowire gas sensor with high sensitivity," Nanotechnology, vol. 19, no. 9, Article ID 095508, 2008.
4. L. You, Y. Sun, J. Ma et al., "Highly sensitive NO_2 sensor based on square-like tungsten oxide prepared with hydrothermal treatment," Sensors and Actuators B, vol. 157, no. 2, pp. 401–407, 2011.
5. X. Chu, S. Liang, W. Sun, W. Zhang, T. Chen, and Q. Zhang, "Trimethylamine sensing properties of sensors based on MoO_3 microrods," Sensors and Actuators B, vol. 148, no. 2, pp. 399–403, 2010.
6. Y. Zhang, X. He, J. Li, H. Zhang, and X. Gao, "Gas-sensing properties of hollow and hierarchical copper oxide microspheres," Sensors and Actuators B, vol. 128, no. 1, pp. 293–298, 2007.
7. H. Steinebach, S. Kannan, L. Rieth, and F. Solzbacher, "H_2 gas sensor performance of NiO at high temperatures in gas mixtures," Sensors and Actuators B, vol. 151, no. 1, pp. 162–168, 2010.
8. R. K. Bedi and I. Singh, "Room-temperature ammonia sensor based on cationic surfactant-assisted nanocrystalline CuO," ACS Applied Materials and Interfaces, vol. 2, no. 5, pp. 1361–1368, 2010.
9. B. J. Hansen, N. Kouklin, G. Lu, I.-K. Lin, J. Chen, and X. Zhang, "Transport, analyte detection, and opto-electronic response of p-type CuO nanowires," Journal of Physical Chemistry C, vol. 114, no. 6, pp. 2440–2447, 2010.
10. C. Yang, X. Su, F. Xiao, J. Jian, and J. Wang, "Gas sensing properties of CuO nanorods synthesized by a microwave-assisted hydrothermal method," Sensors and Actuators B, vol. 158, no. 1, pp. 299–303, 2011.

11. X. Gou, G. Wang, J. Yang, J. Park, and D. Wexler, "Chemical synthesis, characterisation and gas sensing performance of copper oxide nanoribbons," Journal of Materials Chemistry, vol. 18, no. 9, pp. 965–969, 2008.

12. F. Zhang, A. Zhu, Y. Luo, Y. Tian, J. Yang, and Y. Qin, "CuO nanosheets for sensitive and selective determination of H_2S with high recovery ability," Journal of Physical Chemistry C, vol. 114, no. 45, pp. 19214–19219, 2010.

13. Y. Zhang, X. He, J. Li, H. Zhang, and X. Gao, "Gas-sensing properties of hollow and hierarchical copper oxide microspheres," Sensors and Actuators B, vol. 128, no. 1, pp. 293–298, 2007.

14. K. Keis, L. Vayssieres, S.-E. Lindquist, and A. Hagfeldt, "Nanostructured ZnO electrodes for photovoltaic applications," Nanostructured Materials, vol. 12, no. 1, pp. 487–490, 1999.

15. The American Chemical Society, Middle School Chemistry, chapter 5, Water is a Polar Molecule, http://www.middleschoolchemistry.com/lessonplans/chapter5/lesson1.

16. W. Göpel and K. D. Schierbaum, "Electronic conductance and capacitance sensors," in A Comprehensive Survey, W. Göpel, J. Hesse, and J. N. Zemel, Eds., vol. 2, VCH Weinheim, New York, NY, USA, 1991.

17. F. P. Koffyberg and F. A. Benko, "A photoelectrochemical determination of the position of the conduction and valence band edges of p-type CuO," Journal of Applied Physics, vol. 53, no. 2, pp. 1173–1177, 1982.

CHAPTER 15

COMPLEX-MORPHOLOGY METAL-BASED NANOSTRUCTURES: FABRICATION, CHARACTERIZATION, AND APPLICATIONS

Antonella Gentile [1,2], Francesco Ruffino [1,2] and Maria Grazia Grimaldi [1,2]

[1]Department of Physics and Astronomy-University of Catania, via S. Sofia 64, 95123 Catania, Italy
[2]MATIS IMM-CNR, via S. Sofia 64, 95123 Catania, Italy

ABSTRACT

Due to their peculiar qualities, metal-based nanostructures have been extensively used in applications such as catalysis, electronics, photography, and information storage, among others. New applications for metals in areas such as photonics, sensing, imaging, and medicine are also being developed. Significantly, most of these applications require the use of metals in the form of nanostructures with specific controlled properties. The properties of nanoscale metals are determined by a set of physical parameters that include size, shape, composition, and structure. In recent years, many research fields have focused on the synthesis of nanoscale-sized metallic materials with complex shape and composition in order to optimize the optical and electrical response of devices containing metallic nanostructures. The present paper aims to overview the most recent results—in terms of fabrication methodologies, characterization of the physico-chemical properties and applications—of complex-morphology metal-based nanostructures. The paper strongly focuses on the correlation between the complex morphology and the structures' properties, showing how the morphological complexity (and its nanoscale control) can often give access to a wide range of innovative properties exploitable for innovative functional device production. We begin with an overview of the basic concepts on the correlation between structural and optical parameters of nanoscale metallic materials with complex shape and composition, and the possible solutions offered by nanotechnology in a large range of applications (catalysis, electronics, photonics, sensing). The aim is to assess the state of the art, and then show the innovative contributions that can be proposed in this research field. We subsequently report on innovative, versatile and low-cost synthesis techniques, suitable for providing a good control on the size, surface density, composition and geometry of the metallic nanostructures. The main purpose of this study is the fabrication of functional nanoscale-sized materials, whose properties can be tailored (in a wide range) simply by controlling the structural characteristics. The modulation of the structural parameters is required to tune the plasmonic properties of the nanostructures for applications such as biosensors, opto-electronic or photovoltaic devices and surface-enhanced Raman scattering (SERS) substrates. The structural characterization of the obtained nanoscale materials is employed in order to define how the synthesis parameters affect the structural characteristics of the resulting metallic nanostructures. Then, macroscopic measurements are used to probe their electrical and optical properties. Phenomenological growth models are drafted to explain the processes involved in the growth and evolution of such composite systems. After the synthesis and characterization of the metallic nanostructures, we study the effects of the incorporation of the complex morphologies on the optical and electrical responses of each specific device.

Citation: Antonella Gentile, Francesco Ruffino and Maria Grazia Grimaldi, Complex-Morphology Metal-Based Nanostructures: Fabrication, Characterization, and Applications, doi: 10.3390/nano6060110.

1INTRODUCTION

Nanoscale-sized materials are materials with at least one dimension between 1 and 100 nm [1]. The crystalline nanomaterials are also characterized by a single-domain crystalline lattice, without the complicating presence of grain boundaries [2]. Scientific interest in nanomaterials has been growing steadily due to their unique position as a bridge between atoms and bulk solids, as well as their fascinating properties and potential applications [3]. The ability to generate such nanoscale-sized materials is central to advances in many areas of modern science and technology. In principle, the electron confinement by a nano-material provides the most powerful means to manipulate the electronic, optical, and magnetic properties of a solid material. This notion explains why nanomaterials have been the primary source for discovering and studying quantum size effects, with examples of quantized excitation [4,5,6,7], Coulomb blockade [8], metal-insulator transition [9,10], and superparamagnetism [11].

The study of metallic materials is one of the most ancient scientific fields. Among all kinds of inorganic solids, metals deserve special attention because they represent more than two-thirds of the elements in the periodic table. Most metals crystallize in the same cubic close-packed (ccp) structure, a face-centered cubic (fcc) lattice that allows for easy characterization. Properties such as strength, toughness, thermal and electrical conductivities, ductility and high melting point make the metals useful for applications ranging from household items to rocket ships. Traditional applications are mainly based on the bulk metallic properties.

New applications exploit the novel properties of metal nanomaterials [12,13,14,15,16,17,18,19,20,21,22,23,24,25,26,27,28,29]. Nanomaterials exhibit fascinating size-, shape-, and crystal-form-dependent properties. Like the bulk metals, nanomaterials of metals are also going to bring profound changes in many spheres of science, technology, and industry. Though metal nanoparticles have a long history of preparation and applications, the field has undergone explosive growth in recent years. Metal nanoparticles with excess morphologies have already been prepared, such as polyhedrons, plates, prisms, rods, wires, nanoboxes, nanocages, dumbbells, nanoshuttles, stars, branched rods and wires, dendrites, nanorings, nanotubes, *etc*. Such metal nanostructures possess a range of advantageous properties, and many metals have found extensive use in applications that include catalysis [30], electronics [31], photography [32], and information storage [33,34], among others. New applications for metals in areas including photonics [14,16,17,18,19,35,36], sensing [37], imaging [38], and medicine [23,39] are also being developed. Significantly, most of these applications require the use of metals in a finely divided state, preferably in the form of nanocrystals with precisely controlled properties. The properties of a metal nano-crystal are determined by a set of physical parameters that may include its size, shape, composition, and structure. In principle, one can tailor the properties of a metal nanocrystal by controlling anyone of these parameters, but the flexibility and scope of change are highly sensitive to the specific parameter. For example, in the case of localized surface plasmon resonance (LSPR) and surface-enhanced Raman scattering (SERS), both computational and experimental studies have demonstrated that the shape and structure of Au or Ag nanocrystals play the most important role in determining the number, position, and intensity of LSPR modes, as well as the spectral region or polarization dependence for effective molecular detection via SERS [40]. In the case of catalysis, it is well established that the activity of a metal nanocrystal can be enhanced by reducing its size [41]. The selectivity, however, is most sensitive to the packing of atoms on the surface or the exposed facets of a nanocrystal [42]. For example, Pt can selectively catalyze different types of chemical reactions, with the {100} and {210} facets being most active for reactions involving H_2 and CO, respectively [43]. The facets exposed by a nanocrystal clearly have a strong correlation with the shape. These and many other examples definitively illustrate the importance of shape control towards the efficient utilization of metal nanocrystals. The last decade has evidenced the successful synthesis of metal nanocrystals in a variety of shapes. Examples include: sphere; spheroid; cube; cuboctahedron; octahedron; tetrahedron; right bipyramid; decahedron; icosahedron; thin plate with a triangular, hexagonal, or circular profile; and rod or wire with a circular, square, rectangular, pentagonal, or octagonal cross-sections [2,44,45,46,47,48,49,50,51] (Figure 1). The emergence of many novel approaches induces synthesis and synthetic design, control of composition, size, morphology, and assembly structure and impressive advances in the characterization and manipulation techniques of metallic nanoparticles. A number of applications have been realized and multitudes of new applications have been envisaged. Among the many morphologies investigated, the anisotropic metal nanostructures have attracted a great deal of attention over the last decade due to their unique properties originating in their shape and their potential impact on new technologies.

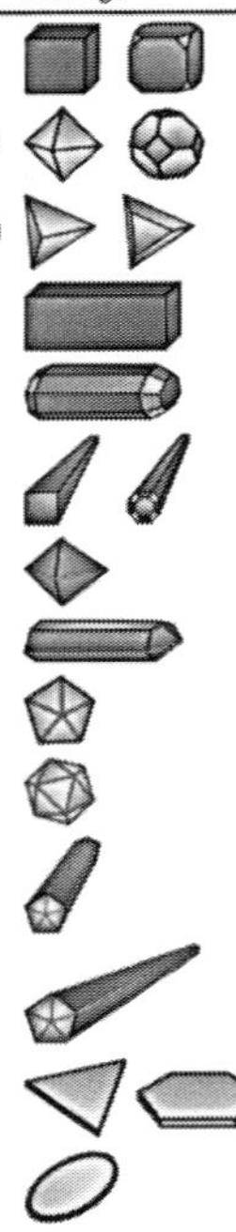

Structures	Shapes	Schematic drawings	Metals
single-crystal	perfect/truncated cube[a]		Pd, Ag, Au, Pt, Cu, Rh, Bi, Fe
	perfect/truncated octahedron[b]		Pd, Ag, Au, Pt
	perfect/truncated tetrahedron[a]		Ag, Au, Pt, Rh
	rectangular bar		Pd, Ag, Pt
	octagonal rod		Pd, Au, Fe, Co, Ni
	rectangular or octagonal wire		Pb, In, Sn, Sb, Fe, Co
singly twinned	right bipyramid		Pd, Ag
	beam		Ag
multiply twinned	decahedron[a]		Pd, Ag, Au
	icosahedron[a]		Pd, Au
	five-fold twinned pentagonal rod		Pd, Ag, Au, Cu
	five-fold twinned pentagonal wire		Ag, Au, Cu
	triangular/hexagonal plate		Pd, Ag, Au, Cu, Pb, Bi, Co, Ni
	disc		Sn, Co

Figure 1: A summary of different shapes that have been achieved for various metal nanocrystals. Reproduced with permission from [2]. Copyright Wiley, 2008.

By controlling the shape of the particle, the crystallographic planes that are exposed at the surfaces are tailored and, as a consequence, the atomic arrangement that makes the interface with the medium can be tuned. This has been shown to be critical for most chemical–physical properties, expanding the already broad range of applications of nanoparticles (NPs). One of the most appealing examples impacted by the shape-dependent properties of NPs is heterogeneous catalysis [52]. Several catalysts are composed by metal NPs dispersed on high surface area supports. In a heterogeneous catalytic reaction, the binding of the reagent molecules and the formation of the new bonds, which will form the product molecules, depend not only on the available orbitals of the reagents but also on the geometry of the atomic arrangement at the metal surface. Hence, controlling the exposed facets of an NP can be a major step toward a deeper understanding of the catalytic reaction and the design of better catalysts. Another key example in regards to the plasmonic-based behavior of noble metal nanostructures: metals such as gold and silver show a resonance in the ultraviolet-visible (UV-vis) region of the spectra known as surface plasmon [53]. This resonance arises from collective oscillations of the conduction electrons of the metal and strongly depends on the size, shape, local environment [14,16,17,18,19,24,54], and assembly of the NPs [55]. When a molecule is attached to the surface of a metal NP, its Raman scattering cross-section can be enhanced by several orders of magnitude, from 10^6 up to 10^9 if the incident light is tuned to the surface plasmon resonance of the particle [56]. This effect, known as SERS, results from an intense local amplification of the electric field near a metal surface when surface plasmon resonates in phase with the incident light. Because the electromagnetic near-field intensity is higher at sharp edges and tips, anisotropic NPs exhibit high Raman scattering enhancement factors that have been exploited for the detection of diluted molecules in analytical applications. For example, Mulvihill *et al.* [57] explored Langmuir–Blodgett monolayers of cubic, cuboctahedral, and octahedral silver NPs to detect trace amounts of arsenic in water. The best results were obtained with octahedral silver NPs, for which a detection level in the ppb range was achieved.

To understand the effects of the complex shape of the metallic nanoparticles on the plasmonic properties, the results of the optical characterizations related to non-regular geometries are shortly presented: for example, dark-field optical microscopy and near-field optical extinction microscopy enable the observation of plasmon resonances of a single particle. In dark-field optical microscopy, only the light scattered by the structure under study is collected in the detection path, while the directly transmitted light is blocked using a dark-field condenser. This enables the study of single particles dilutely dispersed on a substrate. Figure 2a shows an example of the dipolar plasmon line-shapes related to colloidal silver particles of different shapes. Conversely, in near-field optical spectroscopy, a thin (metalized or uncoated) fiber tip with an aperture to the order of 100 nm is brought into close proximity of the particle using an appropriate feedback scheme. The plasmon resonances can then be mapped out using either illumination through the tip and collection in the far-field, or evanescent illumination from the substrate side and light collection via the tip. Figure 2 also shows scattered light images (Figure 2b) and plasmon line-shapes (solid lines in Figure 2c) of a variety of gold nanoparticles [58,59,60].

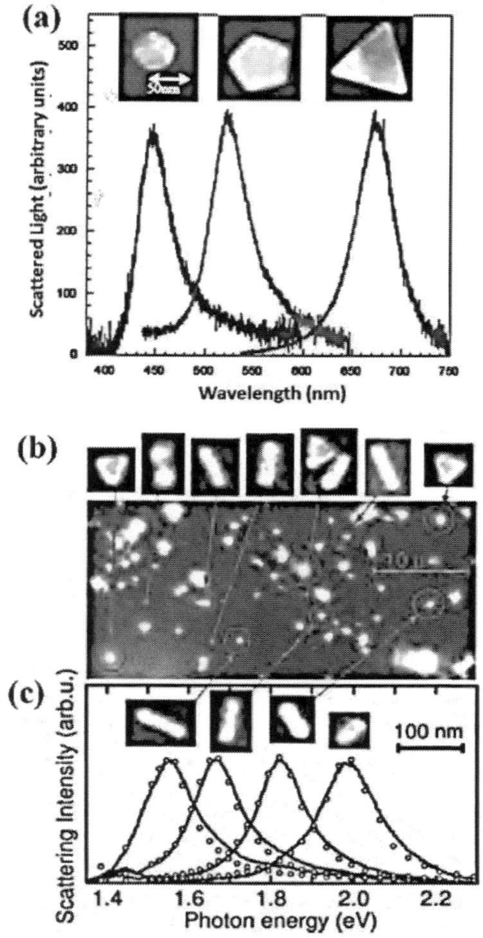

Figure 2: (**a**) Scattering spectra of single silver nanoparticles of different shapes obtained in dark-field configuration. Optical dark field images together with scanning electron microscope (SEM) images of individual gold nanoparticles (**b**) and corresponding scattering spectra (**c**) for an incident light polarization along the long particle axis. (**a**) Reproduced with permission from [59]. Copyright American Institute of Physics, 2002. (**b**,**c**) reproduced with permission from [60]. Copyright American Institute of Physics, 2003.

Therefore, Figure 2 highlights that the localized plasmon resonance of a single metallic nanoparticle can be shifted in frequency from the Fröhlich frequency [58] via alterations in particle shape and size. Furthermore, in particle ensembles, additional shifts are expected to occur due to electromagnetic interactions between the localized modes. For small particles, these interactions are essentially of a dipolar nature, and the particle ensemble can, in a first approximation, be treated as an ensemble of interacting dipoles. Two regimes have to be distinguished, depending on the magnitude of the interparticle distance (d). For closely spaced particles, $d \ll \lambda$, near-field interactions with a distance dependence of d^{-3} dominate, and the particle array can be described as an array of point dipoles interacting via their near-field. In this case, strong field localization in nano-sized gaps between adjacent particles has been observed for regular one-dimensional particle chains [61]. The field localization is due to a suppression of scattering into the far-field via excitation of plasmon modes in particles along the chain axis, mediated by near-field coupling. Figure 3 illustrates this fact by showing the simulated distribution of the electric field above single gold nanoparticles (Figure 3a) and a particle chain (Figure 3b). From the images, it can clearly be seen that scattering is drastically suppressed for closely spaced particles, and that the fields are instead highly localized at interstitial sites. Interparticle junctions such as these therefore serve as hot-spots for field enhancement, which will be discussed later. Therefore, the interparticle coupling will lead to shifts in the spectral position of the plasmon resonance compared to the case of an isolated particle.

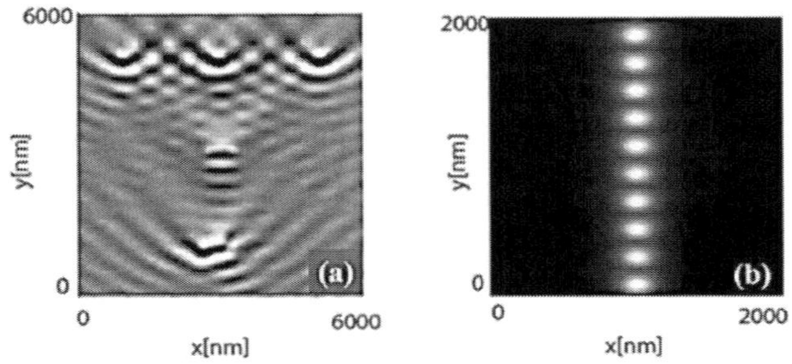

Figure 3: Simulated intensity distribution of the optical near-field above an ensemble of well-separated gold particles (**a**) and a chain of closely spaced gold nanoparticles (**b**). While for separated particles interference effects of the scattered fields are visible, in the particle chain, the fields are closely confined in gaps between adjacent particles. Plasmon resonances were excited using prism coupling with the direction of the in-plane moment component as outlined in the pictures. Reproduced with permission from [61]. Copyright Wiley, 2001.

Maier *et al.* [62], using one-dimensional arrays of 50 nm gold particles with varying interparticle distance, demonstrated experimentally the shifts in resonance energy using far-field extinction spectroscopy. Due to the strong scaling of the interaction strength with d^{-3}, particle separations in excess of 150 nm are sufficient to recover the behavior of essentially isolated particles. For larger particle separations, $d >> \lambda$, far-field dipolar coupling with a distance dependence of d^{-1}dominates. For example, two-dimensional gratings of gold nanoparticles with various lattice constants [63] (Figure 4) shows that far-field coupling has pronounced influences on the plasmon line-shape, both in terms of resonance frequency as well as spectral width. We note that interactions between metal nanoparticles can be further enhanced by providing additional coupling pathways as, for example, in the form of propagating Surface Plasmon Polariton (SPPs) for particle arrays fabricated on a conductive substrate.

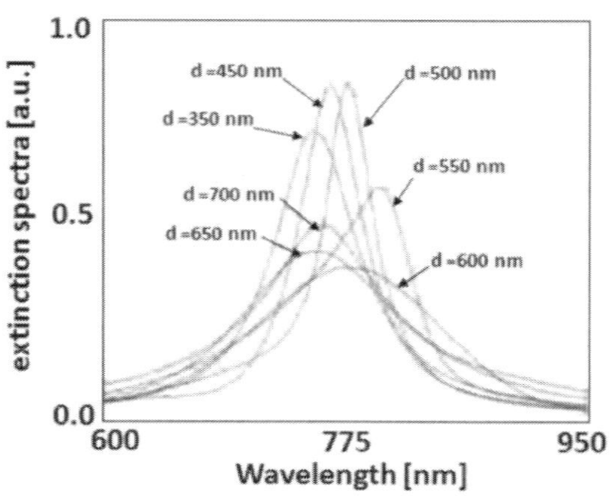

Figure 4: Extinction spectra for square two-dimensional gratings of gold nanoparticles (height 14 nm, diameter 150 nm) with grating constant d situated on a glass substrate. Reproduced with permission from [63]. Copyright American Physical Society, 2000.

These examples show how the optical response of isolated metal nanostructures or metal nanostructure assemblies are strongly dependent on the nanostructure morphology and topology. The rationale design of the morphology and topology of the nanostructures can give the possibility to confer a specific functionality to the nanostructures in relation to its interaction with the electromagnetic radiation. For example, a single nanostructure of complex morphology can be characterized by several hot-spots with respect to the interaction with the electromagnetic radiation. The same number of hot-spots are, instead, obtainable by a large number of spherical nanoparticles assembled in a specific array.

Starting from these general ideas, in the following sections we report synthesis methods and property characterization for some peculiar metal nanostructures with complex shapes and composition in view of advanced applications in the field of electronic, opto-electronic, sensing, bio-sensing and magnetic devices.

GOLD NANORINGS AND RING DIMERS

The field enhancement around nanoscale metallic particles offers an efficient tool to manipulate light–matter interactions at defined optical frequencies [64,65]. For spectroscopic applications, the field confinement around these structures reduces the mismatch between the wavelength of the incident light and size of the molecules under study. In the SERS, this can boost the Raman cross-section of molecules by many orders of magnitude. SERS offers a very sensitive tool for biological and chemical sensing [66], enabling vibrational spectra to be obtained even from single molecules [67]. The signal enhancement in SERS mostly comes from the excitation of localized surface plasmon resonance (LSPRs) and the "lightning rod" effect [68]. The latter is particularly pronounced for sharp tips and nanoparticle dimers with small inter-gap size. The optimum substrate for SERS should have the highest average field enhancement possible, be highly tunable, and offer a broad resonance bandwidth, although satisfying all these criteria is very challenging. Promising candidates as highly tunable nanostructures fulfilling these requirements are metal nanostructure characterized by a ring geometry [13,69]. The optical properties of gold nanorings fabricated by colloidal lithography were reported by Aizpurua *et al.* [69]. They showed that the LSPR frequency strongly depends on the ratio of the thickness to ring outer radius. This tunability, which is similar to that of nanoshells [70], allows the control of the plasmon resonances from visible to near-IR wavelengths via reducing the ring thickness. The ring-like nanoparticles (Figure 5a,c) exhibit tunable plasmon resonances in the near infrared that are not observed for solid particles of similar size (nanodisks in Figure 5b). The volume confined by the nanoring is empty and therefore accessible to various sensing and spectroscopy applications at the nanometer scale. The experimentally observed optical response of these structures is well described by numerical simulations (Figure 5d) and the main features can be qualitatively understood from simple models of the charge oscillation patterns. The predictive character of these calculations allows one to tailor the shape of a particle in order to achieve excitation spectra on demand with controlled field enhancement. The reduced amount of the metal in the ring geometry compared to disks of the same radius results in smaller absorption loss. Fabricating metallic rings with high-resolution methods such as electron beam lithography enables a great control of the periodic arrangement of the nanoparticles. Moreover, the ring geometry has an advantage over the nanoshell pattern in that molecules can attach to the inner wall as well as the outer wall, which is beneficial as substantial field enhancement occurs there. In addition to the conventional SERS, studying the field confinement in metallic rings could also have applications in optical corrals [71] and lasing spacers [72].

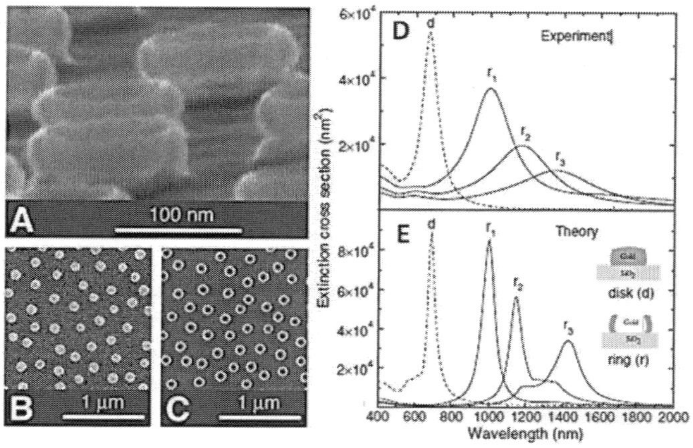

Figure 5: SEM images of gold nanorings and nanodisks prepared by colloidal lithography. (**a**) 80° tilt image of a ring structure. The walls of the rings are thin enough for the 30 keV electrons to pass through; (**b,c**) Top views of disks and rings taken at an acceleration voltage of 1.5 keV. The heights of the disks and rings in the figure are ≈20 and ≈40 nm, respectively, whereas the radius is ≈60 nm in both cases. The thickness of the ring walls was estimated at 14 ± 2 nm from sideview SEM images similar to (**a**); (**d**) Experimental extinction spectra of disks (*d*, dashed line) and rings (*r*, solid lines). The estimated wall thickness d of the rings is $d = 14 \pm 2$ nm (r_1), d = 10 ± 2 nm (r_2), and d = 9 ± 2 nm (r_3), respectively. (**d**) Calculated extinction spectra for the same particles as in (**a**). The rings exhibit near-infrared features at larger wavelengths for thinner walls. The disks show a dipolar excitation at around 700 nm. Reproduced with permission from [69]. Copyright American Physical Society, 2003.

Banaee *et al.* [64] compared the SERS spectra related to single nanoring and nanoring dimmers with that of an array of nanodisk dimers. For SERS measurement, self-assembled monolayers of benzenethiol are formed on the gold nanostructures. To compare SERS enhancements of ring and ring dimer structures with the disk dimmers, similar steps are taken for the measurement of the latter. The experimental SERS enhancement factor (EFs) for 1074 cm^{-1} line of benzenethiol for all patterns are measured and that of the ring and dimer rings results in 3.9×10^6 and 4.2×10^6, respectively, and they are roughly seven times higher than the dimer disks.

Even if the nanoring architecture presents several strengths with respect the nanoshell one, this last resulted of primary importance when its plasmonic properties were combined with those of the nanorod architecture in a single structure, named "nanorice", as discussed in the following section.

GOLD NANORICES

This new hybrid nanoparticle geometry combines the plasmonic properties of both nanorods and nanoshells in a single structure. This dielectric core-metallic shell prolate spheroid nanoparticle bears a remarkable resemblance to a grain of rice, inspiring the name "nanorice". Wang *et al.* [73] report nanorice (see Figure 6) composed of a spindle-shaped hematite core coated with a continuous nanometer-thick Au shell. To apply the plasmon hybridization picture to analyze the resonances of this layered nanoparticle geometry, the authors analyze the plasmon resonances of a solid metallic nanorod and a nanocavity of arbitrary ellipticity as the parent plasmons which, when hybridized, give rise to the plasmon resonances of nanorice. The fabrication of nanorice involves seeded metallization of spindle-shaped hematite nanoparticle cores (Figure 6A. Small Au nanoparticles (~2 nm in diameter) are immobilized onto the surface of (3-aminopropyl) trimethoxysilane (APTMS) functionalized cores at a nominal coverage of ~30%. The immobilized Au colloids act as nucleation sites for electroless Au plating onto the surface of core particles, leading to the gradual formation of a continuous and complete Au shell layer. This is essentially the same metallization procedure used in silica–Au nanoshell synthesis [74] and shows that this approach is readily adaptable to produce uniform metallization layers on the surfaces of other oxide nanoparticles. Further metal deposition onto the nanostructure increases the thickness of the metal layer. Nanorice can be dispersed homogeneously in solvents such as water and ethanol to form colloidal solutions or be dispersed and immobilized on polyvinylpyridine (PVP)-coated substrates as individual, randomly oriented nanoparticles.

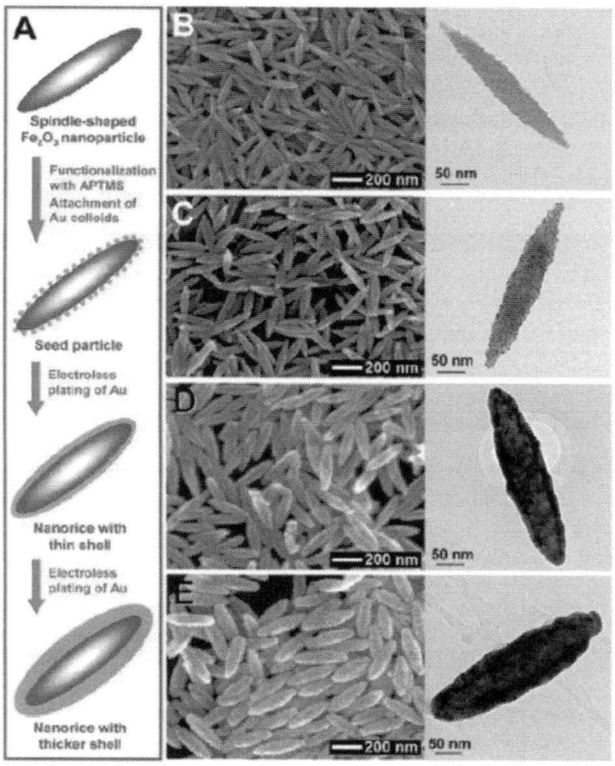

Figure 6: (**A**) Schematics of the fabrication of hematite-Au core-shell nanorice particles. SEM (**left**) and transmission electron microscopy (TEM) (**right**) images of (**B**) hematite core (longitudinal diameter of 340 ± 20 nm, and transverse diameter of 54 ± 4 nm; (**C**) seed particles; (**D**) nanorice particles with thin shells (13.1 ± 1.1 nm); and (**E**) nanorice particles with thick shells (27.5 ± 1.7 nm). Reproduced with permission from [73]. Copyright American Chemical Society, 2006.

Extinction spectra of nanorice with varying thicknesses of the shell layer are shown in Figure 7A. Representative SEM image of the nanorice sample used to obtain these optical measurements is shown in Figure 7B. The strong plasmon resonance feature observed in the spectra in Figure 7A arises because of the longitudinal plasmon of this layered structure and exhibits a highly sensitive structural dependence of its optical resonance, which blueshifts as the metal layer thickness is increased. A finite difference time domain (FDTD) analysis [75] of the far field extinction spectrum of this nanostructure reveals that the transverse plasmon mode (Figure 7C inset) has a much weaker extinction cross-section than the longitudinal mode of this nanostructure (Figure 7C. The weaker, higher energy resonance observed in the spectrum of nanorice is attributable to the far weaker transverse plasmon mode, with a small additional contribution from a higher order longitudinal plasmon mode. The local fields associated with the longitudinal and transverse nanorice plasmons are shown in Figure 7D,E. The asperities of this structure support very strong local field intensity enhancements (>7000 for this specific nanorice geometry) at wavelengths corresponding to the longitudinal plasmon resonance of the nanostructure. This intensity enhancement is several times larger than those reported for nanofabricated bowtie junctions [76] and what has been predicted and measured in scanning probe junctions [77,78]. Such strong, tunable local fields make this geometry highly attractive for use in designing substrates for surface-enhanced spectroscopy-based sensing. These large plasmon-resonant local field enhancements are similar in magnitude to the localized plasmon resonant "hot-spots" occurring in junctions between metallic nanoparticles when a dimer plasmon resonance is excited [79,80]. The nanorice local fields should give rise to intense SERS enhancements with the added advantage that the hot-spots are completely open to the surrounding medium in this geometry. From this point of view, each nanorice particle can potentially serve as a standalone, optically addressable nanoscale substrate for surface-enhanced spectroscopies. Moreover, because the enhanced near-field intensities can extend several tens of nanometers from the surface of the nanorice, these particles may exhibit unique advantages in the spectroscopic sensing and characterization of large biomolecules, such as proteins and DNA, biological samples, or materials placed directly adjacent to the nanoparticle.

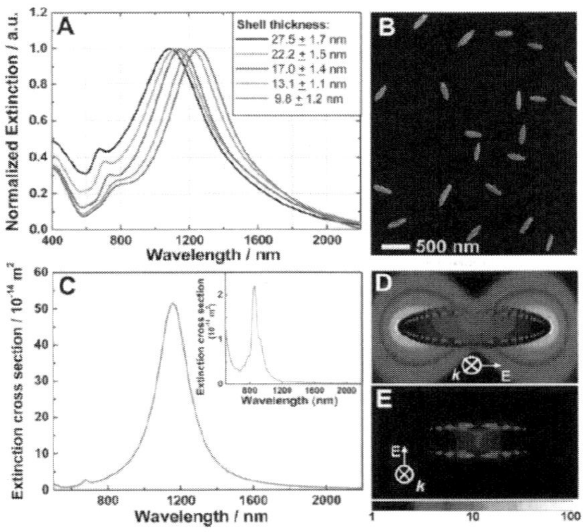

Figure 7: (**A**) Extinction spectra of hematite-Au core-shell nanorice with different shell thicknesses. Two plasmon peaks are observed for each sample. The plasmons at longer and shorter wavelengths are the longitudinal and transverse plasmons, respectively. The samples measured are monolayers of isolated nanoshells immobilized on polyvinylpyridine (PVP)-glass slides; (**B**) A SEM image of a monolayer of nanorice particles (shell thickness of 13.1 (1.1 nm) on a PVP-glass slide; (**C**) Calculated far-field extinction spectra of the nanorice with incident polarization along the longitudinal and (inset) transverse axis of a nanorice particle using finite difference time domain (FDTD). The nanorice particle employed for the FDTD simulations is composed of a hematite core with longitudinal diameter of 340 nm and transverse diameter of 54 nm surrounded by a 13-nm-thick Au shell. Near-field profile of the nanorice under resonance excitations: (**D**) incident polarization along the longitudinal axis, λ_{ex} = 1160 nm; (**E**) incident polarization along the transverse axis, λ_{ex} = 860 nm. Reproduced with permission from [73]. Copyright American Chemical Society, 2006.

The nanorice architecture is the simplest example of how hot-spots can originate in a single nanostructure simply deforming the spherical geometry. The next example, the nanoporous gold architecture, shows as the topological complexity on the nanoscale allows the establishment of a high number of hot-spots in an extended nanostructured system giving rise, at the same time, to a system with an enormous exposed surface area which is beneficial for ultrasensitive devices.

NANOPOROUS GOLD FILMS

Porous materials have recently been attracting considerable attention because of a wide range of applications in catalysis, sensing, micro/nanoelectromechanical systems, and biotechnology [81]. One example is nanoporous gold formed by chemically or electrochemically dealloying of silver–gold alloys [82,83,84,85], which possesses bi-continuous porosity and excellent electrical and thermal conductivities. Nanoporous gold offers an extensive surface where the analytes can connect, thereby increasing sensitivity and efficiency [86]. Nanoporous gold (NPG) may be formed by a spontaneous pattern-forming instability during the chemical etching of silver from gold–silver alloys [87]. The leaching of the less noble metal gives rise to a bi-continuous sponge-like structure of nanopores and gold ligaments [88] whose geometric features depend on the alloy composition and on the experimental conditions of the dealloying process [89]. The nanoporous structure affects also the optical properties: the plasma frequency ω_p exhibits red-shift due to the lower density in comparison with bulk gold and the material shows metallic behavior for wavelengths above the near-IR range [90]. Nanoporous gold with excellent thermal stability and chemical inactivity has recently been exploited as an attractive substrate for SERS applications because of its large surface area and bi-continuous porous structure in three dimensions [91]. These SERS-active substrates combine the self-organized and highly interacting nanoscale morphology of NPG with the advantages of reproducibly nanopatterned periodic structures. Nevertheless, conventional nanoporous gold with a pore size of about tens of nanometers made by room temperature dealloying does not show a promising SERS enhancement although nanostructured gold, such as gold nanoparticles, are known to be SERS active for a number of organic and biological molecules [92]. Thus, optimizing the microstructure is crucial to the improvement of the SERS enhancements and thus the SERS application of nanoporous gold, which has been attempted by a few of research groups [93,94,95]. More recently, Kucheyev *et al.* [94] reported that the optimized SERS enhancement can be achieved in the annealed nanoporous gold with a coarsened pore size of ~250 nm. However, Qian [96] found that the strongest SERS enhancement takes place from the nanoporous gold with an ultrafine nanopore size of ~5–10 nm. The correlation between the structural parameters of the NPG films and the improvement of the optical properties drives the search to optimize the synthesis processes in order to obtain a greater control over the size of the nanopore structures. Subsequently, we report synthesis methods used to obtain porous materials with a greatly enhanced surface-to-volume ratio which is particularly suitable for the realization of plasmonic supports for sensing purposes.

Lang *et al.* [97], report the plasmonic properties of free-standing nanoporous gold films with an intricate bicontinuous nanostructure. $Ag_{65}Au_{35}$ (atomic ratio) leaves with a size of ~20 mm × 20 mm × 100 nm were dealloyed in a 70% (mass ratio) HNO_3 solution at room temperature. Nanopore sizes can be tailored from 10 to 50 nm by controlling the corrosion time in the range of 5 min to 24 h (Figure 8a,b). Intermediate porous structures are quenched by distilled water, and residual acid within nanopore channels is removed by water rinsing. UV-vis extinction spectra of the NPG films with pore sizes of 10, 15, 20, 25, 30, 40, and 50 nm soaked in the water are shown in Figure 8c. Regardless of nanopore sizes, two characteristic resonant bands can be detected in all the NPG samples. One at ~490 nm, resulting from the resonant absorption of gold films, is independent of nanopore sizes and dielectric surroundings. The other at ~550–650 nm, arising from the excitation of localized surface plasmon resonance, shows obvious band shift with the nanopore sizes and dielectric indices of surrounding media, suggesting that NPG is a promising candidate as plasmonic sensors for organic and biologic molecule detection. The spectral features of NPG are apparently different from those of other nanostructured golds. For instance, gold nanoparticles are characterized by an isolate surface plasmon resonance (SPR) band and gold nanorods exhibit one transverse plasmon band at ~520 nm [98,99]. The low wavelength peak of NPG at 490 nm originates from the resonant absorption of gold films [100]. The resonant location mainly relies on film thickness rather than nanopore and ligament sizes. Thus, it is nearly independent of nanopore sizes. In contrast, the high wavelength band, arising from LSPR, represents the significant redshift with the increase in pore sizes, which expectedly results from the effective electron oscillation lengths that are determined by the gold ligament sizes. Due to these intense SPR peaks, the authors tested the produced NPG films as plasmonic sensors. Figure 8a–c show the extinction spectra of the NPG films immersed into a series of organic dielectric media. The SPR peak at 490 nm (λ_1) does not show a detectable shift when the refractive indices of the media (n) increase from 1.33 to 1.495. In contrast, the SPR band at ~540 nm (λ_2) represents evident redshift with the refractive indices (seeFigure 8e–g). The shift magnitude of λ_2 has a linear relationship with the refractive index as shown in Figure 8h. For the λ_2SPR band, the NPG films with larger pore sizes show more significant redshift with the refractive index. Thus, the refractive index sensitivity of NPG is comparable to those of other nanostructured golds, such as gold nanorods [99], hollow sphere and cylindrical holes.

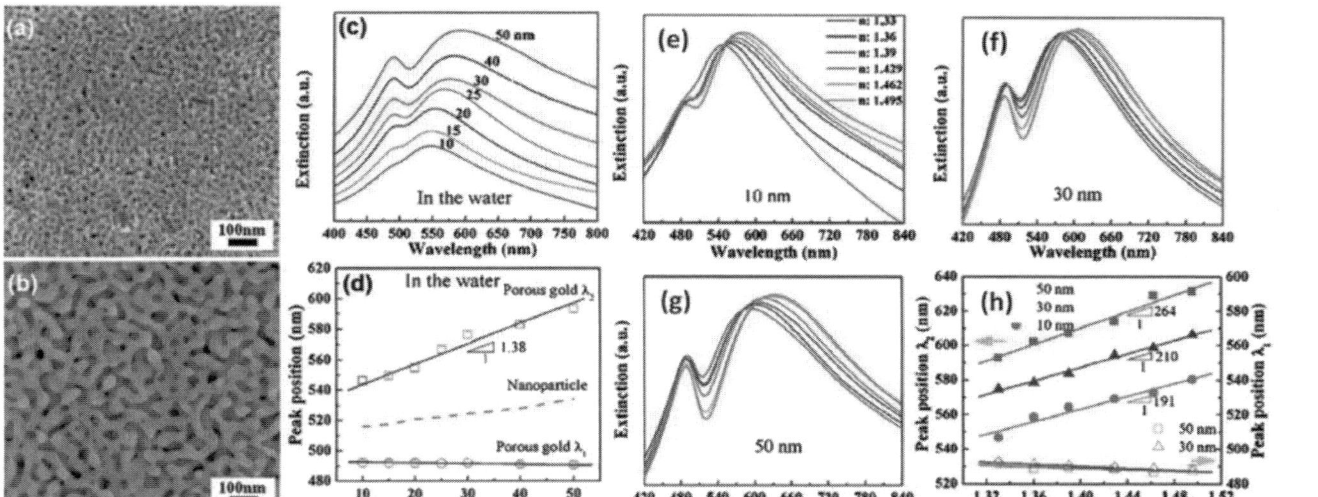

Figure 8: SEM micrographs of NPG synthesized by means of corrosion for 5 min (**a**) and 24 h (**b**); (**c**) ultraviolet (UV)-vis extinction spectra and (**d**) the resonant peak position of λ_1 and λ_2 of NPG with the pore sizes of 10–50 nm in water. For comparison, the dashed line represents the size dependence resonant band of gold nanoparticles. UV-vis extinction spectra of porous gold films with (**e**) 10 nm; (**f**) 30 nm; and (**g**) 50 nm are recorded by immersing them into various dielectric environments. Refractive index of these solutions increase from left to right: water (*n* = 1.33), ethanol (*n* = 1.36), 3:1 ethanol/toluene (*n* = 1.39), 1:1 ethanol/toluene (*n* = 1.429), 1:3 ethanol/toluene (*n* = 1.462), and toluene (*n* = 1.495). (d) Dependence of resonance (λ_1), empty symbols and LSPR (λ_2, solid symbols) peaks of NPG films on refractive index. Reproduced with permission from [97]. Copyright American Institute of Physics, 2011.

Qian *et al.* [96] report, instead, the results of the characterization of two classes of $Ag_{35}Au_{65}$ (atomic %, 50/50 by weight) films with the thicknesses of ~100 nm and ~60 μm, respectively, used to synthesize nanoporous gold by selective chemical and electrochemical dealloying in 70% HNO_3 aqueous solution at the temperatures of −20, 0, and 25 °C, respectively. By controlling the dealloying time and temperatures, the authors prove that the nanopore sizes can be tailored from ~5 to ~33 nm for 100 nm thick films (see SEM images in Figure 9a,b). The 60 μm thick nanoporous gold with a pore size of ~55 nm, produced by room-temperature dealloying for 48 h, is further annealed at 200, 300, 400, 500, and 600 °C for 2 h in air, which results in the coarsening of nanoporous gold with pore sizes ranging from 90 to 700 nm. The nanoporous gold specimens with various nanopore sizes were subjected to SERS experiments with rhodamine 6G (R6G) and crystal violet (CV) 10B as test molecules (see SEM images in Figure 9c–f). SERS enhancements of nanoporous gold for both R6G and CV molecules prove that ultrafine nanopores possess the strongest SERS enhancement. Although high enhancements can be obtained in the annealed samples with a large pore size of 350 nm, similar to the observations of Kucheyev *et al.* [94], this anomalous enhancement originates from the ultrafine pimple-type irregularities on the rough gold surfaces. Thus, the observations of the SERS enhancements of nanoporous gold confirm that the high SERS enhancements result from smaller microstructure features, either smaller pore sizes or the rough gold ligament surfaces, which are believed to promote electromagnetic field enhancements and provide more active sites for molecule adsorption [101,102]. Because the formation and coarsening of nanoporous gold are controlled by surface diffusion, the smaller pore sizes formed by a short dealloying time at low temperatures generally contain more or less residual silver. It is known that the SERS effect of silver exceeds that of gold [103]. Thus, the stronger Raman scattering from the ultrafine nanoporous gold may also include the extra contribution from the residual silver. However, the fact that the annealed nanoporous gold (without detectable residual silver) containing a high density of surface irregularities exhibits the analogous enhancement as the ultrafine nanoporous gold indicates that the ultrafine structures, both small pore sizes and fine surface pimple-type irregularities, play the major role in the strong SERS enhancement of the nanoporous gold.

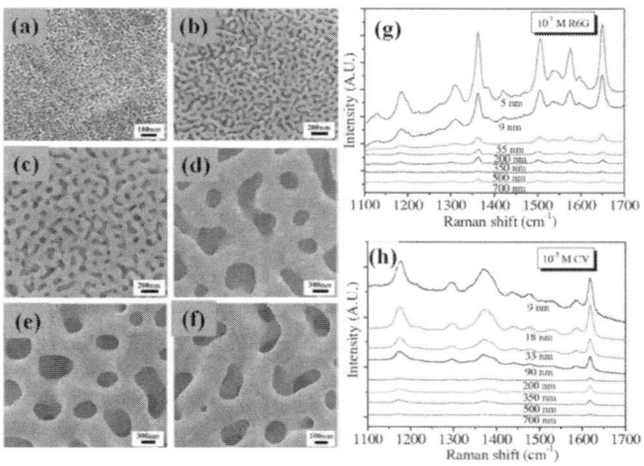

Figure 9: Representative SEM micrographs of nanoporous gold with various nanopore sizes. (**a**) Nanoporous gold film after 5 min dealloying at room temperature; (**b**) dealloyed at room temperature for 48 h; (**c**) thoroughly rinsed nanoporous gold annealed at 200 °C for 2 h; (**d**) 400 °C for 2 h; (**e**) 500 °C for 2 h; and (**f**) 600 °C for 2 h. SERS spectra of nanoporous gold with different pore sizes for (**g**) 10^{-7} mol/L R6G aqueous solution and (**h**) 10^{-5} mol/L crystal violet (CV) methanol solution. Reproduced with permission from [96]. Copyright American Institute of Physics, 2007.

Chen *et al.* [95] studied the plasmonic properties of a 20-nm-thick nanoporous Au film by far-field and near-field optical techniques. The film is prepared sequentially by deposition of gold and copper, thermal annealing, and chemical etching, and has randomly distributed nanopores with sizes ranging between 20 and 350 nm. The nanoporous film has a different transmittance and a lower reflectance when compared with those of a 20-nm-thick plain Au film in the wavelength range between 400 and 1000 nm. As a result, the absorbance of the nanoporous film is much higher and can be attributed to the conversion of incident light into SPPs. In the dark-field scattering spectrum, a broad peak appears at around 630 nm and confirms the presence of HPR (Hole Plasmon resonances) [104] of the nanoporous film. With the aid of transmission mode NSOM (Near Scanning Optical Microscopy), local field distribution on the film is observed and reveals the generation of SPPs. Furthermore, two types of local field enhancement with spatial extents of around 200 nm and 1 nm are observed. The two types of enhancement correlate with strong and weak SPP localizations, respectively.

The peculiar optical properties of this complex-morphology structure are highlighted by the simulation reported by Lang, using a discrete dipole approximation (DDA) [105]. A simplified nanostructure representing the key structural features of NPG is introduced to qualitatively simulate the optical properties of NPG with the bulk dielectric function of Au [106]. Under the plane wave with a wavelength of 514 nm propagating along the direction normal to the top surface of the nanostructure, the total near-field E^2/E_0^2 distributions are calculated with pore size $d = 10$ nm and Au ligaments size $D = 30$ nm, and also with $d = D = 10$ nm, and $d = D = 20$ nm, respectively, exhibiting that E^2/E_0^2 increases with the decrease in the d/D ratio. This is qualitatively in accordance with experimental measurements of the SERS enhancements reported in many works. The electromagnetic field enhancements from near-field coupling dramatically increase as the ratio of d/D decreases compared to the NPG with the same pore size but large d/D ratios. Therefore, for SERS effect of NPG films, in addition to the LSPR, the weakening of plasmon damping with increasing ligament sizes obviously enhances the near-field coupling between neighboring ligaments, which gives rise to the observed SERS improvements with the decrease in the d/D ratios.

For the sake of completeness, we mention that in ideal conditions, nanoporous gold films can be produced without compromising its electrical performance. In fact, also the electrical transport properties of nanoporous gold films are widely studied [107,108,109,110]. Smith *et al.* [107] analyzed samples of nanoporous gold of various morphologies by a combination of electrical and optical data. Concerning the electrical response, the authors reported the resistance of a growing gold film on glass as a function of volume (or area) fraction f of gold (Figure 10), finding a percolation-type behavior. This resistance data is consistent with a percolation threshold of $f_c = 0.31 \pm 0.03$.

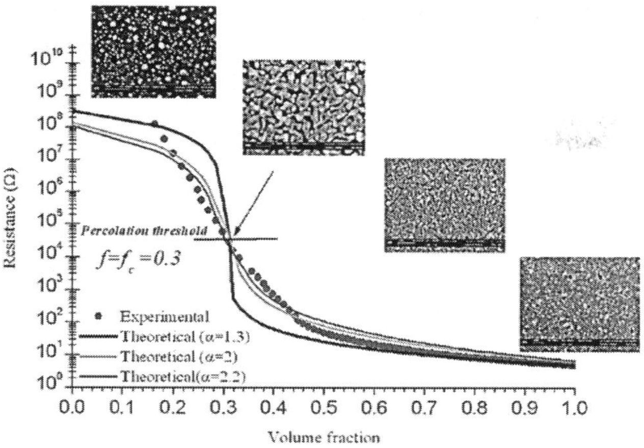

Figure 10: Resistance (dots are the experimental data) of a growing gold film on glass as a function of volume (or area) fraction f of gold. Images included show characteristic morphology at each stage of gold coverage. Reproduced with permission from [107]. Copyright American Physical Society, 2008.

We highlight that Chen *et al.* [111] exploited the electrical properties of nanoporous gold films to fabricate a non-enzyme electrochemical glucose sensor: The free-standing NPG films showed a robust and sensitive current response to glucose. The authors also investigated effects of pore sizes and detecting potentials. Chloride ions on glucose oxidation were systematically investigated finding that NPG with a smaller pore size possesses higher sensitivity.

Recent research has likewise been devoted to the fabrication of porous Au nanostructures such as porous Au nanoparticles [112] or nanowires [113]. For example, Wang *et al.* [112] developed a combination of a "top-down" approach (substrate-conformal imprint lithography) and two "bottom-up" approaches (dewetting and dealloying) for the fabrication of perfectly ordered 2D arrays of nanoporous gold nanoparticles (Figure 11). The dewetting of Au/Ag bilayers on the periodically prepatterned substrates leads to the interdiffusion of Au and Ag and the formation of an array of Au–Ag alloy nanoparticles. The array of alloy nanoparticles is transformed into an array of nanoporous gold nanoparticles by a following dealloying step in HNO_3. This technique allows for the control of particle size, particle spacing, and ligament size (or pore size) by varying the period of the structure, total metal layer thickness, and the thickness ratio of the as-deposited bilayers. Chauvin*et al.* [113] developed a two-step approach allowing creation of highly ordered porous gold nanowire arrays with a length up to a few centimeters. This two-step approach consists of the growth of gold/copper alloy nanowires by magnetron cosputtering on a nanograted silicon substrate, serving as a physical template, followed by a selective dissolution of copper by an electrochemical anodic process in diluted sulfuric acid. The authors demonstrated that the pore size of the nanowires can be tailored between 6 and 21 nm by tuning the dealloying voltage between 0.2 and 0.4 V and the dealloying time within the range of 150–600 s. In addition, they showed that the initial gold content (11 to 26 atomic %) and the diameter of the gold/copper alloy nanowires (135 to 250 nm) are two important parameters that must carefully be selected to precisely control the porosity of the material (Figure 12).

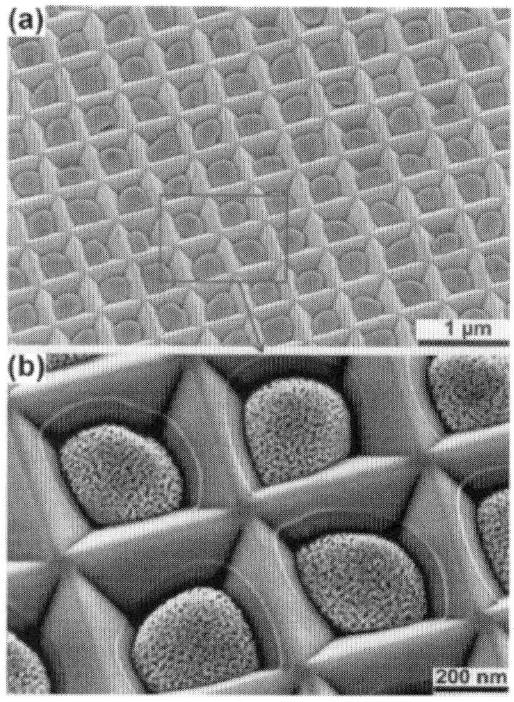

Figure 11: SEM images (false color) at 25° tilt of the perfectly ordered array of the nanoporous gold nanoparticles formed from the 15 nm Au/30 nm Ag bilayers. Reproduced with permission from [112]. Copyright Beilstein-Institut, 2012.

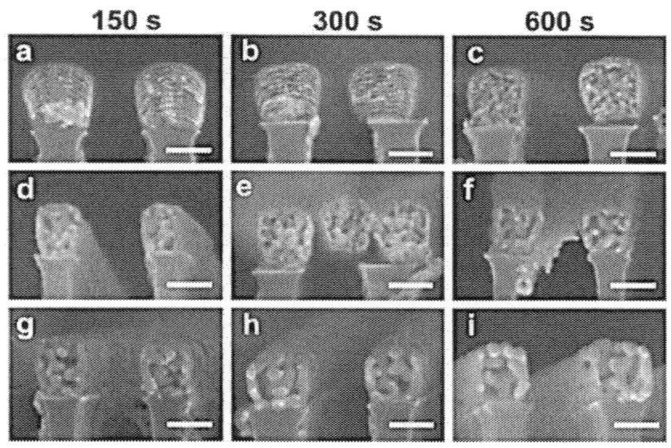

Figure 12: Cross-sectional SEM micrographs showing the morphological evolution of Au–Cu nanowires with an initial Au content of 11 atomic % and a diameter of 200 nm as a function of the dealloying time and potential. Different dealloying potentials were explored: (**a**–**c**) 0.2; (**d**–**f**) 0.3; and (**g**–**i**) 0.4 V. Scale bar: 100 nm. Reproduced with permission from [113]. Copyright American Chemical Society, 2016.

The nanorice and nanoporous architectures highlighted the key role played by the "hot-spots" in determining very high enhancement factors in SERS performances. Now, the question which arises concerns how to maximize the number of hot-spots in a single nanostructure. A nanorice architecture generates a low number of hot-spots. Nanoporous gold can create a very high number of hot-spots which, however, derive in an extended configuration by the interaction of several gold ligaments. Therefore, we are now looking for a single defined nanostructure which, thanks to the complexity of its geometry, can give rise in a confined spatial region to a very high number of hot-spots. As described in the next section, this is due to nanodendritic geometry.

METALLIC NANODENDRITES

The "lightning-rod" effect [68] is particularly pronounced for metal nanostructures with sharp tips and with small inter-gap size. In this sense, one of the most promising candidates for a highly tunable SERS system is the dendritic architecture. Metallic nanodendrites, defined as large fractal aggregates with hyperbranched architectures (see Figure 13) have attracted much attention due to their importance in understanding the fractal growth phenomena (to be distinguished from non-fractal structures such as compact or periodic geometries), and their potential applications in functional devices, plasmonics, biosensing and catalysis.

(a)

(b)

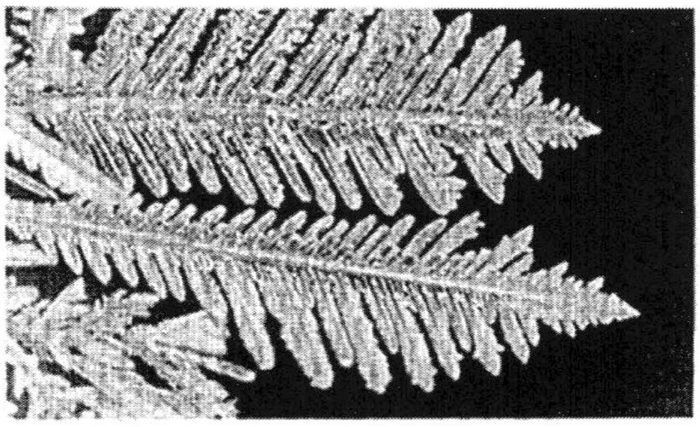

Figure 13: Dendritic structures obtained by electrochemical deposition of copper. (**a**) From 0.5 M $CuSO_4$/0.5 M H_2SO_4 at 350 mV (frame width = 2.5 mm); and (**b**) from 0.5 M $CuCl_2$ /0.5 M H_2SO_4 at 500 mV (frame width = 0.6 mm). Reproduced with permission from [114]. Copyright American Physical Society, 1995.

Fractal nanostructures can be formed by metals (Au [115,116], Pt [117,118], Pd [119,120], Rh [121,122], Ni [123], Ru [124,125]), metal oxides [126,127,128,129], and semiconductors [130,131,132,133,134,135,136]. Because of their higher structural complexity, compared to nanospheres, nanowires and nanodiscs, these branched nanostructures are expected to have a wide range of technological applications [135,137]. For example, they have interesting catalytic [114,115], magnetic [121,123,136], optical [115,135,136], and electronic [135,136] properties, and semiconductor heterostructures have potential uses in solar cells [136]. Branched nanostructures have also been suggested to be interconnections in the bottom-up self-assembly of future nanocircuits and nanodevices. Therefore, the mechanism by which these structures are formed [134,136] is a crucial step toward realizing synthetic control over their structural and functional properties. Synthetic control over the size and shape of these nanostructures is achieved using different chemical approaches. The controlled synthesis of these metal complex morphologies allows the modulation of the optical properties for each specific application.

To understand some optical behavior of metallic nanodendrites, we start from some consideration on the optical response of metal nanoparticles. Au and Ag nanoparticles show a characteristic absorption in the visible region of the electromagnetic spectrum due to the surface plasmon resonance. The position and width of the surface plasmon resonance band are considerably influenced by the size and shape of the nanocrystals. The wavelength dependence of the extinction coefficient is fairly well described by the Mie theory [62] for metal particles with sizes much smaller than the wavelength of light [127]. According to the Mie theory, the extinction coefficient is given by [138]

$$k = \frac{18\pi N V \varepsilon_m^{3/2}}{\lambda} \frac{\varepsilon_2}{[\varepsilon_1 + 2\varepsilon_m]^2 + \varepsilon_2^2} \tag{1}$$

where λ is the wavelength of incident light, and ε_1 and ε_2 are the real and imaginary parts of the complex dielectric function of the metal, and ε_m is the dielectric constant of the surrounding medium assumed to be an invariant with respect to the frequency of the light. The overall dielectric constant is given by

$$\varepsilon = \varepsilon_1 + i\varepsilon_2 \tag{2}$$

One can readily see that the dependence of the extinction coefficient on wavelength is Lorentzian. Moreover, it is known that anisotropic metallic nanoparticles normally exhibit two principal SPR absorption peaks, characteristic of the short (transverse band) and long (longitudinal band) axes. To obtain more information about the changes in the optical response that occurs from a regular to a complex morphology of metallic nanostructures, we report here the results of some experimental works. Huang *et al.* [138] show the UV-vis absorption spectrum of the as-prepared Au nanodendrites suspended in water (see inset in Figure 14). The spectrum displays a gradual increase in absorption from ~500 nm to the near-IR region without indication of leveling off, which could be attributed to the longitudinal plasmon band, indicating a remarkable overlapping between the transverse band and the longitudinal band. It is well known that the position and intensity of the longitudinal band depend largely on the size, aspect ratio and mutual coupling of Au nanocrystals. The authors suggest that the observed overlapping between the transverse band and the longitudinal band for the obtained Au nanodendrites could be attributed to the polydispersity in the length and diameter of the trunks and branches, which may lead to a variety of sizes and aspect ratios. The multiple coupling between neighboring trunks and side branches may also result in a longitudinal plasmon band with the position lying in a wide range of wavelengths. The observed absorption spectrum may thus represent a contour combining the size, aspect ratio and coupling factors of the hierarchical dendritic structures. Conversely, Kaniyankandy *et al.* [139] report the optical properties of silver nanodendrites, synthesized by electrodeposition using $AgNO_3$ as the source in ammoniacal solution. They report that SEM patterns of the samples show that as a consequence of the addition to the starting solution of a small amount of ammonia, the deposits (Ag B) show a dramatic change in morphology from irregularly shaped to long strands or rods. Furthermore, an increase of NH_3 (Ag D) does not greatly affect the morphology. The structural characterizations, shown in this work, indicate that the average value of thickness of the central stem of a dendrite is ~50 nm. Along the central stem (with an average length of ~5 μm), branching is visible for every ~50 nm, and in some cases it is also hyperbranched. The branches of the dendrites are predominantly planar and hyperbranched. To illustrate the changes of the optical properties due to the different morphology of the resulting nanostructures, the authors compare the results of the simulation with the results of the optical measurements (Figure 15).

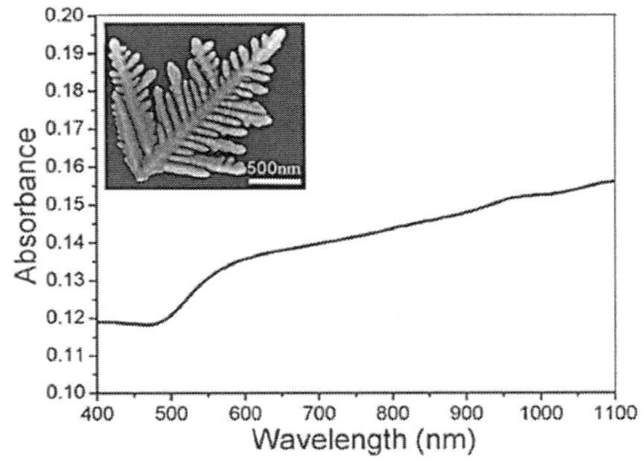

Figure 14: UV-vis absorption spectrum of Au nanodendrites. The inset shows the Au nanodendrites grown in mixed dodecyltrimethylammonium bromide/β-cyclodextrin solution. Reproduced with permission from [138]. Copyright American Chemical Society, 2010.

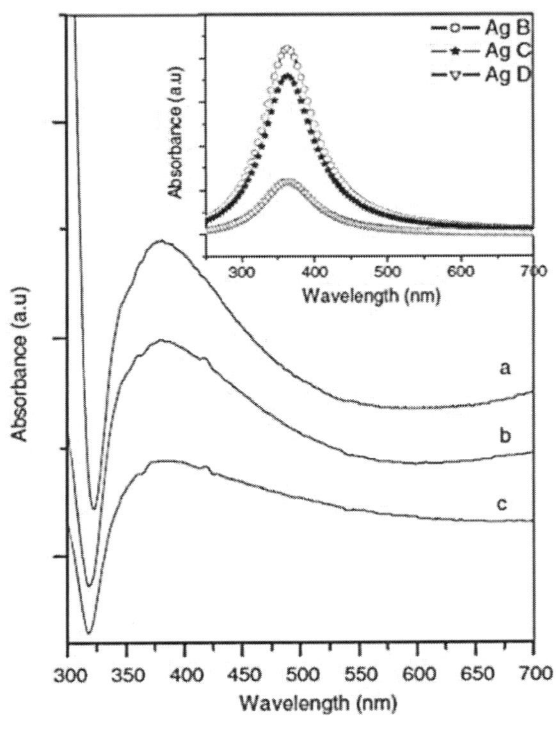

Figure 15: UV-vis absorption spectra of the samples: (**a**) Ag B, 0.3 M aqueous NH_3solution; (**b**) Ag C, 0.6 M aqueous NH_3 solution and (**c**) Ag D, 1.2 M aqueous NH_3solution. The inset reports the simulated absorption spectrum from the Mie theory for the samples. Reproduced with permission from [139]. Copyright Institute of Physics, 2007.

The simulation of the UV-vis spectra related to the metal colloids is shown in the inset of Figure 15. The authors highlight that the spectra show the Lorentzian nature of the band, and that the decrease in the size of nanoparticles leads to a decrease in the peak height and an increase in the peak width. The experimental UV-vis spectra (Figure 15) clearly give a single broad peak with maxima at $\lambda = 380$ nm. This indicates that the particles are predominantly spherical and do not have shape anisotropy. Additionally, the peak is highly asymmetric and has a large tail towards the lower energy region, *i.e.*, towards red. The peak extends to the entire visible region of the electromagnetic spectrum. This is contrary to the expected Lorentzian line shape. Kaniyankandy *et al.* explain that this result is due to the fact that, according to the Mie theory, metal nanoparticles are assumed to be non-interacting spheres. However, in their work, the metal nanoparticles are assembled in the form of a supramolecular dendritic structure. Therefore, a considerable amount of interaction is expected among the nanoparticles. Hence, the contribution from higher order multipoles and the distribution of depolarization factors have to be taken into consideration while simulating the absorption spectra. The distribution of depolarization factors in interacting metal spheres gives rise to a shape asymmetry for the SPR band. This has been previously observed by several authors [140]. The peak position at 380 nm is indicative of the particle sizes in the range 20–30 nm. Moreover, with an increase in the NH_3(hyperbranched nanostructures) content, the peak position shows only an extremely small shift towards the blue region, indicating a small decrease in the particle size, while the peak width increases with increasing ammonia content in the electrolyte. To explain these effects, the authors consider the size effects of metal nanoparticles. Depending on the size, there exist two regimes in the influence of particle size on the UV-vis spectrum [141]. The size regime is defined by the electron mean free path of the metal. For the bulk silver, the electron mean free path is ~52 nm. For particles smaller than the mean free path of the electrons, the influence of the grain boundary scattering cannot be ruled out, and must be taken into consideration. In the scenario when silver particle size is >52 nm, the peak position shows a red shift with increasing size, and the peak width increases with respect to the size (this is defined as an extrinsic size effect). For particle sizes lower than 52 nm, the peak position shows a very little change towards blue, and the peak width increases with decreasing size. Kaniyankandy *et al.* [138] indicate that this is the intrinsic size effect. In this case the size of all samples are in the range where the intrinsic size effect is operational. Therefore, the observation of the increase in the peak width with decreasing particle size can be explained on the basis of the intrinsic size effect. It is interesting to note that the SPR band spreads throughout the entire visible region of the electromagnetic spectrum. This observation is suggestive of aggregated silver nanoparticles, as well as in the present case where the aggregates form an ordered supramolecular structure. This hints at the applicability of these structures in SERS studies, because of an enhancement based on an electromagnetic mechanism, in which the high overlap of incident and scattered light with a metal SPR absorption band is important. The greater the overlap, the higher the enhancement. In the present case, since the SPR band extends throughout the entire visible region, the light of any wavelength in the visible region will be enhanced, thereby giving more flexibility with respect to the application in SERS. Furthermore, the highly branched dendritic structure has a higher surface area, which enables an increase in the sensitivity of SERS [142].

To highlight the good response of the metallic nanodendrites as SERS-active substrates, we report the results of Qiu *et al.*[143]. The authors fabricated highly ordered and regular silver nanodendrites (on a silicon substrate) with tunable inter-nanowire gap size (Figure 16a,b) for placing analyte molecules (C_{60} nanoclusters) in hot-spots between closely spaced nanowires (Figure 16c), leading to tunable SERS enhancement. In order to give experimental evidence for the hot-spots of the form "metal/nanoclusters/metal," the authors position many C_{60} nanoclusters as Raman probes in the junctions between the neighboring arms of silver nano-dendrites. Figure 16c shows the typical TEM image of C_{60} nanoclusters (the almost-circular regions between the two arms of the dendrite) coupled with silver nano-dendrite. C_{60} nanoclusters with sizes varying from 10 to 50 nm were clearly observed on the surface and between the arms of the dendrite, forming the "metal/nanoclusters/metal" structure. The SERS spectrum of such a system is reported in Figure 16d and it shows a variety of Raman bands (identified by the corresponding Raman shifts in Figure 16d). For comparison, the authors couple C_{60}nanoclusters on the silicon wafer, using a 1 mM C_{60} solution. No visible C_{60} Raman bands (see bottom of Figure 16d) are observed. The comparison, first of all, suggests that the strong enhancement for the dendritic pattern should be attributed to the fact that the nano-dendrites are assembled on the Si substrate with a very high density and with many horns and, thus, C_{60} nanoclusters in abundant hot-spots between closely spaced dendrite arms lead to intense SERS enhancement. The Raman spectrum of the C_{60} nanoclusters on the dendritic structures is analyzed by the authors supposing that the SERS process consists of a five-steps (as pictured in Figure 16d): (1) An ingoing photon interacts with the substrate exciting a plasmon; (2) The plasmon polarizes a molecule bound to the surface, creating a large effective dipole moment; (3) If the molecule now changes the vibrational state, then its molecular polarization alters; (4) The change in molecular polarization couples back into an emitted plasmon; (5) Finally, the plasmon couples with an outgoing Raman scattered photon. However, when the C_{60} nanoclusters are adsorbed on silver nano-dendrites, the symmetry of the C_{60} molecule is reduced, resulting in the splitting of the Raman bands from the degenerate modes of C_{60}. As a result, the number of the vibration modes is greatly increased in the SERS spectrum (Figure 16d). A similar argument was invoked by Nikoobakht and El-Sayed [144] who forwarded that SERS intensities are far higher for molecules on aggregated gold nanorod deposits compared with monomeric nanorods, an observation attributed to an enhanced electromagnetic (EM) field in the inter-nanorod region. These studies indicate that the precise control of gaps between nanorods or nanowires on a SERS-active substrate is likely to be critical for the fabrication of substrates with uniformly high EFs, and for understanding collective surface plasmons. This is the critical reason many works carried out theoretical examinations of the local EM properties by the simulation method to assess the EM near-field distributions for a dendrite-like model pattern with different inter-arm gap dimensions.

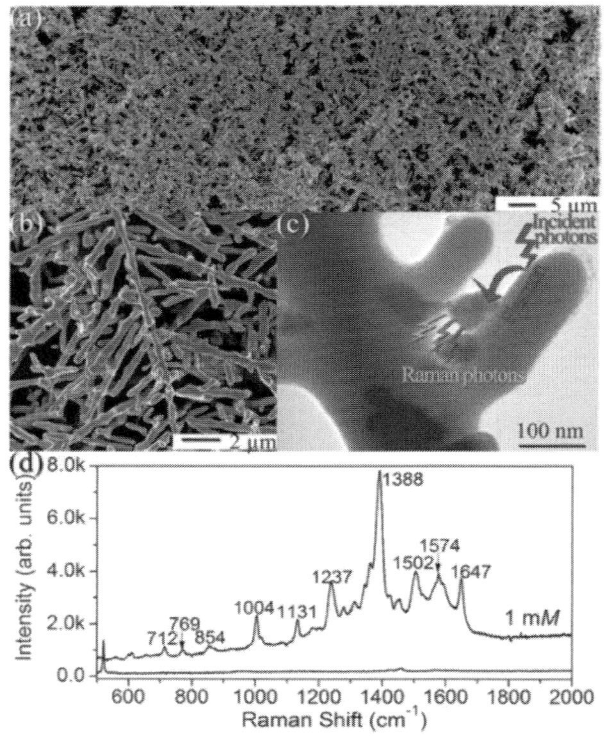

Figure 16: (**a**) Large-scale SEM image and (**b**) magnified image of silver nano-dendrites; (**c**) typical TEM image of C_{60} nanoclusters coupled with silver dendrite. The inset shows a schematic SERS process; light is coupled with a plasmon leading to an interaction and this is then Raman scattered by C_{60} nanoclusters on the surface of silver dendrite, and the outgoing plasmon is then scattered back into a photon; (**d**) SERS spectrum of C_{60} nanoclusters coupled with silver dendrites from a solution of 1 mM C_{60} in toluene. The Raman spectrum at the bottom corresponds to the sample of C_{60} nanoclusters coupled with Si wafer from 1 mM C_{60} toluene solution. Reproduced with permission from [143]. Copyright Institute of Physics, 2009.

The results of Qiu *et al.* therefore indicate that the nanodendritic architecture can be realized by multibranched metallic nanowires that act as efficient active SERS substrates. Regarding the one-dimensional nanostructure field, the next section aims to show that some interesting properties can be obtained from one-dimensional metal-based nanocomposite systems, demonstrating their potential as a basis for new-generation devices. The particular systems described aim to cross the localized surface plasmonic properties of metal nanoparticles with the waveguiding properties of silica nanowires.

GOLD NANOPARTICLE-EMBEDDED DIELECTRIC NANOWIRES

Noble-metal nanoparticles encapsulated in a dielectric matrix have attracted sustained interest over several centuries owing to their unusual optical and electrical properties. Metal-dielectric nanocomposites show non-linear and fast optical response near the SPR frequency due to their enhanced third-order optical susceptibilities [145]. Thus, they have been applied in optical switching devices [146,147]. Several approaches, such as the sol-gel process, metal-dielectric co-sputtering deposition, metal-ion implantation into a dielectric matrix and pulsed laser deposition have been used to prepare metal-dielectric nanocomposites. In addition to the mentioned metal-dielectric nanocomposites that are mainly prepared in thin film or bulk forms, the one-dimensional noble-metal nanoparticle chain has become an intensive research focus because of its potential applications for nanoscale-integrated optics below the diffraction limit of light. It has been suggested theoretically and experimentally that the electromagnetic energy can be transported below the diffraction limit by a coupled-plasmon mode in the linear chains of noble-metal nanoparticles. The successful synthesis of a one-dimensional hybrid nanosystem known as a "peapod" offers a promising opportunity to realize a wide variety of functionalities. The best-known example among them is the all-carbon hybrid peapod structure, consisting of fullerene molecules encapsulated in single-walled carbon nanotubes, which has been demonstrated to show several unusual physical properties.

In the paper of Hu *et al.* [145], the approach for fabricating a self-organized metal nanoparticle chain encapsulated in a dielectric nanowire with SPR-induced conductivity is reported. Cathodoluminescence measurements on the gold nanopeapodded silica nanowires are carried out using a Gatan with an acceleration voltage of 8 keV. Figure 17 shows typical cathodoluminescent spectra of gold nanopeapodded silica nanowires at five different temperatures.

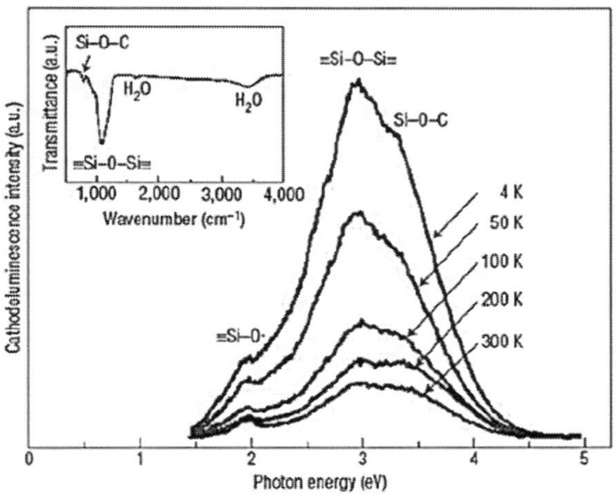

Figure 17: Temperature-dependent cathodoluminescent spectra of gold nano-peapodded silica nanowires. The inset shows the corresponding Fourier transform infrared spectroscopy (FTIR) spectrum. Reproduced with permission from [145]. Copyright Nature Publishing Group, 2006.

The authors highlight that the spectra are composed of three characteristic peaks at 1.9, 3.0, and 3.3 eV. The 1.9 eV band can be ascribed to the non-bridging oxygen hole centers or oxygen dangling bonds (≡Si–O). The 3.0 eV band corresponds to the two-fold coordinated Si lone pair centers originating from the formation of strain bonds (≡Si–O–Si≡), when mechanical stress is applied to the silica glass 22, 23. The 3.3 eV band may be correlated with the formation of the Si–O–C owing to the reaction between ≡Si–O defects and carbon atoms during nanowire growth, which has been reported in carbon-doped silica. As no carbon-containing gas reactants are intentionally introduced during nanowire growth, the authors suppose that the incorporated carbon might originate from the sputtering process of gold thin films under ambient air or the residual contaminant in their reaction chamber. Analyses of bonding configurations from FTIR data also confirm the existence of Si–O–C and ≡Si–O–Si≡ (Figure 17 inset).

Ultraviolet-visible absorption measurements are performed on a set of the composite nanowires grown on quartz substrate with different gold filling in the silica nanowires. The room-temperature ultraviolet-visible absorption spectra for plain silica nanowires (dotted line), gold-peapodded silica nanowires (solid line) and gold-filled silica nanowires, wherein the gold segments show an aspect ratio of 3–5 (dashed line), are shown in Figure 18. The gold-filled and gold-peapodded silica nanowires reveal a pronounced SPR absorption band at 532 nm, whereas the plain silica nanowires do not show any SPR absorption. The SPR band at 532 nm is attributed to the transverse plasmon mode, and the longer wavelength band is associated with the longitudinal plasmon mode, which was observed for composite nanowires containing gold segments of high aspect ratio (see the dashed line in the plot in Figure 18). The absence of the longitudinal mode in the gold-peapodded nanowires (see the solid line in the plot in Figure 18) may be ascribed to the small aspect ratio (<1.2) of the gold nanoparticles. The observed background in the composite nanowires originates from an interband transition of valence electrons to the Fermi surface.

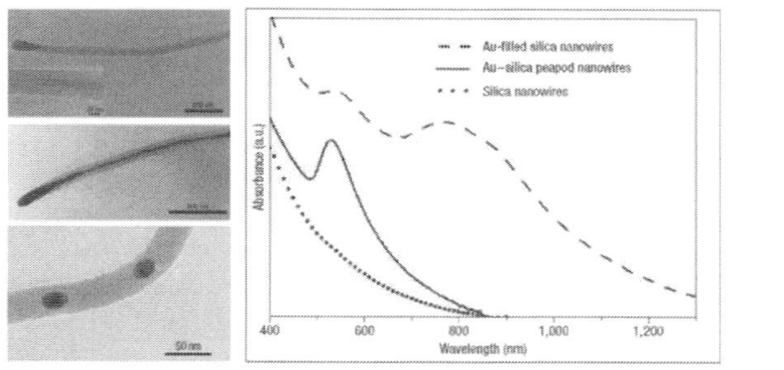

Figure 18: Optical absorption spectra for plain silica nanowires (dotted line), gold peapodded silica nanowires (solid line) and gold-filled silica nanowires wherein the aspect ratio of the gold segment is about 3–5 (dashed line). Reproduced with permission from [145]. Copyright Nature Publishing Group, 2006.

Photoresponse measurements based on the nanowire ensemble devices are conducted under alternate light illumination of different wavelengths and under dark conditions. The resistance of the hybrid peapod nanowire device shows nearly a five-fold smaller value than its counterpart without gold nanoparticles. More interestingly, the hybrid peapod nanowire device presents a wavelength-dependent photoresponse in resistance. The photoresponse behavior at an SPR wavelength of 532 nm shows a larger difference in resistance between the illumination on and off conditions (Figure 19).

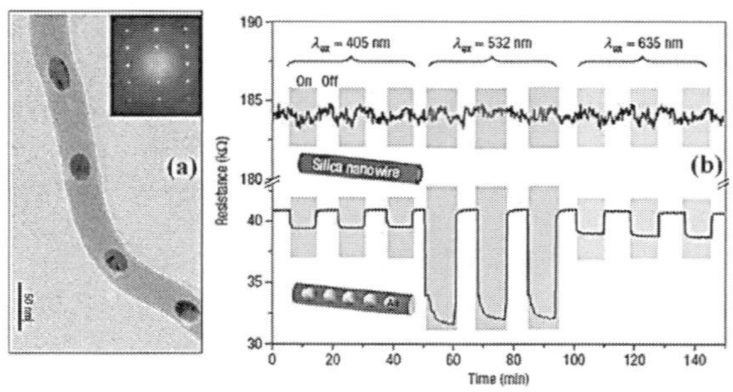

Figure 19: (**a**) TEM images showing the structure of metal nanoparticle-encapsulated silicon dioxide nanowire. The inset in in (**a**) shows

an electron-diffraction pattern recorded along the [120] zone axis; (**b**) Photoresponse measurements. The room-temperature resistance response as a function of time to light illumination for plain silica nanowires (**upper** part) and gold nanopeapodded silica nanowires (**lower** part). Shaded (pink, excitation wavelength lex = 635 nm; green, lex = 532 nm; purple, lex = 405 nm) and unshaded regions mark the light-on and light-off periods. Reproduced with permission from [145]. Copyright Nature Publishing Group, 2006.

The resistance change is nearly five times that observed using 632 and 405 nm light. Furthermore, an excitation-intensity-dependent resistance is measured using 532 nm light at various pumping powers. The authors believe that the enhanced photosensitive behavior observed from the hybrid nanowire device can be attributed to the SPR, as the hybrid nanowires have a pronounced SPR absorption at 532 nm. The enhancement of photoresponse behavior arising from the excitation of the surface plasmon has been reported for metal-oxide-metal (MOM) tunneling junctions, which is attributed to the generation of hot electrons owing to the decay of surface plasmon polaritons. The generated hot electrons drift or diffuse to the oxide barrier, and tunnel to the counter electrode. In the case of their study, one-dimensional gold nanopeapodded silica nanowire can be regarded as an MOM tunneling junction in series. The generation of hot electrons and electron tunneling through the oxide nanowires may contribute to the enhanced photoresponse behavior in our hybrid nanowire. The authors suggest that this simple one-step approach could be extended to the preparation of other metal-dielectric peapod-type nanowires with different functionalities.

Furthermore, SiO_2 nanowire (NW)-Au NPs composite systems are garnering great scientific and technological interest for their potential applications in, for example, electronic devices, biosensors, nanoscale optical devices and sensors. In particular, these pea-podded structures are used as light waveguides or biosensors, in line with the results reported by Wang *et al.* [146]. The authors report SPR-enhanced molecular oxygen sensing by single Au NPs-SiO_x NW under 532 nm illumination (visible light) at room temperature. Figure 20a,b show field-emission scanning electron microscopy images of two terminal single-NW devices consisting of a silica NW without Au NPs (a) and a silica NW embedded with Au NPs (b). To understand the basic electrical and optoelectronic properties of the single-NW devices, the conventional current–voltage (I–V) and photoresponse were measured at room temperature in vacuum by the authors (Figure 20c,d). As shown in Figure 20c, the I–V plots of both NWs are linear, indicating the ohmic nature of the e-beam-defined contacts. The slope of the I–V plot indicates that the electrical conductivity of the Au NP-embedded silica NW is higher than that of the bare silica NW.

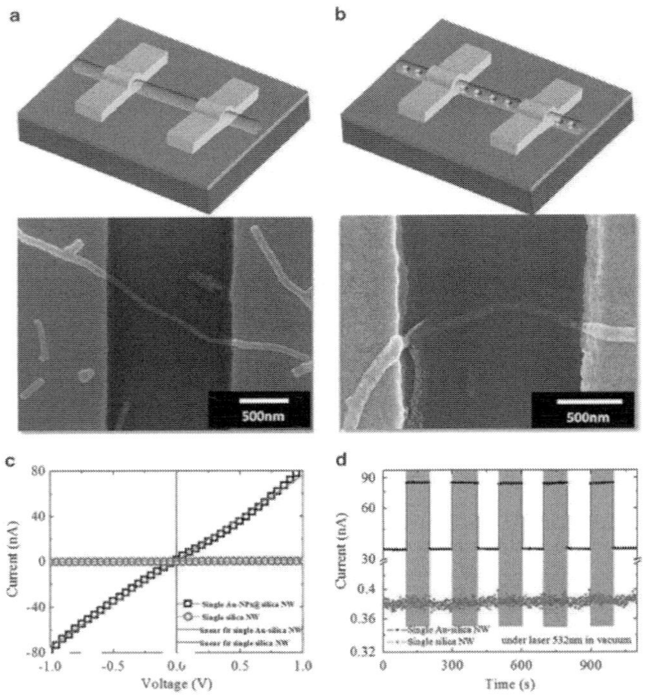

Figure 20: Field-emission scanning electron microscopy images of nanowire (NW) devices consisting of (**a**) plain silica NW (without the Au NPs) and (**b**) Au NPs-silica NW; (**c**) Dark I–V characteristics of single NW devices, with and without Au NP peapods, measured in vacuum, and the corresponding linear fit to the dataset; (**d**) Dark and photocurrent measured under vacuum in single NW devices, with and without Au NP peapods, as a function of time. The color bars indicate the duration of the 532 nm illumination. The measurement has been performed with a 1 V applied bias at room temperature. The bare silica NW showed no photoresponse, whereas the Au NPs-silica NW showed strong photoresponse. Reproduced with permission from [146]. Copyright Nature Publishing Group, 2013.

This trend occurs because the creation of more electron-hopping sites in the Au NPs-silica NW leads to enhanced conductivity [147]. During the growth of Au NP-peapodded silica NWs, residual Au atoms may partially incorporate into the silica matrix of the wire during the peapod formation. The incorporation of Au may produce impurities and/or defect-related hopping sites in the silica matrix [147]. The Au-embedded silica NW has higher conductivity than the plain silica NW, because of its shorter hopping length. The photoresponse measurements of the NW devices, conducted under alternating 532 nm illumination and dark conditions, are shown in Figure 20d. The plain silica NW device does not show any significant photoresponse; the ratio of photo and dark current is ~0.01. In contrast, the Au NPs-silica NW device presents a reversible photoresponse and a stable photocurrent when illuminated by 532 nm wavelength radiation (in Figure 20d the color bars indicate the duration of the 532 nm illumination), with a photo/dark current ratio of ~1.48. Assuming the band gap of silica NWs to be in the range of ~6.8 eV, and denying any absorption of the 532 nm light, these data suggest that the photocurrent can be attributed to the near-resonance SP absorption of the Au NPs (within the silica NWs) that has been reported at ~532 nm. So, under 532 nm illumination, the gold nanoparticles absorb photons so that electrons are (plasmonically) excited becoming charge carriers which can be transmitted through the silica nanowires by hopping processes between neighboring Au

NPs. Taking into account these optoelectronic properties, such complex systems can be used as a light waveguide. Indeed, the resonant coupling between the incident light and the plasmonic resonances of Au NPs can induce an enhancement of the output signal.

CONCLUSIONS AND PERSPECTIVES

In this paper, we reviewed the basic concepts related to some fabrication methods and the physico-chemical properties of complex-morphology metal nanostructures, focusing on the dependence of these properties on the specific nanostructure morphology. The main highlight of the review is that the fine wide-range control of the metal nanostructures' morphological characteristics allow the wide-range tuning of their properties such as the electrical or optical ones. We tried to highlight that assemblies of complex-morphology metal nanostructures present intriguing properties, often more effective and versatile with respect to the properties of assemblies of spherical metal nanostructures. Therefore, in addition to chemical composition and size, the shape of the metal nanostructures is demonstrated to be a further fundamental parameter for reaching a wider range control of the properties of the systems towards functional technological applications in areas such as sensors, energy conversion, flexible electronics, optoelectronics, *etc*. Starting from the description of some peculiar cases (gold nanorings, gold nanorods, nanoporous gold structures, gold and silver nanodendrites, silica nanowires embedded with gold nanoparticles) we highlighted the disruptive role of the symmetry breaking in the metal nanostructure morphology in reaching unexplored nanomaterial behaviors. Due to the many unique characteristics of the complex-morphology metal nanostructures, we believe that the detailed understanding of the basic physical phenomena involved in the materials synthesis and electron transport and interaction with electromagnetic radiation can allow the desired control over properties and applications. The review thus emphasizes the basic microscopic mechanisms and processes and the general physical concepts suitable for the material properties' interpretation and the structure-property correlations. Besides the basic processes and the general concepts, this review aims at a comprehensive schematization of the main technological applications currently in development somewhere in the world so as the future possible ones.

Some perspectives can be drawn on the basis of the described systems and properties throughout the paper. In particular, we believe that the more interesting perspectives can be argued looking at the properties (and to the microscopic origin) of the illustrated complex-morphology metal-based nanostructures and thinking how these properties, presented singularly by the specific complex-morphology structure, can be crossed simultaneously in a unique nanostructure. Consequently, the prospective to realize some type of complex-morphology nanostructure originating from the complex-morphology characteristics of the two or more structures presented above aimed to cross the complex morphologies and meld the corresponding properties and performances. We briefly describe this idea through practical examples and propose the following:

a. **Hybrid nanostructures**: crossing the properties of Au nanorings and Au nanodendrities (see, for example, the possible architecture in Figure 21 (plan-view)). Possibly, such a type of structure could present LSPR frequency strongly dependent on the ratio of the thickness to ring outer radius and could simultaneously take advantage of the hot-spots for the electromagnetic regions originating in the inter-gap spaces between the sharp tips. So what can we expect from such a structure? Reasonably, very high surface-enhanced Raman scattering enhancement factors at a wavelength established by the ratio of the thickness to ring outer radius.

Figure 21: Picture of a possible complex-morphology Au nanostructure (plan-view) crossing the properties of Au nanorings with the properties of sharp Au tips typical of nanodendritic structures.

b. **Nanoporous Au nanodendrites** (see the picture in Figure 22): crossing the properties of nanoporous Au and Au nanodendrites. In this case, the SERS effects characteristic of the nanodendritic Au structures (due to hot-spot engineering) could possibly be enormously enhanced thanks to the very exposed surface of the porous material where the analytes can connect and determine an increased sensitivity and efficiency.

Figure 22: Picture of a possible nanoporous Au nanodendrites.

c. **Branched (*i.e.*, fractal-type) silica nanowires embedded with Au nanoparticles** (see Figure 23): In this case, we believe that properties of the silica nanowires embedded with Au nanoparticles, such as their photoconductivity, take advantage of the branched shape due to the occurrence of the hot-spots and, hence, of the consequent amplification of the electromagnetic field in between a larger number of in-gap regions between nanoparticles.

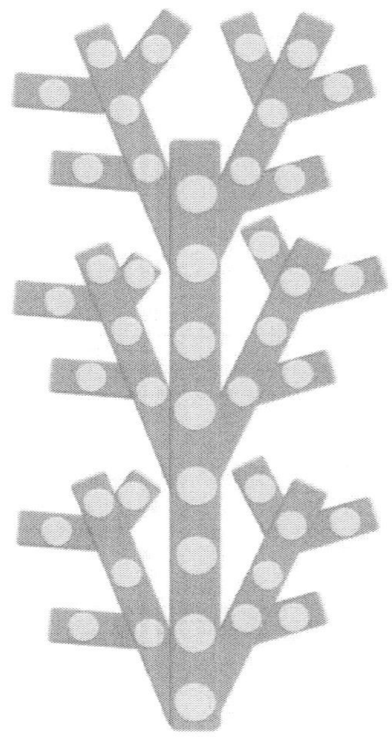

Figure 23: Picture of a possible branched silica nanowires embedded with Au nanoparticles.

In addition to these fascinating complex-morphology metal-based nanostructures proposed as perspectives, nano-architectures to further improve efficiency with respect to electromagnetic-based phenomena, other possibilities include: porous gold nanoparticle- embedded silica nanowires, a system as pictured in Figure 21 but composed by nanoporous gold, multi-sharpened nanorices, *etc*. In general, the perspective idea can be: crossing specific morphologies in a unique structure (a truly hierarchically concept) to cross the properties and efficiencies (with respect to some specific process) to obtain, through mutual combination, un-precedented properties and higher efficiency with respect to the efficiencies of the original structures separately.

REFERENCES

1. Fahlman, B.D. *Materials Chemistry*; Springer: New York, NY, USA, 2007; Volume 1.
2. Xia, Y.; Xiong, Y.; Lim, B.; Skrabalak, S.E. Shape-controlled synthesis of metal nanocrystals: Simple chemistry meets complex physics? *Angew. Chem. Int. Ed. Engl.* 2009, *48*, 60–103.
3. Halperin, W.P. Quantum size effects in metal particles. *Rev. Mod. Phys.* 1986, *58*, 533–607.
4. Bawendi, M.G.; Steigerwald, M.L.; Brus, L.E. The quantum mechanics of larger semiconductor clusters ("quantum dots"). *Annu. Rev. Phys. Chem.* 1990, *41*, 477–496.
5. Weller, H. Quantized Semiconductor Particles: A novel state of matter for materials science. *Adv. Mater.* 1993, *5*, 88–95.
6. Alivisatos, A.P. Semiconductor Clusters, Nanocrystals, and Quantum Dots. *Science* 1996, *271*, 933–937.
7. Murray, C.B.; Kagan, C.R.; Bawendi, M.G. Synthesis and characterization of monodisperse nanocrystals and close-packed nanocrystals assemblies. *Annu. Rev. Mater. Sci.* 2000, *30*, 545–610.
8. Maheshwari, V.; Kane, J.; Saraf, R. Self-Assembly of a Micrometers-Long One-Dimensional Network of Cemented Au Nanoparticles. *Adv. Mater.* 2008, *20*, 284–287.
9. Mott, N.F. Metal-Insulator Transition. *Rev. Mod. Phys.* 1968, *40*, 677–683.
10. Markovich, G.; Collier, C.P.; Henrichs, S.E.; Remacle, F.; Levine, R.D.; Heath, J.R. Architectonic Quantum Dot Solids.*Acc. Chem. Res.* 1999, *32*, 415–423.
11. Jeong, U.; Teng, X.; Wang, Y.; Yang, H.; Xia, Y. Superparamagnetic Colloids: Controlled Synthesis and Niche Applications. *Adv. Mater.* 2007, *19*, 33–60.
12. Xiong, Y.; Lu, X. *Metallic Nanostructures: From Controlled Synthesis to Applications*; Springer: New York, NY, USA, 2015.
13. Mishra, Y.K.; Kabiraj, D.; Sulania, I.; Pivin, J.C.; Avasthi, D.K. Synthesis and characterization of gold nanorings. *J. Nanosci. Nanotechnol.* 2007, *7*, 1878–1881.
14. Mohapatra, S.; Mishra, Y.K.; Ghatak, J.; Kabiraj, D.; Avasthi, D.K. Surface plasmon resonance of Ag nanoparticles embedded in partially oxidized amorphous Si matrix. *J. Nanosci. Nanotechnol.* 2008, *8*, 4285–4289.
15. Mishra, Y.K.; Mohapatra, S.; Chakravadhanula, V.S.K.; Lalla, N.P.; Zaporojtchenko, V.; Avasthi, D.K.; Faupel, F. Synthesis and Characterization of Ag-Polymer Nanocomposites. *J. Nanosci. Nanotechnol.* 2010, *10*, 2833–2837.
16. Mohapatra, S.; Mishra, Y.K.; Warrier, A.M.; Philip, R.; Sahoo, S.; Arora, A.K.; Avasthi, D.K. Plasmonic, Low-Frequency Raman, and Nonlinear Optical-Limiting Studies in Copper–Silica Nanocomposites. *Plasmonics* 2012, *7*, 25–31.
17. Mishra, Y.K.; Adelung, R.; Kumar, G.; Elbahri, M.; Mohapatra, S.; Singhal, R.; Tripathi, A.; Avasthi, D.K. Formation of Self-organized Silver Nanocup-Type Structures and Their Plasmonic Absorption. *Plasmonics* 2013, *8*, 811–815.
18. Kumar, M.; Suchand Sandeep, C.S.; Kumar, G.; Mishra, Y.K.; Philip, R.; Reddy, G.B. Plasmonic and Nonlinear Optical Absorption Properties of Ag:ZrO_2 Nanocomposite Thin Films. *Plasmonics* 2014, *9*, 129–136.
19. Mishra, Y.K.; Mohapatra, S.; Singhal, R.; Avasthi, D.K.; Agarwal, D.C.; Ogale, S.B. Au–ZnO: A tunable localized surface plasmonic nanocomposite. *Appl. Phys. Lett.* 2008, *92*.
20. Singhal, R.; Agarwal, D.C.; Mohapatra, S.; Mishra, Y.K.; Kabiraj, D.; Singh, F.; Avasthi, D.K.; Chawla, A.K.; Chandra, R.; Mattei, G.; *et al.* Synthesis and characterizations of silver-fullerene C_{70} nanocomposite. *Appl. Phys. Lett.* 2008, *93*.

21. Mishra, Y.K.; Chakravadhanula, V.S.K.; Hrkac, V.; Jebril, S.; Agarwal, D.C.; Mohapatra, S.; Avasthi, D.K.; Kienle, L.; Adelung, R. Crystal growth behaviour in Au-ZnO nanocomposite under different annealing environments and photoswitchability. *J. Appl. Phys.* 2012, *112*.

22. Mishra, Y.K.; Mohapatra, S.; Kabiraj, D.; Mohanta, B.; Lalla, N.P.; Pivin, J.C.; Avasthi, D.K. Synthesis and characterization of Ag nanoparticles in silica matrix by atom beam sputtering. *Scr. Mater.* 2007, *56*, 629–632.

23. Mishra, Y.K.; Mohapatra, S.; Avasthi, D.K.; Kabiraj, D.; Lalla, N.P.; Pivin, J.C.; Sharma, H.; Kar, R.; Singh, N. Gold–silica nanocomposites for the detection of human ovarian cancer cells: A preliminary study. *Nanotechnology* 2007, *18*.

24. Chakravadhanula, V.S.K.; Mishra, Y.K.; Kotnur, V.G.; Avasthi, D.K.; Strunkus, T.; Zaporojtchenko, V.; Fink, D.; Kienle, L.; Faupel, F. Microstructural and plasmonic modifications in Ag-TiO_2 and Au-TiO_2 nanocomposites through ion beam irradiation. *Beilstein J. Nanotechnol.* 2014, *5*, 1419–1431.

25. Ruffino, F.; De Bastiani, R.; Grimaldi, M.G.; Bongiorno, C.; Giannazzo, F.; Roccaforte, F.; Spinella, C.; Raineri, V. Self-organization of Au nanoclusters on the SiO_2 surface induced by 200 keV-Ar^+ irradiation. *Nucl. Instrum. Methods Phys. Res. B* 2007, *257*, 810–814.

26. Ruffino, F.; Torrisi, V.; Marletta, G.; Grimaldi, M.G. Kinetic growth mechanisms of sputter-deposited Au films on mica: From nanoclusters to nanostructured microclusters. *Appl. Phys. A* 2010, *100*, 7–13.

27. Ruffino, F.; Grimaldi, M.G. Island-to-percolation transition during the room-temperature growth of sputtered nanoscale Pd films on hexagonal SiC. *J. Appl. Phys.* 2010, *107*.

28. Ruffino, F.; Torrisi, V.; Marletta, G.; Grimaldi, M.G. Atomic force microscopy investigation of the kinetic growth mechanisms of sputtered nanostructured Au film on mica: Towards a nanoscale morphology control. *Nanoscale Res. Lett.* 2011, *6*, 112.

29. Ruffino, F.; Pugliara, A.; Carria, E.; Bongiorno, C.; Spinella, C.; Grimaldi, M.G. Formation of nanoparticles from laser irradiated Au thin film on SiO_2/Si: Elucidating the Rayleigh-instability role. *Mater. Lett.* 2012, *84*, 27–30.

30. Lewis, L.N. Chemical catalysis by colloids and clusters. *Chem. Rev.* 1993, *93*, 2693–2730.

31. Davey, N.M.; Seymour, R.J. The Platinum Metals in Electronics. *Platin. Met. Rev.* 1985, *29*, 2–11.

32. Hamilton, J.F. The silver halide photographic process. *Adv. Phys.* 1988, *37*, 359–441.

33. Murray, C.B.; Sun, S.; Doyle, H.; Betley, T. Monodisperse 3*d* Transition-Metal (Co,Ni,Fe) Nanoparticles and Their Assembly intoNanoparticle Superlattices. *MRS Bull.* 2001, *26*, 985–991.

34. Ditlbacher, H.; Krenn, J.R.; Lamprecht, B.; Leitner, A.; Aussenegg, F.R. Spectrally coded optical data storage by metal nanoparticles. *Opt. Lett.* 2000, *25*, 563–565.

35. Maier, S.A.; Brongersma, M.L.; Kik, P.G.; Meltzer, S.; Requicha, A.A.G.; Atwater, H.A. Plasmonics—A Route to Nanoscale Optical Devices. *Adv. Mater.* 2001, *13*, 1501–1505.

36. Sanders, A.W.; Routenberg, D.A.; Wiley, B.J.; Xia, Y.; Dufresne, E.R.; Reed, M.A. Observation of Plasmon Propagation, Redirection, and Fan-Out in Silver Nanowires. *Nano Lett.* 2006, *6*, 1822–1826.

37. Taton, T.A.; Mirkin, C.A.; Letsinger, R.L. Scanometric DNA array detection with nanoparticle probes. *Science* 2000,*289*, 1757–1760.

38. Chen, J.; Saeki, F.; Wiley, B.J.; Cang, H.; Cobb, M.J.; Li, Z.Y.; Au, L.; Zhang, H.; Kimmey, M.B.; Li, X.; *et al.* Gold nanocages: Bioconjugation and their potential use as optical imaging contrast agents. *Nano Lett.* 2005, *5*, 473–477.

39. West, J.L.; Halas, N.J. Engineered nanomaterials for biophotonics applications: Improving sensing, imaging, and therapeutics. *Annu. Rev. Biomed. Eng.* 2003, *5*, 285–292.

40. Kottmann, J.P.; Martin, O.J.F.; Smith, D.R.; Schultz, S. Plasmon resonances of silver nanowires with a nonregular cross section. *Phys. Rev. B* 2001, *64*.

41. Valden, M.; Lai, X.; Goodman, D.W. Onset of Catalytic Activity of Gold Clusters on Titania with the Appearance of Nonmetallic Properties. *Science* 1998, *281*, 1647–1650.

42. Narayanan, R.; El-Sayed, M.A. Catalysis with Transition Metal Nanoparticles in Colloidal Solution: Nanoparticle Shape Dependence and Stability. *J. Phys. Chem. B* 2005, *109*, 12663–12676.

43. Falicov, L.M.; Somorjai, G.A. Correlation between catalytic activity and bonding and coordination number of atoms and molecules on transition metal surfaces: Theory and experimental evidence. *Proc. Natl. Acad. Sci. USA* 2005, *82*, 2207–2211.

44. Fan, Z.; Huang, X.; Tan, C.; Zhang, H. Thin metal nanostructures: Synthesis, properties and applications. *Chem. Sci.*2015, *6*, 95–111.

45. Elechiguerra, J.L.; Reyes-Gasga, J.; Yacaman, M.J. The role of twinning in shape evolution of anisotropic noble metal nanostructures. *J. Mater. Chem.* 2006, *16*, 3906–3919.

46. Zhang, J.; Langille, M.R.; Mirkin, C.A. Photomediated Synthesis of Silver Triangular Bipyramids and Prisms: The Effect of pH and BSPP. *J. Am. Chem. Soc.* 2010, *132*, 12502–12510.

47. Liu, M.; Leng, M.; Yu, C.; Wang, X.; Wang, C. Selective Synthesis of Hexagonal Ag Nanoplates in a Solution-Phase Chemical Reduction Process. *Nano Res.* 2010, *3*, 843–851.

48. Zhang, Q.; Ge, J.; Pham, T.; Goebl, J.; Hu, Y.; Lu, Z.; Yin, Y. Reconstruction of Silver Nanoplates by UV Irradiation: Tailored Optical Properties and Enhanced Stability. *Angew. Chem. Int. Ed.* 2009, *48*, 3516–3519.

49. Zhou, J.; An, J.; Tang, B.; Xu, S.; Cao, Y.; Zhao, B.; Xu, W.; Chang, J.; Lombardi, J.R. Growth of Tetrahedral Silver Nanocrystals in Aqueous Solution and Their SERS Enhancement. *Langmuir* 2008, *24*, 10407–10413.

50. Zhang, J.; Li, S.; Wu, J.; Schatz, G.C.; Mirkin, C.A. Plasmon-Mediated Synthesis of Silver Triangular Bipyramids.*Angew. Chem. Int. Ed.* 2009, *121*.

51. Sau, T.K.; Rogach, A.L. *Complex-Shaped Metal Nanoparticles: Bottom-up Synthesis and Applications*; Wiley-VCH: Weinheim, Germany, 2012.

52. Bratlie, K.M.; Lee, H.; Komvopoulos, K.; Yang, P.; Somorjai, G.A. Platinum Nanoparticle Shape Effects on Benzene Hydrogenation Selectivity. *Nano Lett.* 2007, *7*, 3097–3101.

53. Pastoriza-Santos, I.; Liz-Marzán, L.M. Colloidal silver nanoplates. State of the art and future challenges. *J. Mater. Chem.* 2008, *18*, 1724–1737.

54. Tao, A.; Sinsermsuksakul, P.; Yang, P. Polyhedral Silver Nanocrystals with Distinct Scattering Signatures. *Angew. Chem. Int. Ed.* 2006, *45*, 4597–4601.

55. Tao, A.; Sinsermsuksakul, P.; Yang, P. Tunable plasmonic lattices of silver nanocrystals. *Nat. Nanotechnol.* 2007, *2*, 435–440.

56. Yoshida, K.-I.; Itoh, T.; Tamaru, H.; Biju, V.; Ishikawa, M.; Ozaki, Y. Quantitative evaluation of electromagnetic enhancement in surface-enhanced response Raman scattering from plasmonic properties and morphologies of individual Ag nanostructures. *Phys. Rev. B* 2010, *81*.

57. Mulvihill, M.; Tao, A.; Benjauthrit, K.; Arnold, J.; Yang, P. Surface-Enhanced Raman Spectroscopy for Trace Arsenic Detection in Contaminated Water. *Angew. Chem. Int. Ed.* 2008, *47*, 6456–6460.

58. Maier, S.A. *Plasmonic: Fundamentals and Applications*; Springer: New York, NY, USA, 2007.

59. Mock, J.J.; Barbic, M.; Smith, D.R.; Schultz, D.A.; Schultz, S. Shape effects in plasmon resonance of individual colloidal silver nanoparticles. *J. Chem. Phys.* 2002, *116*, 6755–6759.

60. Kuwata, H.; Tamaru, H.; Esumi, K.; Miyano, K. Resonant light scattering from metal nanoparticles: Practical analysis beyond Rayleigh approximation. *Appl. Phys. Lett.* 2003, *83*, 4625–4627.

61. Krenn, J.R.; Salerno, M.; Félidj, N.; Lamprecht, B.; Schider, G.; Leitner, A.; Aussenegg, F.R.; Weeber, J.C.; Dereux, A.; Goudonnet, J.P. Light field propagation by metal micro- and nanostructures. *J. Microsc.* 2001, *202*, 122–128.

62. Maier, S.A.; Brongersma, M.L.; Kik, P.G.; Atwater, H.A. Observation of near-field coupling in metal nanoparticle chains using far-field polarization spectroscopy. *Phys. Rev. B* 2002, *65*.

63. Lamprecht, B.; Schider, G.; Ditlbacher, R.T.; Lechner, H.; Krenn, J.R.; Leitner, A.; Aussenegg, F.R.; Weeber, J.C. Metal nanoparticle gratings: Influence of dipolar particle interaction on the plasmon resonance. *Phys. Rev. Lett.* 2000, *84*, 4721–4724.

64. Banaee, M.G.; Crozier, K.B. Gold nanorings as substrates for surface-enhanced Raman scattering. *Opt. Lett.* 2010, *35*, 760–762.

65. Novotny, L. Nano-optics: Optical antennas tuned to pitch. *Nature* 2008, *455*.

66. Kneipp, K.; Moskovits, M.; Kneipp, H. *Surface-Enhanced Raman Scattering: Physics and Applications*; Springer: Berlin, Germany; New York, NY, USA, 2006; Volume 103.

67. Dieringer, J.A.; Lettan, R.B., II; Scheidt, K.A.; Van Duyne, R.P. A Frequency Domain Existence Proof of Single-Molecule Surface-Enhanced Raman Spectroscopy. *J. Am. Chem. Soc.* 2007, *129*, 16249–16256.

68. Kang, J.H.; Kim, D.S.; Park, Q.-H. Local Capacitor Model for Plasmonic Electric Field Enhancement. *Phys. Rev. Lett.*2009, *102*.

69. Aizpurua, J.; Hanarp, P.; Sutherland, D.S.; Kall, M.; Bryant, G.W.; Garcia de Abajo, F.J. Optical Properties of Gold Nanorings. *Phys. Rev. Lett.* 2003, *90*.

70. Talley, C.E.; Jackson, J.B.; Oubre, C.; Grady, N.K.; Hollars, C.W.; Lane, S.M.; Huser, T.R.; Nordlander, P.; Halas, N.J. Surface-Enhanced Raman Scattering from Individual Au Nanoparticles and Nanoparticle Dimer Substrates. *Nano Lett.*2005, *5*, 1569–1574.

71. Babayan, Y.; McMahon, J.M.; Li, S.; Gray, S.K.; Schatz, G.C.; Odom, T.W. Confining Standing Waves in Optical Corrals.*ACS Nano* 2009, *3*, 615–620.

72. Zheludev, N.I.; Prosvirnin, S.L.; Papasimakis, N.; Fedotov, V.A. Lasing spaser. *Nat. Photonics* 2008, *2*, 351–354.

73. Wang, H.; Brandl, D.W.; Fei, L.; Nordlander, P.; Halas, N.J. Nanorice: A Hybrid Plasmonic Nanostructure. *Nano Lett.*2006, *6*, 827–832.

74. Oldenburg, S.J.; Averitt, R.D.; Westcott, S.L.; Halas, N.J. Nanoengineering of optical resonances. *Chem. Phys. Lett.* 1998,*288*, 243–247.

75. Oubre, C.; Nordlander, P. Optical Properties of Metallodielectric Nanostructures Calculated Using the Finite Difference Time Domain Method. *J. Phys. Chem. B* 2004, *108*, 17740–17747.

76. Schuck, P.J.; Fromm, D.P.; Sundaramurthy, A.; Kino, G.S.; Moerner, W.E. Improving the Mismatch between Light and Nanoscale Objects with Gold Bowtie Nanoantennas. *Phys. Rev. Lett.* 2005, *94*.

77. Sanchez, E.J.; Novotny, L.; Xie, S.X. Near-Field Fluorescence Microscopy Based on Two-Photon Excitation with Metal Tips. *Phys. Rev. Lett.* 1999, *82*, 4014–4017.

78. Hartschuh, A.; Sanchez, E.J.; Xie, S.X.; Novotny, L. High-Resolution Near-Field Raman Microscopy of Single-Walled Carbon Nanotubes. *Phys. Rev. Lett.* 2003, *90*.

79. Michaels, A.M.; Jiang, J.; Brus, L. Ag Nanocrystal Junctions as the Site for Surface-Enhanced Raman Scattering of Single Rhodamine 6G Molecules. *J. Phys. Chem. B* 2000, *104*, 11965–11971.

80. Nordlander, P.; Oubre, C.; Prodan, E.; Li, K.; Stockman, M.I. Plasmon Hybridization in Nanoparticle Dimers. *Nano Lett.* 2004, *4*, 899–903.

81. Qian, L.H.; Chen, M.W. Ultrafine nanoporous gold by low-temperature dealloying and kinetics of nanopore formation. *Appl. Phys. Lett.* 2007, *91*.

82. Forty, A.J. Corrosion micromorphology of noble metal alloys and depletion gilding. *Nature* 1979, *282*, 597–598.

83. Pickering, H.W. Characteristic features of alloy polarization curves. *Corros. Sci.* 1983, *23*, 1107–1120.

84. Oppenheim, C.; Trevor, D.J.; Chidsey, C.E.D.; Trevor, P.L.; Sieradzki, K. *In Situ* Scanning Tunneling Microscopy of Corrosion of Silver-Gold Alloys. *Science* 1991, *254*, 687–689.

85. Erlebacher, J.; Aziz, M.J.; Karma, A.; Dimitrov, N.; Sieradzki, K. Evolution of nanoporosity in dealloying. *Nature* 2001,*410*, 450–453.

86. Biener, J.; Nyce, G.W.; Hodge, A.M.; Biener, M.M.; Hamza, A.V.; Maier, S.A. Nanoporous Plasmonic Metamaterials.*Adv. Mater.* 2008, *20*, 1211–1217.

87. Senior, N.A.; Newman, R.C. Synthesis of tough nanoporous metals by controlled electrolytic dealloying. *Nanotechnology* 2006, *17*, 2311–2316.

88. Fujita, T.; Qian, L.H.; Inoke, K.; Erlebacher, J.; Chen, M.W. Three-dimensional morphology of nanoporous gold. *Appl. Phys. Lett.* 2008, *92*.

89. Ruffato, G.; Romanato, F.; Garoli, D.; Cattarin, S. Nanoporous gold plasmonic structures for sensing applications. *Opt. Express* 2011, *19*, 13164–13170.

90. Maaroof, A.I.; Gentle, A.; Smith, G.B.; Cortie, M.B. Bulk and surface plasmons in highly nanoporous gold films. *J. Phys. D* 2007, *40*, 5675–5682.

91. Collinson, M.M. Nanoporous Gold Electrodes and Their Applications in Analytical Chemistry. *ISRN Anal. Chem.* 2013,*2013*.

92. Kneipp, K.; Haka, A.S.; Kneipp, H.; Badizadegan, K.; Yoshizawa, N.; Boone, C.; Shafer-Peltier, K.E.; Motz, J.T.; Dasari, R.R.; Feld, M.S. Surface-Enhanced Raman Spectroscopy in Single Living Cells Using Gold Nanoparticles. *Appl. Spectrosc.* 2002, *56*, 150–154.

93. Dixon, M.C.; Daniel, T.A.; Hieda, M.; Smilgies, D.M.; Chan, M.H.W.; Allara, D.L. Preparation, structure, and optical properties of nanoporous gold thin films. *Langmuir* 2007, *23*, 2414–2422.

94. Kucheyev, S.O.; Hayes, J.R.; Biener, J.; Huse, T.; Talley, C.E.; Hamza, A.V. Surface-enhanced Raman scattering on nanoporous Au. *Appl. Phys. Lett.* 2006, *89*.

95. Chen, H.-A.; Long, J.-L.; Lin, Y.-H.; Weng, C.-J.; Lin, H.-N. Plasmonic properties of a nanoporous gold film investigated by far-field and near-field optical techniques. *J. Appl. Phys.* 2011, *110*.

96. Qian, L.H.; Yan, X.Q.; Fujita, T.; Inoue, A.; Chen, M.W. Surface enhanced Raman scattering of nanoporous gold: Smaller pore sizes stronger enhancements. *Appl. Phys. Lett.* 2007, *90*.

97. Lang, X.; Qian, L.; Guan, P.; Zi, J.; Chen, M. Localized surface plasmon resonance of nanoporous gold. *Appl. Phys. Lett.*2011, *98*.

98. Okamoto, T.; Yamaguchi, I.; Kobayashi, T. Local plasmon sensor with gold colloid monolayers deposited upon glass substrates. *Opt. Lett.* 2000, *25*, 372–374.

99. Marinakos, S.M.; Chen, S.H.; Chilkoti, A. Plasmonic Detection of a Model Analyte in Serum by a Gold Nanorod Sensor. *Anal. Chem.* 2007, *79*, 5278–5283.

100. Kreibig, U.; Vollmer, M. *Optical Properties of Metal Clusters*; Springer: Berlin, Germany, 1995.

101. Gersten, J.; Nitzan, A. Electromagnetic theory of enhanced Raman scattering by molecules adsorbed on rough surfaces. *J. Chem. Phys.* 1980, *73*, 3023–3037.

102. Jensen, T.; Kelly, L.; Lazarides, A.; Schatz, G.C. Electrodynamics of Noble Metal Nanoparticles and Nanoparticle Clusters. *J. Clust. Sci.* 1999, *10*, 295–317.

103. Fleger, Y.; Rosenbluth, M. Surface Plasmon and Surface Enhanced Raman Spectra of Aggregated and Alloyed Gold-Silver Nanoparticles. *Int. J. Opt.* 2009, *2009*.

104. Park, T.H.; Mirin, N.; Lassiter, J.B.; Nehl, C.L.; Halas, N.J.; Nordlander, P. Optical Properties of a Nanosized Hole in a Thin Metallic Film. *ACS Nano* 2008, *2*, 25–32.

105. Lang, X.Y.; Chen, L.Y.; Guan, P.F.; Fujita, T.; Chen, M.W. Geometric effect on surface enhanced Raman scattering of nanoporous gold: Improving Raman scattering by tailoring ligament and nanopore ratios. *Appl. Phys. Lett.* 2009, *94*.

106. Johnson, P.B.; Christy, R.W. Optical Constants of the Noble Metals. *Phys. Rev. B* 1972, *6*, 4370–4379.

107. Smith, G.B.; Maaroof, A.I.; Cortie, M.B. Percolation in nanoporous gold and the principle of universality for two-dimensional to hyperdimensional networks. *Phys. Rev. B* 2008, *78*.

108. Hopkins, P.E.; Serrano, J.R.; Phinney, L.M.; Li, H.; Misra, A. Boundary scattering effects during electron thermalization in nanoporous gold. *J. Appl. Phys.* 2011, *109*.

109. Whal, P.; Trauβnig, T.; Landgraf, S.; Jin, H.-J.; Weissmüller, J.; Würschum, R. Adsorption-driven tuning of the electrical resistance of nanoporous gold. *J. Appl. Phys.* 2010, *108*.

110. Mishra, A.K.; Bansal, C.; Hahn, H. Surface charge induced variation in the electrical conductivity of nanoporous gold.*J. Appl. Phys.* 2008, *103*.

111. Chen, L.Y.; Lang, X.Y.; Fujita, T.; Chen, M.W. Nanoporous gold for enzyme-free electrochemical glucose sensors. *Scr. Mater.* 2011, *65*, 17–20.

112. Wang, D.; Ji, R.; Albrecht, A.; Schaaf, P. Ordered arrays of nanoporous gold nanoparticles. *Beilstein J. Nanotechnol.*2012, *3*, 651–657.

113. Chauvin, A.; Delacôte, C.; Molina-Luna, L.; Duerrschabel, M.; Boujtita, M.; Thiry, D.; Du, K.; Ding, J.; Choi, C.-H.; Tessier, P.-Y.; *et al.* Planar Arrays of Nanoporous Gold Nanowires: When Electrochemical Dealloying Meets Nanopatterning. *ACS Appl. Mater. Interfaces* 2016, *8*, 6611–6620.

114. Barkey, D.; Oberholtzer, F.; Wu, Q. Kinetic Anisotropy and Dendritic Growth in Electrochemical Deposition. *Phys. Rev. Lett.* 1995, *75*, 2980–2983.

115. Hao, E.; Bailey, R.C.; Schatz, G.C.; Hupp, J.T.; Li, S.Y. Synthesis and Optical Properties of "Branched" Gold Nanocrystals. *Nano Lett.* 2004, *4*, 327–330.

116. Chen, S.H.; Wang, Z.L.; Ballato, J.; Foulger, S.H.; Carroll, D.L. Monopod, Bipod, Tripod, and Tetrapod Gold Nanocrystals. *J. Am. Chem. Soc.* 2003, *125*, 16186–16187.

117. Herricks, T.; Chen, J.Y.; Xia, Y.N. Polyol Synthesis of Platinum Nanoparticles: Control of Morphology with Sodium Nitrate. *Nano Lett.* 2004, *4*, 2367–2371.

118. Teng, X.W.; Yang, H. Synthesis of Platinum Multipods: An Induced Anisotropic Growth. *Nano Lett.* 2005, *5*, 885–891.

119. Ramirez, E.; Jansat, S.; Philippot, K.; Lecante, P.; Gomez, M.; Masdeu-Bulto, A.M.; Chaudret, B.-J. Influence of organic ligands on the stabilization of palladium nanoparticles. *J. Organomet. Chem.* 2004, *689*, 4601–4610.

120. Antonietti, M.; Goltner, C. Superstructures of Functional Colloids: Chemistry on the Nanometer Scale. *Angew. Chem. Int. Ed.* 1997, *36*, 910–928.

121. Ewers, T.D.; Sra, A.K.; Norris, B.C.; Cable, R.E.; Cheng, C.H.; Shantz, D.F.; Schaak, R.E. Spontaneous Hierarchical Assembly of Rhodium Nanoparticles into Spherical Aggregates and Superlattices. *Chem. Mater.* 2005, *17*, 514–520.

122. Hoefelmeyer, J.D.; Niesz, K.; Somorjai, G.A.; Tilley, T.D. Radial Anisotropic Growth of Rhodium Nanoparticles. *Nano Lett.* 2005, *5*, 435–438.

123. Ely, T.O.; Amiens, C.; Chaudret, B.; Snoeck, E.; Verelst, M.; Respaud, M.; Broto, J.M. Synthesis of Nickel Nanoparticles. Influence of Aggregation Induced by Modification of Poly(vinylpyrrolidone) Chain Length on Their Magnetic Properties. *Chem. Mater.* 1999, *11*, 526–529.

124. Vidoni, O.; Philippot, K.; Amiens, C.; Chaudret, B.; Balmes, O.; Malm, J.O.; Bovin, J.O.; Senocq, F.; Casanove, M.J. Novel, Spongelike Ruthenium Particles of Controllable Size Stabilized Only by Organic Solvents. *Angew. Chem. Int. Ed.* 1999, *38*, 3736–3738.

125. Pelzer, K.; Vidoni, O.; Philippot, K.; Chaudret, B.; Colliere, V. Organometallic Synthesis of Size-Controlled Polycrystalline Ruthenium Nanoparticles in the Presence of Alcohols. *Adv. Funct. Mater.* 2003, *13*, 118–126.

126. Wang, Z.L. Functional oxide nanobelts: Materials, properties and potential applications in nanosystems and biotechnology. *Annu. Rev. Phys. Chem.* 2004, *55*, 159–196.

127. Gao, P.X.; Wang, Z.L. Nanopropeller arrays of zinc oxide. *Appl. Phys. Lett.* 2004, *84*, 2883–2885.

128. Cheng, Y.; Wang, Y.S.; Chen, D.Q.; Bao, F. Evolution of Single Crystalline Dendrites from Nanoparticles through Oriented Attachment. *J. Phys. Chem. B* 2005, *109*, 794–798.

129. Yan, H.Q.; He, R.R.; Johnson, J.; Law, M.; Saykally, R.J.; Yang, P.D. Dendritic Nanowire Ultraviolet Laser Array. *J. Am. Chem. Soc.* 2003, *125*, 4728–4729.

130. Wang, D.; Qian, F.; Yang, C.; Zhong, Z.H.; Lieber, C.M. Rational Growth of Branched and Hyperbranched Nanowire Structures. *Nano Lett.* 2004, *4*, 871–874.

131. Peng, Z.A.; Peng, X.G. Nearly Monodisperse and Shape-Controlled CdSe Nanocrystals via Alternative Routes: Nucleation and Growth. *J. Am. Chem. Soc.* 2002, *124*, 3343–3353.

132. Manna, L.; Milliron, D.J.; Meisel, A.; Scher, E.C.; Alivisatos, A.P. Controlled growth of tetrapod-branched inorganic nanocrystals. *Nat. Mater.* 2003, *2*, 382–385.

133. Milliron, D.J.; Hughes, S.M.; Cui, Y.; Manna, L.; Li, J.B.; Wang, L.-W.; Alivisatos, A.P. Colloidal nanocrystal heterostructures with linear and branched topology. *Nature* 2004, *430*, 190–195.

134. Grebinski, J.W.; Hull, K.L.; Zhang, J.; Kosel, T.H.; Kuno, M. Solution-Based Straight and Branched CdSe Nanowires.*Chem. Mater.* 2004, *16*, 5260–5272.

135. Lee, S.M.; Jun, Y.W.; Cho, S.N.; Cheon, J. Single-Crystalline Star-Shaped Nanocrystals and Their Evolution: Programming the Geometry of Nano-Building Blocks. *J. Am. Chem. Soc.* 2002, *124*, 11244–11245.

136. Jun, Y.W.; Jung, Y.Y.; Cheon, J. Architectural Control of Magnetic Semiconductor Nanocrystals. *J. Am. Chem. Soc.* 2002,*124*, 615–619.

137. Wang, D.L.; Lieber, C.M. Inorganic materials: Nanocrystals branch out. *Nat. Mater.* 2003, *2*, 355–356.

138. Huang, T.; Meng, F.; Qi, L. Controlled Synthesis of Dendritic Gold Nanostructures Assisted by Supramolecular Complexes of Surfactant with Cyclodextrin. *Langmuir* 2010, *26*, 7582–7589.

139. Kaniyankandy, S.; Nuwad, J.; Thinaharan, C.; Dey, G.K.; Pillai, C.G.S. Electrodeposition of silver nanodendrites. *Nanotechnology* 2007, *18*.

140. Claro, F.; Fuchs, R. Optical absorption by clusters of small metallic spheres. *Phys. Rev. B* 1986, *33*, 7956–7960.

141. Link, S.; El-Sayed, M.A. Spectral Properties and Relaxation Dynamics of Surface Plasmon Electronic Oscillations in Gold and Silver Nanodots and Nanorods. *J. Phys. Chem. B* 1999, *103*, 8410–8426.

142. Chen, M.C.; Tsai, S.D.; Chen, M.R.; Ou, S.Y.; Li, W.-H.; Lee, K.C. Effect of silver-nanoparticle aggregation on surface-enhanced Raman scattering from benzoic acid. *Phys. Rev. B* 1995, *51*, 4507–4515.

143. Qiu, T.; Zhou, Y.; Li, J.; Zhang, W.; Lang, X.; Cui, T.; Chu, P.K. Hot spots in highly Raman-enhancing silver nano-dendrites. *J. Phys. D* 2009, *42*.

144. Nikoobakht, B.; El-Sayed, M.A. Preparation and Growth Mechanism of Gold Nanorods (NRs) Using Seed-Mediated Growth Method. *Chem. Mater.* 2003, *15*, 1957–1962.

145. Hu, M.-S.; Chen, H.-L.; Shen, C.-H.; Hong, L.-S.; Huang, B.-R.; Chen, K.-H.; Chen, L.-C. Photosensitive gold-nanoparticle-embedded dielectric nanowires. *Nat. Mater.* 2006, *5*, 102–106.

146. Wang, S.B.; Huang, Y.F.; Chattopadhyay, S.; Chang, S.J.; Chen, R.S.; Chong, C.W.; Hu, M.S.; Chen, L.C.; Chen, K.H. Surface plasmon-enhanced gas sensing in single gold-peapodded silica nanowires. *NPG Asia Mater.* 2013, *5*.

147. Wang, S.-B.; Hu, M.-S.; Chang, S.J.; Chong, C.-W.; Han, H.-C.; Huang, B.-R.; Chen, L.-C.; Chen, K.-H. Gold nanoparticle-modulated conductivity in gold peapodded silica nanowires. *Nanoscale* 2012, *4*, 3660–3664.

CHAPTER 16

COPPER NANOWIRES AND THEIR APPLICATIONS FOR FLEXIBLE, TRANSPARENT CONDUCTING FILMS: A REVIEW

Vu Binh Nam and Daeho Lee

Department of Mechanical Engineering, Gachon University, Seongnam 13120, Korea

ABSTRACT

Cu nanowires (NWs) are attracting considerable attention as alternatives to Ag NWs for next-generation transparent conductors, replacing indium tin oxide (ITO) and micro metal grids. Cu NWs hold great promise for low-cost fabrication via a solution-processed route and show preponderant optical, electrical, and mechanical properties. In this study, we report a summary of recent advances in research on Cu NWs, covering the optoelectronic properties, synthesis routes, deposition methods to fabricate flexible transparent conducting films, and their potential applications. This review also examines the approaches on protecting Cu NWs from oxidation in air environments.

INTRODUCTION

The demand for transparent screens is huge and increasing rapidly. From large flat-panel displays to small smartphones and smart watches, the devices we use and encounter every day and everywhere need transparent conducting films (TCFs). Also, optoelectronic devices such as thin-film solar cells [1,2], organic light-emitting diodes (OLEDs) [3,4], thermo-electrics [5], and sensors [6] utilize TCFs as their substrates. Moreover, flexibility is becoming an additional requirement for TCFs as the demand for flexible electronics is growing rapidly. In the current transparent conductor industry, indium tin oxide (ITO) films coated on glass occupy a majority of the market. However, current ITO films are deposited using a vacuum process, which requires expensive equipment and generates significant material waste. Furthermore, ITO films cannot withstand repeated bending or rolling because they are brittle and prone to cracking, limiting their use in truly flexible devices [7]. More importantly, indium, which is a rare material, causes fluctuations in ITO prices [8]. As a result, the use of traditional ITO-based TCFs led to delays in terms of new production and innovations. The developed markets for flexible and curved screens require transparent conductor materials that can be processed at low cost on plastic substrates and that have excellent mechanical and optoelectronic performance. Thus, many researchers have investigated alternative materials for ITO for use in flexible transparent conducting electrodes, such as carbon-based nanomaterials (nanotubes [9,10] and graphene [11,12]), zinc oxide (ZnO)-based transparent conducting oxides [13,14,15], noble metals (Au, Ag) [16,17], metal micro grids [18,19], nanowire grids [20,21], and graphene–silver nanowire (NW) hybrids [22]. However, conducting networks of carbon-based nanomaterials, including carbon nanotubes (CNTs) and graphene, exhibit relatively poor optoelectronic performance as compared to ITO [23,24]. The high price of noble metals makes these metals uncompetitive for large-area, low-cost transparent conductors, even though they exhibit

Citation: Vu Binh Nam and Daeho Lee, Copper Nanowires and Their Applications for Flexible, Transparent Conducting Films: A Review, doi: 10.3390/nano6030047.

superior conductivity and transmittance. Recently, noble metal NWs have emerged as promising materials for achieving performance equivalent to that of ITO. Ag NWs achieved a transparency with a low sheet resistance (R_s), which surpasses the ITO performance [25]. Ag NWs also can be deposited on flexible substrates by spin coating or vacuum transfer [18,26,27]. However, the high material cost of Ag still presents a challenge to the mass production of optoelectronic devices.

Copper (Cu) NWs have recently garnered increasing attention as excellent candidates for TCFs. Because Cu has high intrinsic conductivity, it is a potential alternative to ITO and Ag NWs. Moreover, Cu is only 6% less conductive than Ag, very abundant (1000 times more abundant than Ag), and 100 times less expensive than Ag or ITO [28]. Several studies have been reported, demonstrating how Cu NW films can achieve high optical transmittance with high electric conductivity [29,30,31]. However, challenges remain for the stabilization of Cu NWs because they are prone to oxidation under ambient conditions, and thus they cannot retain performance comparable to that of ITO.

In this review, we report recent research progress on Cu NWs as promising transparent conducting materials. This review covers the optoelectronic properties, synthesis routes of Cu NWs, deposition methods to fabricate TCFs, and their potential applications. Firstly, we summarize the relationship between the physical dimensions and optoelectronic properties of metal NWs. In the second part, the synthesis routes of Cu NWs and the fabrication processes of Cu NW thin films are reviewed. In the third part, we discuss how to protect and passivate Cu NWs from oxidation. Finally, the applications of Cu NWs and required future studies are proposed.

RELATIONSHIP OF OPTOELECTRONIC PROPERTIES AND NANOWIRE DIMENSIONS

Electrical Properties of Nanowire Films

High-quality TCFs require both high conductivity and high transmittance [32]. In general, however, a metal thin film with high conductivity leads to low transmittance because the conductivity is proportional to the film thickness [33]. A functional relationship between NW dimensions and the properties of NW-based films is very important and useful for the synthesis of NWs optimized for use in transparent electrodes. In this part, we summarize the structure–property correlations of NW films based on recently published articles. Although many theories and experiments introduced here have been established and performed with materials other than Cu NWs, we can predict the properties of Cu NWs based on these previous studies.

The first theory for predicting the dependence of resistivity on thin-film dimensions was introduced by Fuchs *et al.* [34]. Dingle subsequently extended Fuchs's theory to metal NWs [35]. In this model, the dependence of the resistivity on the diameter of NWs can be expressed as:

$$\rho_d = \frac{\rho_o + \rho_o (1-\rho)\, l_o}{D} \tag{1}$$

where ρd is the resistivity of a NW of diameter *D*, ρo is the bulk or the so-called "intrinsic resistivity" of the NW material, ρ is the specularity factor which denotes the fraction of conduction electrons undergoing specular reflection at the NW surface, and l_o is the free path of the conduction electrons. Figure 1a shows the resistivity dependence of Ag NWs on the diameter [36]. The resistivity decreases as the NW diameter increases, which is attributed to the surface scattering of conduction electrons [33,36,37]. Figure 1b shows the measured R_s (dots) as a function of the diameter of Ag NW networks with network pitches of 500 (blue), 700 (green), and 1,000 nm (orange) [38]. The solid lines represent the fitted plots of R_s, which confirm that R_s decreases for all networks as the diameter of the NW increases. Figure 1c shows experimental (points) and simulative (lines) R_s values for Ag NW thin films as a function of area fraction (AF) for the three L/D values indicated, where AF is defined as the number density of the nanowires in a film multiplied by the projected area of a nanowire. We can verify that the R_s decreases with increasing AF and L/D.

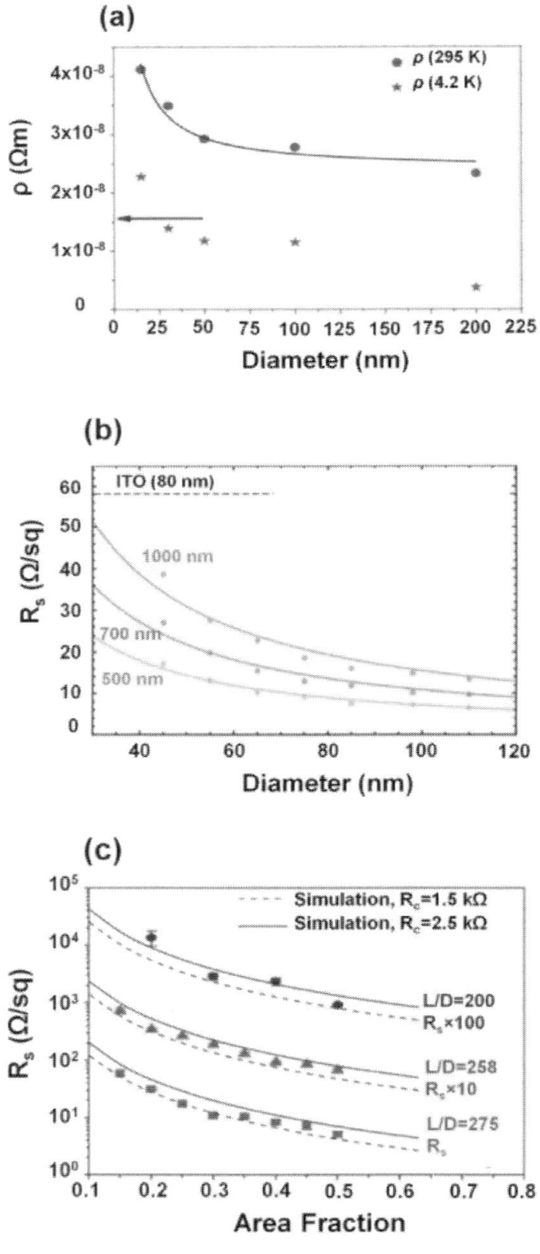

Figure 1: (a) Plot of the resistivity (ρ) of Ag nanowires (NWs) at 295 K and 4.2 K and the residual resistivity as a function of the NW

diameter. The line is the fit to the data using Equation (1). The arrow is the bulk value of the resistivity of Ag, $1.629 \times 10^{-8}\Omega$m. (Reproduced with permission from [36]. Copyright American Physical Society, 2006); (**b**) The dots represent measured sheet resistances (R_s) as a function of NW width for network pitches of 500 (blue), 700 (green), and 1000 nm (orange). The solid lines represent fitted plots for each pitch. (Reproduced with permission from [38]. Copyright American Chemical Society, 2012); (**c**) Experimental R_s values for silver NW thin films (points) as a function of the area fraction (AF) for the L/D values indicated.R_s from quasi-2D simulations of rods with specified L and D values use an effective contact resistance (R_c) to fit the simulations to the experimental data; the best fits correspond to $R_c = 1.5$ kΩ (dashed lines) and 2.5 kΩ (solid lines) (Reproduced with permission from [42]. Copyright American Chemical Society, 2013).

The effect of NW dimensions on the properties of NW-based TCFs can also be understood by the percolation threshold. Li*et al.* [39] used Monte Carlo simulation to verify that the length of the widthless sticks, *L,* and the critical number density of sticks, N_c, achieves percolation conductivity by the following equation:

$$N_c L^2 \approx 5.71 \qquad (2)$$

where 5.71 is the simulation result from extraction of a high-precision N_c value. This equation indicates that double-length NWs require only a quarter of the number density of NWs to achieve percolation. Therefore, if more NWs are added to a film that has just achieved percolation, more conducting pathways formed between NWs increase network conductivity [40].

Recently, Borchert *et al.* integrated simulations and experiments to predict the relationship between R_s and the dimensions of Cu NWs [41]. They reported that the R_s value of the Cu NW films decreases as AF increases. In their study, they assumed that all NWs can be modeled as cylinders with a constant diameter, which is taken to be the experimentally obtained average diameter. The AF of a film covered by NWs was calculated by the following equation [41]:

$$AF = \frac{4\,(Nt)_i^{\mathrm{NW}}\,M_i}{\pi \rho_B D} \qquad (3)$$

where $(Nt)_i^{\mathrm{NW}}$ is the atomic areal density of element i in the NW sample measured by Rutherford backscattering spectrometry, Mi is the molecular weight of the NW component element, ρ_B is the bulk atomic density of the material, and$D$ is the diameter of the NW.

Optical Properties of Nanowire Films

Controlling optical properties of NW thin films is also critical for TCF applications. Although many studies regarding optical properties of NW films have been performed using materials other than Cu NWs, the fundamental relationship between the NW structure and the optical properties can be applied to general NW films, including Cu NW films. The optical properties of NW films can be determined by considering the surface AF of the NW films. The amount of light blocked by a NW network is equivalent to the light blocked by the AF of the film covered by the NWs. Bergin *et al.* predicted how transmittance is affected by the dimensions of NWs of a random NW network [40]. The transmittance of the film is calculated by

$$T(\%) = 100 - a \times \mathrm{AF} \qquad (4)$$

where T is the transmittance of the film in %, the fitting parameter a accounts for the diameter and wavelength-dependent optical properties of the NWs, and AF is the area fraction of the NW films given by Equation (3). The fitted plots with Equation (4) are shown in Figure 2a with $a = 87$ and $a = 61$, which correspond to the diameters of 51 nm and 30 nm of Ag NWs, respectively [43]. The plots indicate that the transmittance of NW films increases with decreasing NW diameter and area coverage. As the area coverage of the NWs in a film approaches zero, the transmittance of the NW film approaches 100%.

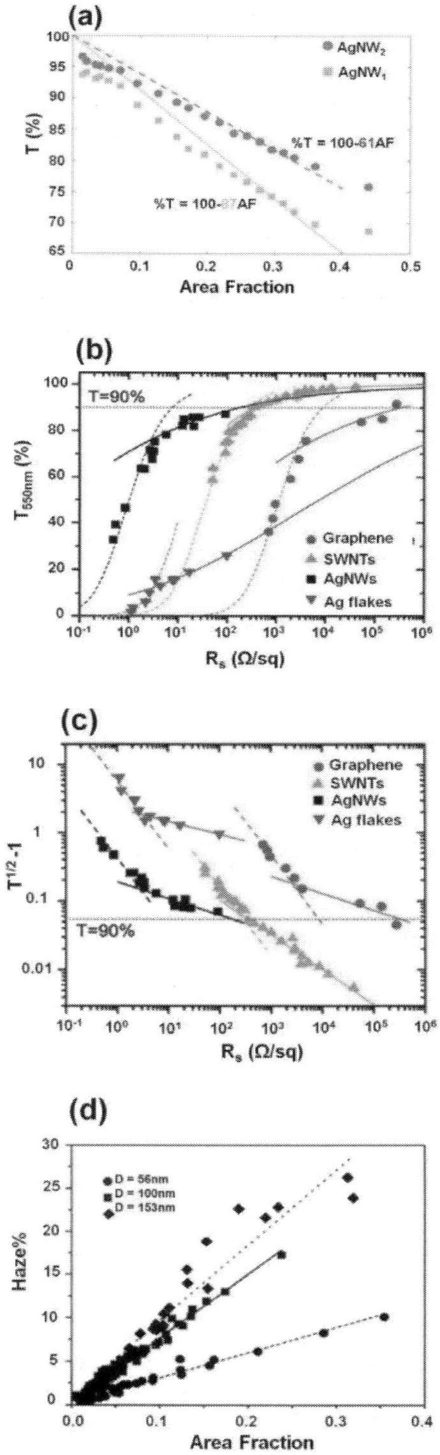

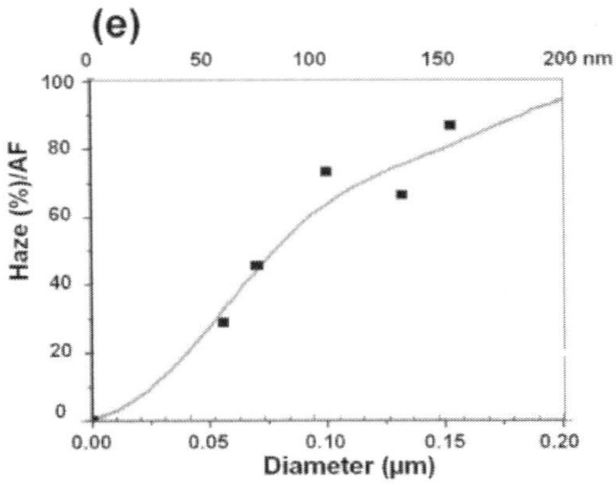

Figure 2: (**a**) The relationship between AF and transmittance (T). Different L/D in AgNW1 and AgNW2 lead to different trends in both T and R_s. (Reproduced with permission from [43]. Copyright Springer Science + Business Media on behalf of the American Coatings Association and the Oil and Colour Chemist Association, 2015); (**b**) Transmittance (550 nm) plotted as a function of sheet resistance (R_s) for thin films prepared from four nanostructured materials, graphene, single-walled carbon-based nanomaterials, Ag nanowires (NWs) and Ag flakes. The dashed lines represent fitted plots of the bulk regime using Equation (5), while the solid lines corresponds to the percolative regime using Equation (6); (**c**) The date in Figure 2a are re-plotted on the axes of ($T^{-1/2} - 1$) and R_s on log-log scale. Note that ($T^{-1/2} - 1$) is proportional to film thickness. The dashed lines represent fitted plot of the bulk regime using Equation (5), while the solid lines corresponds to the percolative regime using Equation (6). (Reproduced with permission from [37]. Copyright American Chemical Society, 2010); (**d**) Haze factor (HF) *vs.* AF of Ag NWs with diameters of 56, 100, and 153 nm; (**e**) Slope of HF(%) *vs.* AF, HF(%)/AF, as a function of Ag NW diameter. (Reproduced with permission from [45]. Copyright AIP Publishing, 2013).

In general, for NW conducting films, De *et al.* [37] proposed a simple model in which the transmittance is related to sheet resistance in the bulk and percolative regimes by the following equations:

$$T = \left(1 + \frac{1}{\Pi_{bulk}} \frac{Z_o}{2R_s}\right)^{-2} \quad (5)$$

$$T = \left[1 + \frac{1}{\Pi_{percolative}} \left(\frac{Z_o}{R_s}\right)^{1/(n+1)}\right]^{-2} \quad (6)$$

where T is the transmittance of the film, Z_o is the impedance of free space, n is the percolation exponent, R_s is the sheet resistance, Π_{bulk} and $\Pi_{percolative}$ are the dimensionless number of the bulk and percolative regimes, respectively. Figure 2b shows T as a function of R_s for thin films prepared from four nanostructured materials which are graphene, single-walled carbon nanotubes, Ag NWs, and Ag flakes. In most cases, the bulk-like behavior begins to occur at relatively low R_s values which correspond to thick films (dashed lines) and does not occur in the technologically relevant regime ($T > 90\%$), whereas the percolative regime can be identified as a straight line on a log-log plot of ($T^{-1/2} - 1$) *versus* R_s (equivalent to a graph of thickness *vs.* R_s) by Equation (6). The data for in Figure 2b are re-plotted with this form in Figure 2c. For each curve, the bulk and percolative regimes are clearly present, where n and Π can be obtained by fitting the data in both regimes. In short, high transmittance of NW films can be obtained with large Π values by decreasing the NW diameter [37].

In addition to transmittance as a critical optical property of TCFs, haze factor (HF) is becoming another important factor that affects the performance of TCFs. The haze factor (HF) is defined as [44]:

$$HF = \frac{I_s}{I_s + I_d} \times 100\ (\%) \quad (7)$$

where I_d is the light flux transmitted directly and I_s is the flux of scattered light. The control of HF is an ongoing challenge for NW-based transparent electronic devices because NWs are strong scattering centers. Figure 2d,e show the HF values of NW films as a function of AF and the diameter of the NW, respectively. HF appears to scale linearly with AF, and it increases significantly with increasing NW diameter [45,46]. Furthermore, the haze value decreases almost linearly with increasing total transmittance [29]. Khanarian *et al.* proved that less than 1% HF of the Ag NW mesh film can be achieved if the diameter of the NW is less than 50 nm and the length is greater than 5 μm [45]. On the other hand, TCFs with different haze factors can be applied to different fields. For instance, large HF values are preferred for thin-film photovoltaic applications because an increase in the amount of light scattered from the electrode can enhance the coupling of light into the absorber material, thereby increasing the efficiency of the device [47], whereas in display applications, HF has to be kept to a minimum (usually <1%) to improve the clarity of a film [48].

Consequently, the optical properties of NW film are strongly dependent on the diameter and area coverage of the NWs on it. The relationship offers guidelines for developing a method of synthesizing NWs for optimal performance.

SYNTHESIS OF CU NWS AND FABRICATION OF CU NW THIN FILMS

Synthesis of Cu NWs

Several methods for the synthesis of Cu NWs, such as chemical vapor deposition [49], vacuum thermal decomposition [50], and electro-spinning [51] have been reported. For the past ten years, chemical-solution methods (CSMs), including hydrothermal synthesis [52], reduction of a precursor solution [53], and catalytic synthesis [54] have emerged as important and mainstream routes to NW synthesis. CSMs have several advantages, such as fewer constraints regarding precursor selection and the conditions of the solvents and reactions, and the feasibility of large-scale production. Cu NW synthesis by CSMs at low temperature under ambient conditions is easier in terms of scaling up and commercialization than that by vacuum-vapor processes. All the CSMs are similar in that they use a reducing agent to transform metal ions into metal atoms and a capping agent to bring metal atoms into the NWs. In this section, we summarize the synthesis of Cu NWs by CSMs for use in TCFs. There are two general approaches: (i) the amine-assisted approach and (ii) the catalyst-assisted approach.

Amine-Assisted Synthesis

Chang *et al.* developed one of the first solution-phase synthesis methods of Cu NWs [55]. The scanning electron microscopy (SEM) image, photographic image of mother liquor, magnified SEM and transmission electron microscopy (TEM) image of Cu NWs synthesized by their method are displayed in Figure

3a–d, respectively. The formation of Cu NWs in that work was based on the following redox reaction:

$$Cu^{2+} + N_2H_4 \xrightarrow{OH^-,\ Heating,\ \mathrm{EDA}} Cu\ NWs \quad (8)$$

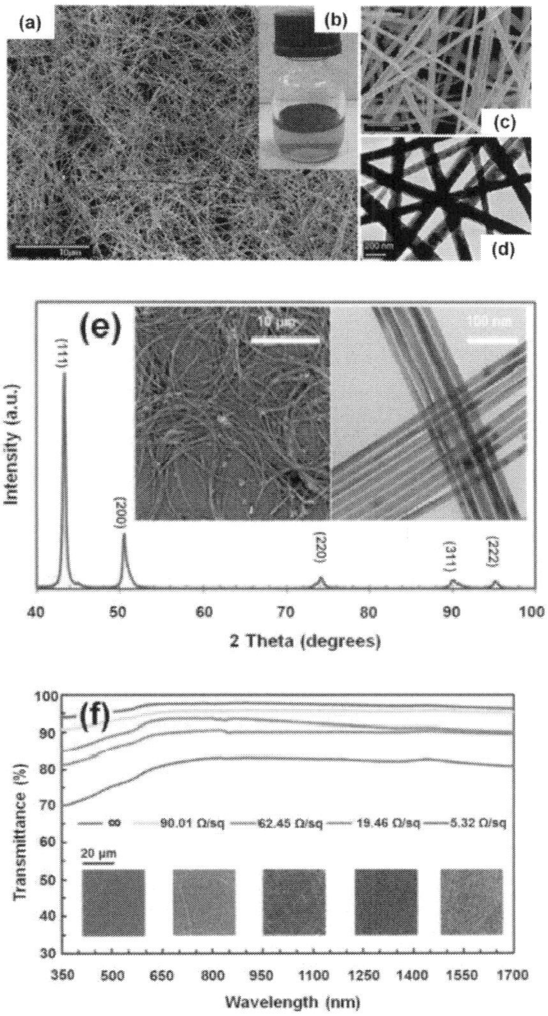

Figure 3: (**a**) Scanning electron microscopy (SEM) image of general views of Cu NWs; (**b**) Photographic image of as-prepared Cu NWs in mother liquor; (**c**) SEM image of detailed views of Cu NWs; (**d**) Transmission electron microscopy (TEM) image of Cu NWs (Reproduced with permission from [55]. Copyright American Chemical Society, 2005); (**e**) X-ray diffraction pattern of Cu NWs. Inset, SEM (**left**) and TEM (**right**) images of as-grown Cu NWs; (**f**) Wavelength-dependent transmittance, sheet resistance, and corresponding SEM images of transparent conductors. Substrate contribution is excluded. (Reproduced with permission from [29]. Copyright American Chemical Society, 2015).

Cu^{2+} ions were reduced by hydrazine (N_2H_4) with ethylenediamine (EDA, $C_2H_8N_2$) as a capping agent when a reactor containing highly basic copper nitrate ($Cu(NO_3)_2$) aqueous precursors was controlled at 25–100 °C. The resulting Cu NWs were straight, with uniform diameters in the range of 60–160 nm and lengths greater than 40 μm (Figure 3a–d). Rathmell *et al.* reported modification of this approach, focusing on large-scale synthesis produced Cu NWs 90 ± 10 nm in diameter and 10 ± 3 μm in length [8]. They also verified that the concentration of EDA is a critical parameter for anisotropic growth of NWs. When EDA was not added to the reaction, only Cu nanoparticles were produced. Shi *et al.* used copper(II) chloride ($CuCl_2$) aqueous solution as a Cu^{2+} ion supplier, and employed octadecylamine (ODA, $C_{18}H_{39}N$) substituting for hydrazine and EDA as both a reductant and a surfactant [56]. The resulting Cu NWs had uniform diameters of 50–100 nm and lengths up to several millimeters. Jin *et al.* reported a procedure to synthesize Cu NWs by reducing Cu^{2+} ions with glucose in water with hexadecylamine (HAD) serving as a capping agent. [57]. They produced Cu NWs with controlled shapes, high purity, and good uniformity in relatively large quantities. The resulting Cu NWs had a uniform diameter of 25 nm and lengths ranging from several tens to hundreds of micrometers; some of them were as long as several millimeters. More recently, Cui*et al.* used oleylamine (OLA, $C_{18}H_{37}N$) both as a coordinating ligand for the $CuCl_2$ precursor and a capping ligand on the Cu NW surface while tris(trimethylsilyl) silane ($C_9H_{28}Si_4$) was introduced as a mild reducing reagent. Their method could produce ultrathin and highly crystalline Cu NWs with an average diameter of 17.5 nm and a length of 17 μm, which resulted in transparent Cu NW TCFs as shown in Figure 3e,f [29]. Another advantage of their approach is that the resulting Cu NW TCFs show a dramatically reduced HF due to the small NW diameter.

Catalyst-Assisted Synthesis

The formation of Cu NWs using a catalyst-assisted synthesis is based on the following redox reaction:

$$Cu^{2+} \xrightarrow{Catalyst\ (Ni,\ Pt),\ Amines,\ Heating} Cu\ NV \quad (9)$$

These synthesis routes also employ amines, but differ from above-mentioned methods in that catalytic reactions occur during synthesis. Guo *et al.* successfully synthesized long and fine Cu NWs in OLA solution by the nickel(II) acetylacetonate ($Ni(acac)_2$) catalytic procedure [58]. The resulting Cu NWs had high aspect ratios with diameters of 16.2 ± 2 nm and lengths up to 40 μm. Firstly, under heating at 175 °C, Ni^{2+} ions in the OLA solution were reduced to Ni atoms. Secondly, Cu atoms were formed by the galvanic replacement reaction of Cu^{2+} ions with Ni atoms; in this reaction, Ni^{2+} ions acted as catalysts. Lastly, Cu NWs grew slowly from Cu nanocrystal seeds and elongated in the presence of Cl^- ions from the $CuCl_2$ precursor. The presence of Cl^- ions was indispensable because the Cl^- ions attached onto the (100) facets, resulting in strongly anisotropic growth. Zhang *et al.* synthesized ultralong crystalline Cu NWs with excellent dispersibility. through the self-catalytic growth of Cu NWs within a liquid crystalline medium of hexadecylamine (HAD, $C_{16}H_{35}N$) and cetyltrimethylammonium bromide (CTAB, $C_{19}H_{42}BrN$) [59]. The average diameter of Cu NWs produced by this method was 78 nm and the lengths varied from tens to hundreds of micrometers. It is noted that platinum (Pt) catalyst was indispensable for their method since no particles or NWs were observed without the catalyst even after long reaction time, which is attributed to the fact that the presence of Pt surface first catalyzes the reduction of Cu ions into metallic clusters or particles, which then migrate to the solution and act as seeds for growth of the NWs.

Fabrication of Cu NW Thin Films

The synthesized Cu NWs should be deposited on a substrate to fabricate TCFs. Vacuum filtration [31,58], spin coating

[60,61,62], Meyer rod coating [30], nitrocellulose-based ink printing [63,64], drop-spray coating [65,66], and roll-to-roll printing [67] are the examples of general NW deposition processes regardless of the types of materials. However, the R_s value of the as-transferred or as-coated TCFs is generally high due to contact resistance between NWs. Several ways to minimize contact resistance to achieve lower R_s values of NW networks have been proposed.

He *et al.* reported a new approach by which tough and excellent-performance Cu NW transparent electrodes were fabricated using electrospun fibers as a mask [60]. As shown in Figure 4, electrospun nanofiber networks provide percolating paths that result in completely connected metal NW patterns, which in turn lead to high-conductivity networks. Interestingly, the performance of the transparent Cu NW electrodes in their work is better than that of ITO, CNTs, and graphene-based electrodes in that the Cu NW films showed excellent adhesion to the substrate, offering the benefit of being usable on polymer substrates. In addition, the low fabrication cost and simplicity of electrospinning make it a promising method to produce large-scale transparent electrodes. Han *et al.* introduced ultrafast plasmonic nanoscale welding by using a polarized laser under ambient conditions. A Cu NW percolation network was first formed on various substrates from a Cu NW solution, followed by laser irradiation, which generated laser-induced heat at the junctions of each NW due to the plasmonic effect. As a consequence, the resistance of Cu NW network can be reduced significantly because the NW junctions were welded together [31]. The electrical performance of NW TCFs can also be improved by NW alignment. Kang *et al.* [68] introduced a capillary printing technique for the fabrication of transparent electrodes using Ag NWs (Figure 5). The Ag NW solution was deposited and dragged to generate uniformly aligned NWs by controlling the nanochannels. The aligned NWs were predicted to have a significantly lower percolation threshold, which would improve the transmittance and conductivity of the film. Although their process was applied to Ag NW alignment, the same process can possibly be used for Cu NWs. In addition, mechanical pressing [69,70], thermal annealing [71], and optothermal heating [72], reported to improve electrical performance of Ag NW electrodes, can be applied to Cu NW electrodes as well.

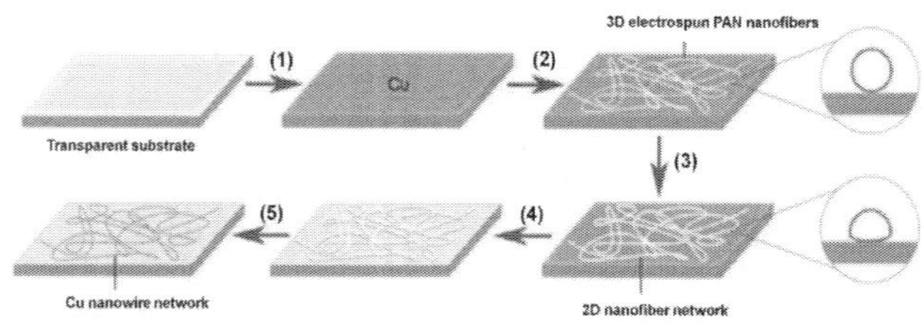

Figure 4: Scheme of Cu NW network electrode fabrication: (**1**) Electron-beam deposition of Cu film on the transparent substrate; (**2**) Electrospun polyacrylonitrile (PAN) nanofibers on the Cu-covered substrate; (**3**) Solvent vapor annealing. The insets show schematics of the fiber cross-sections before and after solvent vapor annealing; (**4**) Metal etching; (**5**) Removal of PAN fibers by dissolution. Cu NW network on transparent substrate after PAN is removed (Reproduced with permission from [73]. Copyright American Chemical Society, 2014).

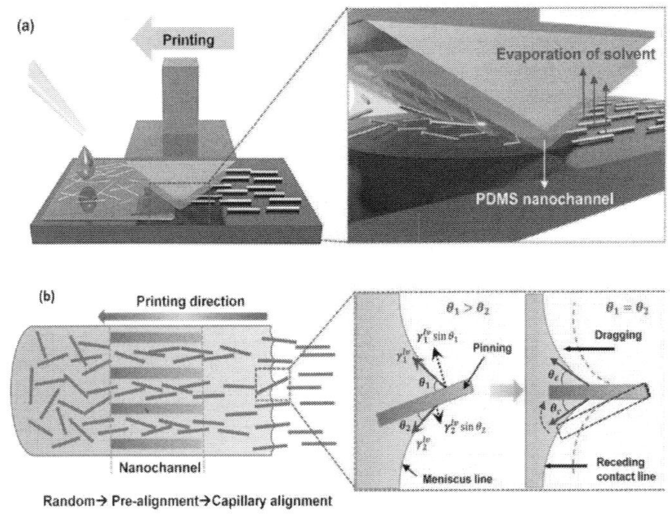

Figure 5: Solution-printed highly aligned Ag nanowire (NW) arrays: (**a**) Schematic of the capillary printing process using a nanopatterned polydimethylsiloxane (PDMS) stamp to produce highly aligned NW arrays; (**b**) Schematic showing the alignment process during capillary printing of unidirectional NW arrays. The solvent-evaporation-induced capillary force produces highly aligned networks by dragging confined NWs at the solid-liquid-vapor contact line. (Reproduced with permission from [68]. Copyright American Chemical Society, 2015).

The roughness of a NW film is another important factor for device applications because high roughness of a TCF surface leads to leakage or short circuits [71]. NW films have a surface roughness of at least two NW diameters because NWs stack in junctions after they are transferred from the solution to the substrate [74]. Researchers have found various ways to address this issue, such as polymer coating [75], thermal annealing [70], and mechanical pressing [63]. These approaches can generally be applied regardless of the types of materials.

Adhesion of NWs to a substrate also affects device performance because NWs with poor adhesion can be wiped off the substrate, resulting in malfunction of the device. The degree of adhesion depends on various factors such as the chemical and physical properties of the NW, the type of the substrate, and post-processing. Poor adhesion can be solved by embedding the NWs into the surface of the polymer matrix [74,76], or by encapsulating NWs within a thin layer of organic materials [65]. Because the electrical properties of NWs are also affected by the embedment and encapsulation, proper treatment should be carefully applied.

PROTECTION OF CU NWS FILMS AGAINST OXIDATION

The most critical issue of Cu NWs is that Cu easily reacts with oxygen under ambient conditions; it thus oxidizes and becomes non-conductive. Furthermore, as the temperature increases, the oxidation rate accelerates. For this reason, protection of Cu NW films from atmospheric corrosion is the most crucial challenge for researchers in their attempts to apply Cu NWs to commercial production. Hsu *et al.* tried to solve this issue by using atomic layer deposition (ALD) to coat a passivation layer of aluminum-doped zinc oxide (AZO) and aluminum oxide (Al_2O_3) onto electrospun Cu NWs, as shown in the inset of Figure 6a [77]. In their work, the AZO layer improved the reliability of Cu NWs owing to its high transparency and conductivity, whereas the outer ultrathin Al_2O_3 layer enhanced the chemical resistance of Cu NWs to corrosion and oxidation. The core-shell structure of Cu_{core}-AZO/$Al_2O_{3\text{-}shell}$ NWs retained the same resistance before and after ALD, and the transmittance decreased by only 1% (Figure 6a). This passivation layer therefore had a remarkable effect in protecting Cu NWs from thermal oxidation at high temperatures up to 160 °C in dry air and up to 80 °C in humid air (Figure 6b). Rathmell*et al.* reported solution-phase synthesis of Cu_{core}–Ni_{shell} NWs with several Cu to Ni ratios [78]. The transmittance of the Cu NWs coated with Ni in a Cu:Ni ratio of 1:1 decreased from 94% to 84% (Figure 6c) at the low R_s of 60 Ω/sq. However, the Cu_{core}–Ni_{shell} NW films show 1,000 times more resistance to oxidation than the Cu NW films. As shown in Figure 6d, without any Ni coating, the R_s value of the bare Cu NW-based TCFs began to increase after 1 day in a dry oven at 85 °C. TheR_s value soared by an order of magnitude after 5 days, whereas the R_s value of the Ag NW films increased by an order of magnitude after 13 days. In comparison, the R_s value of the Ni-coated Cu NW-based TCFs remained remarkably stable over a period of 30 days. On the other hand, Chen *et al.* coated Zn, Sn, and In onto Cu NWs by a solution-phase process in order to protect Cu NW films from oxidation without degrading the optoelectric properties of the film [79]. Coating Cu NWs with a complete shell of Zn, Sn, or In usually leads to a decrease in the transmittance of the film of 4%–9%; however, this transmittance loss can be largely recovered by exposing the core-shell NWs to oxidizing conditions, or H_2O_2 treatment. For instance, treatment of a Cu_{core}–Zn_{shell} NW film in a 2 wt % solution of H_2O_2 in H_2O for 15 min increased the transmittance of the film from 76% to 84% while maintaining the low R_s value of the film. Mehta *et al.* developed a novel method using low-temperature plasma-enhanced chemical vapor deposition (CVD) to coat uniform graphene onto the surfaces of Cu NWs [80]. Graphene has remarkable in-plane stiffness to prevent mechanical deformation and impermeability to protect Cu against chemical reactions. A strong enhancement of both electrical and thermal conductivity in graphene-encapsulated Cu NWs was observed because passivation of the Cu surface with graphene induced partial elastic surface scattering, which consequently resulted in an increase of electron transport through the NWs. Recently, Song *et al.* demonstrated conductive Cu@Cu–Ni NW elastomer composites with ultrahigh performance stability against oxidation, bending, and stretching. The Cu@Cu–Ni NWs were fabricated by one-pot heating of $CuCl_2$ in an OLA solution of $Ni(acac)_2$ to achieve high performance with a transparency of 80% and an R_s value of 62.4 Ω/sq (Figure 6e). During the 30-day measurement, the R_s values of the Cu@Cu–Ni NW composites remained at 20 Ω/sq, and the fluctuation amplitude was within 2 Ω/sq (Figure 6f). Furthermore, the lifetime of the Cu@Cu–Ni NW elastomer composites was estimated to exceed 1200 days, which was one of the longest stable durations of Cu NW-based electrodes [81].

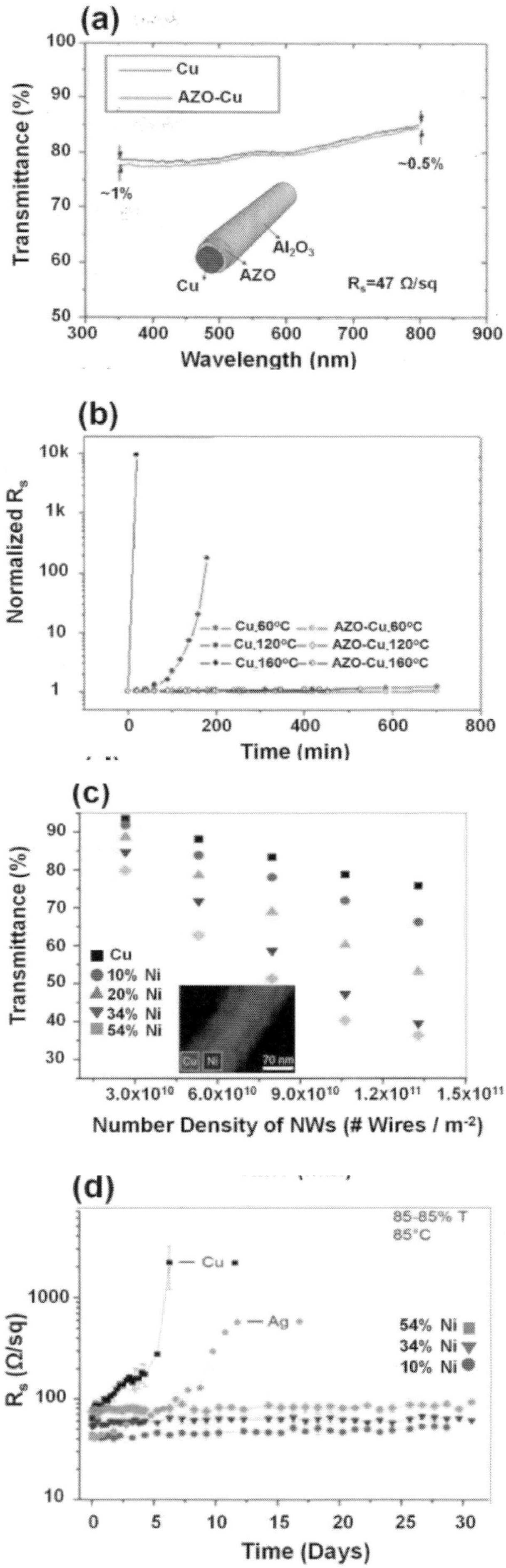

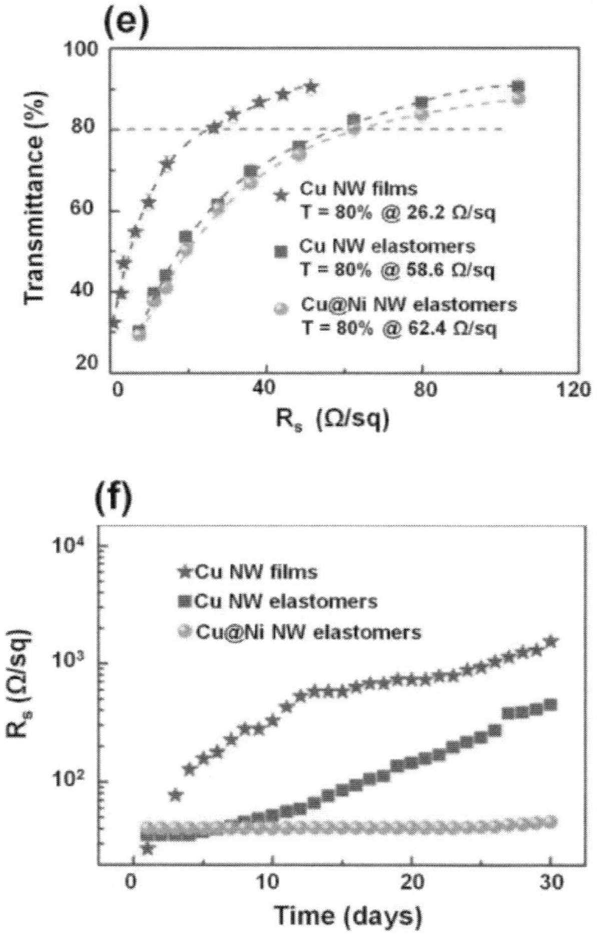

Figure 6: (**a**) Transmittance of electrospun Cu nanofibers before and after atomic layer deposition (ALD); inset is a cartoon (not to scale) of the Cu@AZO/Al_2O_3 nanofiber (**b**) Normalized sheet resistance (R_s) of bare and protected Cu NWs *vs.* baking time at 80, 120, and 160 °C. Solid circles refer to bare Cu nanofibers and hollow circles represent AZO-Cu nanofibers. In the case of bare Cu nanofibers at 120 and 160 °C, the resistance exceeded the multimeter limit (20 MΩ) in the end. (Reproduced with permission from [77]. Copyright American Chemical Society, 2012); (**c**) A plot of specular transmittance*vs.* number density of NWs, which shows the effect of increasing wire diameter on the film transmittance; inset is an energy-dispersive X-ray spectroscopy (EDS) mapping image of a cupronickel NW coated with 54 mol % nickel; (**d**) Plot of R_s *vs.* time for films of Ag NWs, Cu NWs, and cupronickel NWs stored at 85 °C (Reproduced with permission from [78]. Copyright American Chemical Society, 2012); (**e**) Transmittance as a function of R_s for various NW electrodes; (**f**) R_s *vs.* time for various NW electrodes during exposure in the natural environment. (Reproduced with permission from [81]. Copyright American Chemical Society, 2014).

More recently, it was revealed that laser irradiation [82] can turn Cu_xO NWs back into Cu NWs. Those processes can be performed rapidly at room temperature without using any vacuum process. The resulting Cu NWs exhibit superior conductivity with better resistance to the environment due to the enhanced Cu NW junctions. As a result, Cu_xO NWs can be employed as starting materials for Cu NW thin films, thereby facilitating long-term storage and versatile applications of the material.

APPLICATIONS AND PERSPECTIVE

As investigated above, Cu NW-based TCFs have shown excellent performance in terms of optoelectronic properties and mechanical flexibility. Therefore, they have potential application in various optoelectronic devices such as touchscreens [31,83], solar cells [63,84,85,86], OLEDs [81,87,88], and sensors [89,90]. Stewart *et al.* showed that organic solar cells (OSCs) using Cu_{core}–Ni_{shell} NW films as transparent anodes can achieve a device efficiency of 4.9%, nearly equivalent to the 5% efficiency of OSCs utilizing Ag NW anodes [63]. Won *et al.* also successfully adopted Cu NW-based transparent electrodes into $Cu(In_{1-x},Ga_x)(S,Se)_2$ thin-film solar cells, which exhibited a power conversion efficiency of 7.1% [85]. Song *et al.*demonstrated usage of Cu@Cu–Ni NWs as conductors in OLED devices that were stable under extreme bending and stretching conditions [81]. For sensor applications, Cu NWs have been widely investigated in the development of electrochemical biosensors due to their efficient electron transfer and good mechanical strength. Stortini *et al.* prepared an ensemble of Cu NW electrodes with highly controlled morphologies. The electrodes demonstrated efficient sensing performance with respect to nitrate determination at concentration levels as low as a few micromoles [89]. As for touchscreen applications, Han *et al.* reported a resistive-type touchscreen device employing Cu NWs as a transparent conductor [31].

The required optical and electrical properties of TCFs can be different for each application. For example, touchscreens often require an R_s value of 50–100 Ω/sq and a transmittance of more than 85% along with a low HF, whereas solar cell applications require an R_s value below 20 Ω/sq and a transmittance of more than 90% [91].

The excellent mechanical durability of Cu NWs embedded on the surface of a transparent glass-fabric reinforced plastic (GFRHybrimer) was confirmed by cyclic bending (Figure 7a). The long-term thermal and oxidation stability test of CuNW-GFRHybrimer revealed no drastic change in electrical performance after about 15 days (Figure 7b). Accordingly, we can expect to meet the demand of daily-use portable devices by adopting flexible Cu NW touchscreen panels in the near future as illustrated in Figure 7c.

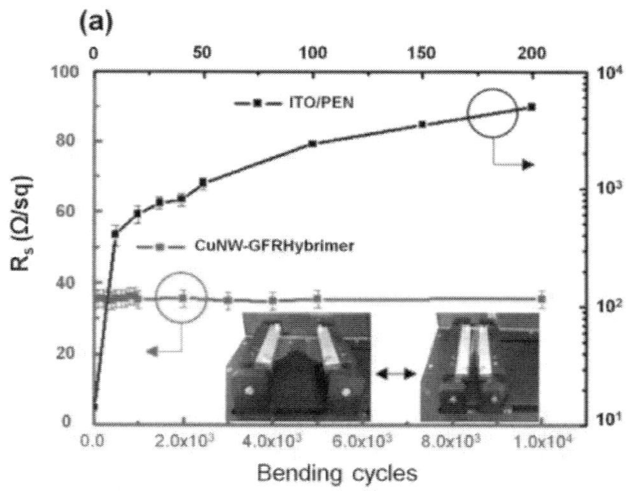

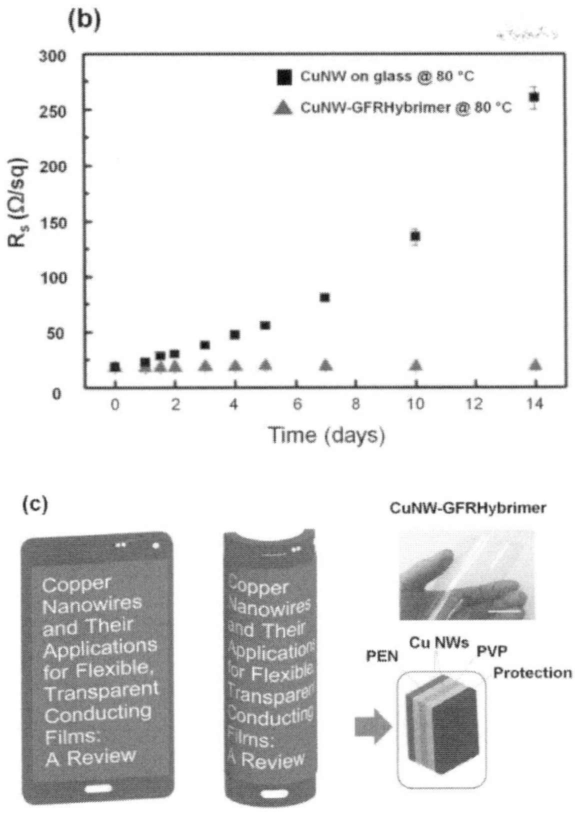

Figure 7: (**a**) Bending test results of a GFRHybrimer film (sheet resistance, R_s = 35 Ω/sq) and a reference indium tin oxide (ITO)/polyethylene naphthalate (PEN) film (R_s = 15 Ω/sq). Top and right axes are for the ITO/PEN (the inset represents the experimental setup for the bending test; bending radius is 5 mm); (**b**) Thermal and oxidation stability tests of bare Cu NW film and CuNW-GFRHybrimer film at 80 °C; (**c**) A schematic example of flexible smartphone using Cu NWs in the future. The inset is a photograph of a Cu NW-GFRHybrimer film (scale bar is 3 cm). (Reproduced with permission from [88]. Copyright American Chemical Society, 2014).

ITO has been the most widely used material for TCFs for a long time, and it will continue to play a vital role in the future. However, Cu NW films that are durable and flexible and that enhance the optoelectronic properties with lower production cost are also a possibility. In this regard, Cu NWs stand as a strong alternative to ITO; therefore, in the future, there will be immense opportunities for exploring new designs using Cu NW-based TCFs, including flexible displays and curved surfaces.

CONCLUSIONS

Cu NWs have emerged as the promising next-generation conducting materials due to their excellent performance with high conductivity, high transmittance, mechanical flexibility, and cost-effectiveness. In terms of transmittance and R_s, Cu NW films are already comparable to commercial ITO films while also exhibiting much better flexibility that allows devices based on Cu NWs to be bent or stretched. These superior characteristics make Cu NWs strong competitors for use in flexible electronics in the future. In this review, we briefly summarized the relationship between the dimensions and optoelectronic properties of NWs, and examined the control of Cu NW structure by various synthesis and fabrication routes for optimal performance of Cu NW films. We also discussed the development of new approaches to eliminating the wire-wire junction resistance between Cu NWs for better electrical performance. There are still challenges that need to be overcome, such as oxidation, haze, and large scalable fabrication, before Cu NWs can be fully integrated into commercial devices. In the future, further studies that cover the following aspects would be required: (a) the long-term stability of Cu NWs against oxygen, high temperatures, chemical etching, and mechanical strains; and (b) the fine-tuning of the structure and facile synthesis of Cu NWs. Once these problems are solved, Cu NWs as the foundation of next-generation TCFs can be expected.

ACKNOWLEDGMENTS

This research was supported by the Basic Science Research Program through the National Research Foundation of Korea (NRF) funded by the Ministry of Science, ICT and Future Planning (NRF-2015R1C1A1A02037541), and by the Gachon University research fund of 2014 (GCU-2014-0107).

REFERENCES

1. Kang, M.G.; Kim, M.S.; Kim, J.; Guo, L.J. Organic solar cells using nanoimprinted transparent metal electrodes. *Adv. Mater.* 2008, *20*, 4408–4413.
2. Lee, D.; Rho, Y.; Allen, F.I.; Minor, A.M.; Ko, S.H.; Grigoropoulos, C.P. Synthesis of hierarchical TiO_2 nanowires with densely-packed and omnidirectional branches. *Nanoscale* 2013, *5*, 11147–11152.
3. Wu, J.; Agrawal, M.; Becerril, H.A.; Bao, Z.; Liu, Z.; Chen, Y.; Peumans, P. Organic light-emitting diodes on solution-processed graphene transparent electrodes. *ACS Nano* 2009, *4*, 43–48.
4. Tak, Y.-H.; Kim, K.-B.; Park, H.-G.; Lee, K.-H.; Lee, J.-R. Criteria for ITO (indium–tin-oxide) thin film as the bottom electrode of an organic light emitting diode. *Thin Solid Films* 2002, *411*, 12–16.
5. Loureiro, J.; Neves, N.; Barros, R.; Mateus, T.; Santos, R.; Filonovich, S.; Reparaz, S.; Sotomayor-Torres, C.M.; Wyczisk, F.; Divay, L.; *et al*. Transparent aluminium zinc oxide thin films with enhanced thermoelectric properties. *J. Mater. Chem. A* 2014, *2*, 6649–6655.
6. Bai, S.L.; Sun, C.Z.; Wan, P.B.; Wang, C.; Luo, R.X.; Li, Y.P.; Liu, J.F.; Sun, X.M. Transparent conducting films of hierarchically nanostructured polyaniline networks on flexible substrates for high-performance gas sensors. *Small*2015, *11*, 306–310.
7. Gordon, R.G. Criteria for choosing transparent conductors. *MRS Bull.* 2000, *25*, 52–57.
8. Rathmell, A.R.; Bergin, S.M.; Hua, Y.L.; Li, Z.Y.; Wiley, B.J. The growth mechanism of copper nanowires and their properties in flexible, transparent conducting films. *Adv. Mater.* 2010, *22*, 3558–3563.
9. Ding, Z.Q.; Zhu, Y.P.; Branford-White, C.; Sun, K.; Um-I-Zahra, S.; Quan, J.; Nie, H.L.; Zhu, L.M. Self-assembled transparent conductive composite films of carboxylated multi-walled carbon nanotubes/

poly(vinyl alcohol) electrospun nanofiber mats. *Mater. Lett.* 2014, *128*, 310–313.

10. McCarthy, M.A.; Liu, B.; Donoghue, E.P.; Kravchenko, I.; Kim, D.Y.; So, F.; Rinzler, A.G. Low-voltage, low-power, organic light-emitting transistors for active matrix displays. *Science* 2011, *332*, 570–573.
11. Sagar, R.U.R.; Zhang, X.Z.; Xiong, C.Y.; Yu, Y. Semiconducting amorphous carbon thin films for transparent conducting electrodes. *Carbon* 2014, *76*, 64–70.
12. Kim, K.S.; Zhao, Y.; Jang, H.; Lee, S.Y.; Kim, J.M.; Kim, K.S.; Ahn, J.H.; Kim, P.; Choi, J.Y.; Hong, B.H. Large-scale pattern growth of graphene films for stretchable transparent electrodes. *Nature* 2009, *457*, 706–710.
13. Girtan, M.; Vlad, A.; Mallet, R.; Bodea, M.A.; Pedarnig, J.D.; Stanculescu, A.; Mardare, D.; Leontie, L.; Antohe, S. On the properties of aluminium doped zinc oxide thin films deposited on plastic substrates from ceramic targets. *Appl. Surf. Sci.* 2013, *274*, 306–313.
14. Chen, T.-H.; Chen, T.-Y. Effects of annealing temperature on properties of Ti-Ga-doped ZnO films deposited on flexible substrates. *Nanomaterials* 2015, *5*, 1831–1839.
15. Lee, D.; Pan, H.; Ko, S.H.; Park, H.K.; Kim, E.; Grigoropoulos, C.P. Non-vacuum, single-step conductive transparent ZnO patterning by ultra-short pulsed laser annealing of solution-deposited nanoparticles. *Appl. Phys. A* 2012, *107*, 161–171.
16. De, S.; Higgins, T.M.; Lyons, P.E.; Doherty, E.M.; Nirmalraj, P.N.; Blau, W.J.; Boland, J.J.; Coleman, J.N. Silver nanowire networks as flexible, transparent, conducting films: Extremely high DC to optical conductivity ratios. *ACS Nano* 2009, *3*, 1767–1774.
17. Zhang, Y.Y.; Ronning, F.; Gofryk, K.; Mara, N.A.; Haberkorn, N.; Zou, G.F.; Wang, H.Y.; Lee, J.H.; Bauer, E.; McCleskey, T.M.; *et al.* Aligned carbon nanotubes sandwiched in epitaxial NbC film for enhanced superconductivity.*Nanoscale* 2012, *4*, 2268–2271.
18. Lee, D.; Paeng, D.; Park, H.K.; Grigoropoulos, C.P. Vacuum-free, maskless patterning of Ni electrodes by laser reductive sintering of NiO nanoparticle ink and its application to transparent conductors. *ACS Nano* 2014, *8*, 9807–9814.
19. Liu, Y.K.; Sie, Y.Y.; Liu, C.A.; Lee, M.T. A novel laser direct writing system integrated with A & F XXY alignment platform for rapid fabrication of flexible electronics. *Smart Sci.* 2015, *3*, 87–91.
20. Ge, Z.B.; Wu, S.T. Nanowire grid polarizer for energy efficient and wide-view liquid crystal displays. *Appl. Phys. Lett.*2008, *93*.
21. Kim, S.H.; Park, J.D.; Lee, K.D. Fabrication of a nano-wire grid polarizer for brightness enhancement in liquid crystal display. *Nanotechnology* 2006, *17*, 4436–4438.
22. Lee, M.S.; Lee, K.; Kim, S.Y.; Lee, H.; Park, J.; Choi, K.H.; Kim, H.K.; Kim, D.G.; Lee, D.Y.; Nam, S.; *et al.* High-performance, transparent, and stretchable electrodes using graphene-metal nanowire hybrid structures. *Nano Lett.*2013, *13*, 2814–2821.
23. Arnold, M.S.; Stupp, S.I.; Hersam, M.C. Enrichment of single-walled carbon nanotubes by diameter in density gradients. *Nano Lett.* 2005, *5*, 713–718.
24. Spitalsky, Z.; Tasis, D.; Papagelis, K.; Galiotis, C. Carbon nanotube-polymer composites: Chemistry, processing, mechanical and electrical properties. *Prog. Polym. Sci.* 2010, *35*, 357–401.
25. Song, M.; You, D.S.; Lim, K.; Park, S.; Jung, S.; Kim, C.S.; Kim, D.H.; Kim, D.G.; Kim, J.K.; Park, J.; *et al.* Highly efficient and bendable organic solar cells with solution-processed silver nanowire electrodes. *Adv. Funct. Mater.* 2013, *23*, 4177–4184.
26. Yao, H.F.; Sun, J.L.; Liu, W.; Sun, H.S. Effect of a magnetic field on the preparation of silver nanowires using solid electrolyte thin films. *J. Mater. Sci. Technol.* 2007, *23*, 39–42.
27. Eom, H.; Lee, J.; Pichitpajongkit, A.; Amjadi, M.; Jeong, J.H.; Lee, E.; Lee, J.Y.; Park, I. Ag@Ni core-shell nanowire network for robust transparent electrodes against oxidation and sulfurization. *Small* 2014, *10*, 4171–4181.
28. Jewell, S.; Kimball, S. *Mineral Commodity Summaries 2014*; US Geological Survey: Reston, VA, USA, 2014.
29. Cui, F.; Yu, Y.; Dou, L.T.; Sun, J.W.; Yang, Q.; Schildknecht, C.; Schierle-Arndt, K.; Yang, P.D. Synthesis of ultrathin copper nanowires using tris(trimethylsilyl) silane for high-performance and low-haze transparent conductors. *Nano Lett.* 2015, *15*, 7610–7615.
30. Ye, S.R.; Rathmell, A.R.; Stewart, I.E.; Ha, Y.C.; Wilson, A.R.; Chen, Z.F.; Wiley, B.J. A rapid synthesis of high aspect ratio copper nanowires for high-performance transparent conducting films. *Chem. Commun.* 2014, *50*, 2562–2564.
31. Han, S.; Hong, S.; Ham, J.; Yeo, J.; Lee, J.; Kang, B.; Lee, P.; Kwon, J.; Lee, S.S.; Yang, M.Y.; *et al.* Fast plasmonic laser nanowelding for a Cu-nanowire percolation network for flexible transparent conductors and stretchable electronics.*Adv. Mater.* 2014, *26*, 5808–5814.
32. Hecht, D.S.; Hu, L.B.; Irvin, G. Emerging transparent electrodes based on thin films of carbon nanotubes, graphene, and metallic nanostructures. *Adv. Mater.* 2011, *23*, 1482–1513.
33. O'Connor, B.; Haughn, C.; An, K.H.; Pipe, K.P.; Shtein, M. Transparent and conductive electrodes based on unpatterned, thin metal films. *Appl. Phys. Lett.* 2008, *93*.
34. Fuchs, K. The conductivity of thin metallic films according to the electron theory of metals. *Math. Proc. Camb. Philos. Soc.* 1938, *34*, 100–108.
35. Dingle, R. The electrical conductivity of thin wires. *Proc. R. Soc. Lond. A* 1950, *201*, 545–560.
36. Bid, A.; Bora, A.; Raychaudhuri, A.K. Temperature dependence of the resistance of metallic nanowires of diameter $\geq$ 15 nm: Applicability of bloch-gruneisen theorem. *Phys. Rev. B* 2006, *74*.
37. De, S.; King, P.J.; Lyons, P.E.; Khan, U.; Coleman, J.N. Size effects and the problem with percolation

in nanostructured transparent conductors. *ACS Nano* 2010, *4*, 7064–7072.

38. Van de Groep, J.; Spinelli, P.; Polman, A. Transparent conducting silver nanowire networks. *Nano Lett.* 2012, *12*, 3138–3144.

39. Li, J.; Zhang, S.-L. Finite-size scaling in stick percolation. *Phys. Rev. E* 2009, *80*.

40. Bergin, S.M.; Chen, Y.H.; Rathmell, A.R.; Charbonneau, P.; Li, Z.Y.; Wiley, B.J. The effect of nanowire length and diameter on the properties of transparent, conducting nanowire films. *Nanoscale* 2012, *4*, 1996–2004.

41. Borchert, J.W.; Stewart, I.E.; Ye, S.R.; Rathmell, A.R.; Wiley, B.J.; Winey, K.I. Effects of length dispersity and film fabrication on the sheet resistance of copper nanowire transparent conductors. *Nanoscale* 2015, *7*, 14496–14504.

42. Mutiso, R.M.; Sherrott, M.C.; Rathmell, A.R.; Wiley, B.J.; Winey, K.I. Integrating simulations and experiments to predict sheet resistance and optical transmittance in nanowire films for transparent conductors. *ACS Nano* 2013, *7*, 7654–7663.

43. Ahn, K.; Kim, D.; Kim, O.; Nam, J. Analysis of transparent conductive silver nanowire films from dip coating flow. *J. Coat. Technol. Res.* 2015, *12*, 855–862.

44. Kim, T.; Canlier, A.; Cho, C.; Rozyyev, V.; Lee, J.Y.; Han, S.M. Highly transparent Au-coated ag nanowire transparent electrode with reduction in haze. *ACS Appl. Mater. Interfaces* 2014, *6*, 13527–13534.

45. Khanarian, G.; Joo, J.; Liu, X.Q.; Eastman, P.; Werner, D.; O'Connell, K.; Trefonas, P. The optical and electrical properties of silver nanowire mesh films. *J. Appl. Phys.* 2013, *114*.

46. Preston, C.; Xu, Y.L.; Han, X.G.; Munday, J.N.; Hu, L.B. Optical haze of transparent and conductive silver nanowire films. *Nano Res.* 2013, *6*, 461–468.

47. Preston, C.; Fang, Z.Q.; Murray, J.; Zhu, H.L.; Dai, J.Q.; Munday, J.N.; Hu, L.B. Silver nanowire transparent conducting paper-based electrode with high optical haze. *J. Mater. Chem. C* 2014, *2*, 1248–1254.

48. Araki, T.; Jiu, J.T.; Nogi, M.; Koga, H.; Nagao, S.; Sugahara, T.; Suganuma, K. Low haze transparent electrodes and highly conducting air dried films with ultra-long silver nanowires synthesized by one-step polyol method. *Nano Res.*2014, *7*, 236–245.

49. Choi, H.; Park, S.H. Seedless growth of free-standing copper nanowires by chemical vapor deposition. *J. Am. Chem. Soc.* 2004, *126*, 6248–6249.

50. Haase, D.; Hampel, S.; Leonhardt, A.; Thomas, J.; Mattern, N.; Buchner, B. Facile one-step-synthesis of carbon wrapped copper nanowires by thermal decomposition of copper(II)-acetylacetonate. *Surf. Coat. Technol.* 2007, *201*, 9184–9188.

51. Khalil, A.; Hashaikeh, R.; Jouiad, M. Synthesis and morphology analysis of electrospun copper nanowires. *J. Mater. Sci.* 2014, *49*, 3052–3065.

52. Zang, W.L.; Li, P.; Fu, Y.M.; Xing, L.L.; Xue, X.Y. Hydrothermal synthesis of Co-ZnO nanowire array and its application as piezo-driven self-powered humidity sensor with high sensitivity and repeatability. *RSC Adv.* 2015, *5*, 84343–84349.

53. Xu, P.; Chen, W.Z.; Wang, Q.; Zhu, T.S.; Wu, M.J.; Qiao, J.L.; Chen, Z.W.; Zhang, J.J. Effects of transition metal precursors (Co, Fe, Cu, Mn, or Ni) on pyrolyzed carbon supported metal-aminopyrine electrocatalysts for oxygen reduction reaction. *RSC Adv.* 2015, *5*, 6195–6206.

54. Panciera, F.; Chou, Y.C.; Reuter, M.C.; Zakharov, D.; Stach, E.A.; Hofmann, S.; Ross, F.M. Synthesis of nanostructures in nanowires using sequential catalyst reactions. *Nat. Mater.* 2015, *14*, 820–825.

55. Chang, Y.; Lye, M.L.; Zeng, H.C. Large-scale synthesis of high-quality ultralong copper nanowires. *Langmuir* 2005, *21*, 3746–3748.

56. Shi, Y.; Li, H.; Chen, L.Q.; Huang, X.J. Obtaining ultra-long copper nanowires via a hydrothermal process. *Sci. Technol. Adv. Mater.* 2005, *6*, 761–765.

57. Jin, M.S.; He, G.N.; Zhang, H.; Zeng, J.; Xie, Z.X.; Xia, Y.N. Shape-controlled synthesis of copper nanocrystals in an aqueous solution with glucose as a reducing agent and hexadecylamine as a capping agent. *Angew. Chem. Int. Ed.*2011, *50*, 10560–10564.

58. Guo, H.Z.; Lin, N.; Chen, Y.Z.; Wang, Z.W.; Xie, Q.S.; Zheng, T.C.; Gao, N.; Li, S.P.; Kang, J.Y.; Cai, D.J.; *et al.* Copper nanowires as fully transparent conductive electrodes. *Sci. Rep.* 2013, *3*.

59. Zhang, D.Q.; Wang, R.R.; Wen, M.C.; Weng, D.; Cui, X.; Sun, J.; Li, H.X.; Lu, Y.F. Synthesis of ultralong copper nanowires for high-performance transparent electrodes. *J. Am. Chem. Soc.* 2012, *134*, 14283–14286.

60. Ahn, Y.; Jeong, Y.; Lee, Y. Improved thermal oxidation stability of solution-processable silver nanowire transparent electrode by reduced graphene oxide. *ACS Appl. Mater. Interfaces* 2012, *4*, 6410–6414.

61. Chen, T.G.; Huang, B.Y.; Liu, H.W.; Huang, Y.Y.; Pan, H.T.; Meng, H.F.; Yu, P.C. Flexible silver nanowire meshes for high-efficiency microtextured organic-silicon hybrid photovoltaics. *ACS Appl. Mater. Interfaces* 2012, *4*, 6856–6863.

62. Chung, C.H.; Song, T.B.; Bob, B.; Zhu, R.; Duan, H.S.; Yang, Y. Silver nanowire composite window layers for fully solution-deposited thin-film photovoltaic devices. *Adv. Mater.* 2012, *24*, 5499–5504.

63. Stewart, I.E.; Rathmell, A.R.; Yan, L.; Ye, S.R.; Flowers, P.F.; You, W.; Wiley, B.J. Solution-processed copper-nickel nanowire anodes for organic solar cells. *Nanoscale* 2014, *6*, 5980–5988.

64. Li, S.J.; Chen, Y.Y.; Huang, L.J.; Pan, D.C. Large-scale synthesis of well-dispersed copper nanowires in an electric pressure cooker and their application in transparent and conductive networks. *Inorg. Chem.* 2014, *53*, 4440–4444.

65. Akter, T.; Kim, W.S. Reversibly stretchable transparent conductive coatings of spray-deposited silver nanowires. *ACS Appl. Mater. Interfaces* 2012, *4*, 1855–1859.

66. Celle, C.; Mayousse, C.; Moreau, E.; Basti, H.; Carella, A.; Simonato, J.P. Highly flexible transparent film heaters

based on random networks of silver nanowires. *Nano Res.* 2012, *5*, 427–433.

67. Deng, B.; Hsu, P.C.; Chen, G.C.; Chandrashekar, B.N.; Liao, L.; Ayitimuda, Z.; Wu, J.X.; Guo, Y.F.; Lin, L.; Zhou, Y.; *et al.* Roll-to-roll encapsulation of metal nanowires between graphene and plastic substrate for high-performance flexible transparent electrodes. *Nano Lett.* 2015, *15*, 4206–4213.

68. Kang, S.; Kim, T.; Cho, S.; Lee, Y.; Choe, A.; Walker, B.; Ko, S.J.; Kim, J.Y.; Ko, H. Capillary printing of highly aligned silver nanowire transparent electrodes for high-performance optoelectronic devices. *Nano Lett.* 2015, *15*, 7933–7942.

69. Hu, L.B.; Kim, H.S.; Lee, J.Y.; Peumans, P.; Cui, Y. Scalable coating and properties of transparent, flexible, silver nanowire electrodes. *ACS Nano* 2010, *4*, 2955–2963.

70. Tokuno, T.; Nogi, M.; Karakawa, M.; Jiu, J.T.; Nge, T.T.; Aso, Y.; Suganuma, K. Fabrication of silver nanowire transparent electrodes at room temperature. *Nano Res.* 2011, *4*, 1215–1222.

71. Lee, J.Y.; Connor, S.T.; Cui, Y.; Peumans, P. Solution-processed metal nanowire mesh transparent electrodes. *Nano Lett.*2008, *8*, 689–692.

72. Garnett, E.C.; Cai, W.S.; Cha, J.J.; Mahmood, F.; Connor, S.T.; Christoforo, M.G.; Cui, Y.; McGehee, M.D.; Brongersma, M.L. Self-limited plasmonic welding of silver nanowire junctions. *Nat. Mater.* 2012, *11*, 241–249.

73. He, T.; Xie, A.; Reneker, D.H.; Zhu, Y. A tough and high-performance transparent electrode from a scalable and transfer-free method. *ACS Nano* 2014, *8*, 4782–4789.

74. Gaynor, W.; Burkhard, G.F.; McGehee, M.D.; Peumans, P. Smooth nanowire/polymer composite transparent electrodes. *Adv. Mater.* 2011, *23*, 2905–2910.

75. Yu, Z.B.; Zhang, Q.W.; Li, L.; Chen, Q.; Niu, X.F.; Liu, J.; Pei, Q.B. Highly flexible silver nanowire electrodes for shape-memory polymer light-emitting diodes. *Adv. Mater.* 2011, *23*, 664–668.

76. Miller, M.S.; O'Kane, J.C.; Niec, A.; Carmichael, R.S.; Carmichael, T.B. Silver nanowire/optical adhesive coatings as transparent electrodes for flexible electronics. *ACS Appl. Mater. Interfaces* 2013, *5*, 10165–10172.

77. Hsu, P.C.; Wu, H.; Carney, T.J.; McDowell, M.T.; Yang, Y.; Garnett, E.C.; Li, M.; Hu, L.B.; Cui, Y. Passivation coating on electrospun copper nanofibers for stable transparent electrodes. *ACS Nano* 2012, *6*, 5150–5156.

78. Rathmell, A.R.; Nguyen, M.; Chi, M.F.; Wiley, B.J. Synthesis of oxidation-resistant cupronickel nanowires for transparent conducting nanowire networks. *Nano Lett.* 2012, *12*, 3193–3199.

79. Chen, Z.F.; Ye, S.R.; Stewart, I.E.; Wiley, B.J. Copper nanowire networks with transparent oxide shells that prevent oxidation without reducing transmittance. *ACS Nano* 2014, *8*, 9673–9679.

80. Mehta, R.; Chugh, S.; Chen, Z. Enhanced electrical and thermal conduction in graphene-encapsulated copper nanowires. *Nano Lett.* 2015, *15*, 2024–2030.

81. Song, J.Z.; Li, J.H.; Xu, J.Y.; Zeng, H.B. Superstable transparent conductive $Cu@Cu_4Ni$ nanowire elastomer composites against oxidation, bending, stretching, and twisting for flexible and stretchable optoelectronics. *Nano Lett.* 2014, *14*, 6298–6305.

82. Han, S.; Hong, S.; Yeo, J.; Kim, D.; Kang, B.; Yang, M.Y.; Ko, S.H. Nanorecycling: Monolithic integration of copper and copper oxide nanowire network electrode through selective reversible photothermochemical reduction. *Adv. Mater.*2015, *27*.

83. Bao, C.X.; Yang, J.; Gao, H.; Li, F.M.; Yao, Y.F.; Yang, B.; Fu, G.; Zhou, X.X.; Yu, T.; Qin, Y.Q.; *et al. In situ* fabrication of highly conductive metal nanowire networks with high transmittance from deep-ultraviolet to near-infrared. *ACS Nano* 2015, *9*, 2502–2509.

84. Ahn, Y.; Jeong, Y.; Lee, D.; Lee, Y. Copper nanowire-graphene core-shell nanostructure for highly stable transparent conducting electrodes. *ACS Nano* 2015, *9*, 3125–3133.

85. Won, Y.; Kim, A.; Lee, D.; Yang, W.; Woo, K.; Jeong, S.; Moon, J. Annealing-free fabrication of highly oxidation-resistive copper nanowire composite conductors for photovoltaics. *NPG Asia Mater.* 2014, *6*.

86. Ko, S.H. Review of the multi-scale nano-structure approach to the development of high efficiency solar cells. *Smart Sci.* 2014, *2*, 54–62.

87. Won, Y.; Kim, A.; Yang, W.; Jeong, S.; Moon, J. A highly stretchable, helical copper nanowire conductor exhibiting a stretchability of 700%. *NPG Asia Mater.* 2014, *6*.

88. Im, H.G.; Jung, S.H.; Jin, J.; Lee, D.; Lee, J.; Lee, D.; Lee, J.Y.; Kim, I.D.; Bae, B.S. Flexible transparent conducting hybrid film using a surface-embedded copper nanowire network: A highly oxidation-resistant copper nanowire electrode for flexible optoelectronics. *ACS Nano* 2014, *8*, 10973–10979.

89. Stortini, A.M.; Moretto, L.M.; Mardegan, A.; Ongaro, M.; Ugo, P. Arrays of copper nanowire electrodes: Preparation, characterization and application as nitrate sensor. *Sens. Actuators B* 2015, *207*, 186–192.

90. Zhao, Y.X.; Fan, L.L.; Zhang, Y.; Zhao, H.; Li, X.J.; Li, Y.P.; Wen, L.; Yan, Z.F.; Huo, Z.Y. Hyper-branched $Cu@Cu_2O$ coaxial nanowires mesh electrode for ultra-sensitive glucose detection. *ACS Appl. Mater. Interfaces* 2015, *7*, 16802–16812.

91. Rowell, M.W.; McGehee, M.D. Transparent electrode requirements for thin film solar cell modules. *Energ Environ. Sci.*2011, *4*, 131–134.

CHAPTER 17

ELECTRONIC TONGUE BASED ON NANOSTRUCTURED HYBRID FILMS OF GOLD NANOPARTICLES AND PHTHALOCYANINES FOR MILK ANALYSIS

Luiza A. Mercante[1] Vanessa P. Scagion[1,2] Adriana Pavinatto[1] Rafaela C. Sanfelice[1] Luiz H. C. Mattoso[1] and Daniel S. Correa[1,2]

[1]National Laboratory for Nanotechnology in Agribusiness (LNNA), Embrapa Instrumentation, 13560-970 São Carlos, SP, Brazil

[2]Center for Exact Sciences and Technology, Federal University of São Carlos (UFSCar), 13565-905 São Carlos, SP, Brazil

ABSTRACT

The use of gold nanoparticles combined with other organic and inorganic materials for designing nanostructured films has demonstrated their versatility for various applications, including optoelectronic devices and chemical sensors. In this study, we reported the synthesis and characterization of gold nanoparticles stabilized with poly(allylamine hydrochloride) (Au@PAH NPs), as well as the capability of this material to form multilayer Layer-by-Layer (LbL) nanostructured films with metal tetrasulfonated phthalocyanines (MTsPc). Film growth was monitored by UV-Vis absorption spectroscopy, atomic force microscopy (AFM), and Fourier transform infrared spectroscopy (FTIR). Once LbL films have been applied as active layers in chemical sensors, Au@PAH/MTsPc and PAH/MTsPc LbL films were used in an electronic tongue system for milk analysis regarding fat content. The capacitance data were treated using Principal Component Analysis (PCA), revealing the role played by the gold nanoparticles on the LbL films electrical properties, enabling this kind of system to be used for analyzing complex matrices such as milk without any prior pretreatment.

INTRODUCTION

During the past few years, hybrid nanomaterials have been widely used for several kinds of applications such as the development of chemical sensors [1–4] and biosensors [5–7]. Although there are many methods for producing such materials, strong emphasis has been given on the fabrication techniques that allow the precise control of molecular architectures, including the Langmuir-Blodgett (LB) and Layer-by-Layer (LbL) methods [8]. The LbL technique [9] is probably the most commonly used technique owing to its efficiency and molecular architecture control, combined with its experimental simplicity and low cost [10]. Besides the architecture of the film, the choice of the materials is also important to determine the film final properties. Among the many possible materials, gold nanoparticles have been used once they present high surface area and electrical conductivity and catalytic properties that make them excellent materials for the detection of a wide range of analytes [11, 12]. However, the use of gold nanoparticles for improving the electrical properties of LbL films in chemical sensor arrays, such as electronic tongues, has not received much attention.

In this study we focus on nanostructured architectures whose properties are varied significantly by the interplay among the constituent materials. Specifically, here we report the synthesis of gold nanoparticles functionalized with poly(allylamine hydrochloride) and its use for the

Citation: Luiza A. Mercante, Vanessa P. Scagion, Adriana Pavinatto, Rafaela C. Sanfelice, Luiz H. C. Mattoso, and Daniel S. Correa, "Electronic Tongue Based on Nanostructured Hybrid Films of Gold Nanoparticles and Phthalocyanines for Milk Analysis," Journal of Nanomaterials, vol. 2015, Article ID 890637, 7 pages, 2015. doi:10.1155/2015/890637.

fabrication of LbL films with copper and nickel tetrasulfonated phthalocyanines. We use the functionalized nanoparticles to enhance the electrical properties of the LbL films, which were applied in chemical sensing experiments using an electronic tongue (e-tongue) system. The e-tongue was composed of an array of platinum interdigitated electrodes coated with LbL films of distinct compositions and used to analyze milk samples via impedance spectroscopy technique. The electronic tongue was able to distinguish milks with different fat contents, showing the practical utility of the employed approach.

EXPERIMENTAL

Materials

Copper(II) tetrasulfonated phthalocyanine (CuTsPc), nickel(II) tetrasulfonated phthalocyanine (NiTsPc), poly(allylamine hydrochloride) (PAH, M_w = 17.500), sodium borohydride ($NaBH_4$), and tetrachloroauric(III) acid ($HAuCl_4$) were purchased from Sigma-Aldrich and used as received. Platinum interdigitated electrodes (IDEs), each one comprising 50 pairs of fingers with finger width and distance between fingers of 10 μm, were designed to exhibit a capacitive profile. The IDEs were fabricated by photolithography at the Brazilian Nanotechnology National Laboratory (LNNano-LNLS).

Synthesis of Au@PAH Nanoparticles

Gold nanoparticles were synthetized following the procedure described previously by Alencar et al. [13]. 1.0 mL of aqueous $NaBH_4$ solution (10^{-2} mol L^{-1}) was added dropwise under vigorous stirring to a HCl solution (pH 3) containing 1.0 mL PAH (10^{-2} mol L^{-1}) and 1.0 mL $HAuCl_4$ (10^{-3} mol L^{-1}). The stoichiometric ratio of $NaBH_4$ and $HAuCl_4$ was kept at 10 to ensure total reduction of Au^{3+} to Au^0. After 30 min, the reaction was complete, with the aqueous solution containing Au@PAH NPs displaying a reddish color, characteristic of gold nanoparticles formation.

Preparation of LbL Films

Solutions of 1 mg mL^{-1} of Au@PAH NPs, PAH, CuTsPc, and NiTsPc were prepared by dispersing the materials in HCl aqueous solution (pH 3). LbL films were deposited on quartz, silicon, and platinum interdigitated electrodes substrates by alternating immersion into cationic (Au@PAH NPs and PAH) and anionic (CuTsPc and NiTsPc) solutions. PAH/CuTsPc, PAH/NiTsPc, Au@PAH/CuTsPc, and Au@PAH/NiTsPc LbL films were prepared varying the number of bilayers by immersing the substrate for 30 min in the Au@PAH NPs solution and for 5 min into the PAH or phthalocyanines solutions.

LbL Film Characterization

The morphology and thickness of LbL films adsorbed on glass substrates were assessed by AFM using a Dimension V (Veeco) microscope. All images were obtained in the Tapping mode with a scan rate of 0.6 Hz, using silicon nitride tips attached to a cantilever of spring constant of 2.5 N m^{-1}. The root-mean square roughness (R_{rms}) was calculated using Gwydion 2.1 data analysis software. The UV-Vis absorption spectra were recorded using a Perkin-Elmer Lambda 25 on quartz slides. FTIR spectroscopy was performed on a Perkin-Elmer 1000 using silicon wafers as substrates for the films. A total of 64 scans were collected at a resolution of 2 cm^{-1}.

Milk Samples

Twenty-four milks from distinct commercial brands with different fat contents were purchased from a local supplier. The 24 samples included 10 fat, 7 semi-skimmed, and 7 skimmed milk samples. The milk samples were prepared by adding 100 μL of each milk sample to 100 mL of phosphate buffer (pH = 6.5), which were analyzed by the e-tongue system without any prior pretreatment.

Electrical Measurements

Impedance spectroscopy measurements of milk samples using the e-tongue were carried out with a Solartron analyzer (model 1260A). All measurements were carried out in phosphate buffer solutions (pH = 6.5) and the data were collected in the range from 1 Hz up to 1 MHz, with an ac applied voltage of 50 mV. The sensor array composing the e-tongue was comprised of three sensing units: a bare platinum interdigitated electrode and two electrodes coated with 7-bilayer LbL film of PAH/CuTsPc and PAH/NiTsPc (named e-tongue 1). For comparison and in order to investigate the effect of gold nanoparticles on the electrical performance of the LbL films, electrodes coated with 7-bilayer LbL film of Au@PAH/CuTsPc and Au@PAH/NiTsPc were also prepared (named e-tongue 2). The electrodes were left soaked into the solutions for 20 min prior to the start of the electrical measurements. Five independent acquisitions using the two e-tongues were carried out for each milk sample. By carefully washing the sensing units with deionized water to remove possible molecules entrapped within the films, the sensors could be reused in further experiments. Electrical measurements in phosphate buffer solution before and after exposing the sensing units to the milk samples yielded similar results (data not shown), indicating that sensor contamination was negligible, reinforcing the reusability of the sensor. The capacitance values obtained were treated using the statistical method known as Principal Component Analysis (PCA), which correlates samples by decreasing the dimensionality of the original data without losing relevant information. Briefly, PCA makes a linear combination of the data, creating another pair of orthogonal axes into the directions where the higher correlations exist [14].

RESULTS AND DISCUSSION

Synthesis of Au@PAH Nanoparticles

For the synthesis of gold nanoparticles, $NaBH_4$ is employed as the reducing agent of gold ions. However, once $NaBH_4$ is not a strong stabilizing agent for the nanoparticles, PAH was used as stabilizer in order to prevent nanoparticle agglomeration and precipitation [13]. The Au@PAH NPs formed display a surface plasmon resonance (SPR) absorption band centered at 520 nm, according to the UV-VIS spectrum (Figure 1(a)) and spherical shape with a mean diameter of 11 nm, as shown in the TEM image (Figure 1(b)).

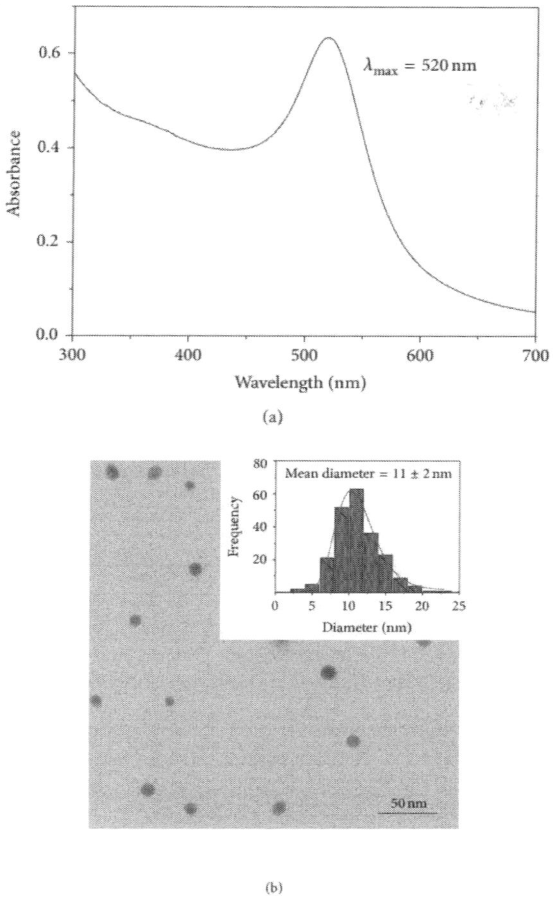

Figure 1: (a) UV-Vis absorption spectrum and (b) TEM image of Au@PAH nanoparticles. The inset in (b) shows the histogram of size distribution.

The colloidal stability of Au@PAH nanoparticles was analyzed by two parameters, the hydrodynamic diameter and zeta potential. The first parameter is associated with the fact that when the nanoparticles are not stable in solution, they tend to agglomerate, yielding higher values of diameters. The zeta potential is related to the effective charge of the particles. The higher the absolute potential value, the higher the stability of the colloidal system, once the charged particles tend to repel each other and therefore overcome the natural tendency of aggregation. Before measuring the colloidal stability of Au@PAH nanoparticles, the dispersion was filtered through a membrane having pore size of 0.22 μM. The values found for the hydrodynamic diameter and zeta potential were 22 nm and 29 mV, respectively, confirming the stability of the synthesized nanoparticles. The positive value of the zeta potential is related to positive charge of PAH due to the protonation of the amino group in the polymer chain. The hydrodynamic diameter value close to the nanoparticles diameter obtained by TEM images (Figure 1(b)) demonstrates that PAH acts as a stabilizer avoiding the agglomeration of the nanoparticles. This information is important because it allows such nanoparticles to be used in the fabrication of LbL films, as will be discussed in the following section.

LbL Films Characterization

Nanostructured LbL films have been widely used as active layers in sensors and biosensors platforms, once they are one of the main approaches used to fabricate highly organized and morphologically controlled nanostructures at molecular level [15–17]. The LbL films architectures were built on quartz substrate and the sequential growth was monitored by UV-Vis absorption spectroscopy, as shown in Figure 2, for films formed by Au@PAH/CuTsPc (Figure 2(a)) and Au@PAH/NiTsPc (Figure 2(b)). The spectra show two distinct absorption regions, one in the visible and the other in the ultraviolet range, assigned to π-π^* electronic transitions from macrocyclic ring of phthalocyanine, referenced as Q and B bands, respectively [18]. More specifically, Q bands arise due to electronic transitions from electrons in the outer rings while B bands arise from electrons in both inner and outer rings [18]. The Q band comprises two different peaks displaying a maximum of absorption: at 617 nm (Figure 2(a)) and 619 nm (Figure 2(b)) assigned to the absorption of dimmers and higher order of aggregates and at 669 nm (Figure 2(a)) and 667 nm (Figure 2(b)) related to the monomers absorption [19]. The B bands are centered around 340 nm for both films and the other absorption regions are attributed to the metal compound in the center of the phthalocyanine structures as reported by Edwards and Gouterman [20]. It is worth noting that the band at 520 nm assigned to the gold nanoparticles surface plasmon resonance could not be observed in the UV-Vis absorption spectra due to the overlapping with Q bands from phthalocyanines and also the low amount of the nanoparticles in the films [13].

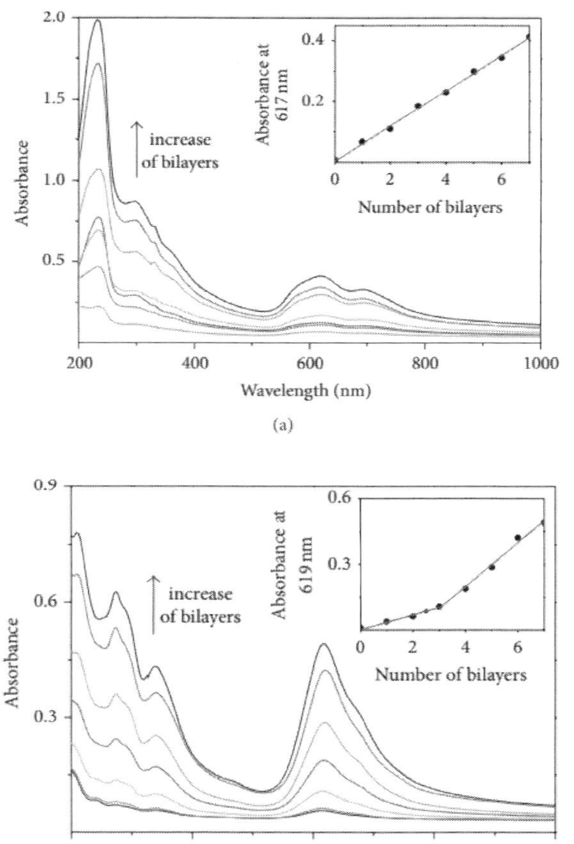

Figure 2: UV-Vis absorption spectra for LbL films ranging from 1 up to 7 bilayers of (a) Au@PAH/CuTsPc and (b) Au@PAH/NiTsPc. The inset shows the absorbance at (a) 617 nm and (b) 619 nm versus number of bilayers for respective architectures.

A linear growth was observed for the film formed by Au@PAH/CuTsPc (inset of Figure 2(a)). The absorbance at 617 nm increased linearly with the number of bilayers, which indicates that the same amount of material was adsorbed for each bilayer deposition. Similar results were reported by Alencar et al. and Silva et al. [13, 21]. However, Au@PAH/NiTsPc film (from 1 to 7 bilayers) presented two linear regions for film growth, as can be observed in the inset of Figure 2(b): one ranging from 1 to 3 bilayers and the other from 3 to 7 bilayers, as displayed. This result suggests reorganization of the materials in the films from the third bilayer leading to irregular (increase) amount of material adsorbed, which can be attributed, for instance, to NiTsPc diffusion/interaction into Au@PAH layers [22].

FTIR measurements were performed in order to investigate the adsorption and supramolecular interaction between the components of the films. Figure 3 shows the spectra for 15-bilayer LbL films formed by Au@PAH/CuTsPc (Figure 3(a)) and Au@PAH/NiTsPc (Figure 3(b)), where the characteristics bands for PAH and phthalocyanines can be viewed in both spectra. The main bands of phthalocyanines are assigned at 1032 and 1034 cm^{-1} (S-O stretching vibration deformation), 1120 and 1126 cm^{-1} (S=O symmetric stretching), and 1624 and 1632 cm^{-1} (C=C axial stretching from aromatic ring). The bands at 1120 and 1130 cm^{-1} (C-N angular deformation) and 1514 and 1534 cm^{-1} (C-N axial deformation) and the bands in the range of 2926–3076 cm^{-1} (C-H axial deformation) can be assigned to both materials, while the strong bands at 3450 and 3456 cm^{-1} are related to the stretching mode of the NH_3^+ from PAH [13, 18]. The bands assigned to S-O stretching are shifted to low energies in the LbL films in comparison with FTIR spectra for CuTsPc [23] and NiTsPc [13] cast films. These findings suggest interaction at supramolecular level between the SO_3^- group of phthalocyanines and NH_3^+ group of PAH, which leads to the formation of the "salt-bridge" reported in the literature [13, 18, 23].

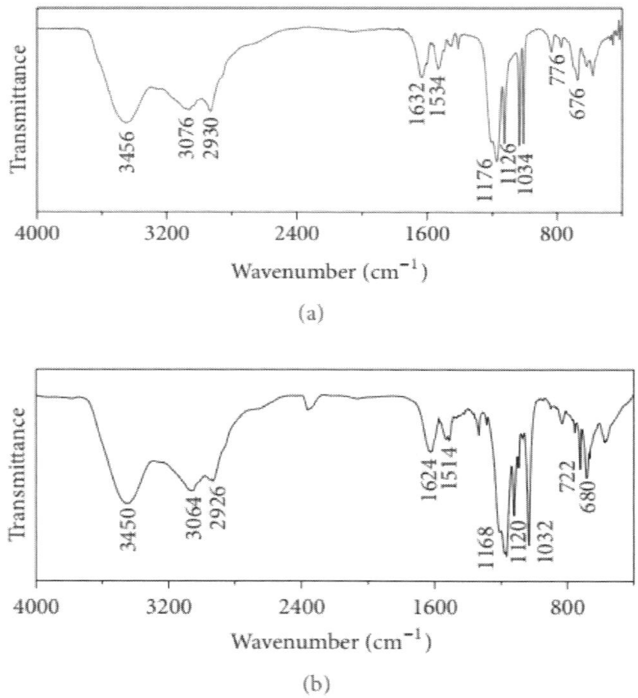

Figure 3: FTIR spectra in the transmission mode for 15-bilayer LbL films of (a) Au@PAH/CuTsPc and (b) Au@PAH/NiTsPc.

The film morphology was investigated by AFM technique using the 7-bilayer LbL films of Au@PAH/CuTsPc and Au@PAH/NiTsPc, as shown in Figures 4(a) and 4(b). Higher roughness (R_{rms}) was observed for Au@PAH/NiTsPc (R_{rms} =8.3 nm) film compared to Au@PAH/CuTsPc (R_{rms} =16.9 nm), which suggests interpenetration of the NiTsPc in the Au@PAH during the film formation. These findings corroborate the difference observed on film growth as shown in the UV-Vis spectra (Figure 2).

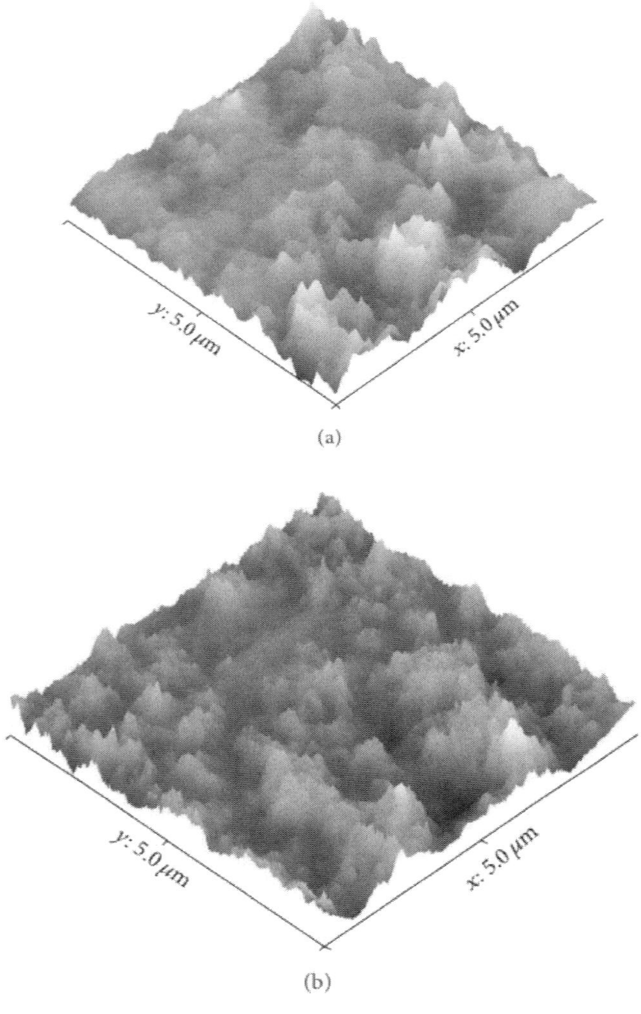

Figure 4: AFM 3D images of 7-bilayer LbL films of (a) Au@PAH/CuTsPc and (b) Au@PAH/NiTsPc.

Electrical Properties

Before using the IDEs modified with the LbL films as the sensing units of the e-tongue system, several measurements were performed with the bare platinum electrodes immersed in distilled water to guarantee similar electrical responses. Therefore, any minor changes observed in the impedance data, after the deposition of the LbL films, are in fact due to the difference between the samples, instead of subtle changes attributed to the design of the electrodes [24]. Capacitance measurements at 1 kHz were performed on the platinum electrodes coated with the LbL films as shown in Figure 5. The frequency of 1 kHz was chosen to analyze the data because normally higher sensitivity is achieved at lower frequencies owing to the importance of the electrical double-layer in the electrical response [25]. In order to evaluate the influence of the gold nanoparticles on the sensing properties of the nanostructured architectures, the LbL films without gold nanoparticles were fabricated and also used for capacitance measurements.

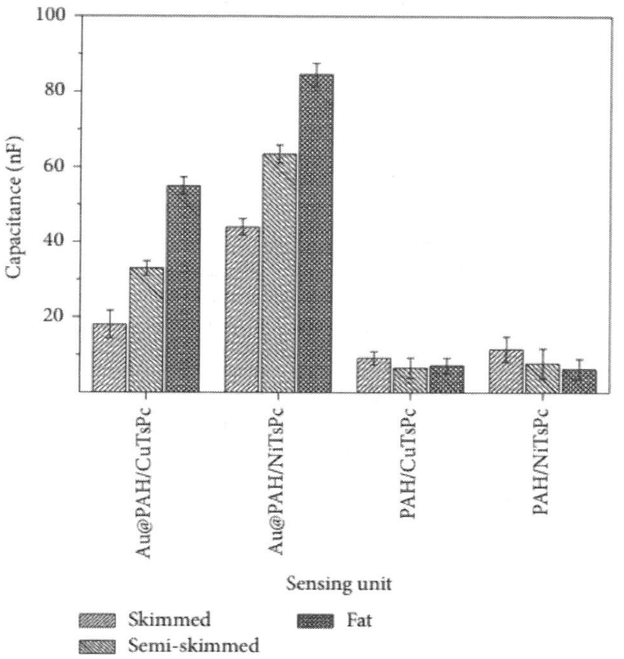

Figure 5: Capacitance measurements of fat, skimmed, and semi-skimmed milk samples at 1 kHz for the four different sensing units containing 7-bilayer LbL films of Au@PAH/CuTsPc, Au@PAH/NiTsPc, PAH/CuTsPc, and PAH/NiTsPc.

From Figure 5, it is clear that the sensing units containing Au@PAH NPs were able to differentiate the milk samples and displayed an increase in the capacitance values at 1 kHz as the fat content increases in the sample. These shifts in the capacitance values are stronger in the LbL films containing the gold nanoparticles, revealing that the gold nanoparticles increase the efficiency of the LbL films as sensing platforms. Possibly, this improvement is due to the good electrical conductivity and the capability of nanoparticles to enhance the electrical contact of the sample molecules with the electrode surface [12].

It is known that the limit of detection of an array of sensors is often lower than that of a single sensor [26]. Therefore, to test the ability of an electronic tongue to classify milk samples, its sensing units (including the blank electrode) formed by interdigitated electrodes modified with 7-bilayer films of PAH/CuTsPc and PAH/NiTsPc (e-tongue 1) were immersed into a set of 24 milk samples (10 fat, 7 skimmed, and 7 semi-skimmed commercial milk samples), and electrical capacitance data at 1 KHz and 50 mV were collected to be used in the PCA analysis. The results in Figure 6(a) show that e-tongue 1 could not satisfactorily distinguish the distinct milk samples, where no correspondence in the data could be related to the principal components. On the other hand, the discriminating capability of the sensor array composed of sensing units with Au@PAH NPs (e-tongue 2) is more evident, as illustrated in the PCA plot of Figure 6(b). For instance, we note that fat milk samples were located on the positive side of PC1. Skimmed milks can be seen in a cluster located on the negative side of PC1 and positive side of PC2, while almost all semi-skimmed milk samples are grouped on the negative sides of PC1 and PC2. Only two samples of semi-skimmed milks are located close to skimmed milks, probably due to the lower fat content for these samples, since semi-skimmed samples can present a range of fat content (1-2%). Moreover, the data variance was mainly accounted using two first principal components, yielding 78% (PC1 + PC2) of the total information collected by the array.

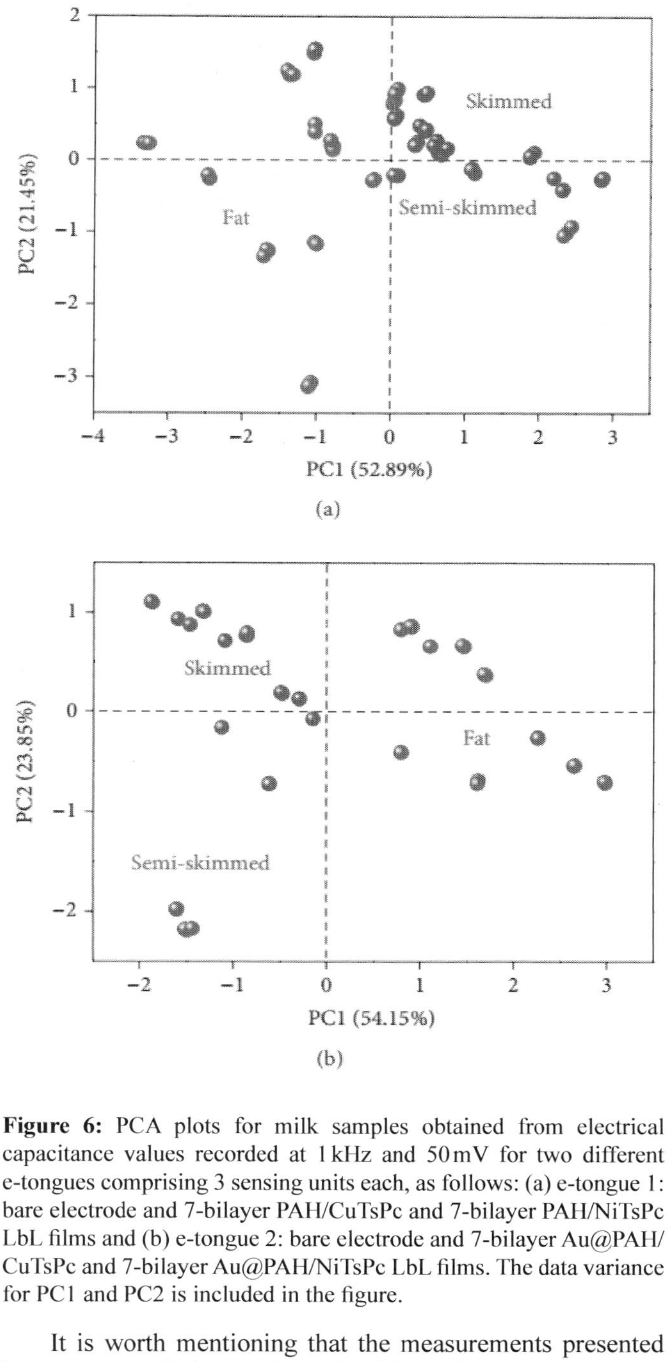

Figure 6: PCA plots for milk samples obtained from electrical capacitance values recorded at 1 kHz and 50 mV for two different e-tongues comprising 3 sensing units each, as follows: (a) e-tongue 1: bare electrode and 7-bilayer PAH/CuTsPc and 7-bilayer PAH/NiTsPc LbL films and (b) e-tongue 2: bare electrode and 7-bilayer Au@PAH/CuTsPc and 7-bilayer Au@PAH/NiTsPc LbL films. The data variance for PC1 and PC2 is included in the figure.

It is worth mentioning that the measurements presented here were carried out using LbL films with a fixed number of bilayers (seven), and therefore no attempt was made to investigate possible effects of the number of film layers on the sensor sensitivity. However, previous studies have demonstrated that sensor optimal performance can usually be attained using from 5-bilayer to 10-bilayer films because they are sufficiently thin to increase sensitivity while providing full coverage of the electrodes, which may not occur for very thin films [27].

CONCLUSION

Gold nanoparticles functionalized with PAH with an average diameter of 11 nm were synthetized and used to grow nanostructured LbL films containing metal tetrasulfonated phthalocyanines. These LbL films were used to modify interdigitated platinum electrodes and were applied as sensing units of an electronic tongue sensor aiming to distinguish milk samples with different fat contents. The presence of the Au@PAH NPs in the LbL films increased the sensor sensitivity when compared to the films without nanoparticles, as shown by the PCA method, revealing that the presence of gold nanoparticles greatly improves the conductivity of the film and enhances the sensitivity of the e-tongue. The results presented here demonstrate that gold nanoparticles can be used to modify LbL films to produce sensing units for an electronic tongue with higher sensitivity to evaluate liquid samples. Such approach can be used in the development of novel sensor platforms for measuring milk quality and other complex liquid matrices in a fast and low-cost way.

ACKNOWLEDGMENTS

The authors acknowledge the financial support by FAPESP (Grants nos. 2012/23880-3, 2013/26712-7, and 2014/16789-5), CNPq, CAPES, MCTI-SisNano, and Embrapa from Brazil.

REFERENCES

1. L. Nicole, C. Laberty-Robert, L. Rozes, and C. Sanchez, "Hybrid materials science: a promised land for the integrative design of multifunctional materials," Nanoscale, vol. 6, no. 12, pp. 6267–6292, 2014.
2. L.-N. Wu, Y.-L. Tan, L. Wang et al., "Dopamine sensor based on a hybrid material composed of cuprous oxide hollow microspheres and carbon black," Microchimica Acta, vol. 182, no. 7, pp. 1361–1369, 2015.
3. L. A. Mercante, A. Pavinatto, L. E. Iwaki et al., "Electrospun polyamide 6/poly(allylamine hydrochloride) nanofibers functionalized with carbon nanotubes for electrochemical detection of dopamine," ACS Applied Materials & Interfaces, vol. 7, no. 8, pp. 4784–4790, 2015.
4. R. S. Andre, A. Pavinatto, L. A. Mercante, E. C. Paris, L. H. Mattoso, and D. S. Correa, "Improving the electrochemical properties of polyamide 6/polyaniline electrospun nanofibers by surface modification with ZnO nanoparticles," RSC Advances, vol. 5, no. 90, pp. 73875–73881, 2015.
5. M. S. Artiles, C. S. Rout, and T. S. Fisher, "Graphene-based hybrid materials and devices for biosensing," Advanced Drug Delivery Reviews, vol. 63, no. 14-15, pp. 1352–1360, 2011.
6. W. Zhang, J. Zheng, C. Tan et al., "Designed self-assembled hybrid Au@CdS core-shell nanoparticles with negative charge and their application as highly selective biosensors," Journal of Materials Chemistry B, vol. 3, no. 2, pp. 217–224, 2015.
7. J. Shen, X. Yang, Y. Zhu, H. Kang, H. Cao, and C. Li, "Gold-coated silica-fiber hybrid materials for application in a novel hydrogen peroxide biosensor," Biosensors and Bioelectronics, vol. 34, no. 1, pp. 132–136, 2012.
8. G. Decher and J. B. Schlenoff, Multilayer Thin Films: Sequential Assembly of Nanocomposite Materials, Wiley, Weinheim, Germany, 2003.
9. G. Decher, "Fuzzy nanoassemblies: toward layered polymeric multicomposites," Science, vol. 277, no. 5330, pp. 1232–1237, 1997.
10. S. Srivastava and N. A. Kotov, "Composite Layer-by-Layer (LBL) assembly with inorganic nanoparticles and nanowires," Accounts of Chemical Research, vol. 41, no. 12, pp. 1831–1841, 2008.
11. K. Saha, S. S. Agasti, C. Kim, X. Li, and V. M. Rotello, "Gold nanoparticles in chemical and biological sensing," Chemical Reviews, vol. 112, no. 5, pp. 2739–2779, 2012.
12. E. Katz, I. Willner, and J. Wang, "Electroanalytical and bioelectroanalytical systems based on metal and semiconductor nanoparticles," Electroanalysis, vol. 16, no. 1-2, pp. 19–44, 2004.
13. W. S. Alencar, F. N. Crespilho, M. V. A. Martins, V. Zucolotto, O. N. Oliveira Jr., and W. C. Silva, "Synergistic interaction between gold nanoparticles and nickel phthalocyanine in layer-by-layer (LbL) films: evidence of constitutional dynamic chemistry (CDC)," Physical Chemistry Chemical Physics, vol. 11, no. 25, pp. 5086–5091, 2009.
14. S. Wold, K. Esbensen, and P. Geladi, "Principal component analysis," Chemometrics and Intelligent Laboratory Systems, vol. 2, no. 1–3, pp. 37–52, 1987.
15. J. R. Siqueira Jr., L. Caseli, F. N. Crespilho, V. Zucolotto, and O. N. Oliveira Jr., "Immobilization of biomolecules on nanostructured films for biosensing," Biosensors and Bioelectronics, vol. 25, no. 6, pp. 1254–1263, 2010.
16. A. Pavinatto, L. A. Mercante, C. S. Leandro, L. H. C. Mattoso, and D. S. Correa, "Layer-by-Layer assembled films of chitosan and multi-walled carbon nanotubes for the electrochemical detection of 17α-ethinylestradiol," Journal of Electroanalytical Chemistry, vol. 755, pp. 215–220, 2015.
17. A. Manzoli, F. M. Shimizu, L. A. Mercante et al., "Layer-by-layer fabrication of AgCl-PANI hybrid nanocomposite films for electronic tongues," Physical Chemistry Chemical Physics, vol. 16, no. 44, pp. 24275–24281, 2014.
18. L. N. Furini, E. Feitosa, P. Alessio, M. H. Shimabukuro, A. Riul Jr., and C. J. L. Constantino, "Tuning the nanostructure of DODAB/nickel tetrasulfonated phthalocyanine bilayers in LbL films," Materials Science and Engineering C, vol. 33, no. 5, pp. 2937–2946, 2013.
19. M. F. Zampa, A. C. F. de Brito, I. L. Kitagawa et al., "Natural gum-assisted phthalocyanine immobilization in electroactive nanocomposites: physicochemical characterization and sensing applications," Biomacromolecules, vol. 8, no. 11, pp. 3408–3413, 2007.
20. L. Edwards and M. Gouterman, "Porphyrins. XV. Vapor absorption spectra and stability:

phthalocyanines," Journal of Molecular Spectroscopy, vol. 33, no. 2, pp. 292–310, 1970. · ·

21. J. R. Silva, N. C. de Souza, and O. N. Oliveira Jr., "Adsorption kinetics and charge inversion in layer-by-layer films from nickel tetrasulfonated phthalocyanine and poly(allylamine hydrochloride)," Journal of Non-Crystalline Solids, vol. 356, no. 18-19, pp. 937–940, 2010. · ·
22. J. M. Garza, P. Schaaf, S. Muller et al., "Multicompartment films made of alternate polyelectrolyte multilayers of exponential and linear growth," Langmuir, vol. 20, no. 17, pp. 7298–7302, 2004.
23. V. Zucolotto, M. Ferreira, M. R. Cordeiro, C. J. L. Constantino, W. C. Moreira, and O. N. Oliveira Jr., "Nanoscale processing of polyaniline and phthalocyanines for sensing applications," Sensors and Actuators B: Chemical, vol. 113, no. 2, pp. 809–815, 2006. · ·
24. P. H. B. Aoki, D. Volpati, A. Riul Jr., W. Caetano, and C. J. L. Constantino, "Layer-by-layer technique as a new approach to produce nanostructured films containing phospholipids as transducers in sensing applications," Langmuir, vol. 25, no. 4, pp. 2331–2338, 2009. · ·
25. D. M. Taylor and A. G. MacDonald, "AC admittance of the metal/insulator/electrolyte interface," Journal of Physics D: Applied Physics, vol. 20, no. 10, pp. 1277–1283, 1987.
26. A. V. Legin, A. M. Rudnitskaya, Y. G. Vlasov, C. Di Natale, and A. D'Amico, "Features of the electronic tongue in comparison with the characteristics of the discrete ion-selective sensors," Sensors and Actuators B: Chemical, vol. 58, no. 1–3, pp. 464–468, 1999.
27. M. Ferreira, A. Riul Jr., K. Wohnrath, F. J. Fonseca, O. N. Oliveira Jr., and L. H. C. Mattoso, "High-performance taste sensor made from Langmuir–Blodgett films of conducting polymers and a ruthenium complex," Analytical Chemistry, vol. 75, no. 4, pp. 953–955, 2003.

CHAPTER 17

TEXTILE-BASED ELECTRONIC COMPONENTS FOR ENERGY APPLICATIONS: PRINCIPLES, PROBLEMS, AND PERSPECTIVE

Vishakha Kaushik, Jaehong Lee, Juree Hong, Seulah Lee, Sanggeun Lee, Jungmok Seo, Chandreswar Mahata and Taeyoon Lee

Nanobio Device Laboratory, School of Electrical and Electronic Engineering, Yonsei University, 50 Yonsei-ro, Seodaemun-Gu, Seoul 120-749, Korea

ABSTRACT

Textile-based electronic components have gained interest in the fields of science and technology. Recent developments in nanotechnology have enabled the integration of electronic components into textiles while retaining desirable characteristics such as flexibility, strength, and conductivity. Various materials were investigated in detail to obtain current conductive textile technology, and the integration of electronic components into these textiles shows great promise for common everyday applications. The harvest and storage of energy in textile electronics is a challenge that requires further attention in order to enable complete adoption of this technology in practical implementations. This review focuses on the various conductive textiles, their methods of preparation, and textile-based electronic components. We also focus on fabrication and the function of textile-based energy harvesting and storage devices, discuss their fundamental limitations, and suggest new areas of study.

INTRODUCTION

Electronic devices have the ability to address basic needs [1,2,3,4,5,6]. When electronic devices were first being developed, scientists imagined great leaps forward such as the development of artificial intelligence, and realized the innovation of the Bell telephone in 1876 [7]. A complex number calculator was later developed by the Bell laboratory in 1939 [8]. Seven years after this valuable development, the first fully-functional digital computer, known as ENIAC, came into existence [9]. Although ENIAC was the fastest machine at that time, it was difficult to physically transport due to size ($2.4 \times 0.9 \times 30$ m^3) and power requirements (150 kW). Thus, the primary goals of researchers at this time were to decrease the size and power consumption of electronic devices, which they did with the miniaturization of electronic devices and developed the first personal computer, Kenbak-1, and the first laptop computer, Epson HX-20 [10,11,12]. The increasing pace of technological developments since the late 20th century has resulted in exploration of mobile and wearable devices [13,14]. Early stage wearable computers were composed of backpack systems with uncomfortable wiring networks. Currently, researchers aim to develop flexible wearable devices that integrate seamlessly into regular clothing such as Google glass, the Apple watch, and electronic skins [15,16,17,18]. An illustration of the history of these electronic components from Bell's telephone to smart wearable devices is presented in Figure 1 [7,8,9,10,11,12,13,14,15,16,17,18,19]. Recent market demand has focused on small electronic components for smart wearable device applications, leading to research and products featuring technology that can directly integrate

Citation: Vishakha Kaushik, Jaehong Lee, Juree Hong, Seulah Lee, Sanggeun Lee, Jungmok Seo, Chandreswar Mahata and Taeyoon Lee, Textile-Based Electronic Components for Energy Applications: Principles, Problems, and Perspective, doi:10.3390/nano5031493.

electronic components into textiles or fabrics for a variety of applications including sensors, medical devices, safety instruments, and planar waveguides [4,20,21,22,23,24,25,26, 27,28,29,30,31].

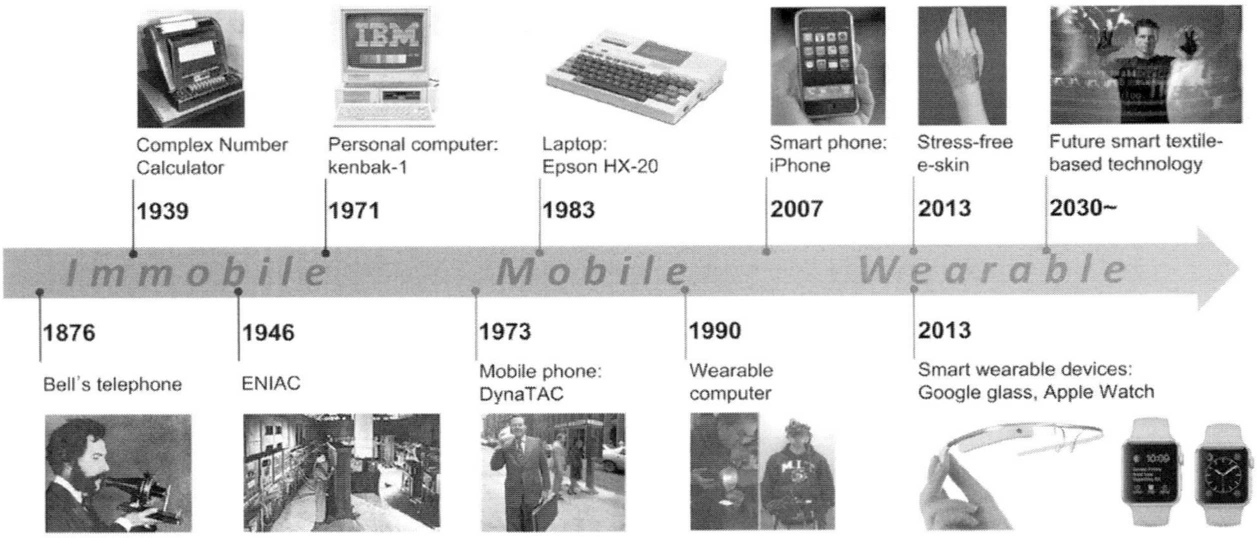

Figure 1: A chronological illustration presenting the journey of electronic components from immobile to mobile, and finally to wearable textile-based electronics.

Textiles have been used by mankind since prehistoric times, and they are essential to daily life. Fashion dictates that the textiles must look attractive and fit comfortably. Traditional textiles are composed of natural or synthetic fibers and filaments. Recent advances in electronics have led to the development of a new generation known as electronic textiles (E-textiles) or smart textiles. E-textiles consist of fabric that provides advanced functions in the form of electronic features or interconnection [32,33]. E-textiles can be categorized into two types: classical, where the electronic components are embedded into garments and integrated, where the electronic components are directly integrated into textile substrates. E-textiles have attracted attention in technical and socioeconomic fields because of their ability to sense and respond to environmental stimuli [33,34]. Due to this ability, they are ideal components for smart wearable electronics, and represent a new branch of flexible multifunctional fabric electronics. Many research groups have explored the integration of flexible devices such as solar cells, field-effect transistors (FETs), light emitting diodes (OLEDs), and photovoltaic devices into textiles; however, these components cannot operate without sufficient electrical energy, making the development of energy harvesting and storage devices essential [35,36,37,38,39,40,41,42,43,44,45,4 6].

This paper reviews the development of various key components that are required for E-textiles. The most fundamental element is the conductive textile, which is required for interconnection to transfer power and data. Energy harvesting and storing applications are necessary for powering the electrical components such as transistors, LEDs, and sensors for their utilization in various functional E-textiles. A systematic view is suggested and elucidated with examples.

CONDUCTIVE TEXTILE: MATERIALS AND FABRICATION

It is ideal for electronic interconnects to be realized on fibers or fabrics for the fabrication of textile-based wearable devices. The development of highly-conductive textiles for interconnection is crucial to minimize energy loss. Three types of materials have been explored for the fabrication of high-conductivity textiles: polymer-based, carbon-based, and metallic materials. In this section, the features of conductive textiles consisting of organic and inorganic materials will be discussed.

Polymer-Based Conductive Textiles

Conductive polymers have attracted significant scientific interest for their interesting and unique electrical, optical, and mechanical properties [47,48,49]. They have been explored in a number of applications such as chemical and biological sensors, drug delivery, corrosion protection, nanoelectronic devices, biomedicine, microwave absorption, and electrocatalysis [50 ,51,52,53,54,55,56,57,58,59]. Shirakawa *et al.* were the first to demonstrate conductivity in polymers via doping, showing that exposure of polyacetylene (PA) polymer to iodine vapor increases conductivity by up to seven orders of magnitude [60]. The enhanced conductivity of PA is attributed to redox reactions (charge transfer complexes) between the PA and iodine vapor. The search for highly conductive polymers has resulted in the expansion of this area, and a number of conductive polymers have been used in the last 35 years. Various applications such as protective clothing, sportswear, health monitoring, actuators, and sensors for wearable E-textiles have now been explored via these conductive polymer fabrics [61,62,63,64,65].

Conjugated polymer polypyrrole (PPy) has received much attention due to its high conductivity paired with chemical and environmental stability [66,67]. PPy also has the advantage of being easy to synthesize at large areas with different porosities at room temperature. Conductive polymer textiles can be fabricated via polymerization, wet spinning, or dip coating processes [68,69,70]. *In situ* polymerization or chemical oxidation polymerization are commonly used [71,72]. The experimental steps used to fabricate conductive PPy via in-situ chemical polymerization are illustrated in Figure 2a. Highly-conductive PPy nanofibers with diameters between 60 and 100 nm that are synthesized in p-hydroxyazobenzene sulfonic acid act as a functional dopant. PPy is deposited when the fabric and polymer solutions come into contact due to the liquid-solid interface interaction. This is considered a physical adsorption followed by a polymerization process, and results in a layer of conducting polymer with a smooth surface over the fibers. PPy can also be deposited onto other complex structures.Figure 2b shows the scanning electron microscope (SEM) image of a conductive fiber composed of carbon nanotube bundles coated with PPy [73]; however, brittleness and rigidity are two potential drawbacks to using PPy.

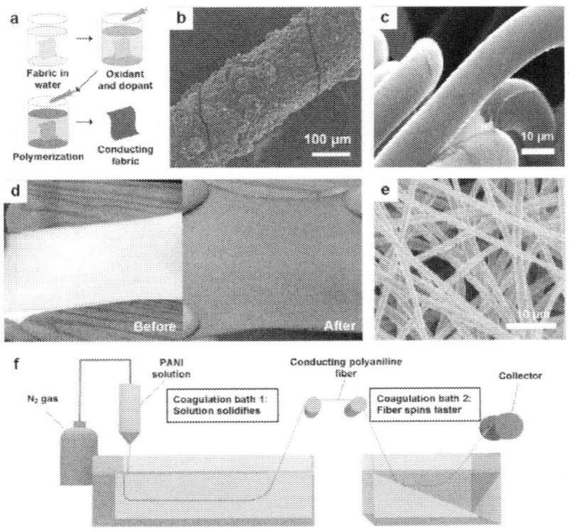

Figure 2: (a) Schematic illustration of the *in situ* chemical polymerization for conducting polymer-coated fabric; (b) SEM image of a conductive fiber composed of carbon nanotube bundles coated with polypyrrole; (c) SEM micrograph of stretchable spandex fabric after a dip coating process with polystyrene sulfonate-doped poly-(3,4-ethylenedioxythiophene) (PEDOT:PSS); (d) Optical image of stretchable fabric before and after dip coating with PEDOT:PSS; (e) Surface morphology of polyaniline (PANI): polyethylene oxide (PEO) electrospun fiber; (f) Wet fiber spinning technique process. Reproduced from [68,73,74] with the permission by Royal Society of Chemistry, Copyright 2015, and by ACS Publications, Copyright 2010, 2012.

Another interesting conductive polymer is a polythiophene derivative, poly-(3,4-ethylenedioxythiophene) (PEDOT), which shows high electrochemical stability in oxidized form due to its planar structure and delocalized π electrons [75]. Its unique structure contains dioxyalkylene bridging groups at positions 3 and 4 of its heterocycle ring, making it the most valuable among all conducting polymers. Thus, it shows high conductivity and good electrical, thermal, and chemical stability when compared to PPy. Although PEDOT itself is a conductive polymer, it has low solubility, which limits usage. Generally, polystyrene sulfonate (PSS)-doped PEDOT (PEDOT:PSS) is used to obtain a stable dispersion in a water solvent. PEDOT:PSS has been used extensively in the last few years as a conducting polymer [76]. The simplest technique for coating is dip coating, where the fabric is simply dipped into the conductive solution, resulting in a conductive polymer fabric. Ding*et al.* have reported the preparation of PEDOT:PSS conductive fabrics using the dip coating process [68]. Figure 2c shows the surface morphology of a spandex fabric after soaking in a commercially-available PEDOT:PSS dispersion. It is clear from theFigure 2c that the surface of the fabric is smooth and uniform after conductive coating. The conductivity of the single soaked fabric was reported to be 0.1 S/cm, reaching up to 2 S/cm after multiple soaking steps. The optical image of the PEDOT:PSS-soaked textile is shown in Figure 2d. Thus, conductive fabrics can be produced in a simple manner.

Polyaniline (PANI), known as aniline black, has also been studied as a conductive polymer. PANI revolutionized polymer chemistry due to its many advantages including stability, cost effectiveness, and switching characteristics between conductive and resistive states. Its electrical conductivity is due to the partial oxidation or reduction process and can be tuned to achieve the required conductivity for a given application. It is fabricated via a chemical oxidative polymerization process of aniline, which is a multistep and slightly complex process. Figure 2e shows the surface morphology of electrospun PANI fibers that exhibit high electrical conductivity between 70 and 150 S/cm [74]. However, a more simplified, single-step fabrication method is reported by Pomfret *et al.* where polyaniline fibers are fabricated using a wet spinning process [77]. The resulting fibers exhibited high conductivity (~1000 S/cm for bulk material) and good mechanical strength (young modulus ~2 GPa). Figure 2f shows the schematic design of the wet spinning process utilized for the production of conductive polyaniline polymer. A mixture of organic solvents and the emeraldine base form of polyaniline is used to make a spinning solution, which then enters the first coagulation bath where it solidifies and forms fibers. In the second coagulation bath, the fibers spin rapidly, resulting in stretching. The stretched fibers are then doped with an acid suitable for the preparation of conducting fibers, as shown in Figure 2f. As discussed earlier, conductive polymers are considered useful in wearable textiles; however, the conductivity values should be modified for their optimization in the production of smart wearable electronics. A comparison of the properties of these conductive fibers is given in Table 1.

Table 1: Summary of materials and their electric properties

Structure of conductive textile	Material use to prepare or to coat on fibers	Merits of Conductive textiles	Technique used to fabricate conductive textile	Advantages of growth technique	Disadvantages of growth technique	Electrical property	Ref.
Polymer-based	Polypyrrole	Chemical and environment stable	Functional dopant induced process	Novel & Easy process, High Yield easily deposit on fabric surface	Need to optimize dopant concentration for better conductivity results	120–130 S/cm	[78]
	Polyaniline	Cost effective, stable	Wet Spinning	Produced fibers show high electronic and mechanical strength, fabricated thick fibers	Need to combine Individual fibers	140-750 S/cm	[79–81]
	PEDOT:PSS	Highly conductive good thermal and chemical stability	Dip Coating	Simple process	Coating is slow, tide lines can forms	0.4–2.0 S/cm	[68,82]
Carbon-based	Graphene	Highly conductive & stable, high strength	Electrostatic self-assembly with BSA followed by chemical reduction	Easy to attach graphene oxide films to textiles	Multistep process, wrinkles observed on surface of samples	10–20 S/cm	[83,84]
	Carbon nanotubes	Conductive, high mechanical strength	Wet spinning	Continuous long length (cm) CNT yarns	Slow processing	125–3000 S/cm	[85–87]
Metal-based	Al-coated conductive paper	Stable, conductive, stretchable	Chemical solution process	Low cost, Vacuum not required, Direct deposition of Al on fibers	Need a suitable catalyst for decomposition process	19 mΩ/□	[88]
	AgNP/Ag nanowires	Conductive	Wet spinning	Simple, highly conductive material	Need optimization of suitable solvents and coagulation liquids	2200–5400 S/cm	[89,90]

Carbon-Based Conductive Textile

Carbon-based materials including carbon nanotubes (CNTs), carbon fibers, carbon nanoparticles, and graphene present extraordinary properties such as high mechanical strength, light weight, environmental stability, and superior thermal and electrical conductivity [91,92,93,94,95,96,97,98]. These intriguing properties make carbon-based materials important candidates in wearable electronic textiles. Among all carbon allotropes, the most widely explored are CNTs and graphene. While carbon fibers have been explored in the textile industry for a long time, CNTs are increasingly chosen over commercial carbon fibers because of their lower density, and higher tensile and compressive strength [99,100,101]. CNT can be coated on various fibers such as cellulose or polyester yarns via a simple and cost effective dipping and drying method, as shown in Figure 3a [87]. Here, cotton fibers are simply dipped into a single-walled carbon nanotube (SWCNT) solution, resulting in a black conductive fabric. Figure 3b shows the surface morphology of the SWCNT-coated cotton fibers. The as-prepared CNT-cotton yarns show high electrical conductivity (125 S/cm) with outstanding stretchability and flexibility. The coating of CNTs results in conductivity of fibers and a CNT-cotton thread can emit the light of an LED, as shown in Figure 3c [102].

CNT fibers without any supportive fiber can be made using wet or dry spinning methods. Zhang *et al.* reported a wet spinning method for the production of long CNTs; a schematic of their wet spinning process is shown in Figure 3d [103]. The electrical conductivity of CNT fibers was enhanced up to ~6.7 × 10^4 S/cm by iodine-doping [104]. The enhancement in the conductivity of CNT fibers via doping is shown in Figure 3e [104]. Jiang *et al.* have fabricated CNT yarns using a super-aligned array of CNTs [105]. The way to produce continuous CNT yarns 30 cm in length from a CNT array is similar to drawing a thread from a silkworm cocoon. Figure 3f shows an optical image of the yarn pulled from a CNT array [105]. The authors have constructed a light bulb filament using these CNT yarns and observed that their strength and conductivity is enhanced by heating at high temperatures. The CNT-yarn-based filament emits incandescent light for 3 h, and after this time span the conductivity of the filament increases by up to 13%, while the tensile strength of the CNT yarn increased from 1 mN to 6.4 mN. Thus, these CNT yarns can be woven into fabrics [105]. However, this process has some disadvantages in terms of slow processing and limited to scaling for thick or multistrand forms. Other issues include high junction resistance.

Graphene oxide (GO) has also been widely studied for conductive fibers. Yun *et al.* presented an electrostatic method for fabricating textiles that are wrapped with reduced graphene oxide (RGO), which makes them conductive and stable (Figure 3g) [83]. Nylon-6 yarns were functionalized with bovine serum albumin molecules to obtain positively-charged surfaces. GO sheets were then electrostatically assembled over the fibers via a simple immersing and shaking process in a GO solution [83]. Finally, the GO are chemically reduced to RGO. They have also demonstrated the lightning of an LED using the synthesized conductive fibers, a process that works well even after folding, as shown in Figure 3h [83]. Freestanding RGO fibers can also be synthesized using wet or dry spinning methods. Xu *et al.* have reported porous graphene oxide with good electrical conductivity (~4.9 × 10^3 S/m) and high specific surface area (884 $m^2 \cdot g^{-1}$) [106]. This graphene-based hybrid structure presents great potential in fiber-based electrochemical super-capacitors [84]. The above described properties of carbon-based materials open a new path in electronic textiles for wearable electronics. However, further investigation is needed to look into the possible side effects of wearable electronics, such as toxicity with skin contact and stability in air.

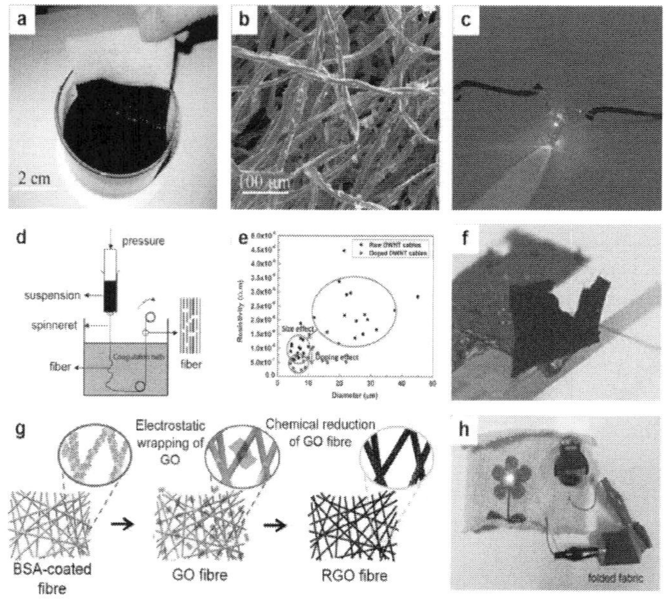

Figure 3: (a) Optical image shows the dip-coating process of fabric into single-walled carbon nanotube (SWCNT)s ink for SWCNT-based conductive fabric; (b) SEM image of a cotton sheet coated with SWNTs on the fibrous surface; (c) Demonstration of LED emission with CNT-based conductive fibers; (d) Schematic of the experimental setup of the wet spinning method for CNT-based conductive fibers; (e) Resistivity *versus*diameter plot for raw DWNT (double walled CNT) and iodine doped DWNT cables; (f) Image shows a free standing CNT array through which a yarn is pulling out; (g) Schematic illustration of the fabrication process of graphene oxide-based conductive fiber using electrostatic force; (h) Photograph of a LED light using a folded, reduced graphene-oxide-based conductive fabric. Reprinted from [83,87,102,103,104,105] with the permission by ACS Publications, Copyright 2010, 2008 and by John Wiley and Sons, Copyright 2008, 2013 and by Nature Publishing Group, Copyright 2002, 2011.

Metal-Based Conductive Textiles

Metallic materials are considered a better choice for wearable electronics because of their high conductivity [89,90]. Vacuum deposition is one of the most widely used techniques for metal deposition and is also frequently used for microfabrication processes. Sputtering or evaporation methods can be used to directly form a thin metal layer on the exposed surface of the fibers [107,108]; however, these techniques have huge drawbacks such as expensive instruments, limited sample size, and non-compatibility for batch process. Also, these methods are not suitable for depositing metal on complex structures such as porous fibers, and the deposited films are weak against repeated deformations like folding or bending. In order to avoid these issues, alternative methods must be used to synthesize metal-based conductive textiles.

Electroplated or electroless plated metals, liquid metals, and metallic nanomaterials such as nanoparticles or nanowires are considered ideal candidates in the growing field of wearable electronics [109,110]. Metal plating methods have been widely used to deposit metal for printed circuit boards, metallization processes in microfabrication, and corrosion protection. Figure 4a shows an SEM image of a gold-plated conductive fiber, in which Kevlar fibers were first treated with Sn^{2+} and Pd^{2+} ions followed by electroless Ni deposition for conductivity [109]. Au was then electroplated onto the surface to achieve a conductive fiber with a conductivity of ~6 S/cm and power output of 1 W. However, the metal films are weakly attached to the surface and may separate from the fiber during further processing. To overcome these obstacles, a recent study presents a new, simple, and inexpensive chemical solution method, as shown in Figure 4b [88]. In this process, an Al precursor was used to deposit Al on the surface of fibrous materials. Figure 4c shows the electrical resistances verses immersion time in the Al precursor solution, and the inset shows an LED connected to these fibers [88].

Another method of synthesizing metal-based conductive textiles is the use of liquid metals [111]. Eutectic gallium indium (EGaIn) is a liquid metal with high conductivity (σ = 3.4 × 10^4 S/cm) and a low melting point (T_m = 15.7 °C) that spontaneously forms a thin oxide layer that can be used to mold the metal into complex shapes [111]. When sufficient force is applied, the oxide layer can be damaged and the shape of the EGaIn can be changed. Zhu *et al.* has filled a hollow elastomer fiber with EGaIn to obtain stretchable conductors, as shown in Figure 4d [112]. The authors observed that the metallic continuity is maintained even after the applying a 700% strain.

Various metallic nanomaterials such as silver, copper, aluminum, gold, and tin are used as conductive materials for textiles [113]. Atwa *et al.* demonstrated nylon, polyester, and cotton threads dip-coated with Ag nanowires (AgNWs), which exhibit high mechanical flexibility and a small increase in resistance after 200 bending cycles as compared to commercial conductive fibers [113]; however, metallic nanomaterials mixed with polymers can lead to better stability. Recently, Lee *et al.* reported the fabrication of silver nanoparticle (AgNP)/polymer-coated Kevlar fiber with a silver content of more than 80 wt % [20]. Figure 4e shows the surface morphology of the as-prepared fibers along with an inset showing the AgNPs on the surface [20]. In this work, the $AgCF_3COO$ solution was absorbed by a poly (styrene-block-butadienstyrene) (SBS) polymer and then reduced to AgNPs using hydrazine hydrate. Figure 4f shows the small change in resistance during 3000 folding tests [20]. In another study, Lee *et al.* inserted AgNWs into a stretchable AgNP/SBS structure to obtain increased conductance under strain [90]. The fiber shows high electrical conductivity, 2450 S/cm, and can stretch up to 900% strain as shown in Figure 4g [90]. Table 1 presents a comparison of the properties of these metallic conductive fibers.

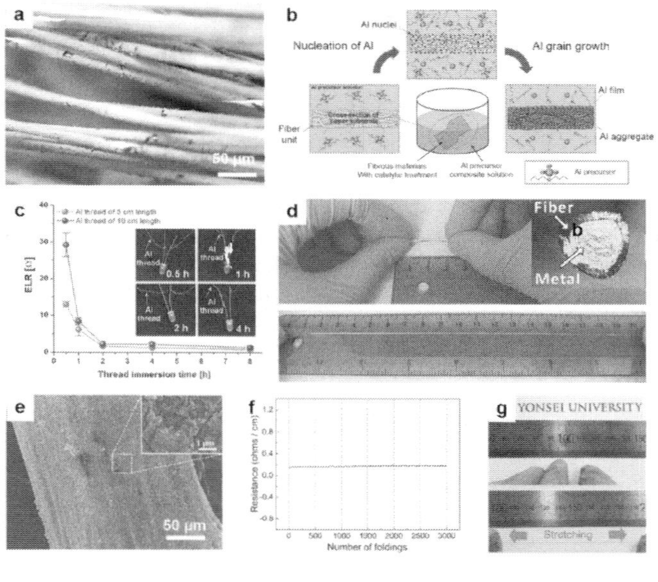

Figure 4: (a) SEM image of conductive Kevlar fibers coated with Ni/Au; (b) Schematic representation of the chemical solution process for the fabrication of conductive Al textiles; (c) Plot of the electrical linear resistance of Al-based conductive fiber synthesized by the chemical solution process and inset shows the functioning LEDs; (d) Optical image of unstretched and stretched (up to 20 cm) conductive fiber; (e) SEM image showing the surface of the Ag nanoparticles and elastomer composite-based conductive fiber; (f) Electrical resistance change in the Ag-based conductive fiber during the 3000 cyclic folding test; (g) Optical image of pre- and 900%-strained Ag nanoparticles and elastomer composite fibers. Reproduced from [20,88,90,109,112] with the permission by ACS Publications, Copyright 2011 and by John Wiley and Sons, Copyright 2013, 2013, 2015, 2015.

TEXTILE ELECTRONIC COMPONENTS

Textile-based electronic devices require electronic functionalities on fiber/textile materials with robust mechanical properties. In this section, we will discuss electronic devices based on fibers and their use in textile electronics. The operation of electronic devices and their stability in a textile system will be explored in detail. Specifically, we will demonstrate recent developments in textile-based electronic devices such as transistors, light emitting diodes (LEDs), and sensors.

Transistor

The integration of electronic devices into textiles at the fiber level is vital to the future of wearable electronics. There are various aspects to be considered before employing electronic devices into a textile-based system such as materials, structures, and the compatibility of electronic devices with textile fabrics. As a fundamental building block of electronic devices, transistors must be integrated into fabric or textiles. In this section, we briefly introduce fiber or textile-based transistors and their operation. In general, an electronic device that can amplify and convert electronic signal to electrical power is known as transistor. Since transistors are basic components in electronic devices, the integration of transistors into fabric or textile systems is highly important. However, conventional inorganic FETs cannot be applied to fabric or textile-based electronics due to flexibility issues and limitations in the fabrication process. Given the technological limitations of inorganic materials, organic thin film transistors (OTFTs) have received significant attention owing to their flexibility, compatibility with flexible electronics, light weight, biocompatibility, low cost, and low-temperature process capability [114,115,116,117].

Organic-based transistors due to their flexibility, easy fabrication process and comparable electrical performance (as compared to inorganic TFTs) are desirable in textile industry [118,119,120]. Two types of fiber-based OTFTs have been reported; fiber organic field effect transistors (F-OFETs) and organic electrochemical transistors (OECTs) [121]. The working principle of F-OFETs is similar to that of commercial FETs. In O-FETs, organic thin-film semiconductors are used as an active layer instead of inorganic materials. Figure 5a shows a schematic diagram of F-OFETs. Here, a commercially available cylindrical metal fiber (Al & stainless steel) is insulated using a uniform layer of polyimide, followed by the evaporation of pentacene as an active material PEDOT:PSS is then deposited on the pentacene surface using a soft lithographic process to form source and drain electrodes. Figure 5b,c shows an optical microscope image of the channel length (~45 μm) and corresponding I_D-V_D curve of an F-OFET [122]. In spite of the low value of the width/length ratio of the F-OFET, the on current and on-off ratio are reasonably high (~10^4), which is fully comparable with those of planar devices. Although F-OFETs exhibit decent electrical properties, they suffer from significant disadvantages such as high operating voltages, high equipment cost, and a complicated fabrication process. OECTs have been considered as an alternative to F-OFETs, since they can be fabricated on the basis of fiber structure. The working principle of OECTs is based on the electro-chemical process, where a reversible process of doping and de-doping of electronic polymers allows the transistor to operate. Figure 5d shows a schematic picture of OECTs on a planar substrate [123]. Here, source (S), drain (D), and gate (G) electrodes are demonstrated on flat substrates such as glass, plastic, and paper. The conducting polymer PEDOT/PSS films are first micro-patterned on a flat substrate, and are followed by the pattering of an electrolyte (micro-channel of ~3 μm height and ~15 μm width) over the patterned PEDOT/PSS film.

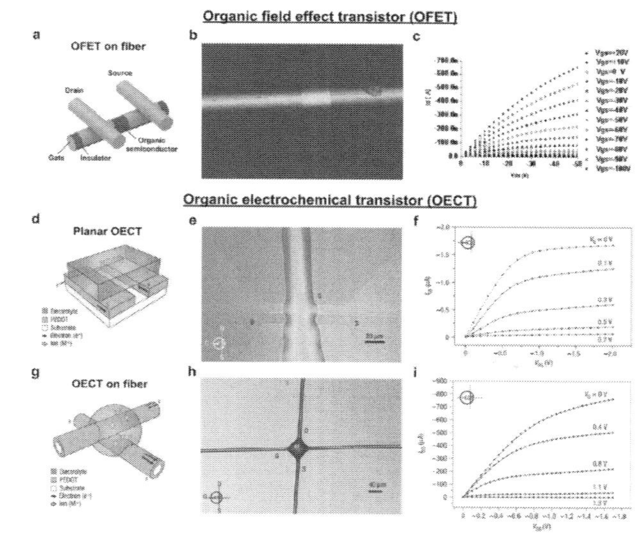

Figure 5: (a) Schematic picture of the organic field effect transistor (O-FET); (b) Optical image showing channel area; (c) The output characteristic of O-FET; (d) Schematic picture of a planar OECT; (e) Optical micrograph of a planar organic electrochemical transistor (OECT); (f) Output characteristic of a planar OECT; (g) Schematic picture of a fiber OECT; (h) Optical micrograph of a fiber OECT; (i) Output characteristic of a fiber OECT. Reprinted from [122,123] with permissions by AIP Publishing LLC, Copyright 2006 and by Nature Publishing Group, Copyright 2007.

An optical micrograph of this configuration is shown in Figure 5e, and the corresponding output characteristic of a planar ECT is shown in Figure 5f [123]. The output characteristics clearly indicate that the current is saturated with increasing drain voltage, and that the gate voltage is low compared to that of F-OFETs. When the gate voltage is zero, the transistor is in the ON state, while the gate voltage transistor moves to the OFF state when the gate voltage is further increased, as shown inFigure 5f. In order to fabricate OECTs on a fiber structure, a simple new method was performed by Hamedi *et al.* [123].Figure 5g shows a schematic picture of a fiber OECT constructed using PEDOT-coated polyamide fibers (~10 μm diameter). Two PEDOT-coated fibers are kept in a cross geometry and an electrolyte is dropped at the fiber junction, as shown in the optical image in Figure 5h [123]. In this cross geometry, one PEDOT-coated fiber acts as source and drain and the other as gate electrode. An active material (electrolyte with 33 wt % PSS, 12 wt % glycol, 8 wt % sorbitol, water and 0.1 M $NaClO_4$) acts as a conducting channel, which is controlled by the gate electrode. In this way, micrometer-sized fiber OECTs can be easily integrated into textiles. The output characteristics of fiber OECTs are similar to planer OECTs, as shown in Figure 5i [123]. The similar electrical characteristics between fiber OECTs and planer OECTs are attributed to the fact that the operation of electrochemical transistor is dominated by the interface effect between the conducting polymer and electrolyte. Therefore, fiber OECTs can be used as effective transistors in spite of completely different geometry when compared to planar OECTs. Theoretically, the fiber OECTs could be used for large scale production: ~100,000 transistors/cm^2 can be constructed using 10 μm fibers. Also, the authors have fabricated a binary tree multiplexer structure by using these fiber OECTs. This multiplexer is able to encode the information from a large number of data sources into a single channel. Hamedi *et al.* fabricated a fiber-based OECT using poly 3-hexylthiophene (P3HT) with 100 μm channel length and showed that the fiber can be operated below ~1 V [124].

Since fiber OECTs have many advantages such as low operating voltages, large amounts of current densities, and low cost production, they can easily replace OFETs and can be integrated to wearable electronics by directly weaving them into fabrics. Bonderover *et al.* have demonstrated textile inverters made by weaving fiber-based transistors [125]. They have shown that E-textiles with electronic components can be designed by simply woven fibers. Thus, in the future it is possible to fabricate such E-textiles. However, their long response time and low switching frequency remain significant hurdles. In addition, the exact charge transport mechanism of fabric OECTs must be revealed to fully understand the operation process.

Light-Emitting Diodes

A light emitting diode (LED) is a semiconductor device that emits visible light when electron hole recombination takes places inside the device. Since the first commercial infrared inorganic LED was presented by Holonyak *et al.* in 1962 [126], various LEDs have been developed around the world such as visible, ultraviolet, and infrared LEDs [127,128,129,130,131]. Since a solid semiconductor material is generally used to generate light in LEDs, LEDs were originally considered solid-state devices. A breakthrough occurred in 1987 when Tang and Van Slyke reported a new type of LED based on organic semiconductors [132]. Organic LEDs (OLEDs) have several advantages, including lower manufacturing cost, faster response time, potential flexibility, low power consumption, and less heat dissipation [133]. Since OLEDs are malleable, they can be compatible with flexible substrates and can be applied to wearable electronics. Current research into OLEDs focuses primarily on two-dimensional (2D) devices, however, one-dimensional (1D) OLEDs have begun to receive attention for future wearable electronics. In this section, various 1D fiber-based OLEDs will be discussed.

Connor *et al.* fabricated a fiber-based OLED using molecular organic compound as an active layer [134]. The inherently-flexible nature of molecular organic compounds is compatible with fiber structure. Figure 6a shows a schematic illustration of the fiber-based OLED structure used in this study [134]. Here, a polyimide-coated silica fiber was prepared as a substrate, and emission layers and organic charge transport layers were deposited between the anode and cathode electrodes via a vacuum thermal evaporation technique. A photograph of a 1 mm segment of green-light OLED on fiber clearly shows the emitting characteristics of green light. For a performance comparison of electricity and luminescence between a fiber-based OLED and planar OLED, they fabricated a planar OLED using a planar silicon/polyimide substrate and investigated the electrical and luminescence characteristics. The electrical and luminescence characteristics of fiber-based OLEDs showed comparable results to planar OLEDs; however, a slightly larger leakage current was measured at low bias due to the surface roughness of the fiber structure. In addition, the fiber is rather thick at ~480 μm, which can hinder practical use. Yang *et al.*recently fabricated single core-shell fiber OLEDs based on an ionic transition-metal complex (iTMC) using a coaxial electrospinning process [135]. A galinstan liquid metal core and a ruthenium tris (bipyridine) with poly (ethyl-oxide) (PEO) mixture shell were co-electrospun to from the cathode and electroluminescent layer, respectively. The anode was formed via thermal evaporation of indium tin oxide. Figure 6b shows a schematic illustration and optical image of an emitting iTMC [135]. The device was initially turned on at 4.2 V and bright electroluminescent was observed at 6.8 V. This work shows the possibility of direct fabrication of 1D flexible OLEDs without flexible substrates; however, the efficiency or brightness of the devices is not clearly shown.

Vohra *et al.* fabricated luminescent F8BT/PEO nanofibers using electrospinning [43]. Figure 6c shows a schematic illustration and fluorescence microscopy image of the conjugated polymer electrospun nanofiber-based OLED [43]. To fabricate the device, a thin layer of PEDOT-PSS was spin coated onto the ITO/glass substrates, and the F8BT/PEO nanofibers were directly collected onto the resulting layer. Then, a PVK (poly *N*-vinylcarbazole) solution was spin coated on top, and finally a thin layer of calcium and aluminum were evaporated. Here, the PVK-coated F8BT/PEO nanofibers functioned as an active layer. This work solved the charge injection issue in organic nanofibers by annealing the device at 150 °C, which changed the active layer into ribbon-like morphology. The measured external quantum efficiencies and luminance were 0.5% and 2300 cd·m^{-2} at 6 V, respectively.

Most recently, Zhang *et al.* developed a color-tunable and wearable fiber-shaped polymer light emitting electrochemical cell (PLEC), which allows for the integration of wearable LEDs in textiles [42]. The fiber-shaped PLEC was fabricated using an all-solution processing method, which increases the chance for practical use. Figure 6d presents a schematic diagram of a flexible fiber-based PLEC [42]. To fabricate the fiber-shaped PLEC, a stainless steel wire was dip coated with ZnO nanoparticles to form an electron transfer layer and decrease leakage current. Then, an electroluminescent polymer layer (a blend of PF-B, ethoxylated trimethylopropane triacrylate and lithium trifluoromethane sulphonate) was grafted over the ZnO-coated stainless steel wire, followed by wrapping the modified steel wire with the aligned carbon nanotube (CNT) sheet. The aligned CNT provided high electrical conductivity and flexibility in the device. Since the fiber-shaped PLEC is a 1D structure, the brightness was the same in all directions. In addition, the authors successfully integrated PLECs into fabrics and showed the tunable colors. This work shows promising benefits such as flexibility, light weight, high brightness that is maintained after bending, and tunable colors, as well as the possibility of integration into wearable electronic devices.

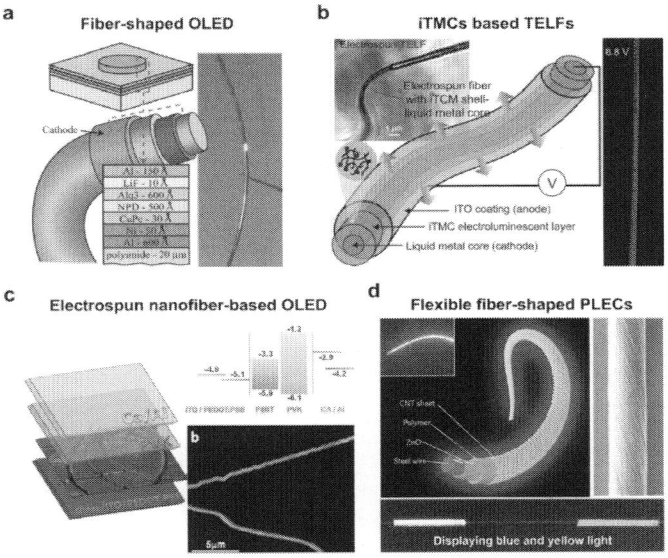

Figure 6: (a) Schematic illustration of a fiber-shaped organic light emitting diodes (OLED) structure compared with a typical OLED structure and a photograph of a flexed fiber with a green light-emitting pixel; (b) Schematic illustration and optical image of ionic transition-metal complex (iTMCs)-based electro-luminescent nanofibers (TELFs) fabricated via coelectrospinning and luminescence response of TELF under an applied voltage of 6.8 V; (c) Schematic of the conjugated polymer electrospun nanofiber-based OLED structure (left), energy level diagram (right top), and fluorescence microscopy image of an electrospun nanofiber, F8BT-PEO (right bottom); (d) Schematic illustration showing the structure of flexible fiber-shaped polymer light-emitting electrochemical cells (PLECs). Left inset: photograph of a fiber-shaped PLEC biased at 10 V, right inset: aligned CNT sheet wrapped around the modified stainless steel wire, and bottom inset: PLEC displaying blue and yellow light. Reproduced from [42,43,134,135] with permissions by John Wiley & Sons, Copyright 2007 and by ACS Publications, Copyright 2012, 2011 and by Nature Publishing group, Copyright 2015.

High Sensitivity Sensors

Sensors are widely used electronic devices which detect specific the characteristics of their surroundings and are present in a wide variety of applications such as electronics, aerospace, medicine, robotics, and machinery [136,137,138,139,140,141,142]. Types of sensors include temperature, pressure, humidity, gravity, sound, and vibration sensors [143,144,145,146,147,148,149,150]. In this review, we focus on textile-based pressure sensors, as sensing pressure is one of the most important functions of wearable electronics.

In 1996, a full-body, soft, flexible sensing suit was developed for a small robot by Japanese researchers [151]. They used an electrically-conductive fabric sheet and conductive strings with a spacer, and measured the contact resistance after applying pressure, demonstrating a promising remote robot sensor capable of controlling robotic body interactions. Shimojo *et al.*fabricated a tactile sensor using pressure-conductive rubber [152] via the development of a thin, flexible, attachable, tactile sensor that can be operated on curved surfaces. They measured the change in resistance under constant strain, observing a change in resistance during a period of 100 s. Their tactile sensor, when properly operated on a robot hand, showed good durability under repeated bending tests. Takamatsu *et al.* developed a large-area pressure sensor on a meter-scale using dye-coated conductive polymer (PEDOT:PSS) and perfluoropolymer dielectric film [153]. The sensor was constructed by weaving fibers into an area of 16 × 16 cm^2. The sensitivity of this design ranged from 0.98 to 9.8 N/cm^2, sensitive enough to detect human touch and therefore suitable for wearable keyboards and healthcare systems.

Recently, Lee *et al.* fabricated a textile-based pressure sensor using Ag-SBS composite conductive fibers [20]. A capacitive type of textile-based sensor was fabricated using a poly(dimethyl-siloxane) (PDMS) coating on the conductive fiber and cross-stacking two PDMS-coated fibers, as shown in Figure 7a [20]. Figure 7b shows the relative changes in capacitance of fiber-based pressure sensors via increasing loads of applied pressure [20]. In this operation, the conductive fiber-based pressure sensor showed stable responses to the various loads (Figure 7c), and detected the loading and unloading of grains of extremely low weight. This work further demonstrates e-textile applications with conductive fiber-based pressure sensors. The textile-based pressure sensor was pixelated through weaving techniques and detected the spatial distribution of external pressures, as shown in Figure 7d [20]. This textile-based pressure sensor was also sewn into a smart glove and clothes and showed the control of drone and spider robots, as shown in Figure 7e [20]. This work demonstrates that human-machine interfaces can be successfully operated using wearable textile-based pressure sensors with ultrahigh sensitivity and stability.

Ma *et al.* presented a new means of fabricating stretchable fabric [154]. To make stretchable knitted fabric, conductive stretchable fibers were first fabricated via wet-spinning. Figure 8a shows the surface morphology of conductive stretchable fibers composed of Ag nanoparticles, nAg-MWCNTs, and poly (vinylidene fluoride-co-hexafluoropropylene) (PVDF-HFP) matrix [154]. Stretchable fabric was then made by knitting the conductive stretchable fibers. The knitted fabric showed reversible characteristics under 100% tensile strain with mechanical and electrical stability (Figure 8b) [154]. In addition, the stretchable fabric was applied to a robotic finger system and demonstrated feasibility for practical application.

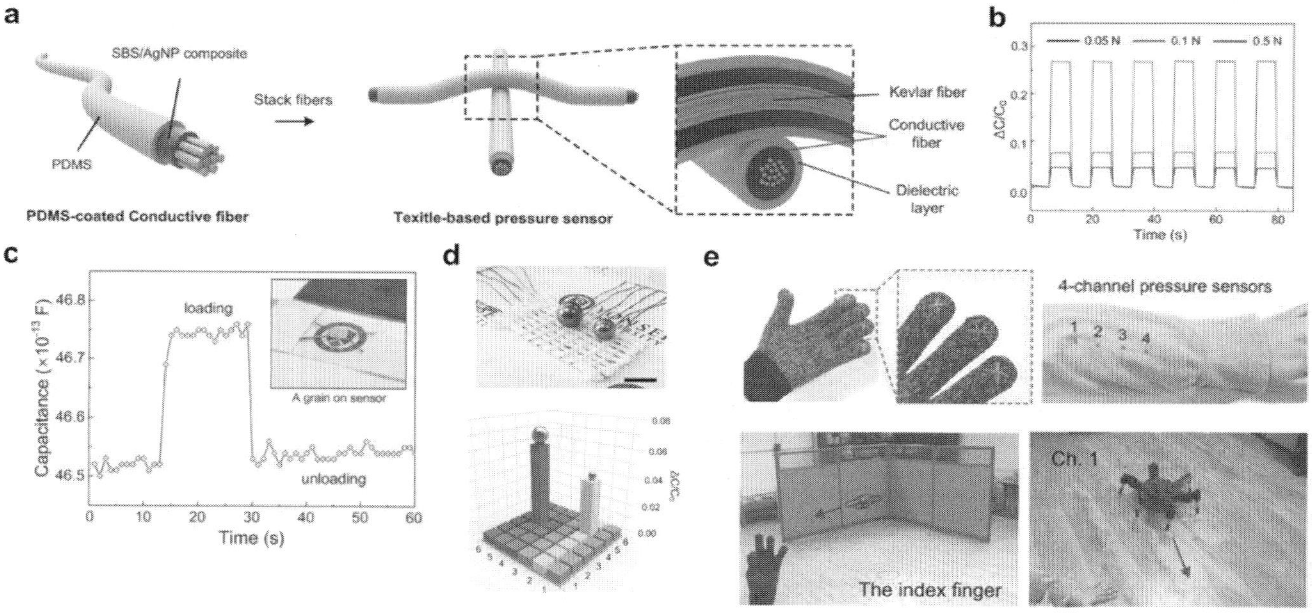

Figure 7: (a) Schematic structure of a textile-based pressure sensor fabricated with PDMS-coated conductive fibers; (b) Capacitive response of the pressure sensor for the various applied loads; (c) Capacitive response of the pressure sensor to the placing and removal of a grain (8 mg); (d) Fabrication of pixelated pressure sensor array; (e) Photographs of the smart glove and clothes with textile-based pressure sensors and operation using smart glove and clothes to control the drone and spider robot. Reprinted from [20] with permissions by John Wiley & Sons, Copyright 2015.

More recently, Lee *et al.* fabricated highly-stretchable conductive fibers composed of Ag nanowires (Ag NWs), Ag nanoparticles (Ag NPs), and SBS polymer using the wet spinning method [90]. Ag NWs allow for significant stretching capability in the fabric because they are aligned along the applied uniaxial strain and bridge the disconnected networks of Ag NPs inside the fiber, as shown in Figure 8c [90].

Figure 8d shows changes in conductivities with regard to various applied tensile strains [90]. As tensile strain increased, conductivities initially decreased, but improved again with increased strain. The reliability test showed stable operation under a repeatable bending test (Figure 8e) [90]. The authors further demonstrated the capability of detecting human motion using the fabricated stretchable sensor. The smart glove was integrated with fibers on each finger and successfully detected finger sign language. Figure 8f shows the detection of the letters "Y" and "S" via the bending of fingers [90].

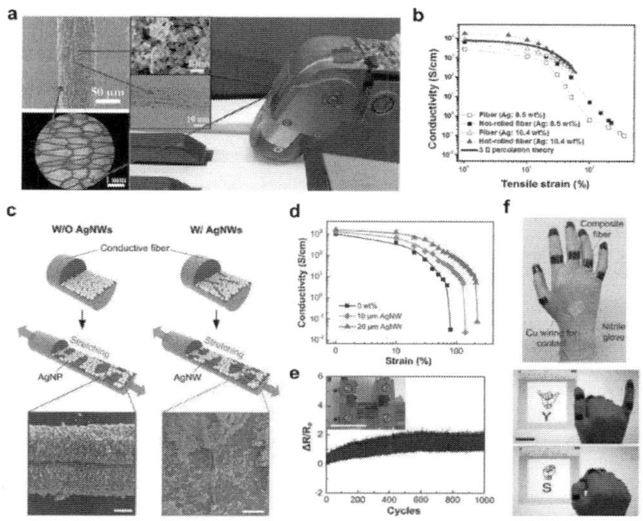

Figure 8: (a) FE-SEM and HRTEM images of knitted fabrics made from highly-conductive stretchable fibers. Bottom of this image shows an optical image of 2 ply rope. The knitted fabric was attached to a robot finger (as shown by arrow) for touch sensing; (b) Plot shows the conductivities of fibers as a function of tensile strain; (c) Schematic illustration of silver nanowires (AgNWs) and silver nanoparticles (AgNPs) in the composite fiber. (Left bottom image in navy colour line: SEM image of AgNP-mixed styrene-block-butadienstyrene (SBS) fiber without AgNWs at 50% strain and and Right bottom image in red colour line: SEM images of the AgNW-AgNP embedded SBS fiber at 50% strain); (d) Changes in conductivity of the composite fiber depending on the length of the AgNWs; (e) Reliability test measured the changes in the normalized resistance of the composite fiber; (f) Photograph of a smart glove made with composite fibers and motion detection of English letters using a sign language. Reproduced from [90,154] with permissions by ACS Publications, Copyright 2014 and by John Wiley & Sons, Copyright 2015.

TEXTILE-BASED ENERGY APPLICATIONS: ENERGY HARVESTING & STORAGE

Although commercial energy harvesting and storage devices using semiconducting thin film exhibit high performance, they also have limitations including their complex fabrication process and high cost. Recent novel materials investigated for energy related devices include dye, organic and inorganic polymers with high efficiency, large capacity, abundant material sources, easy fabrication, and flexibility. Along with the development of materials for energy devices, there are several studies on novel methods for energy harvesting using the human body, such as piezoelectric, triboelectric, and thermoelectric nanogenerators. Nanogenerators can provide self-powered systems that are useful in wireless devices. In particular, fiber-based nanogenerators have been primarily studied in the form of twisted, conjugated, or weaved multiple fiber electrodes integrated with supercapacitors.

Energy Harvesting from Human Motion

As the number of applications related to mobile electronic devices increases drastically, the development of long-lasting power sources for these devices is increasingly in demand. Sustainable self-powered sources have received significant attention as alternatives to conventional rechargeable batteries. Recently, energy harvesting from human body motions has proven a new means of operating many mobile electronic devices without the need for external power sources [155].

The piezoelectric effect is a unique ability of materials with polarization domains or non-centrosymmetric structures. These materials can generate an electrical charge in response to mechanical stresses such as compression, twisting, or distortion. One novel approach using the piezoelectric effect converts human mechanical motion into electrical energy. Wang *et al.* demonstrated that ultrasonic waves were converted into electricity using a Pt-coated zigzag electrode with vertically-aligned ZnO nanowires (NWs) [156]. They optimized the size and shape of the ZnO NWs and obtained high-output power per unit of area reaching ~10 μW/cm^2. The generated energy can potentially facilitate long-term sensors that are applicable to mobile electronic devices and systems [157]. Zhang *et al.* reported a flexible nanogenerator (FNG) fabricated using $BaTiO_3$ nanowire and polyvinyl chloride (PVC) composite piezoelectric fiber [158]. After optimizing their process parameters, they placed FNG on a bent human arm, producing voltage and current outputs up to 1.9 V and 24 nA, respectively. Several methods have been developed for harvesting electrical energy by converting mechanical energy into electricity using piezoelectric materials [159]. A prototype electrostatic non-resonant generator has been designed and optimized by Mitcheson *et al.* [160]. Using flexible polymer membranes, they can generate an output voltage of 250 V when the generator is pre-charged to 30 V. Qin *et al.* have described the concept of textile-based piezoelectric nanogenerators that can harvest electrical energy from the friction between two fibers [161]. Although only a small amount of output current (~±5 pA) was obtained during the pulling/releasing cycles of the nanogenerator, they established a novel method of energy harvesting using fabrics.

Further, Jeong *et al.* reported a novel approach to fabricating a high-performance hyper-stretchable nanogenerator (SEG) with extremely long Ag NWs [162]. The hyper-stretchable piezoelectric elastic composite (PEC) rubber polymer was composed of very long nanowire percolation (VLNP) electrodes and a well-dispersed mixture of lead magnesioniobate lead titanate (PMN-PT) particles and multi-walled carbon nanotubes (MWCNTs). The composite exhibited high stretchability over 200% strain without any mechanical cracking; this could be attributed to variation in the piezoelectric dipoles in the PEC caused by stretching, and helps accumulate electrons, generating voltage and current. An open-circuit voltage (~4 V) and short-circuit current (~500 nA) were obtained and the generated energy can sufficiently operate commercial electronic units. The SEG was tested in biological motion after being sewn onto the knee of a stretchable stocking, as shown in Figure 9a,b [162]. However, textile-type energy harvesters seem to face some prevalent problems such as low energy conversion efficiency, wide spectra of vibration frequencies, and time-dependent amplitudes.

Recently, triboelectric nanogenerators (TENG) have been developed that can harvest electrical energy from mechanical friction, exhibiting high performance, a simple fabrication process, cost-effectiveness and green technology [163,164,165,166]. The principle of TENG operation is the combination of triboelectric and electrostatic induction [166,167]. When two different polymer surfaces are rubbed or compressed, opposite sign electrostatic charges are generated on each surface of the polymer films. Due to the formation of a dipole layer, triboelectric potential in TENG is generated and drives electrical charges through the external load, resulting in the generation of an open circuit voltage and a short circuit current. Lee *et al.* reported a textile, substrate-based triboelectric nanogenerator (T-TENG) fabricated using an assembly of Au-coated fabric and polydimethylsiloxane (PDMS) [165]. To enhance the surface friction energy, thermally-evaporated Al nanoparticles (NPs) were conjugated with Au-coated textile top electrodes. Triboelectric energy was generated when two textile electrodes were repeatedly compressed and released. The validation of T-TENG in energy harvesting using human motion was further examined on a human arm. To utilize this device in practice, they demonstrated that the alternating current (ac) output signals obtained from the T-TENG were converted to direct current (dc) using a bridge rectifier. Using the rectified dc current, a commercial capacitor was charged and a light-emitting diode (LED) was turned on. When the T-TENG attached to the human arm was bent and released, the generated output voltage and current were 139 V and 39 μA, respectively. The high efficiency can be attributed to enhanced surface roughness induced by uniformly-distributed Al NPs on the textile electrodes.

The abovementioned nanogenerators were produced on flexible, metallic-based electrodes. However, to realize the true potential of wearable electronics, it is necessary to study textile-based flexible energy harvesters without metal electrodes. Metal-free, fiber-based TENG capable of converting biomechanical motion into electricity were introduced by Zhong *et al.* [163]. The TENG were produced using two kinds of pretreated cotton threads, each coated with multi-walled carbon nanotubes (CNT) and polytetrafluoroethylene-carbon nanotubes. These two fibers were entangled with each other to form a light-weight and flexible fiber-based generator (FBG). When the entangled fibers were stretched, the inter-fiber gap was reduced and the CNT layer on the CNT-coated cotton thread (CCT) became positively charged due to induction, creating free electrons. Woven FBG was applied to a "power shirt" with a wireless temperature sensor capable of charging to a capacity of 10 nF, as shown in Figure 9c [163]. Figure 9d shows that the rectified DC voltage obtained from the FBG was 2.4 V with the charging time of 27 s, showing that the FBG is capable of lighting a red LED. Since the fabricated power shirt can monitor body temperature cost-effectively due to the self-powered TENG system, it makes viable the possibility of woven textiles in the field of mobile medical systems.

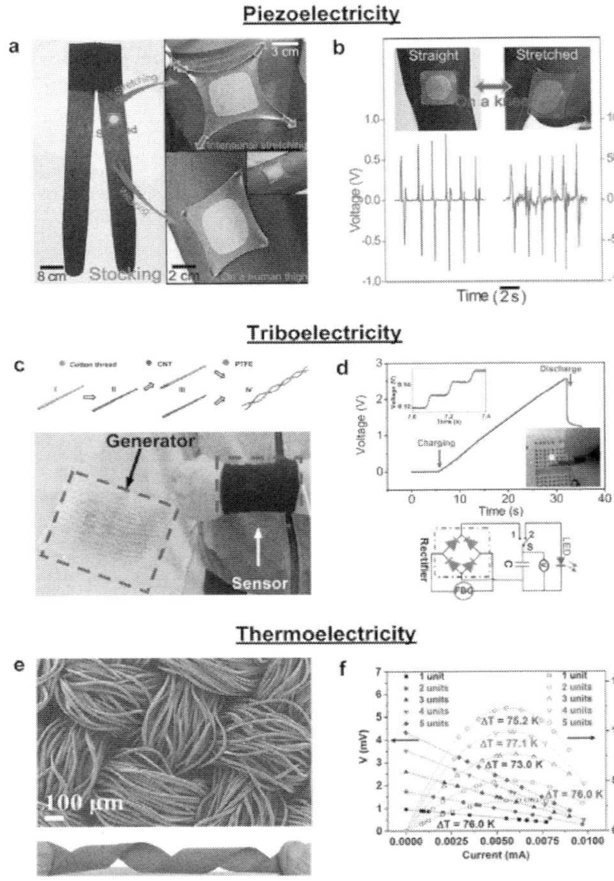

Figure 9: (a) Digital photographs of the stretchable elastic-composite generator (SEG) stitched on a nylon stocking. Photograph of the SEG intentionally stretched and fitted over a human thigh; (b) Stretching on human knees and the generated voltage and current from the SEG on the stocking by bending and straightening the knee; (c) Schematic diagram illustrating the detailed fabrication process of a fiber-based generator (above) and woven fabric FBG sewed on a lab coat (below); (d) The voltage charging curve of the 2.2 μF commercial capacitor by the "power shirt" and typical bridge rectifier circuit for transforming alternating current to direct current; (e) SEM image (above) and digital photography (below) of the polyester fabric after PEDOT:PSS coating; (f) Experimental results of output voltage and power as a function of current for the fabricated devices. Current–power curves are parabolic for all devices. Reproduced from [162,163,168] with permissions by John Wiley & Sons, Copyright 2015 and by ACS Publications, Copyright 2014 and by Nature Publishing Group, Copyright 2014.

Recently, interest in thermoelectric materials has increased due to their ability to generate energy from waste heat. In the presence of temperature gradients, charges in the thermoelectric material are diffused from the hotter to the colder side, inducing electrochemical potential inside the material. In wearable electronics, fiber-based thermoelectric generators (TEGs) are a unique energy harvesting device converting body heat into a reliable power source [168,169,170]. Since human body heat is a steady, plentiful, and convenient source of heat, it is meaningful and useful that TEG can harvest waste body heat. Kim *et al.* described a wearable TEG utilizing a combination of fabric and printed thermocouples that can be easily embedded in clothing to harvest low-temperature energy [169]. To prevent skin irritation, polymer-based fabric was used as a substrate and a conductive silver thread was used for electrical connection. When the TEG was integrated in a shirt, the temperature of the skin was maintained at 32 °C under cold (5 °C) or hot (25 °C) environments. Interestingly, the output power under cold environment was significantly higher at ~147 nW than that of the hot ambient environment, at ~8.1 nW.

To investigate the suitability of thermoelectricity in wearable power generators, Du *et al.* reported a textile-based TEG featuring a long operating time, easy maintenance, and high reliability [168]. This TEG overcame the common problems of toxicity, poor processing ability, and impermeability by introducing a PEDOT:PSS conducting polymer. The inset of Figure 9e shows the morphology of polyester fabric coated with PEDOT:PSS, which had excellent flexibility in any shape [168]. The PEDOT:PSS-coated fabrics were connected using silver wires, as shown in Figure 9f. Using multiple strips of the fabricated device, the output voltage was improved to 4.3 mV due to the Seebeck effect. Using a fabric thermoelectric generator containing 5 thermoelectric strip units, the maximum output of electrical power (P_{max}) was 12.29 nW at a $\Delta T = 75.2$ K, as shown in Figure 9f [168].

Despite the various advantages of TEGs, their limitations include a low energy conversion efficiency rate, slow technology progression, limited application possibilities, and the requirement of a constant heat source. The fabrication process is also complex because TEGs should be flexible and tightly adhere to the human body.

Solar Cells on Textiles

In the past decades, next-generation solar cells such as dye-sensitized solar cells (DSSC), organic solar cells (OSC), and perovskite solar cells (PSC) have attracted much attention due to their low cost and high efficiency [171,172,173]. These solar cells are based on light absorption using dye, organic, or perovskite materials that generate mobile electron-hole pairs [174]. The generated electrons and holes are separated and collected into cathodes and anodes, respectively. Recently, fiber-based solar cells have shown great potential for use in textile electronics and represent a new direction in solar energy harvesting since they can be woven into various types of flexible structures at low-cost. However, these devices would benefit from further study into the development of facile fabrication methods for large-scale production and realization of actual products as well as the feasibility of integration in cloth. Several researchers reporting wearable solar cells have primarily focused on inorganic, dye-sensitized, perovskite, and polymer-based solar cells with lightweight, flexible, and ease of manufacture [38,99,175,176].

Recent studies on dye-sensitized photovoltaic fibers have used primarily copper, steel wire, or polymer fibers coated with an indium tin oxide (ITO) electrode [177,178]. Cai *et al.* demonstrated a new method of fabricating 0.3 mM cis-bis(isothiocyanato)bis(2,2`-bipyridyl-4,4`-dicarboxylato)-ruthenium(II) bis-tetrabutylammonium (N719)-absorbed CNT fibers as a working electrode and CNT fibers coated with poly-vinylidene fluoride (PVDF) as a counter electrode for dye-sensitized photovoltaic application [178]. The fiber-based solar cell was integrated into textiles via conventional weaving techniques. The left side of Figure 10a shows a schematic illustration of the photovoltaic wire, while the middle image inFigure 10a shows an SEM image of the twisted CNT/N719 and CNT/PVDF fibers [178]. The open-circuit voltage and short-circuit current density of the fiber-based dye-sensitized solar cell were 0.69 V and 9.84 mA·cm^{-2}, respectively, with a power conversion efficiency of 3.90%, as shown in the right image in Figure 10a.

Zhang *et al.* demonstrated CdSe NWs and CNT yarn-based photovoltaics, with power conversion efficiency up to 3% [179]. The left image in Figure 10b presents a schematic illustration of fiber-based CdSe-grafted Ti wire fabricated via chemical vapor deposition (CVD) with twisted CNT yarn. The SEM image in Figure 10b shows that the twisted CdSe-Ti wire and CNT fiber make smooth contact without the addition of electrolyte due to the flexibility of the yarn. The performance of CNT yarn and CdSe-Ti wire heterojunction with polysulfide electrodes was characterized by measuring the J-V curve of the fiber cells. The open-circuit voltage (V_{oc}), short-circuit current density (J_{sc}), fill factor (*FF*), and efficiency were 0.65 V, 8.25 mA·cm^{-2}, 55%, and 2.9%, respectively.

Recent successes in perovskite ($CH_3NH_3PbX_3$, X = Cl, Br, I) solar cells provide encouragement for flexible solar cells in wearable electronics [180,181]. Perovskite materials have many attractive properties including broad spectral absorption and high carrier mobility. The efficiency of thin film photovoltaic cells using perovskite materials have been recorded at up to 20% [182,183]. Some attempts have been made to fabricate fiber-based perovskite solar cells for wearable electronics [38,184]; however, it is difficult to integrate the fibers into clothes due to the insufficient flexibility and low surface area of perovskite materials. To solve this issue, Li *et al.* described a CNT fiber-supported double-twisted perovskite solar cell featuring high stability where the perovskite nanocrystals on the CNT fiber sustain the interfacial adhesion under stress [38]. The left image in Figure 10d shows a schematic illustration of the CNT fiber-supported double-twisted perovskite solar cell using twisted*n*-TiO_2 added CNT, meso-TiO_2, $CH_3NH_3PbI_{3-x}Cl_x$, P3HT/SWNT, and an Ag nanowire network [38]. The P3HT/SWNT was used as a hole transport layer, and the Ag nanowire was added to increase the surface area between electrodes. The fabricated perovskite solar cell fiber can be seen in the middle of Figure 10d; the optimized maximum power conversion efficiency (PCE) was 3.03%. Although the performance of the abovementioned textile-based photovoltaic harvesters are, some limitations such as mass production, high cost, and low energy conversion efficiency remained as obstacles to utilization in real fabrics. Further research into textile-based solar cells is needed for the realization of future wearable solar cells.

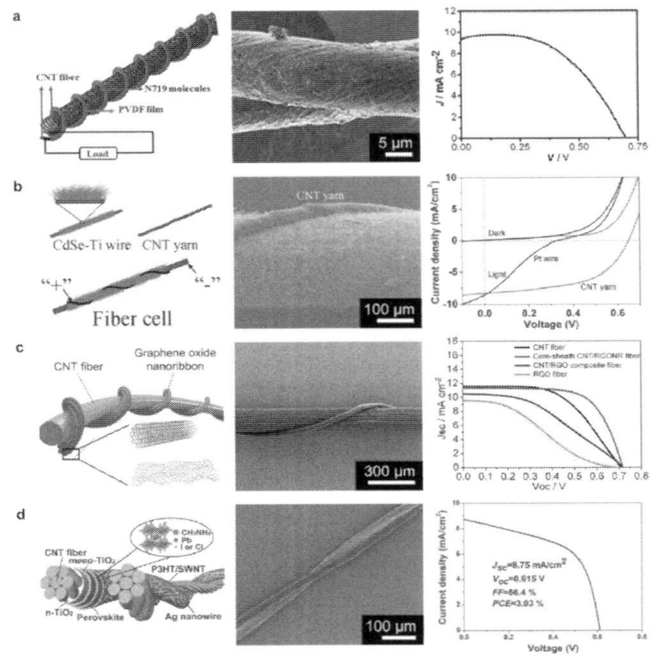

Figure 10: (a) Schematic (left) and SEM image (middle) of the twisted CNT/N719 working electrode and carbon nanotube/poly-vinylidene fluoride (CNT/PVDF) counter electrode. A typical density–voltage (J–V) curve (right) with improved performance after modification of the working electrode; (b) Schematic fabrication process (left) and SEM image (middle) of photo electrochemical cells with a primary electrode of Ti wire, light absorption layer of CdSe nanowire, and counter electrode of CNT. J–V curves under dark and light conditions of different fiber cells using a CNT yarn and Pt wire counter electrode; (c) Schematic (left) and SEM image (middle) of wire shaped dye-sensitized solar cells with twisted CNT fiber. Current density-voltage curves for a wire-shaped DSC (right); (d) Schematic diagram of double twisted solar cell supported by the CNT fiber and the chemical structure of the perovskite material (left). The SEM image shows twisted fibrous solar cells (middle). The current J–V curve is shown for electrical performance (right). Reprinted from [38,99,175,178] with permissions by Royal Society of Chemistry, Copyright 2012, 2012 and by ACS Publications, Copyright 2013 and by John Wiley & Sons, Copyright 2015.

Fabric-Based Supercapacitors

Supercapacitors (SCs) have attracted significant attention as outstanding storage components due to their high power density, operational safety, long cycle life, and fast charge-discharge rates [185,186,187]. Depending on the mechanism of charge storage, SCs can be categorized into two types: the Electrical Double Layer Capacitor (EDLC) and pseudocapacitor. EDLCs are typically fabricated using two electrodes separated by a carbon-based electrolyte and an electrical double layer is formed between the electrodes and electrolyte based dielectric. Pseudocapacitors function based on reversible faradaic-type charging and discharging. The supercapacitor is made of conducting polymers and transition metal oxides, and the mechanism of energy storage depends on the redox reactions between materials. To integrate these supercapacitors into textiles, their components, including electrodes and current collectors, will have to be flexible.

Recently, Yang *et al.* demonstrated a pseudocapacitor-based supercapacitor with improved redox reaction and good reversibility [188]. To further improve the performance, one study introduced a hybrid architecture using MnO_2-coated hydrogenated zinc oxide nanowires (HZnO) grown on a flexible carbon cloth [189,190]. High electrochemical performance was exhibited by amorphous ZnO-doped MnO_2 core-shell nanocable (HZM) electrodes with a polyvinyl alcohol (PVA)/LiCl neutral electrolyte. Figure 11a shows a schematic illustration of the synthesis process of core-shell structures and an SEM image of HZM. The cyclic voltammetry (CV) measurement shown in Figure 11b indicates that the HZnO electrode had a higher free carrier concentration and better electrical stability when compared to untreated ZnO and air-annealed ZnO electrodes which is attributed to the large surface area of the HZnO electrode.

This improvement was further explained by the insertion of hydrogen acting as a shallow donor for the neutral and electrically-inactive oxygen vacancies in *n*-type ZnO. Figure 11c shows the combination of the energy harvesting module, dye-sensitized solar cells (DSSCs), and energy storage component that could illuminate a commercial blue LED from charged SCs when the solar source is off. Figure 11d shows a significant initial drop in the leakage current; this measurement stabilized after 4500 s, indicating the presence of few impurities in the electrode and electrolyte. This research contributed to both the fundamental advancement of this technology as well as the advancement of potential of SCs applications.

Zhang *et al.* also developed a flexible energy fiber composed of titania nanotube-modified Ti wire and MWCNT that can store energy as well as harvest solar energy [191]. P3HT: 1-(3-methoxycarbonyl) propyl-1-phenyl [6,6] C_{61} (PCBM) and PEDOT:PSS with MWCNTs were used as photoactive materials and a PVA/H_3PO_4 gel electrolyte was utilized for capacitor storage, as shown in Figure 11e. When the P3HT: PCBM heterojunction generates excitons and separate carriers (electrons and holes), the carriers are transported through MWCNTs and PEDOT:PSS, and then stored in the energy storage area.Figure 11f shows the energy fibers that were successfully woven into textiles. Figure 11g describes the charge-discharge curves, which indicate a fast charging process. The maximum efficiency and energy storage capacity of the fibers are ~1% and ~65%, respectively. These advances represent various new opportunities for the future of wearable electronics.

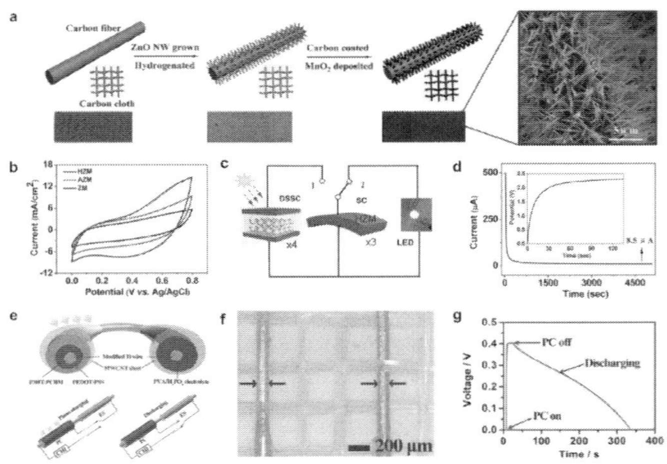

Figure 11: (a) Schematic fabrication process and SEM image of a hydrogenated single crystal ZnO (HZM) and amorphous ZnO-doped

MnO_2 core-shell electrodes; (b) Current-potential characterization with different electrodes with scan rates of 100 mV/s; (c) Design of a stand-alone self-powered system consisting of dye-sensitized solar cells (DSSCs), SCs, and LEDs; (d) Leakage current characteristics of three SCs in series and voltage profile (inset) charged by four DSSCs in series; (e) Schematic illustration and circuit connection mechanism in the process of charging and discharging of photovoltaic conversion (PC) and energy storage (ES) with titania nanotube-modified Ti wire and aligned MWCNT sheet as two electrodes; (f) Photography of energy fibers woven with each other to form flexible textiles; (g) Charging-discharging curve with a current of 0.1 μA during the discharging process with the ES part. Reproduced from [188,191] with permissions by ACS Publications, Copyright 2013 and by John Wiley & Sons, Copyright 2014.

CURRENT LIMITATIONS OF WEARABLE E-TEXTILES

Aforementioned discussion on the wearable E-textiles presents them as a good futuristic candidate in daily life of human outfits. However, for the true realization of wearable E-textiles, some basic issues must be addressed.

One of the main limitations of wearable E-textiles which require more concern is toxicity due to hazards materials used in electronic components. If, we integrate the electronic components in the wearable textiles they must be safe and non-toxic. Electronic components contain toxic material for example batteries have toxic and liquid electrolyte material like Li-ion batteries and tetraethylammonium tetrafluoroborate in acetonitrile [186,192]. Therefore, garments integrated with these electronic components are not suitable to wear because they will be in directly contact with skin. To resolve this issue some reports are available to use a non-toxic material. Jost *et al.* have fabricated flexible and durable textile electrodes for supercapacitors by using non-toxic electrolyte material (sodium and lithium sulfate) [193]. The as prepared fabric electrodes have shown high areal capacitance but the problem with these fabric electrodes is the requirement of an integrated device to seal the liquid electrolyte to avoid its leakage. Further, a non-toxic electrolyte based wearable supercapacitor has been suggested by pasta *et al.* in their well-arranged study [194]. They have reported a simple fabrication technique for supercapacitor showing high specific capacitance and cycling stability. But the drawback of as prepared supercapacitor is the hydrogen evolution overvoltage prevents the operative voltage range as similarly reported by Prosini *et al.* [195]. Several other reports are also available on non-toxic electrolyte materials. However, further improvements are required in order to use these components in wearable textiles and toxicity issue should be carefully monitored before employ them in wearable fabrics.

Other drawback of these wearable E-textiles is related to their multiple uses, *i.e.*, washability and durability. To the best of our knowledge, there is no study on multiple washable textiles integrated with electronic components. In general, multiple washing and special care is required to worn these garments in daily life. Also, the long term usage of these wearable E-textiles is again a matter of concern. It is required to know that the repairing as well as exchanging of electronic device is possible during their long term use by wearer. The other challenge for researchers is to focus on the packaging of electronic components and device performance for their usage over the longer time. Due to crack and poor bending properties of fiber–based electronics, failure of packaging is observed [196]. These wearable electronic fabrics must hold their sensing, actuating and responding properties in qualitative way during their everyday wear and off-course maintain comfort fit to human body. This can be achieved by replacing hard printed circuit board to flexible electronics.

For the practically implementation of these wearable fabrics the mass production is required and for this an interdisciplinary work must be involved between electronic as well as textile peoples. But to the best of knowledge there is a lack of combination of electronics and textiles industries. Thus, for future developments of E-textiles it is required that the experts of these two firms must work together. Other issues include the biocompatibility of these textiles, development of sophisticated multi-functionalities, stable performance of devices and light weight fabrics for future applications. Thus, any realistic commercial implementation can be achieved after addressing all pointed issues.

CONCLUSIONS AND PERSPECTIVES

In this review article, a brief overview of textile electronics is presented based on electronic components, in which the various types of conductive fibers and fabrication techniques are summarized using different materials such as polymers, CNTs, graphene, and metal nanowires. Fiber-based electronic components (transistors, LEDs, and sensors) have also been described. Furthermore, textile-based energy harvesting and storage devices were explained for energy applications in textile electronics. Although considerable performance in textile-based electronic devices has already been achieved, further efforts to improve performance are necessary. A significant body of theoretical and experimental research has been carried out to understand the mechanism and characteristics of E-textiles in wearable electronics. To commercialize textile-based devices in wearable electronics, some important issues will have to be addressed, including mass production, integration into clothes, non-toxic technology, and long-term usage. If the E-textiles can overcome the aforementioned issues, a new era in wearable electronics will begin.

ACKNOWLEDGMENTS

This work was supported by the LG Yonam Foundation. We are thankful to the LG group for the support based on Academic Support Programs of LG Yonam Foundation.

AUTHOR CONTRIBUTIONS

All authors discussed the contents of the review manuscript. V.K. wrote the text of the manuscript. S.L., J.H. and J.L. made the Figures for paper. C.M. also contributed to write the manuscript and revised by V.K. and S.L. Modifications and the proof reading of the manuscript has been done by J.S., S.L., J.H., S.L. and V.K. The manuscript is prepared under the supervision of T.L.

REFERENCES

1. Csapó, Á.; Wersényi, G.; Nagy, H.; Stockman, T. A survey of assistive technologies and applications for

blind users on mobile platforms: A review and foundation for research. *J. Multimodal User Interfaces* 2015.

2. Lewis, Z.H.; Lyons, E.J.; Jarvis, J.M.; Baillargeon, J. Using an electronic activity monitor system as an intervention modality: A systematic review. *BMC Public Health* 2015, *15*, 585–600. [PubMed]
3. Vanveerdeghem, P.; Van Torre, P.; Thielens, A.; Knockaert, J.; Joseph, W.; Rogier, H. Compact personal distributed wearable exposimeter. *IEEE Sens. J.* 2015, *15*, 4393–4401.
4. Yun, J.Y.; Hong, W.G.; Choi, N.-J.; Kim, H.B.; Jun, Y.; Lee, H.-K. Ultrasensitive and highly selective graphene-based single yarn for use in wearable gas sensor. *Sci. Rep.* 2015, *5*, 10904–10910. [PubMed]
5. Bai, P.F.; Hayes, R.A.; Jin, M.; Shui, L.; Yi, Z.C.; Wang, L.; Zhang, X.; Zhou, G. Review of paper-like display technologies (invited review). *Prog. Electromagn. Res.* 2014, *147*, 95–116.
6. Patel, S.; Park, H.; Bonato, P.; Chan, L.; Rodgers, M. A review of wearable sensors and systems with application in rehabilitation. *J. Neuroeng. Rehabil.* 2012, *9*, 1–17. [PubMed]
7. Bell, A.G. Phone and Other Inventions. Available online: http://www.carnetdevol.org/Bell/Phone.html (accessed on 3 August 2015).
8. Tropp, H. Computer report VII: The effervescent years: A retrospective: Behind the computer revolution lay the vision and perseverance of a handful of pioneers committed to open information exchange. *IEEE Spectr.* 1974, *11*, 70–79.
9. Goldstine, H.H.; Goldstine, A. The electronic numerical integrator and computer (ENIAC). *IEEE Ann. Hist. Comput.* 1996, *18*, 10–16.
10. Myslewski, R. File:Ibm pcjr with display.jpg. Available online: https://commons.wikimedia.org/wiki/File:Ibm_pcjr_with_display.jpg (accessed on 14 July 2015).
11. Gilbert, D. Mobile Phone Celebrates 40th Birthday. Available online: http://www.ibtimes.co.uk/mobile-phone-40-years-motorola-dynatac-history-453057 (accessed on 3 August 2015).
12. Steven, S. Epson-hx-20. Available online: https://upload.wikimedia.org/wikipedia/commons/8/8d/Epson-hx-20.jpg(accessed on 14 July 2015).
13. Mann, S. Smart clothing: The wearable computer and wearcam. *Pers. Technol.* 1997, *1*, 21–27.
14. Ritchie, R. History of iPhone: Apple reinvents the phone. Available online: http://www.imore.com/history-iphone-2g(accessed on 3 August 2015).
15. Ackerman, E. Google gets in your face. *IEEE Spectr.* 2013, *50*, 26–29. [Google Scholar]
16. Apple. Apple Watch. Available online: http://www.apple.com/kr/watch/gallery/ (accessed on 3 August 2015).
17. Popat, K.A.; Sharma, P. Wearable computer applications a future perspective. *Int. J. Eng. Innov. Technol.* 2013, *3*, 213–217.
18. Someya Group Organic Transistor Lab Press Kits. Available online: http://www.ntech.t.u-tokyo.ac.jp/en/press/press_for_media/index.html (accessed on 3 August 2015).
19. Caron, N. Meet the Rest of the Cast for Fox's Minority Report TV series. Available online: http://www.blastr.com/2015-3-3/meet-rest-cast-foxs-minority-report-tv-series (accessed on 3 August 2015).
20. Lee, J.; Kwon, H.; Seo, J.; Shin, S.; Koo, J.H.; Pang, C.; Son, S.; Kim, J.H.; Jang, Y.H.; Kim, D.E.; Lee, T. Conductive fiber-based ultrasensitive textile pressure sensor for wearable electronics. *Adv. Mater.* 2015, *27*, 2433–2439.
21. Seesaard, T.; Lorwongtragool, P.; Kerdcharoen, T. Development of fabric-based chemical gas sensors for use as wearable electronic noses. *Sensors* 2015, *15*, 1885–1902.
22. Skrzetuska, E.; Puchalski, M.; Krucińska, I. Chemically driven printed textile sensors based on graphene and carbon nanotubes. *Sensors* 2014, *14*, 16816–16828.
23. Lanata, A.; Valenza, G.; Nardelli, M.; Gentili, C.; Scilingo, E.P. Complexity index from a personalized wearable monitoring system for assessing remission in mental health. *IEEE J. Biomed. Health Inform.* 2015, *19*, 132–139.
24. Zheng, Y.-L.; Ding, X.-R.; Poon, C.C.Y.; Lo, B.P.L.; Zhang, H.; Zhou, X.-L.; Yang, G.-Z.; Zhao, N.; Zhang, Y.-T. Unobtrusive sensing and wearable devices for health informatics. *IEEE Trans. Biomed. Eng.* 2014, *61*, 1538–1554.
25. Zheng, N.; Wu, Z.; Lin, M.; Yang, L.T. Enhancing battery efficiency for pervasive health-monitoring systems based on electronic textiles. *IEEE Trans. Inf. Technol. Biomed.* 2010, *14*, 350–359.
26. Paradiso, R.; Loriga, G.; Taccini, N. A wearable health care system based on knitted integrated sensors. *IEEE Trans. Inf. Technol. Biomed.* 2005, *9*, 337–344.
27. Quinlan, M.; Mayhew, C. The effects of outsourcing on occupational health and safety: A comparative study of factory-based workers and outworkers in the Australian clothing industry. *Int. J. Health Serv.* 1999, *29*, 83–107.
28. Rofouei, M.; Potkonjak, M.; Sarrafzadeh, M. Energy efficient collaborative sensing-based design: Soft keyboard case study. *IEEE Trans. Hum. Mach. Syst.* 2014, *44*, 115–124.
29. Yoo, H.-J. Your heart on your sleeve: Advances in textile-based electronics are weaving computers right into the clothes we wear. *IEEE Solid State Circuits Mag.* 2013, *5*, 59–70.
30. Li, W.; Zhu, L.; Gu, J.; Liu, H. Microwave absorption properties of fabric coated absorbing material using modified carbonyl iron power. *Compos. Part B Eng.* 2011, *42*, 626–630.
31. Moro, R.; Agneessens, S.; Rogier, H.; Dierck, A.; Bozzi, M. Textile microwave components in substrate integrated waveguide technology. *IEEE Trans. Microw. Theory Tech.* 2015, *63*, 422–432.
32. Castano, L.M.; Flatau, A.B. Smart fabric sensors and E-textile technologies: A review. *Smart Mater. Struct.* 2014, *23*, 053001–053027.

33. Stoppa, M.; Chiolerio, A. Wearable electronics and smart textiles: A critical review. *Sensors* 2014, *14*, 11957–11992.

34. Zięba, J.; Frydrysiak, M. Textronics–electrical and electronic textiles. Sensors for breathing frequency measurement.*Fibres Text. East. Eur.* 2006, *14*, 43–48.

35. Sun, H.; You, X.; Yang, Z.; Deng, J.; Peng, H. Winding ultrathin, transparent, and electrically conductive carbon nanotube sheets into high-performance fiber-shaped dye-sensitized solar cells. *J. Mater. Chem. A* 2013, *1*, 12422–12425.

36. Weintraub, B.; Wei, Y.; Wang, Z.L. Optical fiber/nanowire hybrid structures for efficient three-dimensional dye-sensitized solar cells. *Angew. Chem. Int. Ed.* 2009, *48*, 8981–8985.

37. Liu, D.; Zhao, M.; Li, Y.; Bian, Z.; Zhang, L.; Shang, Y.; Xia, X.; Zhang, S.; Yun, D.; Liu, Z.; Cao, A.; Huang, C. Solid-state, polymer-based fiber solar cells with carbon nanotube electrodes. *ACS Nano* 2012, *6*, 11027–11034.

38. Li, R.; Xiang, X.; Tong, X.; Zou, J.; Li, Q. Wearable double-twisted fibrous perovskite solar cell. *Adv. Mater.* 2015, *27*, 3831–3835.

39. Lee, J.B.; Subramanian, V. Weave patterned organic transistors on fiber for E-textiles. *IEEE Trans. Electron Devices* 2005,*52*, 269–275.

40. Tao, X.; Koncar, V.; Dufour, C. Geometry pattern for the wire organic electrochemical textile transistor. *J. Electrochem. Soc.* 2011, *158*, H572–H577.

41. Kylberg, W.; de Castro, F.A.; Chabrecek, P.; Sonderegger, U.; Chu, B.T.-T.; Nüesch, F.; Hany, R. Woven electrodes for flexible organic photovoltaic cells. *Adv. Mater.* 2011, *23*, 1015–1019.

42. Zhang, Z.; Guo, K.; Li, Y.; Li, X.; Guan, G.; Li, H.; Luo, Y.; Zhao, F.; Zhang, Q.; Wei, B.; Pei, Q.; Peng, H. A colour-tunable, weavable fibre-shaped polymer light-emitting electrochemical cell. *Nat. Photonics* 2015, *9*, 233–238.

43. Vohra, V.; Giovanella, U.; Tubino, R.; Murata, H.; Botta, C. Electroluminescence from conjugated polymer electrospun nanofibers in solution processable organic light-emitting diodes. *ACS Nano* 2011, *5*, 5572–5578.

44. Bedeloglu, A.; Demir, A.; Bozkurt, Y.; Sariciftci, N.S. A flexible textile structure based on polymeric photovoltaics using transparent cathode. *Synth. Met.* 2009, *159*, 2043–2048.

45. Jost, K.; Dion, G.; Gogotsi, Y. Textile energy storage in perspective. *J. Mater. Chem. A* 2014, *2*, 10776–10787.

46. Wang, Z.L. Towards self-powered nanosystems: From nanogenerators to nanopiezotronics. *Adv. Funct. Mater.* 2008,*18*, 3553–3567.

47. Das, T.K.; Prusty, S. Review on conducting polymers and their applications. *Polym. Plast. Technol. Eng.* 2012, *51*, 1487–1500.

48. Wegner, G. Polymers with metal-like conductivity—A review of their synthesis, structure and properties. *Angew. Chem. Int. Ed. Engl.* 1981, *20*, 361–381.

49. Patil, A.O.; Heeger, A.J.; Wudl, F. Optical properties of conducting polymers. *Chem. Rev.* 1988, *88*, 183–200.

50. Gerard, M.; Chaubey, A.; Malhotra, B.D. Application of conducting polymers to biosensors. *Biosens. Bioelectron.* 2002,*17*, 345–359.

51. Janata, J.; Josowicz, M. Conducting polymers in electronic chemical sensors. *Nat. Mater.* 2003, *2*, 19–24.

52. Svirskis, D.; Travas-Sejdic, J.; Rodgers, A.; Garg, S. Electrochemically controlled drug delivery based on intrinsically conducting polymers. *J. Controlled Release* 2010, *146*, 6–15.

53. Pernaut, J.-M.; Reynolds, J.R. Use of conducting electroactive polymers for drug delivery and sensing of bioactive molecules. A redox chemistry approach. *J. Phys. Chem. B* 2000, *104*, 4080–4090.

54. Tallman, D.E.; Spinks, G.; Dominis, A.; Wallace, G.G. Electroactive conducting polymers for corrosion control. *J. Solid State Electrochem.* 2001, *6*, 73–84.

55. Hohnholz, D.; Okuzaki, H.; MacDiarmid, A.G. Plastic electronic devices through line patterning of conducting polymers. *Adv. Funct. Mater.* 2005, *15*, 51–56.

56. Green, R.A.; Lovell, N.H.; Wallace, G.G.; Poole-Warren, L.A. Conducting polymers for neural interfaces: Challenges in developing an effective long-term implant. *Biomaterials* 2008, *29*, 3393–3399.

57. Olmedo, L.; Hourquebie, P.; Jousse, F. Microwave absorbing materials based on conducting polymers. *Adv. Mater.* 1993, *5*, 373–377.

58. Malinauskas, A. Electrocatalysis at conducting polymers. *Synth. Met.* 1999, *107*, 75–83.

59. Kingsborough, R.P.; Swager, T.M. Electrocatalytic conducting polymers: Oxygen reduction by a polythiophene-cobalt salen hybrid. *Chem. Mater.* 2000, *12*, 872–874.

60. Shirakawa, H.; Louis, E.J.; MacDiarmid, A.G.; Chiang, C.K.; Heeger, A.J. Synthesis of electrically conducting organic polymers: Halogen derivatives of polyacetylene, $(CH)_x$. *J. Chem. Soc. Chem. Commun.* 1977, 578–580.

61. Schreuder-Gibson, H.L.; Truong, Q.; Walker, J.E.; Owens, J.R.; Wander, J.D.; Jones, W.E. Chemical and biological protection and detection in fabrics for protective clothing. *MRS Bull.* 2003, *28*, 574–578.

62. Cochrane, C.; Koncar, V.; Lewandowski, M.; Dufour, C. Design and development of a flexible strain sensor for textile structures based on a conductive polymer composite. *Sensors* 2007, *7*, 473–492.

63. De Rossi, D.; Carpi, F.; Lorussi, F.; Mazzoldi, A.; Paradiso, R.; Scilingo, E.P.; Tognetti, A. Electroactive fabrics and wearable biomonitoring devices. *AUTEX Res. J.* 2003, *3*, 180–185.

64. Smith, D. Smart clothes and wearable technology. *AI Soc.* 2007, *22*, 1–3.

65. Park, S.; Jayaraman, S. Smart textiles: Wearable electronic systems. *MRS Bull.* 2003, *28*, 585–591.

66. Kim, M.S.; Kim, H.K.; Byun, S.W.; Jeong, S.H.; Hong, Y.K.; Joo, J.S.; Song, K.T.; Kim, J.K.; Lee, C.J.; Lee, J.Y. PET fabric/polypyrrole composite with high electrical conductivity for EMI shielding. *Synth. Met.* 2002, *126*, 233–239.

67. Varesano, A.; Dall'Acqua, L.; Tonin, C. A study on the electrical conductivity decay of polypyrrole coated wool textiles. *Polym. Degrad. Stab.* 2005, *89*, 125–132.

68. Ding, Y.; Invernale, M.A.; Sotzing, G.A. Conductivity trends of PEDOT-PSS impregnated fabric and the effect of conductivity on electrochromic textile. *ACS Appl. Mater. Interfaces* 2010, *2*, 1588–1593.

69. Gospodinova, N.; Terlemezyan, L. Conducting polymers prepared by oxidative polymerization: Polyaniline. *Prog. Polym. Sci.* 1998, *23*, 1443–1484.

70. Jianming, J.; Wei, P.; Shenglin, Y.; Guang, L. Electrically conductive PANI-DBSA/Co-PAN composite fibers prepared by wet spinning. *Synth. Met.* 2005, *149*, 181–186.

71. Dallas, P.; Niarchos, D.; Vrbanic, D.; Boukos, N.; Pejovnik, S.; Trapalis, C.; Petridis, D. Interfacial polymerization of pyrrole and *in situ* synthesis of polypyrrole/silver nanocomposites. *Polymer* 2007, *48*, 2007–2013.

72. Müller, D.; Rambo, C.R.; Recouvreux, D.O.S.; Porto, L.M.; Barra, G.M.O. Chemical *in situ* polymerization of polypyrrole on bacterial cellulose nanofibers. *Synth. Met.* 2011, *161*, 106–111.

73. Xu, R.; Wei, J.; Guo, F.; Cui, X.; Zhang, T.; Zhu, H.; Wang, K.; Wu, D. Highly conductive, twistable and bendable polypyrrole–carbon nanotube fiber for efficient supercapacitor electrodes. *RSC Adv.* 2015, *5*, 22015–22021.

74. Zhang, Y.; Rutledge, G.C. Electrical conductivity of electrospun polyaniline and polyaniline-blend fibers and mats.*Macromolecules* 2012, *45*, 4238–4246.

75. Balint, R.; Cassidy, N.J.; Cartmell, S.H. Conductive polymers: Towards a smart biomaterial for tissue engineering. *Acta Biomater.* 2014, *10*, 2341–2353.

76. Moulé, A.J.; Jung, M.-C.; Rochester, C.W.; Tress, W.; LaGrange, D.; Jacobs, I.E.; Li, J.; Mauger, S.A.; Rail, M.D.; Lin, O.; Bilsky, D.J.; Qi, Y.; Stroeve, P.; Berben, L.A.; Riede, M. Mixed interlayers at the interface between PEDOT:PSS and conjugated polymers provide charge transport control. *J. Mater. Chem. C* 2015, *3*, 2664–2676.

77. Pomfret, S.J.; Adams, P.N.; Comfort, N.P.; Monkman, A.P. Electrical and mechanical properties of polyaniline fibres produced by a one-step wet spinning process. *Polymer* 2000, *41*, 2265–2269.

78. Huang, K.; Wan, M.; Long, Y.; Chen, Z.; Wei, Y. Multi-functional polypyrrole nanofibers via a functional dopant-introduced process. *Synth. Met.* 2005, *155*, 495–500.

79. Wallace, G.G.; Campbell, T.E.; Innis, P.C. Putting function into fashion: Organic conducting polymer fibres and textiles. *Fibers Polym.* 2007, *8*, 135–142.

80. Chacko, A.P.; Hardaker, S.S.; Gregory, R.V.; Samuels, R.J. Viscoelastic characterization of concentrated polyaniline solutions: New insights into conductive polymer processing. *Synth. Met.* 1997, *84*, 41–44.

81. Mottaghitalab, V.; Spinks, G.M.; Wallace, G.G. The development and characterisation of polyaniline—Single walled carbon nanotube composite fibres using 2-acrylamido-2 methyl-1-propane sulfonic acid (AMPSA) through one step wet spinning process. *Polymer* 2006, *47*, 4996–5002.

82. Takahashi, T.; Ishihara, M.; Okuzaki, H. Poly(3,4-ethylenedioxythiophene)/poly(4-styrenesulfonate) microfibers.*Synth. Met.* 2005, *152*, 73–76.

83. Yun, Y.J.; Hong, W.G.; Kim, W.-J.; Jun, Y.; Kim, B.H. A novel method for applying reduced graphene oxide directly to electronic textiles from yarns to fabrics. *Adv. Mater.* 2013, *25*, 5701–5705.

84. Meng, Y.; Zhao, Y.; Hu, C.; Cheng, H.; Hu, Y.; Zhang, Z.; Shi, G.; Qu, L. All-graphene core-sheath microfibers for all-solid-state, stretchable fibriform supercapacitors and wearable electronic textiles. *Adv. Mater.* 2013, *25*, 2326–2331.

85. Jang, E.Y.; Kang, T.J.; Im, H.; Baek, S.J.; Kim, S.; Jeong, D.H.; Park, Y.W.; Kim, Y.H. Macroscopic single-walled-carbon-nanotube fiber self-assembled by dip-coating method. *Adv. Mater.* 2009, *21*, 4357–4361.

86. Chen, T.; Cai, Z.; Yang, Z.; Li, L.; Sun, X.; Huang, T.; Yu, A.; Kia, H.G.; Peng, H. Nitrogen-doped carbon nanotube composite fiber with a core-sheath structure for novel electrodes. *Adv. Mater.* 2011, *23*, 4620–4625.

87. Hu, L.; Pasta, M.; Mantia, F.L.; Cui, L.; Jeong, S.; Deshazer, H.D.; Choi, J.W.; Han, S.M.; Cui, Y. Stretchable, porous, and conductive energy textiles. *Nano Lett.* 2010, *10*, 708–714.

88. Lee, H.M.; Choi, S.-Y.; Jung, A.; Ko, S.H. Highly conductive aluminum textile and paper for flexible and wearable electronics. *Angew. Chem. Int. Ed.* 2013, *52*, 7718–7723.

89. Park, M.; Im, J.; Shin, M.; Min, Y.; Park, J.; Cho, H.; Park, S.; Shim, M.-B.; Jeon, S.; Chung, D.-Y.; *et al.* Highly stretchable electric circuits from a composite material of silver nanoparticles and elastomeric fibres. *Nat. Nanotechnol.*2012, *7*, 803–809.

90. Lee, S.; Shin, S.; Lee, S.; Seo, J.; Lee, J.; Son, S.; Cho, H.J.; Algadi, H.; Al-Sayari, S.; Kim, D.E.; Lee, T. Ag nanowire reinforced highly stretchable conductive fibers for wearable electronics. *Adv. Funct. Mater.* 2015, *25*, 3114–3121.

91. Wu, A.S.; Nie, X.; Hudspeth, M.C.; Chen, W.W.; Chou, T.-W.; Lashmore, D.S.; Schauer, M.W.; Tolle, E.; Rioux, J. Strain rate-dependent tensile properties and dynamic electromechanical response of carbon nanotube fibers. *Carbon* 2012, *50*, 3876–3881.

92. Koziol, K.; Vilatela, J.; Moisala, A.; Motta, M.; Cunniff, P.; Sennett, M.; Windle, A. High-performance carbon nanotube fiber. *Science* 2007, *318*, 1892–1895.

93. Chen, H.; Müller, M.B.; Gilmore, K.J.; Wallace, G.G.; Li, D. Mechanically strong, electrically conductive, and biocompatible graphene paper. *Adv. Mater.* 2008, *20*, 3557–3561.

94. Baughman, R.H.; Zakhidov, A.A.; de Heer, W.A. Carbon nanotubes—The route toward applications. *Science* 2002, *297*, 787–792.

95. Kaushik, V.; Sharma, H.; Girdhar, P.; Shukla, A.K.;

Vankar, V.D. Structural modification and enhanced electron emission from multiwalled carbon nanotubes grown on Ag/Fe catalysts coated Si-substrates. *Mater. Chem. Phys.* 2011,*130*, 986–992.

96. Cao, A.; Dickrell, P.L.; Sawyer, W.G.; Ghasemi-Nejhad, M.N.; Ajayan, P.M. Super-compressible foamlike carbon nanotube films. *Science* 2005, *310*, 1307–1310.

97. Jakubinek, M.B.; Johnson, M.B.; White, M.A.; Jayasinghe, C.; Li, G.; Cho, W.; Schulz, M.J.; Shanov, V. Thermal and electrical conductivity of array-spun multi-walled carbon nanotube yarns. *Carbon* 2012, *50*, 244–248.

98. Ericson, L.M.; Fan, H.; Peng, H.; Davis, V.A.; Zhou, W.; Sulpizio, J.; Wang, Y.; Booker, R.; Vavro, J.; Guthy, C.; *et al.* Macroscopic, neat, single-walled carbon nanotube fibers. *Science* 2004, *305*, 1447–1450.

99. Zhang, L.; Shi, E.; Ji, C.; Li, Z.; Li, P.; Shang, Y.; Li, Y.; Wei, J.; Wang, K.; Zhu, H.; *et al.* Fiber and fabric solar cells by directly weaving carbon nanotube yarns with CdSe nanowire-based electrodes. *Nanoscale* 2012, *4*, 4954–4959.

100. Davis, V.A.; Parra-Vasquez, A.N.G.; Green, M.J.; Rai, P.K.; Behabtu, N.; Prieto, V.; Booker, R.D.; Schmidt, J.; Kesselman, E.; Zhou, W.; *et al.* True solutions of single-walled carbon nanotubes for assembly into macroscopic materials. *Nat. Nanotechnol.* 2009, *4*, 830–834.

101. Vigolo, B.; Pénicaud, A.; Coulon, C.; Sauder, C.; Pailler, R.; Journet, C.; Bernier, P.; Poulin, P. Macroscopic fibers and ribbons of oriented carbon nanotubes. *Science* 2000, *290*, 1331–1334.

102. Shim, B.S.; Chen, W.; Doty, C.; Xu, C.; Kotov, N.A. Smart electronic yarns and wearable fabrics for human biomonitoring made by carbon nanotube coating with polyelectrolytes. *Nano Lett.* 2008, *8*, 4151–4157.

103. Zhang, S.; Koziol, K.K.K.; Kinloch, I.A.; Windle, A.H. Macroscopic fibers of well-aligned carbon nanotubes by wet spinning. *Small* 2008, *4*, 1217–1222.

104. Zhao, Y.; Wei, J.; Vajtai, R.; Ajayan, P.M.; Barrera, E.V. Iodine doped carbon nanotube cables exceeding specific electrical conductivity of metals. *Sci. Rep.* 2011, *1*, 83–87.

105. Jiang, K.; Li, Q.; Fan, S. Nanotechnology: Spinning continuous carbon nanotube yarns. *Nature* 2002, *419*, 801–801.

106. Xu, Z.; Zhang, Y.; Li, P.; Gao, C. Strong, conductive, lightweight, neat graphene aerogel fibers with aligned pores. *ACS Nano* 2012, *6*, 7103–7113.

107. Scholz, J.; Nocke, G.; Hollstein, F.; Weissbach, A. Investigations on fabrics coated with precious metals using the magnetron sputter technique with regard to their anti-microbial properties. *Surf. Coat. Technol.* 2005, *192*, 252–256.

108. Hegemann, D.; Amberg, M.; Ritter, A.; Heuberger, M. Recent developments in Ag metallised textiles using plasma sputtering. *Mater. Technol.* 2009, *24*, 41–45.

109. Little, B.K.; Li, Y.; Cammarata, V.; Broughton, R.; Mills, G. Metallization of kevlar fibers with gold. *ACS Appl. Mater. Interfaces* 2011, *3*, 1965–1973.

110. Zhang, Z.; Dai, S.; Blom, D.A.; Shen, J. Synthesis of ordered metallic nanowires inside ordered mesoporous materials through electroless deposition. *Chem. Mater.* 2002, *14*, 965–968.

111. Zrnic, D.; Swatik, D.S. On the resistivity and surface tension of the eutectic alloy of gallium and indium. *J. Common Met.* 1969, *18*, 67–68.

112. Zhu, S.; So, J.-H.; Mays, R.; Desai, S.; Barnes, W.R.; Pourdeyhimi, B.; Dickey, M.D. Ultrastretchable fibers with metallic conductivity using a liquid metal alloy core. *Adv. Funct. Mater.* 2013, *23*, 2308–2314.

113. Atwa, Y.; Maheshwari, N.; Goldthorpe, I.A. Silver nanowire coated threads for electrically conductive textiles. *J. Mater. Chem. C* 2015, *3*, 3908–3912.

114. Horowitz, G. Organic field-effect transistors. *Adv. Mater.* 1998, *10*, 365–377.

115. Muccini, M. A bright future for organic field-effect transistors. *Nat. Mater.* 2006, *5*, 605–613.

116. Małachowski, M.J.; Żmija, J. Organic field-effect transistors. *Opto-Electron. Rev.* 2010, *18*, 121–136.

117. Sun, Y.; Liu, Y.; Zhu, D. Advances in organic field-effect transistors. *J. Mater. Chem.* 2005, *15*, 53–65.

118. Lee, J.B.; Subramanian, V. Organic Transistors on Fiber: A First Step Towards Electronic Textiles. In Proceedings of IEEE International Electron Devices Meeting (IEDM '03), Technical Digest, Washington, DC, USA, 8–10 December 2003; pp. 8.3.1–8.3.4.

119. Garnier, F.; Horowitz, G.; Peng, X.; Fichou, D. An all-organic "soft" thin-film transistor with very high carrier mobility. *Adv. Mater.* 1990, *2*, 592–594.

120. Tsumura, A.; Koezuka, H.; Ando, T. Macromolecular electronic device: Field-effect transistor with a polythiophene thin film. *Appl. Phys. Lett.* 1986, *49*, 1210–1212.

121. De Rossi, D. Electronic textiles: A logical step. *Nat. Mater.* 2007, *6*, 328–329.

122. Maccioni, M.; Orgiu, E.; Cosseddu, P.; Locci, S.; Bonfiglio, A. Towards the textile transistor: Assembly and characterization of an organic field effect transistor with a cylindrical geometry. *Appl. Phys. Lett.* 2006, *89*, 143515–143517.

123. Hamedi, M.; Forchheimer, R.; Inganäs, O. Towards woven logic from organic electronic fibres. *Nat. Mater.* 2007, *6*, 357–362.

124. Hamedi, M.; Herlogsson, L.; Crispin, X.; Marcilla, R.; Berggren, M.; Inganäs, O. Fiber-embedded electrolyte-gated field-effect transistors for e-textiles. *Adv. Mater.* 2009, *21*, 573–577.

125. Bonderover, E.; Wagner, S. A Woven Inverter Circuit for E-Textile Applications. *IEEE Electron Device Lett.* 2004, *25*, 295–297.

126. Holonyak, N. News - First LED by the GE Engineer, Nick Holonyak. Available online: http://www.gelighting.com/LightingWeb/emea/news-and-media/news/First-LED-by-the-GE-engineer-Nick-Holonyak.jsp (accessed on 16 July 2015).

127. Tan, J.J.; Zou, C.Q.; Du, S.H.; Tan, J.T. Simulation of MIMO channel characteristics for indoor visible light communication with LEDs. *Opt. Int. J. Light Electron Opt.* 2014, *125*, 44–49.
128. Yu, H.; Deng, D.; Chen, L.; Chen, D.; Zhong, J.; Zhao, H.; Xu, S. Luminescent properties of novel $Ca_{2.89}Mg_{0.11}(PO_4)_2$:$Eu^{2+}$ single-phase white light-emitting phosphor for white LEDs. *Ceram. Int.* 2015, *41*, 3800–3805.
129. Zhang, Z.; Wang, J.; Yin, Y.; Zhou, J.; Zhou, S.; Pan, Y. A novel green-to-yellow emitting phosphor: $BaSi_2SN_{2.67}$:$Eu^{2+}$for potential application in UV-LEDs. *Opt. Mater.* 2013, *35*, 1273–1275.
130. Park, S.; Choi, I.-M.; Kim, S.-S.; Kim, S.-M. A portable mid-range localization system using infrared LEDs for visually impaired people. *Infrared Phys. Technol.* 2014, *67*, 583–589.
131. Lourenço, M.A.; Homewood, K.P. Crystalline-silicon-based infra-red LEDs and routes to laser diodes. *Thin Solid Films*2011, *519*, 8441–8445.
132. Tang, C.W.; VanSlyke, S.A. Organic electroluminescent diodes. *Appl. Phys. Lett.* 1987, *51*, 913–915.
133. Misra, A.; Kumar, P.; Kamalasanan, M.N.; Chandra, S. White organic LEDs and their recent advancements. *Semicond. Sci. Technol.* 2006, *21*, R35–R47.
134. O'Connor, B.; An, K.H.; Zhao, Y.; Pipe, K.P.; Shtein, M. Fiber shaped light emitting device. *Adv. Mater.* 2007, *19*, 3897–3900.
135. Yang, H.; Lightner, C.R.; Dong, L. Light-emitting coaxial nanofibers. *ACS Nano* 2012, *6*, 622–628.
136. Wise, K.D. Integrated sensors: Interfacing electronics to a non-electronic world. *Sens. Actuators* 1981, *2*, 229–237.
137. Nese, M.; Seeberg, B.E. Silicon MEMS pressure sensors for aerospace applications. *ASME Proc.* 2006, 195–198.
138. Lin, L.; Yun, W. MEMS Pressure Sensors for Aerospace Applications. In Proceedings of 1998 IEEE Aerospace Conference, Aspen, CO, USA, 28 March 1998; Volume 1, pp. 429–436.
139. Mishra, V.; Singh, N.; Tiwari, U.; Kapur, P. Fiber grating sensors in medicine: Current and emerging applications. *Sens. Actuators Phys.* 2011, *167*, 279–290.
140. Girão, P.S.; Ramos, P.M.P.; Postolache, O.; Miguel Dias Pereira, J. Tactile sensors for robotic applications. *Measurement*2013, *46*, 1257–1271.
141. Wu, W.; Liang, L.; Xu, R.; Liu, X.; Qin, Z. The application of fiber grating sensor network in port-machinery monitoring. *Opt. Lasers Eng.* 2010, *48*, 427–434.
142. Kinnunen, M.; Mian, S.Q.; Oinas-Kukkonen, H.; Riekki, J.; Jutila, M.; Ervasti, M.; Ahokangas, P.; Alasaarela, E. Wearable and mobile sensors connected to social media in human well-being applications. *Telemat. Inform.* 2016, *33*, 92–101.
143. Mansoor, M.; Haneef, I.; Akhtar, S.; De Luca, A.; Udrea, F. Silicon diode temperature sensors—A review of applications. *Sens. Actuators Phys.* 2015, *232*, 63–74.
144. Della Lucia, F.; Zambrozi, P., Jr.; Frazatto, F.; Piazzetta, M.; Gobbi, A. Design, fabrication and characterization of SAW pressure sensors for offshore oil and gas exploration. *Sens. Actuators Phys.* 2015, *222*, 322–328.
145. Yamazoe, N.; Shimizu, Y. Humidity sensors: Principles and applications. *Sens. Actuators* 1986, *10*, 379–398.
146. Zhu, Z.-T.; Mason, J.T.; Dieckmann, R.; Malliaras, G.G. Humidity sensors based on pentacene thin-film transistors.*Appl. Phys. Lett.* 2002, *81*, 4643–4645.
147. Merkys, A.J.; Laurinavichius, R.S.; Rupainene, O.J.; Savichene, E.K.; Jaroshius, A.V.; Shvegzhdene, D.V.; Bendoraityte, D.P. The state of gravity sensors and peculiarities of plant growth during different gravitational loads. *Adv. Space Res.*1983, *3*, 211–219.
148. Pasterkamp, H.; Kraman, S.S.; DeFrain, P.D.; Wodicka, G.R. Measurement of respiratory acoustical signals. Comparison of sensors. *Chest* 1993, *104*, 1518–1525.
149. Turner, R.C.; Fuierer, P.A.; Newnham, R.E.; Shrout, T.R. Materials for high temperature acoustic and vibration sensors: A review. *Appl. Acoust.* 1994, *41*, 299–324.
150. Papadimitriou, C.; Fritzen, C.-P.; Kraemer, P.; Ntotsios, E. Fatigue predictions in entire body of metallic structures from a limited number of vibration sensors using Kalman filtering. *Struct. Control Health Monit.* 2011, *18*, 554–573.
151. Inaba, M.; Hoshino, Y.; Nagasaka, K.; Ninomiya, T.; Kagami, S.; Inoue, H. A Full-Body Tactile Sensor Suit Using Electrically Conductive Fabric and Strings. In Proceedings of the 1996 IEEE/RSJ International Conference on Intelligent Robots and Systems '96 (IROS 96), Osaka, Japan, 4–8 November 1996; Volume 2, pp. 450–457.
152. Shimojo, M.; Namiki, A.; Ishikawa, M.; Makino, R.; Mabuchi, K. A tactile sensor sheet using pressure conductive rubber with electrical-wires stitched method. *IEEE Sens. J.* 2004, *4*, 589–596.
153. Takamatsu, S.; Kobayashi, T.; Shibayama, N.; Miyake, K.; Itoh, T. Fabric pressure sensor array fabricated with die-coating and weaving techniques. *Sens. Actuators Phys.* 2012, *184*, 57–63.
154. Ma, R.; Lee, J.; Choi, D.; Moon, H.; Baik, S. Knitted fabrics made from highly conductive stretchable fibers. *Nano Lett.*2014, *14*, 1944–1951.
155. Starner, T. Human-powered wearable computing. *IBM Syst. J.* 1996, *35*, 618–629.
156. Wang, X.; Song, J.; Liu, J.; Wang, Z.L. Direct-current nanogenerator driven by ultrasonic waves. *Science* 2007, *316*, 102–105.
157. Paradiso, J.; Starner, T. Energy scavenging for mobile and wireless electronics. *IEEE Pervasive Comput.* 2005, *4*, 18–27.
158. Zhang, M.; Gao, T.; Wang, J.; Liao, J.; Qiu, Y.; Yang, Q.; Xue, H.; Shi, Z.; Zhao, Y.; Xiong, Z.; Chen, L. A hybrid fibers based wearable fabric piezoelectric nanogenerator for energy harvesting application. *Nano Energy* 2015, *13*, 298–305.

159. Bouendeu, E.; Greiner, A.; Smith, P.J.; Korvink, J.G. A low-cost electromagnetic generator for vibration energy harvesting. *IEEE Sens. J.* 2011, *11*, 107–113.

160. Mitcheson, P.D.; Miao, P.; Stark, B.H.; Yeatman, E.M.; Holmes, A.S.; Green, T.C. MEMS electrostatic micropower generator for low frequency operation. *Sens. Actuators Phys.* 2004, *115*, 523–529.

161. Qin, Y.; Wang, X.; Wang, Z.L. Microfibre–nanowire hybrid structure for energy scavenging. *Nature* 2008, *451*, 809–813.

162. Jeong, C.K.; Lee, J.; Han, S.; Ryu, J.; Hwang, G.-T.; Park, D.Y.; Park, J.H.; Lee, S.S.; Byun, M.; Ko, S.H.; *et al.* A hyper-stretchable elastic-composite energy harvester. *Adv. Mater.* 2015, *27*, 2866–2875.

163. Zhong, J.; Zhang, Y.; Zhong, Q.; Hu, Q.; Hu, B.; Wang, Z.L.; Zhou, J. Fiber-based generator for wearable electronics and mobile medication. *ACS Nano* 2014, *8*, 6273–6280.

164. Jung, S.; Lee, J.; Hyeon, T.; Lee, M.; Kim, D.-H. Fabric-based integrated energy devices for wearable activity monitors.*Adv. Mater.* 2014, *26*, 6329–6334.

165. Lee, S.; Ko, W.; Oh, Y.; Lee, J.; Baek, G.; Lee, Y.; Sohn, J.; Cha, S.; Kim, J.; Park, J.; *et al.* Triboelectric energy harvester based on wearable textile platforms employing various surface morphologies. *Nano Energy* 2015, *12*, 410–418.

166. Fan, F.-R.; Tian, Z.-Q.; Lin Wang, Z. Flexible triboelectric generator. *Nano Energy* 2012, *1*, 328–334.

167. Zhu, G.; Pan, C.; Guo, W.; Chen, C.-Y.; Zhou, Y.; Yu, R.; Wang, Z.L. Triboelectric-generator-driven pulse electrodeposition for micropatterning. *Nano Lett.* 2012, *12*, 4960–4965.

168. Du, Y.; Cai, K.; Chen, S.; Wang, H.; Shen, S.Z.; Donelson, R.; Lin, T. Thermoelectric fabrics: Toward power generating clothing. *Sci. Rep.* 2015, *5*, 6411–6416.

169. Kim, M.-K.; Kim, M.-S.; Lee, S.; Kim, C.; Kim, Y.-J. Wearable thermoelectric generator for harvesting human body heat energy. *Smart Mater. Struct.* 2014, *23*, 105002–105008.

170. Liang, D.; Yang, H.; Finefrock, S.W.; Wu, Y. Flexible nanocrystal-coated glass fibers for high-performance thermoelectric energy harvesting. *Nano Lett.* 2012, *12*, 2140–2145.

171. O'Regan, B.; Grätzel, M. A low-cost, high-efficiency solar cell based on dye-sensitized colloidal TiO_2 films. *Nature*1991, *353*, 737–740.

172. Kippelen, B.; Brédas, J.-L. Organic photovoltaics. *Energy Environ. Sci.* 2009, *2*, 251–261.

173. Liu, D.; Kelly, T.L. Perovskite solar cells with a planar heterojunction structure prepared using room-temperature solution processing techniques. *Nat. Photonics* 2013, *8*, 133–138.

174. Chen, S.-G.; Stradins, P.; Gregg, B.A. Doping highly ordered organic semiconductors: Experimental results and fits to a self-consistent model of excitonic processes, doping, and transport. *J. Phys. Chem. B* 2005, *109*, 13451–13460.

175. Pan, S.; Yang, Z.; Chen, P.; Fang, X.; Guan, G.; Zhang, Z.; Deng, J.; Peng, H. Carbon nanostructured fibers as counter electrodes in wire-shaped dye-sensitized solar cells. *J. Phys. Chem. C* 2014, *118*, 16419–16425.

176. Cai, X.; Hou, S.; Wu, H.; Lv, Z.; Fu, Y.; Wang, D.; Zhang, C.; Kafafy, H.; Chu, Z.; Zou, D. All-carbon electrode-based fiber-shaped dye-sensitized solar cells. *Phys. Chem. Chem. Phys.* 2012, *14*, 125–130.

177. Hou, S.; Cai, X.; Fu, Y.; Lv, Z.; Wang, D.; Wu, H.; Zhang, C.; Chu, Z.; Zou, D. Transparent conductive oxide-less, flexible, and highly efficient dye-sensitized solar cells with commercialized carbon fiber as the counter electrode. *J. Mater. Chem.* 2011, *21*, 13776–13779.

178. Cai, F.; Chen, T.; Peng, H. All carbon nanotube fiber electrode-based dye-sensitized photovoltaic wire. *J. Mater. Chem.*2012, *22*, 14856–14860.

179. Zhang, L.; Shi, E.; Li, Z.; Li, P.; Jia, Y.; Ji, C.; Wei, J.; Wang, K.; Zhu, H.; Wu, D.; *et al.* Wire-supported CdSe nanowire array photoelectrochemical solar cells. *Phys. Chem. Chem. Phys.* 2012, *14*, 3583–3588.

180. Grätzel, M. The light and shade of perovskite solar cells. *Nat. Mater.* 2014, *13*, 838–842.

181. McGehee, M.D. Perovskite solar cells: Continuing to soar. *Nat. Mater.* 2014, *13*, 845–846.

182. Zhou, H.; Chen, Q.; Li, G.; Luo, S.; Song, T.-b.; Duan, H.-S.; Hong, Z.; You, J.; Liu, Y.; Yang, Y. Interface engineering of highly efficient perovskite solar cells. *Science* 2014, *345*, 542–546.

183. Jeon, N.J.; Noh, J.H.; Yang, W.S.; Kim, Y.C.; Ryu, S.; Seo, J.; Seok, S.I. Compositional engineering of perovskite materials for high-performance solar cells. *Nature* 2015, *517*, 476–480.

184. Qiu, L.; Deng, J.; Lu, X.; Yang, Z.; Peng, H. Integrating perovskite solar cells into a flexible fiber. *Angew. Chem. Int. Ed.*2014, *53*, 10425–10428.

185. Liu, J.; Mirri, F.; Notarianni, M.; Pasquali, M.; Motta, N. High performance all-carbon thin film supercapacitors. *J. Power Sources* 2014, *274*, 823–830.

186. Wang, G.; Zhang, L.; Zhang, J. A review of electrode materials for electrochemical supercapacitors. *Chem Soc Rev* 2012,*41*, 797–828.

187. Yu, Z.; Tetard, L.; Zhai, L.; Thomas, J. Supercapacitor electrode materials: Nanostructures from 0 to 3 dimensions.*Energy Env. Sci* 2015, *8*, 702–730.

188. Yang, P.; Xiao, X.; Li, Y.; Ding, Y.; Qiang, P.; Tan, X.; Mai, W.; Lin, Z.; Wu, W.; Li, T.; *et al.* Hydrogenated ZnO core–shell nanocables for flexible supercapacitors and self-powered systems. *ACS Nano* 2013, *7*, 2617–2626.

189. Jiang, H.; Yang, L.; Li, C.; Yan, C.; Lee, P.S.; Ma, J. High–rate electrochemical capacitors from highly graphitic carbon–tipped manganese oxide/mesoporous carbon/manganese oxide hybrid nanowires. *Energy Environ. Sci.* 2011, *4*, 1813–1819.

190. Li, Q.; Wang, Z.-L.; Li, G.-R.; Guo, R.; Ding, L.-X.; Tong, Y.-X. Design and synthesis of MnO_2/Mn/MnO_2 sandwich-structured nanotube arrays with high supercapacitive performance for electrochemical energy storage. *Nano Lett.*2012, *12*, 3803–3807.

191. Zhang, Z.; Chen, X.; Chen, P.; Guan, G.; Qiu, L.; Lin, H.; Yang, Z.; Bai, W.; Luo, Y.; Peng, H. Integrated polymer solar cell and electrochemical supercapacitor in a flexible and stable fiber format. *Adv. Mater.* 2014, *26*, 466–470.

192. Jiang, D.; Jin, Z.; Henderson, D.; Wu, J. Solvent effect on the pore-size dependence of an organic electrolyte supercapacitor. *J. Phys. Chem. Lett.* 2012, *3*, 1727–1731.

193. Jost, K.; Perez, C.R.; McDonough, J.K.; Presser, V.; Heon, M.; Dion, G.; Gogotsi, Y. Carbon coated textiles for flexible energy storage. *Energy Environ. Sci.* 2011, *4*, 5060–5067.

194. Pasta, M.; La Mantia, F.; Hu, L.; Deshazer, H.D.; Cui, Y. Aqueous supercapacitors on conductive cotton. *Nano Res.*2010, *3*, 452–458.

195. Prosini, P.P.; Pozio, A.; Botti, S.; Ciardi, R. Electrochemical studies of hydrogen evolution, storage and oxidation on carbon nanotube electrodes. *J. Power Sources* 2003, *118*, 265–269.

196. Zou, D.; Lv, Z.; Cai, X.; Hou, S. Macro/microfiber-shaped electronic devices. *Nano Energy* 2012, *1*, 273–281.

INDEX

M

O

P

R

S

T

V

W

X